Water Quality: Monitoring and Treatment

Water Quality: Monitoring and Treatment

Editor: Herbert Lotus

Callisto Reference

www.callistoreference.com

Callisto Reference,
118-35 Queens Blvd., Suite 400,
Forest Hills, NY 11375, USA

Visit us on the World Wide Web at:
www.callistoreference.com

ISBN: 978-1-63239-935-9 (Hardback)

Printed in China.

Cataloging-in-Publication Data

Water quality : monitoring and treatment / edited by Herbert Lotus.
p. cm.
Includes bibliographical references and index.
ISBN 978-1-63239-935-9
1. Water quality. 2. Water--Purification. 3. Water quality--Measurement.
4. Water quality management. I. Lotus, Herbert.
TD370 .W38 2018
628.161--dc23

Table of Contents

Preface

Water quality is increasingly becoming a concern for environmentalists. It is measured to assess if the water is safe for human consumption or for industrial use. The various parameters considered for water quality are temperature, conductivity, pH, turbidity, etc. It studies the characteristics of water such as its chemical, physical and radiological properties. This book on water quality unravels the recent studies in this field. It is compiled in such a manner, that it will provide in-depth knowledge about the theory and applications of accessing water quality. From theories to research to practical applications, case studies related to all contemporary topics of relevance to this field have been included in this book.

The information shared in this book is based on empirical researches made by veterans in this field of study. The elaborative information provided in this book will help the readers further their scope of knowledge leading to advancements in this field.

Finally, I would like to thank my fellow researchers who gave constructive feedback and my family members who supported me at every step of my research.

Editor

Longitudinal Variability of Phosphorus Fractions in Sediments of a Canyon Reservoir Due to Cascade Dam Construction

Qi Liu, Shiliang Liu*, Haidi Zhao, Li Deng, Cong Wang, Qinghe Zhao, Shikui Dong

School of Environment, State Key Laboratory of Water Environment Simulation, Beijing Normal University, Beijing, China

Abstract

Dam construction causes the accumulation of phosphorus in the sediments of reservoirs and increases the release rate of internal phosphorus (P) loading. This study investigated the longitudinal variability of phosphorus fractions in sediments and the relationship between the contents of phosphorus fractions and its influencing factors of the Manwan Reservoir, Lancang River, Yunnan Province, China. Five sedimentary phosphorus fractions were quantified separately: loosely bound P (ex-P); reductant soluble P (BD-P); metal oxide-bound P (NaOH-P); calcium-bound P (HCl-P), and residual-P. The results showed that the total phosphorus contents ranged from 623 to 899 μg/g and were correlated positively with iron content in the sediments of the reservoir. The rank order of P fractions in sediments of the mainstream was HCl-P>NaOH-P>residual-P>BD-P>ex-P, while it was residual-P>HCl-P>NaOH-P>BD-P>ex-P in those of the tributaries. The contents of bio-available phosphorus in the tributaries, including ex-P, BD-P and NaOH-P, were significantly lower than those in the mainstream. The contents of ex-P, BD-P, NaOH-P showed a similar increasing trend from the tail to the head of the Manwan Reservoir, which contributed to the relatively higher content of bio-available phosphorus, and represents a high bio-available phosphorus releasing risk within a distance of 10 km from Manwan Dam. Correlation and redundancy analyses showed that distance to Manwan Dam and the silt/clay fraction of sediments were related closely to the spatial variation of bio-available phosphorus.

Editor: Syuhei Ban, University of Shiga Prefecture, Japan

Funding: This research was funded by the Non-profit Environment Protection Specific Project of China (No. 201209029-4) and the National Natural Sciences Foundation of China (No. 50939001). The funders had no role in study design, data collection and analysis, decision to publish, or preparation of the manuscript.

Competing Interests: The authors have declared that no competing interests exist.

* E-mail: shiliangliu@bnu.edu.cn

Introduction

Dam construction can change rivers' configurations and flow regimes [1,2,3], which will have conspicuous direct effects on nutrient loading in the rivers [4,5]. For example, after the closure of the Three-Gorges Dam in China, the nutrient concentrations and ratios in the water declined dramatically during the high flood season [4]. In the upper stream of the Yellow River, nutrient pollutants varied greatly because of the construction of cascade dams [6]. Dams reduce the transportation of nutrients to marine waters, which indicates that more nutrients settle in the sediments [5].

Among different nutrient pollutants, phosphorus (P) has attracted much attention in sediment research as a key nutrient for phytoplankton growth, controlling the primary productivity of reservoirs [7,8,9,10,11]. When external loading of P occurs, such as from an increase in the drainage of intensively cultivated areas and from sewage, the rate of accumulation of P in the sediments exceeds its ability to release P into water, so sediments act as P sinks. While the external P is reduced, the sediments still release P into the water, which is called internal phosphorus (P) loading [8,9,11,12,13,14,15,16]. Previous research has found that, even when the external phosphorus loading was reduced, P concentrations in lakes either did not change or decreased slightly because of internal phosphorus loading [17,18]. Further, P released by internal P loading contributes to the pool of the P used easily by algae in sediments [19]. Hence, internal P loading has been of great concern in recent years because it can be a potential hazard to aquatic ecosystems [20]. For example, a study in Taihu Lake in China indicated that more than 50% of the inorganic P could be released into the water and used by algae under certain conditions [21]. Although there are several forms of P in sediments, not all of them are released easily from the sediments into the water [22], as this depends on the characteristics of the sediment, environmental factors and the concentration of P in water as well [8,14,22,23,24]. Studies have shown that grain size distribution of the sediments influences the element composition in the sediments, including metal and nutrient contents [25,26,27]. Stone and English (1993) [19] found that fine grained sediment had a different relationship with different P fractions. Further, in river systems influenced by hydropower dams, the spatial grain size distribution of the sediments might be affected by the formative reservoirs and prolonged water renewal time, which eventually has an effect on the heterogeneity of different P fractions in the sediments. Recently, determining P fractions in sediments and the releasing capacity of P fractions from the sediments to the water has been studied and reviewed extensively [12,28,29,30,31]. However, the study of the mechanism of the longitudinal variability of phosphorus fractions in the sediments of canyon reservoirs affected by cascade dams is insufficient and needs further research.

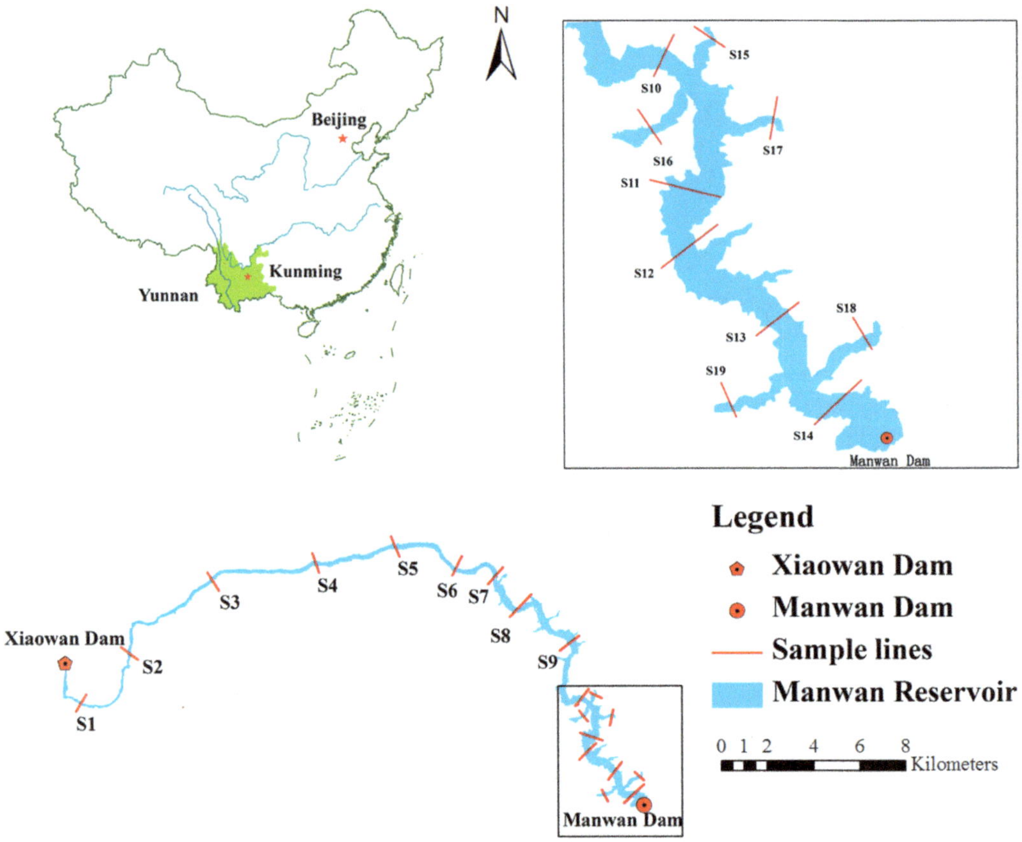

Figure 1. Location of the Manwan Reservoir and the 19 cross-sectional sediment samples, Yunnan Province, China.

Studies have shown that, in the Lower Mekong River mainstream, countries through which the Mekong flows are currently threatened by accelerated eutrophication caused by both hydropower development and climate change [32]. Moreover, the Mekong River Commission (MRC) Water Quality Report (2008) noted that almost one-third of the total P in the Lower Mekong Basin exists as soluble orthophosphate (PO_4-P), which indicated that approximately two-thirds of the total P loading was associated with sediments. In the Upper Mekong River, the external loading of P is lower than that of the Lower Mekong River. Therefore, even slight changes in sediment conditions are likely to affect the P concentration of the water column [33]. However, there are few studies regarding the nutrient condition of the sediments in the Upper Mekong section in China (Lancang River), where a chain of fourteen cascade hydroelectric dams have been planned since the early 1980s, with some completed or currently under construction. Further, human activities are becoming more and more intense, with industrialization and fertilization causing an increased P loading in the river. The Manwan Dam was the first dam constructed in the cascade development project along the Lancang River mainstream. After the Manwan Reservoir began operation, siltation led to increasing concern from researchers and the public because it had increased the elevation of the reservoir bottom by as much as 30 m caused by dam construction [34]. Studies have shown that there was great spatial and temporal variation of heavy metals in the sediments of the Manwan Reservoir, and some metals, such as As, Cd, Cr, Cu, Pb and Zn, have reached contamination levels [35,36]. Further, the study of Zhao, et al. [37], revealed that the potential ecological risk (RI) of multiple heavy metals was related to the grain size of sediments and correlated negatively with the distance from Manwan Dam, a valuable finding for understanding heavy metal contamination in a canyon reservoir. However, the possible spatial variation of P fractions in sediments remains unknown, and this information is useful in understanding the dynamic of trophic conditions of the Manwan Reservoir induced by dam construction.

The aims of our study were to (1) investigate the spatial variation of different P fractions in river sediments of the mainstream and the tributaries of the Manwan Reservoir; (2) estimate the contents of bio-available P in the sediments of Manwan Reservoir, and (3) explore the relationship among P fractions and influencing factors, including metals and sediment grain size.

Materials and Methods

Study Site

The Mekong River is the largest international river in Asia, flowing through seven climatic zones and five countries; it is considered to be one of the most important cradles of human civilization in Southeast Asia [38,39]. The Upper Mekong Section in China (Lancang River), with almost 91% of the drop in elevation of the Mekong River, produces plentiful hydraulic resources [40,41,42]. To date, fourteen cascade dams have been planned there since the early 1980s [35], four of which have been constructed (Xiaowan, Manwan, Dachaoshan and Jinghong) in Yunnan Province. The Manwan Dam, completed in 1993, is the first multimillion kilowatt hydropower station in Yunnan Province [39,42]. The dam is 418 m long and 132 m high with a backwater

Table 1. Extraction procedure used in this work.

step	sequential extraction method	P fraction
1	1 g sediments added to 25 ml 1 M NH_4Cl at pH = 7 shaken for 4 h.	ex-P
2	Residual sample added to 0.11 M $Na_2S_2O_4/NaHCO_3$* shaken for 1 h at 40°C.	BD-P
3	Residual sample added to 0.1 M NaOH shaken for 16 h.	NaOH-P
4	Residual sample added to 0.5 M HCl shaken for 16 h	HCl-P

*Both $Na_2S_2O_4$ and $NaHCO_3$ were the same concentration.

of 70 km near the Xiaowan Dam. The Manwan Reservoir was a canyon reservoir located in a gorge flanked by high mountains, most of whose peaks are higher than 2,200 m above sea level, and deep valleys with a gradient ratio over 15% [36] (Fig. 1). The area of Manwan Reservoir is 23.6 km^2, and the width of the water surface is, on average, 337.1 m, 2.8 times larger in area and 2 times wider than it was before dam construction. The total reservoir capacity is $1{,}060 \times 10^6$ m^3, with a normal water level of 994 m; the effective capacity is 257×10^6 m^3 depending upon seasonal discharge regulation [42].

Sediment Sample Collection

Manwan Dam was constructed and is managed by China Huaneng Group Corporation, a state-owned enterprise. The corporation gave us permission to conduct this field study, which did not involve any endangered or protected species. It was environmentally neutral and did not threaten the welfare of any species or that of the local population. Therefore, it was not related to ethical issues and no specific permissions were required for such activities.

In June, 2012, nineteen surface (15 cm) sediment samples were collected, using cable operated sediment samplers (Van Veen grabs) to investigate the spatial variation in P fractions in the sediments of the Manwan Reservoir. Fourteen cross-sectional samples (S1–S14) were selected to examine the variations in the mainstream of the reservoir, and the remaining five cross-sectional samples (S15–S19) were located at the major tributaries of the reservoir near the dam (Fig. 1). Each sample in a cross-section was a mixture of three sampling sites: the left, middle and right of each section. Both the right and left sampling sites were selected at the same distance to the shore. All of the samples were placed in sealed plastic bags and maintained at 4°C until analysis. After transportation to the laboratory, the samples were kept frozen, and before the analysis, were freeze-dried and ground until all of the particles passed through a 2 mm nylon sieve after removal of the coarse debris [27,43,44].

Analytical Methods

For P fractionation of the sediments from the reservoir, we used the chemical sequential extraction method of Psenner et al. [45], slightly modified by Hupfer et al. [46]. This method fractionates the phosphorus of the sediments into four fractions–loosely bound P (ex-P), reductant soluble P (BD-P), metal oxide-bound P (NaOH-P), calcium-bound P (HCl-P; Table 1) and residual-P–which was the difference between total phosphorus and the four P fractions extracted. All procedures were carried out in triplicate to yield reliable results.

Table 2. Statistical analysis of P fraction concentrations, metals in the sediments and grain size of the sediments of the Manwan Reservoir.

	Range (μg/g)	Average percentage (%)	Average concentration (μg/g)	SD	CV
ex-P	0–1.4	0.1	0.5	0.4	0.8
BD-P	8.7–80.4	4.9	35.5	21.7	0.6
NaOH-P	57.3–270.2	19.6	155.3	67.7	0.4
HCl-P	137.5–403.0	43.6	315.6	91.5	0.3
Residual-P	66.2–369.8	31.9	228.8	97.4	0.4
Al	3.89×10^4–7.57×10^4	49.4	6.07×10^4	1.17×10^4	0.2
Ca	2.37×10^3–4.61×10^4	31.6	2.25×10^4	1.14×10^4	0.5
Fe	2.65×10^4–4.31×10^4	18.4	3.71×10^4	4.62×10^4	0.1
Mn	2.71×10^2–1.09×10^3	0.6	6.76×10^2	1.81×10^2	0.3
S/C	–	47.9	–	33.3	0.7
VFS	–	14.5	–	8.8	0.6
FS	–	12.1	–	15.2	1.3
CS	–	18.6	–	29.9	1.5

S/C: slit/clay fraction (<74 μm).
VFS : very fine sand (74–147 μm).
FS: fine sand (147–246 μm).
CS: coarse sand (246–840 μm).

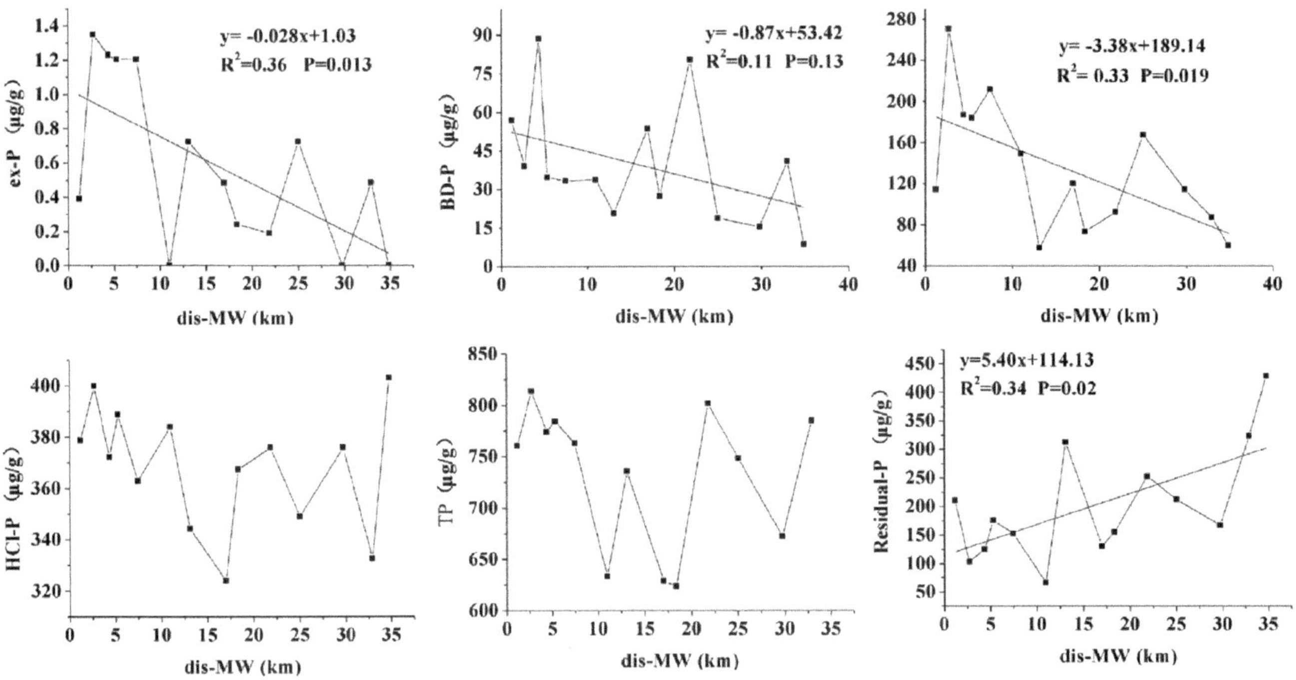

Figure 2. Spatial variation of P fractions in the sediments of the mainstream in Manwan Reservoir.

The extracts in every step were centrifuged at 4500 r/min for 20 minutes, and the soluble reactive phosphorus (SRP) in each fraction was determined by the molybdenum blue/ascorbic acid method (APHA, 1985). For the NaOH extracts, the supernatants were filtered through a 0.45-μm poly-amide filter.

The concentration of total phosphorus was determined by ICP-AES after acid digestion of the freeze-dried samples. Residual phosphorus was the difference between total phosphorus and the sum of the four P fractions extracted above [12]. Total concentrations of Ca, Fe, Mn and Al were determined by the SEPAC method (HJ/T 166-2004) using ICP-AES after wet digestion [47]. The results were calculated on the basis of dry weight sediment.

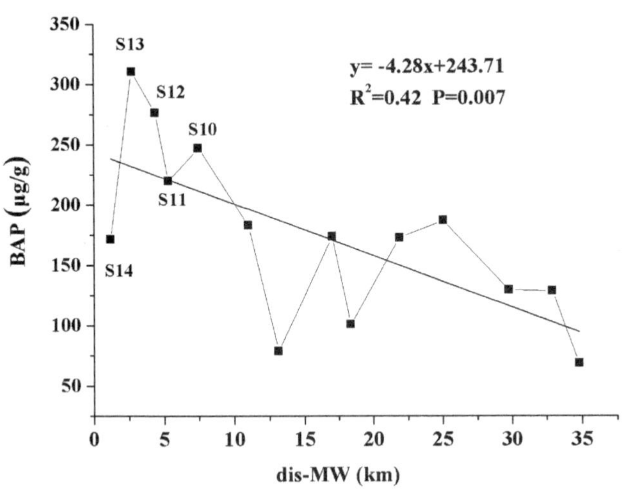

Figure 3. Spatial variation of BAP in the sediments of the mainstream in Manwan Reservoir.

According to the Unified Soil Classification System (USCS), the sediment particles were classified into four grain sizes: coarse/medium sand fraction (246–840 μm); fine sand fraction (147–246 μm); very fine sand fraction (74–147 μm), and silt/clay fraction (<74 μm). The grain size of the sediments was analyzed by an LS 230 laser diffraction particle analyzer (Microtrac S3500).

Statistical Analysis

Redundancy analysis (RDA) is a multivariate direct gradient analysis that enables the identification of variables that best explain the variance pattern of the P fractions [48,49,50]. The data were log (x+1) transformed, centered and standardized before

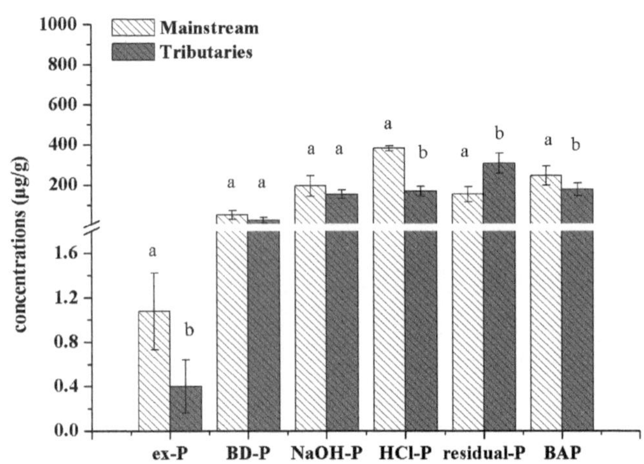

Figure 4. Average concentrations of P fractions in the sediments of mainstream and tributaries. (Different letters indicate significant differences among the mainstream and the tributaries at the 0.05 level).

Table 3. Comparison of TP and the proportion of BAP in IP in different surface sediments.

Sediment source	TP(μg/g)	BAP in IP (%)	Trophic classification	Reference
Lake Koronia, Greece	1156–1305	40–60	hypereutrophic	Christophoros Christophoridis et al, 2006
Lake Volvi, Greece	776–1044	60–80	meso-to-eutrophic	Christophoros Christophoridis et al, 2006
Lake Erken, Sweden	1814	61.4	mesotrophic	Emile Rydin, 2000
Haihe River, China	968–2017	42.2–65.3	mesotrophic	SUN Shujuan et al, 2009
Manwan Reservoir	623–899	37.7	–	Present study

a forward selection procedure combined with Monte Carlo permutation tests (499 permutations) were used to identify which factors contributed significantly to the variation ($p<0.05$). The analysis was performed in CANOCO, Version 4 for Windows, and the results are presented in an ordination diagram in which all of the variables are represented by arrows. A smaller angle between arrows represents a high correlation between variables, and the direction of the arrows represents positive or negative correlations. Pearson correlation analyses were used to provide a further quantitative explanation of the correlation between P fractions and metal content. The P fractions contents in the sediments from different sampling sites were subjected to one-way ANOVA to detect significant differences. The correlation and variance analyses were performed in SPSS 18.0.

Results and Discussion

The Contents of P Fractions and Related Metals in the Manwan Reservoir

The statistical results with respect to different P fractions, related metals and grain size of the sediment particles of the Manwan Reservoir are presented in Table 2. The contents of different P fractions varied greatly. Ex-P is dissolved P, which is absorbed lightly onto the surface of sediment particles or is released from leached P or $CaCO_3$-associated P from organic debris [9,51]. It can estimate the amounts of phosphorus immediately available. In Manwan Reservoir, ex-P consisted of the minimum part of the P pool (0.1% on average). BD-P is mainly redox sensitive P that binds to Fe-hydroxides and Mn compounds and is considered to be potentially available to algae [9,28]. Under the anaerobic conditions of the water-sediment interface, BD-P will be released into the water by reductive Fe dissolution. In Manwan Reservoir, BD-P constituted a minor part of total phosphorus (TP) of the sediments (4.9% on average) which was relatively lower than in other reservoirs. Using the same extracting methods in Lake Simcoe, which is the largest lake in South Ontario and is considered mesotrophic (20%–42% TP), Dittrich et al. found that BD-P was the dominant fraction, accounting for 40% and 57%, respectively, of the long- and short-term sediment P release [52]. In Manwan Reservoir, reactive NaOH-P was the third most abundant P fraction in the sediments and accounted for 19.6% of the total phosphorus, on average. Reactive NaOH-P mainly presented as P bound to the surface of aluminium oxides released at high pH levels because of the ligand exchange reactions of hydroxide ions replacing orthophosphate [12,53], as well as some interior Fe oxides that were not extracted in the BD step [8]. Further, in previous studies, NaOH-P has been used to estimate both short- and long-term available P in the sediments and was verified to be an indicator of algal-available P [54,55]. HCl-P mainly represents calcium-bound P [56], which appears to be non-motile and is not easily bio-available in the sediments [24,57,58]. HCl-P was the most abundant P fraction in the sediments of Manwan Reservoir, constituting 43.6% of TP on average, almost equal to the percentage in Lake Simcoe [52]. Residual-P includes organic phosphorus and refractory P compounds. In Manwan Reservoir, residual-P was the second most abundant P fraction, accounting for 31.9% of TP on average. The rank order of the average concentrations of the four metals in the sediments was Al>Fe>Ca>Mn. The analytical results also showed that there was great spatial variation in the distribution of sediment particles in Manwan Reservoir. According to the average percentage, the silt/clay fraction composed the largest part of all fractions (47.9% on average). The sand fraction, including very fine, fine, and coarse sand, accounted for 45.2% total. Table 2 also shows that Ex-P, Ca, and the coarse sand fraction had relatively high coefficients of variation, which indicated that their contents varied greatly in Manwan Reservoir.

Table 4. P-related metals concentrations in the Manwan Reservoir ($\times10^4$ μg/g).

		Al	Ca	Fe	Mn
Mainstream	Range	3.67–7.15	1.55–4.18	2.98–4.31	0.05–0.11
	Mean	5.97	2.43	3.73	0.07
Tributaries	Range	3.89–7.58	0.22–4.61	2.65–4.21	0.03–0.07
	Mean	6.41	1.65	3.65	0.05

Spatial Variation of P Fractions in the Mainstream and Tributaries

The relationship between concentrations of P fractions and the distance from the sampling site to Manwan Dam (dis-MW) in the mainstream is presented in Fig. 2. In the mainstream, the changing trends of ex-P, BD-P and NaOH-P were similar from the tail to the head of the Manwan Reservoir. The average content of these three P-fractions increased with decreasing distance to the Manwan Dam. Especially within 9 km from the Manwan Dam, the average content of these three P fractions was relatively higher than it was in other parts of the Manwan Reservoir. Particularly, NaOH-P increased faster than ex-P and BD-P. Unlike ex-P, BD-P and NaOH-P, there was not a consistent pattern in the distribution of HCl-P and TP with respect to the distance to the Manwan Dam; further, there was an upward trend in the concentration of residual-P from Manwan to Xiaowan Dam.

A previous study has shown that in the Haihe River, China, the amount of P released from sediments was related closely to the ex-P and BD-P fractions, indicating that the P from these two fractions can be released easily [29]. Furthermore, previous studies

Table 5. The correlation analysis among P fractions, metals and grain size of sediments.

	ex-P	BD-P	NaOH-P	HCl-P	BAP	TP	S/C	VFS	FS	CS	Al	Ca	Fe	Mn	Dis-MW
ex-P	1														
BD-P		1													
NaOH-P	0.550*		1												
HCl-P				1											
BAP	0.613*		0.942**		1										
TP				0.522*		1									
S/C	0.791**				0.497*		1								
VFS								1							
FS						−0.537*			1						
CS					−0.504*					1					
Al	0.470*										1				
Ca			−0.524*		−0.494*							1			
Fe						0.501*	0.544*	−0.629**	−0.723**	−0.488*	0.584*		1		
Mn	0.536*					0.529*							0.527*	1	
dis-MW			−0.631**		−0.634**					0.581*	−0.486*				1

**$p<0.01$;
*$p<0.05$.
BAP: bio-available phosphorus FS: fine sand (147–246 μm) CS: coarse sand (246–840 μm).
S/C: slit/clay fraction (<74 μm) VFS : very fine sand (74–147 μm) Dis-Manwan: distance to Manwan Dam.

have estimated the bio-available phosphorus (BAP) in sediments according to the sequential chemical extraction method by the sum of ex-P, BD-P and NaOH-P [21]. By virtue of this theory, Fig. 3 presents the spatial variation of BAP contents in the mainstream of Manwan Reservoir. The values of BAP were at a maximum in the section between S10 and S14, indicating a higher release risk of bio-available P in the sediment. Further, five tributaries were situated within a distance of 10 km from Manwan dam. Samples in the mainstream (S10–S14) and tributaries (S15–S19) within a distance of 10 km to Manwan Dam were selected to compare the spatial variation of P fractions and BAP using a one-way ANOVA (Fig. 4). The results revealed that there were

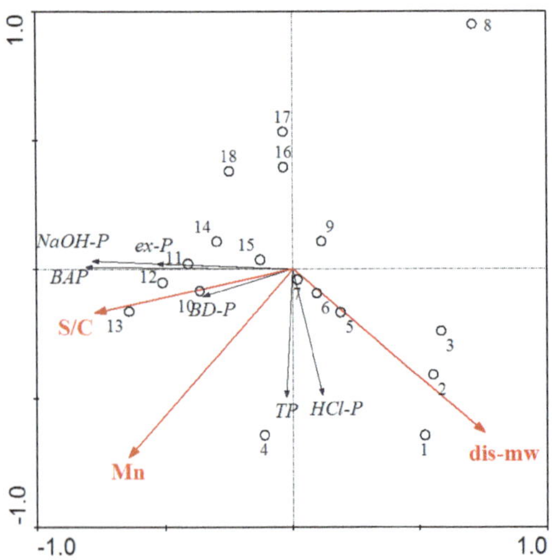

Figure 5. RDA results for P fractions, metals and grain size of sediments.

significant differences in the concentrations of ex-P, HCl-P, residual-P and BAP values between the mainstream and the tributaries at the head of Manwan Reservoir. In the mainstream, the average concentrations of ex-P, HCl-P and residual-P were 1.08 μg/g, 380.50 μg/g and 153.47 μg/g, respectively, whereas in the tributaries, the concentrations of ex-P and HCl-P were smaller by 63% and 56%, respectively, than those in the mainstream. Additionally, the average concentration of residual-P was significantly higher in tributaries than in the mainstream. Also in the mainstream, the contents of BAP were significantly higher than those in the tributaries. Because there are more tributaries at the head of Manwan Reservoir, the untreated sewage and drainage of phosphorus from the agricultural land near the tributaries will merge into the mainstream, causing an increase in inorganic P in the sediments.

The total phosphorus (TP) and the proportion of BAP contents in inorganic P (IP) in Manwan Reservoir compared with other freshwater systems is presented in Table 3. The concentration of TP in Manwan Reservoir ranged from 623–899 μg/g, which was lower than other mesotrophic lakes, indicating that Manwan Reservoir may be low-to-mesotrophic. Therefore, phosphorus released by sediments may influence the water quality greatly. The percentage of BAP in IP indicated that the potential risk of P releasing in the Manwan Reservoir was relatively lower than that in other lakes and rivers. However, constant monitoring and further study are still required to understand the future internal loading in Manwan Reservoir.

Relationship between P Fractions and Influencing Factors

Table 4 shows the concentrations of Al, Ca, Fe and Mn in sediments both in the mainstream and the tributaries of the Manwan Reservoir. The rank order of the concentrations of these four metals both in mainstream and tributaries was Al>Fe>Ca>Mn. The Al content in the tributaries was higher than that in the mainstream, whereas the content of Fe, Ca and Mn in the tributaries was 20%, 47% and 48% lower, respectively, than those in the mainstream. Al and Fe are lithogenous materials and are conservative in the migration process, which contributes to the high content of these metals in the sediments [35,59].

Correlation analysis was used to examine the relationship between the contents of P fractions and related metals (Table 5). The results indicated that there were close relationships among P fractions, metals and the grain size fraction of the sediments. Both ex-P and BAP were correlated positively with the contents of the silt/clay fraction in the sediments; BAP and TP were correlated negatively with the contents of the coarse/medium sand fraction (246–840 μm) and the fine sand fraction (147–246 μm). A significant correlation between the concentration of iron and the silt/clay fraction of sediments was also observed ($r = 0.544$, $p<0.05$). Previous researchers found that grain size has an effect on the chemical composition of the sediments, including the metal contents and the P sorption-desorption ability [25,55]. As a result, the silt/clay fraction may contain more elements, such as iron, which are very important in adsorbing different P fractions, especially NaOH-P and BD-P compositions of BAP [12,60,61]. Ex-P is another fraction of BAP that includes phosphorus lightly absorbed onto the surface of sediment particles; therefore, it is associated closely with the surface physical characteristics of sediments. As the silt/clay fraction has more surface area, so it can absorb more ex-P. The results were consistent with studies performed in the mainstream of Haihe River and Keelung River in China, where a significant correlation between ex-P and silt/clay fraction was found and iron showed a positive linear relationship with fine-grained sediments (grain size<63 μm) [55,60]. Distance to Manwan Dam was correlated significantly with the contents of NaOH-P and BAP, and was also correlated with the contents of coarse sand ($r = 0.581$, $p<0.05$) and aluminium ($r = -0.486$, $p<0.05$) in the sediments (Table 5), indicating that the sediments farther from Manwan Dam contain more coarse sand with fewer aluminium oxides bound to the surface of sediments, which leads to the reduction of NaOH-P and BAP.

Flow variability affects the ecological patterns and processes in river systems, such as the nutrients dynamic [26]. Dams can manipulate the flow regime and consequently, in the upstream of the dam, fine suspended particles are captured and accumulated from the floodplain, while the downstream channel becomes eroded, leading to the coarsening of the sediments [62,63,64] Because of the combined effect of Xiaowan and Manwan dams, the flow velocity downstream from Xiaowan Dam is much higher than in the upstream of Manwan Dam, leading to coarser sediments fractions downstream of Xiaowan Dam. As it approaches the Manwan Dam, the river surface widens and the flow velocity slows, resulting in the accumulation of more slit/clay sediment fractions. According to correlation analysis, the contents of total phosphorus (TP) were correlated positively with Fe ($r = 0.501$, $p<0.05$). Previous studies found that there was an apparent relationship between TP and Fe [12,44]. In the Manwan Reservoir, the HCl-P content was not related to Ca. This finding is consistent with other studies [43,55,65], and can be attributed to the different sources of HCl-P, including sedimentary processes, the exchanges between calcium- and iron-bound phosphate [66] and P from fertilizer caused by runoff [27,43].

Redundancy analysis (RDA) was performed, using four P fractions, bio-available P, and total P as response variables and using metals, distance to Manwan hydropower station and grain size distribution as explanatory variables. The percentage of silt/clay fraction in sediments, distance to Manwan hydropower station and the concentration of Mn in sediments explained the variation significantly ($p = 0.014$, 0.018 and 0.05, respectively; Fig. 5). These three variables accounted for 54% of the total variance, of which the percentage of silt/clay fraction accounted for 20%, distance to Manwan hydropower station explained 19% and the concentration of Mn explained 15%. The percentage of silt/clay fraction correlated positively with the contents of ex-P, BD-P, NaOH-P and BAP in sediments, and the contents of Mn correlated positively with BD-P. RDA analysis also revealed the relationship between P fractions and other factors. Further, RDA can present the relationship between different sampling sites. Concentrations of NaOH-P were correlated positively with ex-P, which was mainly due to its association with the silt/clay fraction of the sediments. The samples from the study sites were distributed in different quarters of the RDA biplot. Samples in the mainstream at the head of the Manwan Reservoir (S10–S14) were related to ex-P, BD-P, BAP and silt/clay fraction of the sediments, which represented a high release risk of P.

Conclusions

Because of the construction of dams, reservoirs have become sensitive regions that have attracted more and more attention in recent years. The release of phosphorus from sediments can influence water quality and water purification capacity, which will endanger the beneficial uses of the river [67]. Thus, spatial analysis of P fractions is important in understanding nutrient pollution and potential ecological risks in the reservoir system. The results of our study are summarized below.

In the mainstream, the rank order of P fractions was HCl-P>NaOH-P>residual-P>BD-P>ex-P. Further, the concentration of ex-P, BD-P and NaOH-P showed an upward trend from Xiaowan to Manwan dam, where the contents of bio-available P (BAP)–the sum of ex-P, BD-P and NaOH-P–were relatively higher from S10 to S14 in the mainstream, especially within approximately 10 km of the Manwan Dam. Moreover, the concentration of BAP in the mainstream was significantly higher than it was in the tributaries; this indicated a greater release risk of phosphorus and more immediately available P in the water. In contrast, there was no trend in the contents of HCl-P and total phosphorus (TP) along the longitudinal direction of Manwan Reservoir. Hence, it may be better to estimate the internal phosphorus loading of the Manwan Reservoir by the contents of BAP rather than TP in the sediments.

In this study, correlation and redundancy analyses (RDA) revealed that silt/clay fraction contents of sediments and distance to Manwan Reservoir influenced the spatial variation of P fractions strongly, especially the bio-available fractions including ex-P, BD-P and NaOH-P. The contents of total phosphorus in the sediments of Manwan Reservoir was 623–899 μg/g, which were lower than some mesotrophic lakes in the world and depended primarily on the contents of iron in the sediments.

The techniques used in this study can be employed widely in freshwater systems and can contribute to the investigation of the tropic conditions, as well as evaluate the potentially mobile P pool stored in canyon reservoir sediments. Especially for an international river like the Mekong nearly half the length of which was in Yunnan Province in China, the construction and operation of cascade dams upstream may cause significant ecological risks downstream [68], risks that are of international concern. Our studies demonstrated that there exist "longitudinal effects" of dams on some phosphorus fractions. Therefore, a management plan was recommended that will intensify the monitoring of the internal P loading of the reservoir, as well as reduce the external P loading that includes the non-point loading from tributaries.

Author Contributions

Conceived and designed the experiments: CW QZ SD LD SL QL. Performed the experiments: HZ QL. Analyzed the data: QL. Contributed reagents/materials/analysis tools: QL SL. Wrote the paper: QL SL.

References

1. Topping DJ, Rubin DM, Vierra L Jr (2000) Colorado River sediment transport 1. Natural sediment supply limitation and the influence of Glen Canyon Dam. Water Resources Research 36: 515–542.
2. Kileshye Onema JM, Mazvimavi D, Love D, Mul ML (2006) Effects of selected dams on river flows of Insiza River, Zimbabwe. Physics and Chemistry of the Earth, Parts A/B/C 31: 870–875.
3. Wei G, Yang Z, Cui B, Li B, Chen H, et al. (2008) Impact of dam construction on water quality and water self-purification capacity of the Lancang River, China. Water Resources Management 23: 1763–1780.
4. Gong G-C, Chang J, Chiang K-P, Hsiung T-M, Hung C-C, et al. (2006) Reduction of primary production and changing of nutrient ratio in the East China Sea: Effect of the Three Gorges Dam? Geophysical Research Letters 33.
5. Warner RF (2012) Environmental impacts of hydroelectric power and other anthropogenic developments on the hydromorphology and ecology of the Durance channel and the Etang de Berre, southeast France. Journal of Environmental Management 104: 35–50.
6. Ouyang W, Hao F, Song K, Zhang X (2010) Cascade Dam-Induced Hydrological Disturbance and Environmental Impact in the Upper Stream of the Yellow River. Water Resources Management 25: 913–927.
7. Sanclements MD, Fernandez IJ, Norton SA (2009) Soil and sediment phosphorus fractions in a forested watershed at Acadia National Park, ME, USA. Forest Ecology and Management 258: 2318–2325.
8. Christophoridis C, Fytianos K (2006) Conditions affecting the release of phosphorus from surface lake sediments. Journal of Environment Quality 35: 1181–1192.
9. Kaiserli A, Voutsa D, Samara C (2002) Phosphorus fractionation in lake sediments – Lakes Volvi and Koronia, N. Greece. Chemosphere 46: 1147–1155.
10. Vollenweider RA (1982) Eutrophication of waters: Monitoring, assessment and control: Organisation for Economic Co-operation and Development.
11. Farmer JG, Bailey-Wattes AE, Kirika A, Scott C (1994) Phosphorus fractionation and mobility. Aquatic Conservation: Marine And Freshwater Ecosystem 4: 45–56.
12. Fytianos K, Kotzakioti A (2005) Sequential fractionation of phosphorus in lake sediments of Northern Greece. Environmental Monitoring and Assessment 100: 191–200.
13. Ramm K, Scheps V (1997) Phosphorus balance of a polytrophic shallow lake with consideration of phosphorus release. Hydrobiologia 342/343: 43–53.
14. Zhang B, Fang F, Guo J, Chen Y, Li Z, et al. (2012) Phosphorus fractions and phosphate sorption-release characteristics relevant to the soil composition of water-level-fluctuating zone of Three Gorges Reservoir. Ecological Engineering 40: 153–159.
15. Jacoby JM, Lynch DD, Welch EB, Perkins MA (1982) Internal phosphorus loading in a shallow eutrophic lake. Water Research 16: 911–919.
16. Cooke GD, McComas MR, Waller DW, Kennedy RH (1977) The occurrence of internal phosphorus loading in two small, eutrophic, glacial lakes in northeastern Ohio. Hydrobiologia 56: 129–135.
17. Boström B, Andersen JM, Fleischer S, Jansson M (1988) Exchange of phosphorus across the sediment-water interface. Hydrobiologia 170: 229–244.
18. Spears BM, Carvalho L, Perkins R, Kirika A, Paterson DM (2006) Spatial and historical variation in sediment phosphorus fractions and mobility in a large shallow lake. Water Research 40: 383–391.
19. Stone M, English M (1993) Geochemical composition, phosphorus speciation and mass transport of fine-grained sediment in two Lake Erie tributaries. Hydrobiologia 253: 17–29.
20. Kopáček J, Borovec J, Hejzlar J, Ulrich K-U, Norton SA, et al. (2005) Aluminum Control of Phosphorus Sorption by Lake Sediments. Environmental Science & Technology 39: 8784–8789.
21. Zhou A, Wang D, Hongxiao T (2005) Phosphorus fraction and bioavailability in Taihu Lake(China) sediment. Journal of Environmental Science 17: 384–388.
22. Xiang S, Zhou W (2011) Phosphorus forms and distribution in the sediments of Poyang Lake, China. International Journal of Sediment Research 26: 230–238.
23. Danen-Louwerse H, Lijklema L, Coenraats M (1993) Iron content of sediment and phosphate adsorption properties. Hydrobiologia 253: 311–317.
24. Jin X, Wang S, Pang Y, Chang Wu F (2006) Phosphorus fractions and the effect of pH on the phosphorus release of the sediments from different trophic areas in Taihu Lake, China. Environmental Pollution 139: 288–295.
25. Zhang C, Wang L, Li G, Dong S, Yang J, et al. (2002) Grain size effect on multi-element concentrations in sediments from the intertidal flats of Bohai Bay, China. Applied Geochemistry 17: 59–68.
26. Maazouzi C, Claret C, Dole-Olivier M-J, Marmonier P (2013) Nutrient dynamics in river bed sediments: effects of hydrological disturbances using experimental flow manipulations. Journal of Soils and Sediments 13: 207–219.
27. Andrieux-Loyer F, Aminot A (2001) Phosphorus forms related to sediment grain size and geochemical characteristics in French coastal areas. Estuarine, Coastal and Shelf Science 52: 617–629.
28. Wang S, Jin X, Zhao H, Wu F (2006) Phosphorus fractions and its release in the sediments from the shallow lakes in the middle and lower reaches of Yangtze River area in China. Colloids and Surfaces A: Physicochemical and Engineering Aspects 273: 109–116.
29. Sun S, Huang S, Sun X, Wen W (2009) Phosphorus fractions and its release in the sediments of Haihe River, China. Journal of Environmental Sciences 21: 291–295.
30. Hong Y, Geng J, Qiao S, Zhang Y, Ding L, et al. (2010) Phosphorus fractions and matrix-bound phosphine in coastal surface sediments of the Southwest Yellow Sea. Journal of Hazardous Materials 181: 556–564.
31. Ruttenberg KC (1992) Development of a sequential extraction method for different forms of phosphorus in marine sediments. Limnology and Oceanography 37: 1460–1482.
32. Lanza GR (2011) Accelerated Eutrophication in the Mekong River Watershed: Hydropower Development, Climate Change, and Waterborne Disease. Eutrophication: causes, consequences and control. 373–386.
33. Vicente I, Andersen FØ, Hansen HCB, Cruz-Pizarro L, Jensen HS (2010) Water level fluctuations may decrease phosphate adsorption capacity of the sediment in oligotrophic high mountain lakes. Hydrobiologia 651: 253–264.
34. He D-m, Zhao W-j, Chen L-h (2004) The ecological changes in Manwan reservoir area and its causes Journal-Yunnan University Natural Sciences 26: 220–226 (in Chinese).
35. Wang C, Liu S, Zhao Q, Deng L, Dong S (2012) Spatial variation and contamination assessment of heavy metals in sediments in the Manwan Reservoir, Lancang River. Ecotoxicology and Environmental Safety 82: 32–39.
36. Zhao Q, Liu S, Deng L, Yang Z, Dong S, et al. (2012) Spatio-temporal variation of heavy metals in fresh water after dam construction: a case study of the Manwan Reservoir, Lancang River. Environmental Monitoring and Assessment 184: 4253–4266.
37. Zhao Q, Liu S, Deng L, Dong S, Wang C (2012) Longitudinal distribution of heavy metals in sediments of a canyon reservoir in Southwest China due to dam construction. Environmental Monitoring and Assessment: 1–10.
38. Liu S, Cui B, Dong S, Yang Z, Yang M, et al. (2008) Evaluating the influence of road networks on landscape and regional ecological risk–A case study in Lancang River Valley of Southwest China. Ecological Engineering 34: 91–99.
39. Zhao Q, Liu S, Deng L, Dong S, Cong, etal. (2012) Landscape change and hydrologic alteration associated with dam construction. International Journal of Applied Earth Observation and Geoinformation 16: 17–26.
40. Heng L, Jiufu L, Haixing T (1998) Present and future of water resources development in Lancang River basin in Yunnan Province. Advances in Water Science 9: 70–76.
41. He D, Wu S, Peng H, Yang Z, Ou X, et al. (2005) A study of ecosystem changes in longitudinal Range-Gorge region and transboundary eco-security in southwest China. Advances in Earth Science 20: 338–344.
42. Fu KD, He DM, Lu XX (2008) Sedimentation in the Manwan reservoir in the Upper Mekong and its downstream impacts. Quaternary International 186: 91–99.
43. Wang P, He M, Lin C, Men B, Liu R, et al. (2009) Phosphorus distribution in the estuarine sediments of the Daliao river, China. Estuarine, Coastal and Shelf Science 84: 246–252.
44. Lin C, Wang Z, He M, Li Y, Liu R, et al. (2009) Phosphorus sorption and fraction characteristics in the upper, middle and low reach sediments of the Daliao river systems, China. Journal of Hazardous Materials 170: 278–285.
45. Psenner R, Pucsko R, Sage M (1984) Fractionation of Organic and Inorganic Phosphorus Compounds in Lake Sediments, An Attempt to Characterize Ecologically Important Fractions (Die Fraktionierung Organischer und Anorganischer Phosphorverbindungen von Sedimenten, Versuch einer Definition Okologisch Wichtiger Fraktionen). Archiv fur Hydrobiologie 1.
46. Hupfer M, Gächter R, Giovanoli R (1995) Transformation of phosphorus species in settling seston and during early sediment diagenesis. Aquatic Sciences-Research Across Boundaries 57: 305–324.
47. SEPAC (2004) The technical specification for soil environmental monitoring. Beijing: Environmental Press of China.
48. Gonzalez C, Orellana L, Casanovas S, Pignata M (1998) Environmental conditions and chemical response of a transplanted lichen to an urban area. Journal of environmental management 53: 73–81.
49. Birks H (1995) Quantitative palaeoenvironmental reconstructions. Statistical modelling of quaternary science data Technical guide 5: 161–254.
50. Van Dobben H, Wolterbeek HT, Wamelink G, Ter Braak C (2001) Relationship between epiphytic lichens, trace elements and gaseous atmospheric pollutants. Environmental Pollution 112: 163–169.
51. Pettersson K (2001) Phosphorus characteristics of settling and suspended particles in Lake Erken. Science of the total environment 266: 79–86.
52. Dittrich M, Chesnyuk A, Gudimov A, McCulloch J, Quaizi S, et al. (2012) Phosphorus retention in a mesotrophic lake under transient loading conditions:

Insights from a sediment phosphorus binding form study. Water Research 17: 1433–1447.

53. Kozerski HP, Kleeberg A (2007) The Sediments and Benthic-Pelagic Exchange in the Shallow Lake Müggelsee (Berlin, Germany). International Review of Hydrobiology 83: 77–112.
54. Zhou Q, Gibson CE, Zhu Y (2001) Evaluation of phosphorus bioavailability in sediments of three contrasting lakes in China and the UK. Chemosphere 42: 221–225.
55. Han L, Huang S, Stanley C, Osborne T (2011) Phosphorus Fractionation in Core Sediments from Haihe River Mainstream, China. Soil and Sediment Contamination 20: 30–53.
56. Zhang Z, Wang Z, Joseph H, Xu X, Wang H, et al. (2012) The release of phosphorus from sediment into water in subtropical wetlands: a warming microcosm experiment. Hydrological Processes 26: 15–26.
57. Ribeiro DC, Martins G, Nogueira R, Cruz JV, Brito AG (2008) Phosphorus fractionation in volcanic lake sediments (Azores – Portugal). Chemosphere 70: 1256–1263.
58. Psenner R, Pucsko R (1988) Phosphorus fractionation: advantages and limits of the method for the study of sediment P origins and interactions. Arch Hydrobiol Beih Ergebn Limnol 30: 43–59.
59. Rubio B, Nombela M, Vilas F (2000) Geochemistry of major and trace elements in sediments of the Ria de Vigo (NW Spain): an assessment of metal pollution. Marine Pollution Bulletin 40: 968–980.
60. Huang K-M, Lin S (2003) Consequences and implication of heavy metal spatial variations in sediments of the Keelung River drainage basin, Taiwan. Chemosphere 53: 1113–1121.
61. Pan K, Wang W-X (2012) Trace metal contamination in estuarine and coastal environments in China. Science of the total environment 421: 3–16.
62. Luo X, Yang S, Zhang J (2012) The impact of the Three Gorges Dam on the downstream distribution and texture of sediments along the middle and lower Yangtze River (Changjiang) and its estuary, and subsequent sediment dispersal in the East China Sea. Geomorphology 179: 126–140.
63. Heath S, Plater A (2010) Records of pan (floodplain wetland) sedimentation as an approach for post-hoc investigation of the hydrological impacts of dam impoundment: The Pongolo river, KwaZulu-Natal. Water Research 44: 4226–4240.
64. Williams GP, Wolman MG (1984) Downstream effects of dams on alluvial rivers. Washington, DC, USA: US Government Printing Office.
65. Morse JW, Cook N (1978) The distribution and form of phosphorus in North Atlantic Ocean deep-sea and continental slope sediments. Limnology and Oceanography 23: 825–830.
66. Golterman H (1988) The calcium-and iron bound phosphate phase diagram. Hydrobiologia 159: 149–151.
67. Kim LH, Choi E, Stenstrom MK (2003) Sediment characteristics, phosphorus types and phosphorus release rates between river and lake sediments. Chemosphere 50: 53–61.
68. Jacobs JW (2002) The Mekong River Commission: transboundary water resources planning and regional security. The Geographical Journal 168: 354–364.

2

Gene Expression Variation in Duplicate *Lactate dehydrogenase* Genes: Do Ecological Species Show Distinct Responses?

Melania E. Cristescu[1,3]*, Bora Demiri[1], Ianina Altshuler[3], Teresa J. Crease[2]

1 Great Lakes Institute for Environmental Research, University of Windsor, Windsor, Ontario, Canada, **2** Department of Integrative Biology, University of Guelph, Guelph, Ontario, Canada, **3** Department of Biology, McGill University, Montreal, Quebec, Canada

Abstract

Lactate dehydrogenase (LDH) has been shown to play an important role in adaptation of several aquatic species to different habitats. The genomes of *Daphnia pulex*, a pond species, and *Daphnia pulicaria*, a lake inhabitant, encode two L-LDH enzymes, LDHA and LDHB. We estimated relative levels of *Ldh* gene expression in these two closely related species and their hybrids in four environmental settings, each characterized by one of two temperatures (10°C or 20°C), and one of two concentrations of dissolved oxygen (DO; 6.5–7 mg/l or 2–3 mg/l). We found that levels of *Ldh*A expression were 4 to 48 times higher than *Ldh*B expression ($p<0.005$) in all three groups (the two parental species and hybrids). Moreover, levels of *LdhB* expression differed significantly ($p<0.05$) between *D. pulex* and *D. pulicaria,* but neither species differed from the hybrid. Consistently higher expression of *Ldh*A relative to *Ldh*B in both species and the hybrid suggests that the two isozymes could be performing different functions. No significant differences in levels of gene expression were observed among the four combinations of temperature and dissolved oxygen ($p>0.1$). Given that *Daphnia* dwell in environments characterized by fluctuating conditions with long periods of low dissolved oxygen concentration, we suggest that these species could employ regulated metabolic depression to survive in such environments.

Editor: Diego Fontaneto, Consiglio Nazionale delle Ricerche (CNR), Italy

Funding: This work was supported by NSERC Canada Discovery grants to TJC and MEC and by an NSERC CREATE grant to MEC. The funders had no role in study design, data collection and analysis, decision to publish, or preparation of the manuscript.

Competing Interests: The authors have declared that no competing interests exist.

* Email: melania.cristescu@mcgill.ca

Introduction

Identifying the molecular changes that contribute to the adaptation of a species to its local environment is an important goal of evolutionary studies [1]. Adaptation to a new environment may be accomplished by molecular changes in enzymes that alter the intrinsic enzymatic rate constants, by a heritable change in the control region of the genome influencing transcript level and consequently enzyme concentration, or by epigenetic factors [2]. Studies investigating the genetic basis of fitness differences in natural populations are often based on protein coding genes involved with glucose metabolism [3], [4], [5], [6]. One such group of enzymes is the NAD-dependent L-lactate dehydrogenases (LDH), which are required by most organisms to reduce pyruvate to L(−) lactate during glycolysis and oxidize L(−) lactate to pyruvate during gluconeogenesis [7].

LDH has been shown to contribute to the adaptation of natural populations in different habitats [2], [8], [9], [10], [11]. For instance, populations of *Fundulus heteroclitus* in Maine and Georgia are fixed for two different alleles at the heart-type *Ldh*B locus. The isozymes encoded by this locus have different temperature-dependent kinetic characteristics due to a few amino acid replacements making these two populations more adapted to their different thermal environments [8]. In addition, an increase in the activity of LDH in the killifish, *F. heteroclitus*, is observed during prolonged exposure to hypoxia [12] indicating the importance of this enzyme in the adaption to different environmental conditions.

A strong association between *Ldh* genotype and habitat has also been observed in members of the *Daphnia pulex* species complex. The pond species, *D. pulex* and the closely related lake species, *D. pulicaria* are fixed for different alleles at the *Ldh*A locus. *Daphnia* in ephemeral ponds are homozygous for the *Ldh*A slow allele (S), whereas *Daphnia* in permanent lakes are homozygous for the *Ldh*A fast allele (F), where S and F refer to relative mobility through gels during protein electrophoresis [13], [14], [15], [16]. Hybrids of the two species, with the heterozygous SF (Slow/Fast) genotype at the *Ldh*A locus, can be found in disturbed, deforested ponds [17] or intermediate habitats. Detailed phylogenetic studies revealed that the main differences between the S and F alleles of the *Ldh*A locus are amino acid substitutions at two sites [18]. This includes the substitution of Aspartic acid in the S allele for Glutamic acid in the F allele at amino acid position 6, and the substitution of Glutamine in S for Glutamic acid in F at position 229, which likely causes the slow/fast mobility shift. Considerably more amino acid variation was observed among LDHB proteins [18].

Several studies have found that *Daphnia* in lakes and ponds differ considerably in their life history traits [19], [20], [21] since these habitats are characterized by different physical and biotic conditions [22]. *Daphnia pulex* populations inhabit fishless, ephemeral ponds for a short period of time in the spring and early summer where they are exposed to anoxia, and eventually desiccation or complete freezing [23]. Conversely, *D. pulicaria* populations can be found in the cold hypolimnetic region of stratified lakes in order to avoid fish predation [24]. It is likely that these two species have diverged due to distinct environmental conditions in their specific habitats such as temperature, dissolved oxygen concentration, biotic interactions, and food sources. However, despite advances in understanding the ecology of *Daphnia* and the availability of new genomic tools, we lack an understanding of the molecular mechanisms used by these species to adapt to ephemeral pond versus permanent lake habitats.

In addition to LDHA, the genomes of *D. pulicaria* and *D. pulex* encode a more genetically variable, second isoenzyme, LDHB [25]. Based on phylogenetic studies, the two loci (*Ldh*A and *Ldh*B) seem to be the product of a relatively recent gene duplication event in *Daphnia*, after its divergence from other crustaceans [25]. The fixation of different *Ldh*A alleles in these two sister species, which occur in different aquatic habitats, suggests that LDH variation could be adaptive, or *LdhA* is linked to loci that are under selection [14], [25]. Therefore, it is of interest to estimate variation in gene expression among these sister species in controlled laboratory conditions.

This study determines the expression patterns of both isoenzymes, *Ldh*A and *Ldh*B, in clones of *D. pulicaria*, *D. pulex* and *D. pulex/pulicaria* hybrids that have been acclimatized to four different environmental settings. Each setting was characterized by one of two temperatures (10°C or 20°C), and one of two concentrations of dissolved oxygen (DO, 6.5–7 mg/l or 2–3 mg/l).

Materials and Methods

Sample collection

This study involved collection of *Daphnia* from ponds and lakes. No specific permissions are required to sample *Daphnia* as they are not endangered or protected species. Lakes were accessed via public boat ramps while ponds were sampled from public roadsides or on private land with the permission of the land owner. A total of eight habitats (three lakes and five ponds) were sampled in Michigan and southern Ontario (Figure S1) during early spring, soon after hatching, when genetic diversity is high [26]. Individual female *Daphnia* were taken to the laboratory and allowed to reproduce parthenogenetically in separate beakers. After genotyping (see below), three genetically unique clonal lines were selected from each of the three pond populations of *D. pulex* (Canard 3E, Solomon, Disputed), and each of the three lake populations of *D. pulicaria* (Warner, Three Lakes II, Lawrence) giving nine clones per species. In addition, we isolated nine *D. pulex/pulicaria* hybrid lines. Four lines were established from pond Canard 1B, three lines were established from pond Canard 3L and one line was established from each of ponds Canard 3 and Disputed. The isolates were maintained in filtered river water at 15−18°C with a 14-h light, 10-h dark photoperiod and fed every 3–4 days with a combination of the microalgae species *Nannochloropsis* and *Tetraselmis* (Reed Mariculture).

Genotyping and sexuality tests

Isolates from both pond and lake populations were initially genotyped at 21 previously mapped microsatellite markers by Cristescu et al. [27] and Xu et al. [28]. The 12 *D. pulex* linkage groups were each represented by one to three loci [29]. Isolates with unique multilocus genotypes were selected for the gene expression experiment. In order to identify *D. pulex/pulicaria* hybrids from the pond populations, the *Ldh*A genotype of each isolate was also determined by Cristescu et al. [27] and Xu et al. [28]. Previous surveys of allozyme variation have shown that lake populations are fixed for an electrophoretically "fast" (F) allele at the *Ldh*A locus [26], [30] while pond populations are either fixed for the "slow" (S) allele or contain SF heterozygotes. The heterozygotes are considered to be hybrids of *D. pulex* and *D. pulicaria* [15], [28] and are reported to reproduce by obligate parthenogenesis [26]. We used S and F allele-specific primers based on *Ldh*A sequences from both species [25] to determine the genotype of each isolate. Sexuality tests were conducted on the pond isolates to determine their mode of reproduction (cyclical parthenogenesis or obligate parthenogenesis), using the method of Innes *et al.* [26], which involves rearing female *Daphnia* in the absence of males and observing their ability, or lack thereof, to deposit eggs (dormant embryos) into their ephippial egg case.

Acclimation and experimental methods

The 27 clonal lines were cultured under standardized conditions, as described above, for a period of one year to eliminate maternal effects. Progenitor females from each line were selected to establish the mother generation of the acclimation experiments.

The four treatments were characterized by a combination of two temperatures, 10°C or 20°C, and two concentrations of dissolved oxygen (DO), 6.5–7 mg/l or 2–3 mg/l. The high DO concentration was obtained by bubbling air into the aquaria using a Topfin Airpump 8000, for 30 min twice a day. The low DO concentration was obtained by bubbling nitrogen gas in the same manner. Each of the four aquaria contained 50-ml chambers each inoculated with one juvenile female from each of the 27 genetically distinct clones. The floor of the chambers was 60-micron mesh, which allowed water and food to circulate freely throughout the aquarium and the chambers (Figure 1). This experimental design allowed the maintenance of uniform experimental conditions across the *Daphnia* clones while keeping them physically separated from one another. The *Daphnia* were exposed to a 14 h light: 10 h dark cycle and were fed daily, at libitum with the unicellular, previously frozen, green algae *Scenedesmus* to maintain an equivalent of 1 mg of C/L in the aquaria. The first clutch of offspring produced by each female was removed from the chamber. Once a female produced the 2nd and 3rd clutch, she was removed and the offspring were left in the chamber to mature. When these offspring reached maturity and eggs could be seen in their brood pouches, they were collected, placed in 400 µL of RNA later (Ambion), and frozen instantly in liquid nitrogen to prevent RNA degradation. The samples were then stored at −80°C until total RNA was extracted.

RNA isolation and cDNA synthesis

Total RNA was extracted from four to six individuals from each chamber using the RNA extraction and purification protocol developed by the *Daphnia* Genomic Consortium [31]. RNA purification was performed using the RNeasy Mini Kit (QIAGEN, Toronto, Canada). During RNA purification the samples were exposed to RNase-free DNAse I for 15 min prior to RNA elution. The purified nucleic acids were re-suspended in 30 µl of RNAse-free water. The concentration of the RNA samples was checked with the NanoVue spectrophotometer (GE Healthcare Technologies) and the ratio of absorbance at 260 nm/280 nm was measured. All samples used for the experiment had a ratio of ~2.0, which is generally accepted to represent RNA that is

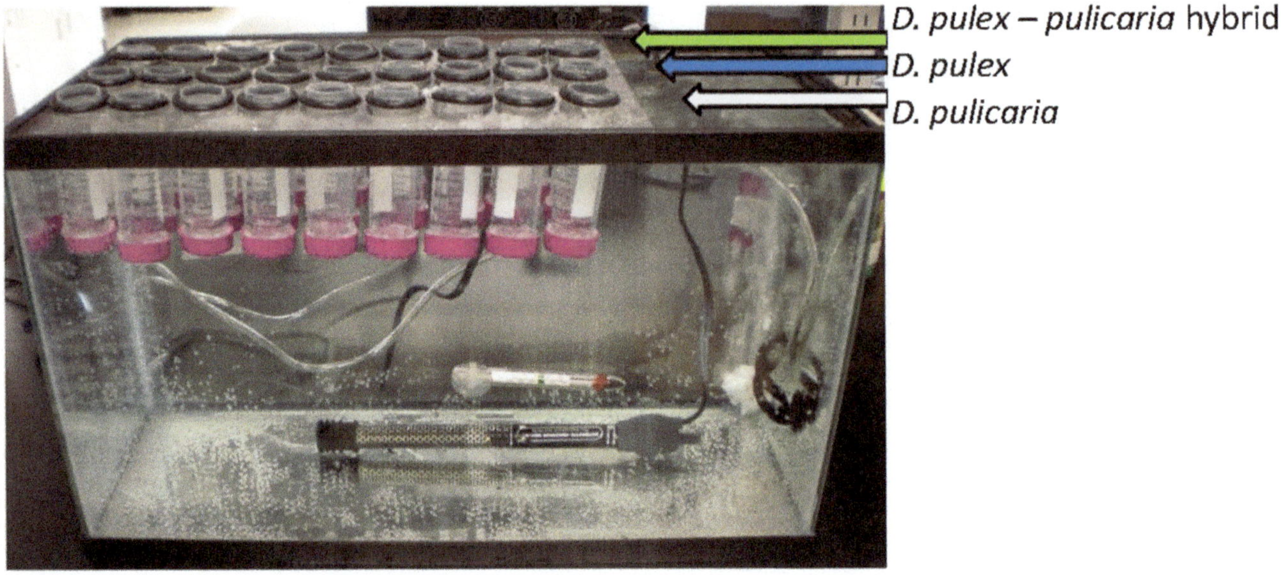

Figure 1. Aquarium set up. Each of the four aquaria contained 27 chambers that were filled with one of nine *D. pulex* clones (pond), nine *D. pulicaria* clones (lake), and nine hybrid clones. Each chamber had a mesh bottom allowing food and air/nitrogen to circulate freely within the aquarium while keeping the clones isolated from each other. There were four aquaria each characterized by different environmental conditions. Aquarium 1 corresponds to 20°C/High [6.5–7 mg/L] DO; aquarium 2 (20°C/Low [2–3 mg/L] DO); aquarium 3 (10°C/High DO); and aquarium 4 (10°C/Low DO).

contaminated with little or no DNA. PCR reactions lacking Reverse Transcriptase were run on the RNA samples to confirm that there was no DNA contamination.

Forty ng of RNA from each extraction was reverse-transcribed with the Sensiscript Reverse Transcription Kit (QIAGEN) according to the manufacturer's protocol, using Oligo-dT primers from Ambion at a final concentration of 1 μM, and Sensiscript reverse transcriptase enzyme. The reverse transcription reaction was carried out at 37°C for 60 minutes with 20 μl total volume.

Quantitative PCR

Primers for quantitative PCR (qPCR) analysis of *Ldh*A and *Ldh*B expression (Table 1) were designed from sequences downloaded from the DOE Joint Genome Institute's *Daphnia pulex* database using Integrated DNA Technologie's PrimerQuest^{SM} program, which is based on the Primer3 code [32]. The primer sequences (Table 1) for the reference gene, *Glyceraldehyde-3-phosphate dehydrogenase* (*Gapdh*), were obtained from Spanier et al. [33]. The primers were designed to generate 90–216 base pair (bp) amplicons and the primer pair efficiency ranged from 94.9% to 96.6%. Primer pair efficiency was estimated using LinRegPCR v. 11.0 [34], [35], which determines the mean PCR efficiency of a primer pair by a linear regression fit of the data in the exponential phase of the qPCR reaction. Three qPCR reactions were performed on each RNA sample in a total volume of 20 μl that contained 10 μl of 2X SYBR Premix Ex Taq^{TM} master mix (Takara Bio USA, Madison, WI), 2 μl of cDNA template (made from 40 ng of RNA in 20 μl volume), 0.4 μl of ROX^{TM} Reference Dye, 6.8 μl of H_2O, and 0.4 μl (10 μM) each of the forward and reverse primers. Negative control reactions without cDNA template were also included to check for non-specific signal due to PCR artifacts such as primer dimerization. Reactions were carried out in clear, qPCR compatible, 96 well plates (Axygen, Waltham, MA). Amplification was performed on an ABI 7500 Real-Time thermal cycler (Applied Biosystems, Foster City, CA) using the thermal cycling parameters recommended for the SYBR Premix Ex Taq^{TM}: 95°C for 10 sec, followed by 40 cycles of 95°C for 5 sec and 60°C for 34 sec. The threshold was set to 0.0276. A melting curve analysis was performed at the end of each qPCR reaction. Only one peak was observed, and there were no primer dimers detected in the qPCR reactions.

Data analysis and statistics

The value of $2^{-\Delta CT}$ was used to determine the level of *Ldh*A and *Ldh*B expression relative to *Gapdh* expression, which is assumed to be constant across treatments [36], in each clone in each of the four treatments. Triplicate C_T values for each RNA sample were averaged before performing the $-\Delta C_T$ calculations. A univariate analysis of variance (ANOVA) was done on the ΔC_T values using IBM SPSS STATISTICS 19.0.0, to determine significant differences in *Ldh* relative to *Gapdh* gene expression in samples from different treatments. Because the overall comparisons of groups (*D. pulex*, *D. pulicaria*, hybrids) within genes were significant, pairwise group comparisons of the C_T values were also done using a t-test with the Bonferroni correction.

Results

Ldh genotyping and sexuality tests

Our sexuality tests confirmed that the hybrid isolates reproduced by obligate parthenogenesis and the *D. pulex* and *D. pulicaria* isolates reproduced by cyclic parthenogenesis. These results were consistent with the *LdhA* results reported by Cristescu et al. [27] and Xu et al. [28]. Lake isolates (*D. pulicaria*) were fixed for an electrophoretically "fast" (F) allele at the *Lactate dehydrogenase* A (*Ldh*A) locus while sexual pond isolates (*D. pulex*) were fixed for a "slow" (S) allele. All SF heterozygote isolates were considered to be hybrids between *D. pulex* and *D. pulicaria*.

Table 1. Primers used for qPCR analyses of *Ldh* expression in *D. pulex* and *D. pulicaria*.

Gene Code	Gene Name	Function	Gene ID*	Primer Sequences [5′ to 3′]	Amplicon size [bp]	Localization in gene	PCR Efficiency
LdhA	L-Lactate dehydrogenase A	GlycolyticEnzyme	Dappu- 230172	ATCCAGACTCCTGTTGCCCATTCA	216	Forward primer in 2nd exon, reverse in 3rd exon	94.9%
				TTCGCCCTTGAGTTTGTCCTCCAT			
LdhB	L-lactate dehydrogenase B	GlycolyticEnzyme	Dappu- 61140	TTGTCCAATACAGTCCCGACACCA	90	4th exon	96.6%
				GCAACCCACTGAGTTTCCAAGCAA			
Gapdh	glyceraldehyde-3-phosphate dehydrogenase	Glycolytic Enzyme	Dappu-302823	TGGGATGAGTCACTGGCATAC	136	3rd exon	95.2%
				GAAAGGACGACCAACAACAAAC			

*The gene ID refers to the *Daphnia pulex* draft genome annotation (dappu v1.1) at www.jgi.doe.gov.

Quantitative PCR

Due to the loss of some clones during the experiments, the analyses are based on the six clones from each of the three groups (for a total of 18 clones) for which we obtained data from all four treatment tanks (Tables S1, S2, S3 in File S1). When more than six clones survived, we selected six clones that maximized the number of shared clones between treatments. The expression of *Ldh*A was significantly different from *Ldh*B in all three groups; *D. pulex, D. pulicaria*, and the hybrid *D. pulex/pulicaria* (source = Gene, Table 2), although both genes were expressed at a much lower level than the reference gene, *Gapdh* (Tables S1, S2, S3 in File S1). In particular, *Ldh*A was expressed 26 to 40 times more than *Ldh*B in *D. pulex*, 4 to 15 times more than *Ldh*B in *D. pulicaria*, and 20 to 48 times more than *Ldh*B in *D. pulex-pulicaria* hybrids (Figure 2). The ANOVA also showed a significant difference in gene expression among the three *Daphnia* groups for each gene (source = Group, Table 2). *Ldh*A expression was similar in *D. pulex* and *D. pulicaria* but higher in the hybrids (Figure 2, Tables S1, S2, S3 in File S1). However, none of the pairwise group differences were significant, although p-values for comparisons between the hybrid and each parent species were less than 0.1 (Table 3). Conversely, *Ldh*B expression was significantly higher in *D. pulicaria* than in *D. pulex* and although expression in the *D. pulex/pulicaria* hybrids was more similar to that in *D. pulex* than *D. pulicaria*, it was not statistically significantly different from either parental species (Figure 2, Table 3, Tables S1, S2, S3 in File S1). Despite significant differences in *Ldh* expression between the two genes and the different groups, there was no significant variation in expression of either gene due to temperature or dissolved oxygen (Table 2).

Discussion

Previous studies revealed that the genes encoding the LDHA and LDHB isozymes are under purifying selection with *Ldh*A under stronger evolutionary constraint than *Ldh*B [25]. This suggests that while both isozymes likely produce functional enzymes, *Ldh*A may be more important in *Daphnia* metabolism, which is also consistent with the observation that *Ldh*A expression is significantly higher than *Ldh*B expression in both species and their hybrids, in all four environmental conditions. Substantially higher levels of *Ldh*A than *Ldh*B expression were also observed in a previous study of *Ldh* in these two species under a single set of environmental conditions [37]. The pairwise comparisons show that mean levels of *Ldh*B expression differ significantly between *D. pulicaria* and *D. pulex* with the former group having higher expression of the gene compared to the latter. However, neither of these species was significantly different from the *D. pulex/pulicaria* hybrids, which showed an intermediate level of *LdhB* expression.

It is generally assumed that the duplicate gene is redundant in function immediately after a duplication event. Thus, deleterious mutations in the control region that reduce or eliminate expression of one duplicate may have little effect on overall fitness, and could ultimately drift to fixation [38], [39]. Alternatively, the duplicated genes could undergo subfunctionalization [39] such that they perform only a subset of functions performed by the original protein, and/or the genes are only expressed in specific tissues. Such specialization does occur in jawed vertebrates in which there are three LDH isozymes found predominantly in three different types of tissue; anaerobic tissues (e.g. white skeletal muscle), aerobic tissues (e.g. heart, brain), and spermatozoa of mammals and birds [40]. Our results suggest that LDH specialization could also have occurred in *Daphnia*, such that the relative expression of *Ldh*A and *Ldh*B is correlated with the quantity of tissues in which

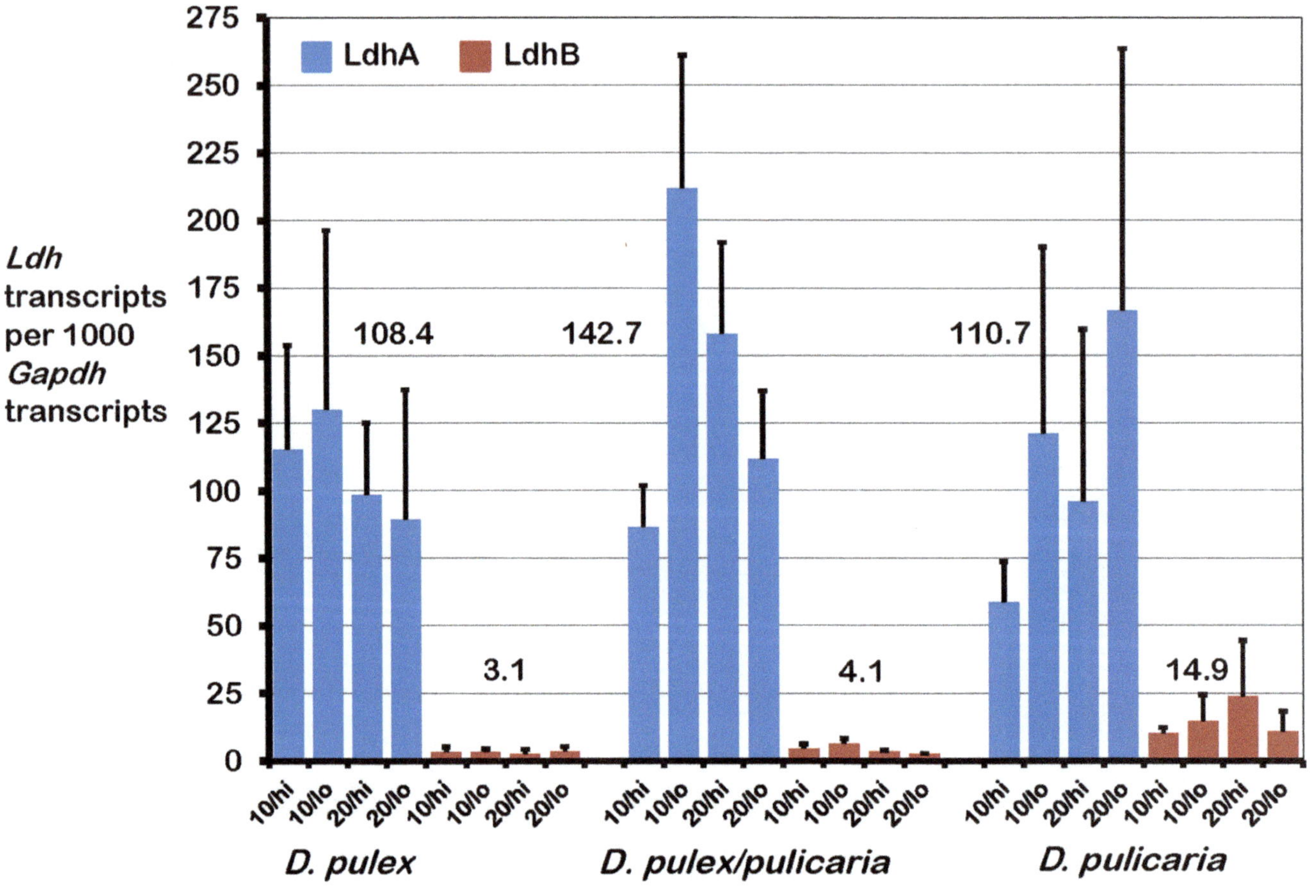

Figure 2. Expression of *Ldh*A and *Ldh*B relative to *Gapdh* in *Daphnia* under four combinations of temperature and dissolved oxygen. Each *Daphnia* group (*D. pulex*, *D. pulicaria* and hybrids) was represented by six clones in the final analysis. The Y-axis shows the number of *Ldh*A and *Ldh*B transcripts per 1000 *Gapdh* transcripts calculated using the $2^{-\Delta CT}$ method. The four treatments consisted of two temperatures, 10°C or 20°C each combined with two levels of dissolved oxygen, 6.5–7 mg/L (hi) or 2–3 mg/L (lo). Blue bars represent *LdhA* and red bars represent *LdhB*. Error bars represent standard error of the mean (SEM). The numbers on the graph are the mean value of relative *Ldh* expression for each treatment in each group. Only the difference between mean relative *Ldh*B expression in *D. pulex* (3.1) and *D. pulicaria* (14.9) is significant (Table 3).

Table 2. Analysis of variance of relative *Ldh* expression in 18 clones of *Daphnia pulex, D. pulicaria* and hybrids.

Gene	Source	Df	F	p-value
	Gene			**0.000**
*Ldh*A	Dissolved Oxygen (DO)	1	1.089	0.301
	Temperature (Temp)	1	0.366	0.547
	Group	**2**	**3.286**	**0.044**
	DO * Temp	1	0.421	0.519
	DO * Group	2	0.905	0.410
	Temp * Group	2	0.247	0.782
*Ldh*B	Dissolved Oxygen (DO)	1	0.384	0.538
	Temperature (Temp)	1	2.881	0.095
	Group	**2**	**9.948**	**0.000**
	DO * Temp	1	0.047	0.829
	DO * Group	2	0.228	0.797
	Temp * Group	1	0.014	0.986

The number of *Ldh* transcripts per *Gapdh* transcript was calculated using the $2^{-\Delta CT}$ method and is the dependent variable in this analysis. Gene is *Ldh*A or *Ldh*B; Group refers to *D. pulex*, *D. pulicaria*, and *D. pulex-pulicaria* hybrids; temperature is 10°C or 20°C; dissolved oxygen is high (6.5–7 mg/l) or low (2–3 mg/l). Interactions are indicated by an asterisk (*). Sources of variation that show a significant difference in gene expression ($p<0.05$) are in bold-face type.

Table 3. Pairwise comparison of *Ldh*A *and Ldh*B expression between genotypes in *Daphnia.*

Gene	Group	Group	p-value
*Ldh*A	*D. pulex*	*D. pulicaria*	1.000
	D. pulex	*D. pulex - pulicaria* hybrid	0.093
	D. pulicaria	*D. pulex - pulicaria* hybrid	0.088
*Ldh*B	*D. pulex*	*D. pulicaria*	**0.000**
	D. pulex	*D. pulex - pulicaria* hybrid	0.056
	D. pulicaria	*D. pulex - pulicaria* hybrid	0.139

The results are based on post-hoc t-test with the Bonferroni correction. The number of *Ldh* transcripts per *Gapdh* transcript was calculated using the $2^{-\Delta CT}$ method and is the dependent variable in this analysis. Genotype refers to *D. pulex, D. pulicaria,* and *D. pulex-pulicaria* hybrids; Gene is *Ldh*A or *Ldh*B. P-values that show a significant difference in gene expression ($p<0.05$) are in bold-face type.

each is expressed. However, further investigation is needed to confirm this hypothesis. The overall expression of *Ldh*B was extremely low in all three groups, in all four environmental conditions, and further analysis will be required to determine if the mRNA transcripts are actually translated into protein. It is possible that this gene is no longer expressed at a functional level, although analysis of sequences from 85 isolates from 9 different *Daphnia* species suggests that selection is maintaining a functional *Ldh*B protein-coding sequence [18].

LDH catalyzes the final step of the anaerobic metabolic pathway, glycolysis [41] and *Ldh* gene expression has been linked with adaptation of species to environments characterized by different dissolved oxygen concentrations and temperatures. For example, gene expression differences are greater between northern and southern populations of *Fundulus heteroclitus* than between southern *F. heteroclitus* and its sister species *F. grandis* [9]. These patterns of gene expression may be a result of the northern *F. heteroclitus* population adapting to colder water and the southern *F. heteroclitus* and *F. grandis* adapting to warmer water [9]. However, temperature did not affect the expression of either *Ldh* gene in *D. pulex, D. pulicaria*, or the hybrids in our study despite the fact that temperature regimes can be very different in ponds and lakes. For example, shallow ponds may be very cold in the spring when the *Daphnia* first hatch from diapausing eggs, but then warm rapidly throughout the spring and early summer. Conversely, the hypolimnion of deep stratified lakes, where *Daphnia* seek refuge from fish predation during the day [42], can be close to 4°C all year round. Thus, it is possible that the temperature range to which *Daphnia* were exposed in our experiments was not broad enough to elicit a stress response.

In low oxygen conditions, glycolysis rate should be enhanced in order to compensate for oxidative energetic metabolism decay. Moreover, pyruvate to lactate transformation must be efficient in order to keep a high $NAD^+/NADH$ ratio and therefore allow continued glycolytic flux [41]. In mammals [43] and fish [12], genes encoding proteins involved in the homeostatic response to hypoxia, such as LDH are over-expressed in acute hypoxic conditions. However, Daphnia *Ldh* expression was not found to be up- regulated under low oxygen supply. This could be due to the fact that *Daphnia* is a hypoxia-tolerant invertebrate in which a response to prolonged periods of hypoxia includes regulated metabolic depression [44], [45], which is a stress-coping mechanism that relies on energy conservation and involves the suppression of every major ATP-utilizing function in the cell [45]. *Daphnia pulex* and the hybrids inhabit shallow temporary ponds with fluctuating DO concentrations that range from 5.2 mg/L to 1.1 mg/L. Conversely, *D. pulicaria* inhabits the hypoxic hypolimnion of lakes. Therefore, these *Daphnia* species dwell in environments characterized by long periods of low DO concentration and could be employing regulated metabolic depression to survive in such environments. Further studies are needed to elucidate the mechanism involved in this adaptation.

Conclusions

We studied whether the expression of the two *Ldh* loci (*Ldh*A and *Ldh*B) varies under different environmental conditions in *D. pulicaria, D. pulex* and their hybrids. We observed significant differences in the level of *Ldh*B expression compared to *Ldh*A ($p<0.001$), with *Ldh*A being expressed 4 to 48 times more than *Ldh*B in all three groups and in all combinations of temperature and oxygen concentration. These results suggest that the function of these two genes might be undergoing specialization such that *Ldh*B is expressed only in a particular set of tissues or organs. Furthermore, although mean *Ldh*A expression was higher in the hybrids, none of the differences between the three *Daphnia* groups were significant. However, mean *Ldh*B expression was significantly higher in *D. pulicaria* than *D. pulex*, and although the level of expression in the hybrid was closer to that in *D. pulex*, it was not significantly different from expression in *D. pulicaria*. Future research on LDHA and LDHB enzyme activity, thermo-stability and rate of protein synthesis under different environmental settings could help clarify the strong association between *Ldh*A genotype and habitat that has been observed in previous studies.

Supporting Information

Figure S1 Distribution of *Daphnia* sampling sites in Michigan and Ontario. The five ponds are Disputed, Solomon and three ponds in the Canard area. The three lakes are Lawrence, Three Lakes II, and Warner.

File S1 Contains Table S1, CT values for all *Daphnia* clones and qPCR replicates included in this study. Table S2, Gene expression of *Ldh*A and *Ldh*B relative to the reference gene, *Gapdh* in 18 clones of *Daphnia* across four combinations of temperature and dissolved oxygen. Table S3, Gene expression of *Ldh*A and *Ldh*B relative to the reference gene, *Gapdh* in three groups of *Daphnia* clones across four combinations of temperature and dissolved oxygen.

Acknowledgments

We thank A. Pecani and P. Sahota for technical assistance with *Daphnia* culturing. We also thank A. Constantin and Sen Xu for providing genotyping data. C. Caceres provided logistical support. D. Heath, A. Hubberstey, R. Williams, S. Xu, A. Zhan and two anonymous reviewers provided valuable advice.

Author Contributions

Conceived and designed the experiments: MEC BD TJC. Performed the experiments: BD IA. Analyzed the data: BD IA TJC. Contributed reagents/materials/analysis tools: MEC. Contributed to the writing of the manuscript: MEC BD IA TJC.

References

1. Karl I, Schmitt T, Fischer K (2009) Genetic differentiation between alpine and lowland populations of a butterfly is related to PGI enzyme genotype. Ecography 32: 488–496.
2. Crawford DL, Powers DA (1992) Evolutionary adaptation to different thermal environments via transcriptional regulation. Molecular Biology and Evolution 9: 806–813.
3. Eanes WF (1999) Analysis of selection on enzyme polymorphisms. Annual Review of Ecology, Evolution and Systematics 30: 301–326.
4. Dahlhoff EP, Rank NE (2000) Functional and physiological consequences of genetic variation at phosphoglucose isomerase: Heat shock protein expression is related to enzyme genotype in a montane beetle. Proceedings of the National Academy of Sciences USA 97: 10056–10061.
5. Haag CR, Saastamoinen M, Marden JH, Hanski I (2005) A candidate locus for variation in dispersal rate in a butterfly metapopulation. Proceedings of the Royal Society B Biological Sciences 272: 2449–2456.
6. Ellegren H, Sheldon BC (2008) Genetic basis of fitness differences in natural populations. Nature 452: 13.
7. Everse J, Kaplan NO (1973) Lactate dehydrogenases: structure and function. Advances in Enzymology and Related Subjects of Biochemistry 37: 61–133.
8. Powers DA, Lauerman T, Crawford DL, Dimichele L (1991) Genetic mechanisms for adapting to a changing environment. Annual Review of Genetics 25: 629–659.
9. Oleksiak MF, Churchill GA, Crawford DL (2002) Variation in gene expression within and among natural populations. Nature Genetics 32: 261–266.
10. Schulte PM, Glemet HC, Fiebig AA, Powers DA (2000) Adaptive variation in lactate dehydrogenase-B gene expression: Role of a stress-responsive regulatory element. Proceedings of the National Academy of Sciences of the USA 97: 6597–6602.
11. Johns GC, Somero GN (2003) Evolutionary convergence in adaptation of proteins to Temperature: A4-Lactate Dehydrogenases of Pacific Damselfishes (*Chromis spp*.). Molecular Biology and Evolution 21: 314–320.
12. Rees BB, Bowman JA, Schulte PM (2001) Structure and sequence conservation of a putative hypoxia response element in the Lactate dehydrogenase-B gene of *Fundulus*. The Biological Bulletin 200: 247–251.
13. Crease TJ, Lee SK, Yu SL, Spitze K, Lehman N et al. (1997) Allozyme and mtDNA variation in populations of the *Daphnia pulex* complex from both sides of the Rocky Mountains. Heredity 79: 242–251.
14. Pfrender ME, Spitze K, Lehman N (2000) Multi-locus genetic evidence for rapid ecologically based speciation in *Daphnia*. Molecular Ecology 9: 1717–1735.
15. Hebert PDN, Finston TL (2001) Macrogeographic patterns of breeding system diversity in the *Daphnia pulex* group from the United States and Mexico. Heredity 87: 153–161.
16. Heier CR, Dudycha JL (2009) Ecological speciation in a cyclic parthenogen: Sexual capability of experimental hybrids between *Daphnia pulex* and *Daphnia pulicaria*. Limnology and Oceanography 54: 492–502.
17. Hebert PDN, Crease TJ (1983) Clonal diversity in populations of *Daphnia pulex* reproducing by obligate parthenogenesis. Heredity 51: 353–369.
18. Crease TJ, Floyd R, Cristescu ME, Innes D (2011) Evolutionary factors affecting Lactate dehydrogenase A and B variation in the *Daphnia pulex* species complex. BMC Evolutionary Biology 11: 212.
19. Dudycha JL, Tessier AJ (1999) Natural genetic variation of life span, reproduction, and Juvenile growth in Daphnia. Evolution 53: 1744–1756.
20. Dudycha JL (2003) A multi-environment comparison of senescence between sister species of Daphnia. Oecologia 135: 555–563.
21. Dudycha JL (2004) Mortality dynamics of Daphnia in contrasting habitats and their role in ecological divergence. Freshwater Biology 49: 505–514.
22. Welborn GA, Skelly DK, Werner EE (1996) Mechanisms creating community structure across a freshwater habitat gradient. Annual Review of Ecology and Systematics 27: 337–363.
23. Colbourne JK, Hebert PDN, Taylor DJ (1997) Evolutionary origins of phenotypic diversity in Daphnia. In: Givnish TJ, Sytsma KJ (Eds.), Molecular Evolution and Adaptive Radiation. Cambridge University Press, Cambridge, 163–188.
24. Wright D, Shapiro J (1990) Refuge availability: a key to understanding the summer disappearance of Daphnia. Freshwater Biology 24: 43–62.
25. Cristescu ME, Innes DJ, Stillman JH, Crease TJ (2008) D- and L-lactate dehydrogenases during invertebrate evolution. BMC Evolutionary Biology 8: 268.
26. Innes DJ, Schwartz SS, Hebert PDN (1986) Genotypic diversity and variation in mode of reproduction among populations in the *Daphnia pulex* group. Heredity 57: 345–355.
27. Cristescu ME, Constantin A, Bock DG, Caceres CE, Crease TJ (2012) Speciation with gene flow and the genetics of habitat transitions. Molecular Ecology 21: 1411–1422.
28. Xu S, Innes DJ, Lynch M, Cristescu ME (2013) The role of hybridization in the origin and spread of asexuality in Daphnia. Molecular Ecology 22: 4549–4561.
29. Cristescu ME, Colbourne JK, Radivojac J, Lynch M (2006) A microsatellite-based genetic linkage map of the water flea *Daphnia pulex*: On the prospect of crustacean genomics. Genomics 88: 415–430.
30. Hebert PDN, Beaton MJ, Schwartz SS, Stanton DJ (1989) Polyphyletic origins of asexuality in *Daphnia pulex*. I. Breeding-system variation and levels of clonal diversity. Evolution 43: 1004–1015.
31. Lopez JA, Bohuski E (2007) Total RNA extraction with TRIZOL reagent and purification with QIAGEN RNeasy Mini Kit. ©DGCIndiana University.
32. Rozen S, Skaletsky H (2000) Primer3 on the WWW for general users and for biologist rogramers. Methods Molecular Biology 132: 365–86.
33. Spanier KI, Leese F, Mayer C, Colbourne JK, Gilbert D et al. (2010) Predator-induced defences in *Daphnia pulex*: Selection and evaluation of internal reference genes for gene expression studies with real-time PCR. BMC Molecular Biology 11: 50.
34. Ramakers C, Ruijter JM, Lekanne Deprez RH, Moorman AFM (2003) Assumption-free analysis of quantitative real-time polymerase chain reaction (PCR) data. Neuroscience Letters 339: 62–66.
35. Ruijter JM, Ramakers C, Hoogaars WMH, Karlen Y, Bakker O et al. (2009) Amplification efficiency: linking baseline and bias in the analysis of quantitative PCR data. Nucleic Acids Research 37: 45.
36. VanGuilder HD, Vrana KE, Freeman WM (2008) Twenty-five years of quantitative PCR for gene expression analysis. BioTechniques 44: 619–62.
37. Cristescu ME, Egbosimba EE (2009) Evolutionary history of d-Lactate dehydrogenases: a phylogenomic perspective on functional diversity in the FAD binding oxidoreductase/transferase type 4 family. Journal of Molecular Evolution 69: 276–287.
38. Hughes AL (1994) The evolution of functionally novel proteins after gene duplication. The Royal Society Proceedings: Biological Sciences 256: 119–124.
39. Lynch M, Force A (2000) The probability of duplicate gene preservation by subfunctionalization. Genetics 154: 459–473.
40. Markert CL, Shaklee JB, Whitt GS (1975) Evolution of a gene. Multiple genes for LDH isozymes provide a mode of the evolution of gene structure, function and regulation. Science 189: 102–114.
41. Rossignol F, Solares M, Balanza E, Coudert J, Clottes E (2003) Expression of *Lactate dehydrogenase* A and B genes in different tissues of rats adapted to chronic hypobaric hypoxia. Journal of Cellular Biochemistry 89: 67–79.
42. Threlkeld ST (1979) The midsummer dynamics of two *Daphnia* species in Wintergreen Lake, Michigan. Ecological Society of America 60: 165–179.
43. Semenza GL (1999) Regulation of mammalian O2 homeostasis by hypoxia-inducible factor 1. Annual Review of Cell and Developmental Biology 15: 551–578.
44. Paul RJ, Colmorgen M, Pirow R, Chen Y, Tsai M (1998) Systemic and metabolic responses in *Daphnia magna* to anoxia. Comparative Biochemistry and Physiology Part A 120: 519–530.
45. Gorr TA (2004) *Daphnia* and *Drosophila*: two invertebrate models for O_2 responsive and HIF- mediated regulation of genes and genomes. International Congress Series 1275: 55–62.

3

Seasonal Variations and Resilience of Bacterial Communities in a Sewage Polluted Urban River

Tamara García-Armisen[1]*, Özgül İnceoğlu[1]*, Nouho Koffi Ouattara[1], Adriana Anzil[1], Michel A. Verbanck[2], Natacha Brion[3], Pierre Servais[1]*

1 Ecology of Aquatic Systems, Université Libre de Bruxelles, Campus de la Plaine, Brussels, Belgium, **2** Department of Water Pollution Control, Université Libre de Bruxelles, Campus Plaine, Brussels, Belgium, **3** Analytical and Environmental Chemistry, Vrije Universiteit Brussels, Brussels, Belgium

Abstract

The Zenne River in Brussels (Belgium) and effluents of the two wastewater treatment plants (WWTPs) of Brussels were chosen to assess the impact of disturbance on bacterial community composition (BCC) of an urban river. Organic matters, nutrients load and oxygen concentration fluctuated highly along the river and over time because of WWTPs discharge. Tag pyrosequencing of bacterial 16S rRNA genes revealed the significant effect of seasonality on the richness, the bacterial diversity (Shannon index) and BCC. The major grouping: -winter/fall samples versus spring/summer samples- could be associated with fluctuations of *in situ* bacterial activities (dissolved and particulate organic carbon biodegradation associated with oxygen consumption and N transformation). BCC of the samples collected upstream from the WWTPs discharge were significantly different from BCC of downstream samples and WWTPs effluents, while no significant difference was found between BCC of WWTPs effluents and the downstream samples as revealed by ANOSIM. Analysis per season showed that allochthonous bacteria brought by WWTPs effluents triggered the changes in community composition, eventually followed by rapid post-disturbance return to the original composition as observed in April (resilience), whereas community composition remained altered after the perturbation by WWTPs effluents in the other seasons.

Editor: Patrick C.Y. Woo, The University of Hong Kong, Hong Kong

Funding: This study was mainly performed in the scope of the GESZ research project (Towards the Good Ecological Status of River Zenne: Reevaluating Brussels wastewater management) from the "Impulse Environment 2008" program of the Brussels Institute for Research and Innovation (Innoviris). A part of the work was also performed in the scope of the project "Tracing and Integrated Modelling of Natural and Anthropogenic effects on Hydrosystems" (TIMOTHY), an Interuniversity Attraction Pole (IAP6.13) funded by the Belgian Federal Science Policy Office (BELSPO). Ö Ýnceoðlu benefits from a FNRS (Fonds National de la Recherche Scientifique, Belgium) grant. The funders had no role in study design, data collection and analysis, decision to publish, or preparation of the manuscript.

Competing Interests: The authors have declared that no competing interests exist.

* E-mail: pservais@ulb.ac.be

These authors contributed equally to this work.

Introduction

Rivers flowing through cities are often used as receiving body for treated and untreated urban wastewaters all over the world [1,2]. Some of these sewage-contaminated rivers are amongst the most extreme examples of ecosystems disturbed by human activities. The impact of wastewater release on the receiving river depends primarily on the size of the city, the type of treatment applied to wastewaters and the flow of the river; the higher the river flow is, the higher its dilution capacity is. The UN predicted that, for 2050, 70% of the world population will be living in cities and this urban growth will mainly occur in less developed countries, where wastewater treatment facilities are scarce [2]. By 2025, it is expected that 27 megacities of more than 10 million inhabitants will exist, 21 of which in less developed countries [2]. Considering the importance, scarcity and fragility of freshwater ecosystems, research about the impact of the release of high amount of sewage on rivers functioning and ecological health is essential to preserve and have a rational management of this resource over a long term period.

It is now well known that microbial community inhabiting aquatic systems are one of the key players in the biogeochemical cycling of organic matter and nutrients and thus in the recovery and maintenance of ecosystem health and balance [3,4]. Sewage brings high loads of organic and inorganic pollutants that strongly modify the habitat of native freshwater bacteria. For instance, basic parameters such as temperature, pH and dissolved oxygen which are known as drivers of the bacterial community composition (BCC) [5,6,7] in aquatic systems can be influenced by urban wastewater discharges. Additionally, the loads of toxic trace metals from industries, of Persistent Organic Pollutants (POPs) and pharmaceuticals resulting from human activities could also affect the BCC [8,9,10,11]. Additionally, high concentrations of particulate matter within wastewater can modify light penetration in the water column and prevent phototrophic organisms' growth.

Sewage also brings high loads of allochthonous microorganisms [12]. If the allochthonous bacteria remain active after their release in the environments, as it has been previously suggested [13,14,15], they can have a major impact on the BCC of the river and could influence biogeochemical cycles. Besides, utilization of waters contaminated with allochthonous pathogenic microorganisms causes health risk.

Investigating the microbial population dynamic and composition of sewage polluted rivers and evaluating the resistance (insensitivity to disturbance) and resilience (the rate of recovery after disturbance (of microbial communities are essential for understanding how ecosystem health and functioning could be affected by the disturbances related to major sewage discharges. Despite the importance of these ecosystems and the increasing interest and advances in freshwater microbial ecology [16,17,18,19], papers devoted to BCC of urban rivers are still underrepresented in the scientific literature. Until now, the majority of the studies dealing with microbial community in urban aquatic settings have focused on indicator bacteria related to fecal and organic pollution [20,21,22,23]. Although some studies were done on polluted river sediments [24,25,26,27], to the best of our knowledge only few papers have examined the BCC of urban rivers waters [7,28,29,30].

The objectives of the present study were thus: **(1)** to study the spatial and temporal variations of the BCC along a sewage polluted urban river; **(2)** to identify the main environmental factors driving the microbial population dynamics in the watercourse; **(3)** to compare the river BCC before and after the discharge of major wastewater treatment plants (WWTPs) effluents; and **(4)** to analyze the fate and dynamics of different functional groups of sewage related bacteria once discharged into the river.

For this purpose, the Zenne River in Brussels (Belgium), a river highly disturbed by urban wastewater, was chosen as a model. This small river, with an annual average discharge of 4 $m^3\ s^{-1}$ upstream from Brussels, crosses the city and receives the effluents released by the two WWTPs of Brussels (treating together more than 1.5 million inhabitant equivalents); the average flow of treated wastewaters is in the same order of magnitude than the Zenne River flow rate upstream from Brussels. The river waters downstream from Brussels are thus roughly half composed of treated wastewaters, this proportion being even higher during the low flow periods of the river [31]. Seven stations along the river and the treated effluents of the two Brussels's WWTPs were sampled during four sampling campaigns at different seasons in 2010. The BCC was analyzed using 16S rRNA gene amplicon 454 pyrosequencing; in parallel, a large set of environmental parameters were measured.

Materials and Methods

Ethics Statement

No specific ethical or institutional permits were required to sample the Zenne River and the experimental studies did not involve endangered or protected species. We also obtained permission from Vivaqua (www.vivaqua.be) to sample the effluent of Brussels South WWTP and from Aquiris (www.aquiris.be) to sample the effluent of Brussels North WWTP.

Study Site and Sampling Strategy

The Zenne River is located in the Scheldt watershed and flows in the three Belgian regions (Wallonia, Flanders and Brussels-Capital); it is a tributary of the Dijle River (Fig. 1). The Zenne watershed (surface 991 km^2) is characterized by agricultural activities in its upstream part and an important urbanization in its downstream part. The annual average discharge of the Zenne River arriving in Brussels from the upstream part of the watershed is 4 $m^3\ s^{-1}$ (deduced from the last 20 years water flow data monitored by the Hydrologisch Informatie Centrum – HIC). Before the river reaches the Brussels area, it already receives several effluents of small scale WWTPs. The population density in the watershed is very high (on average 1260 inhabitants km^{-2}) and mostly located in Brussels city and suburbs. The Zenne River has a total length of 103 km and crosses Brussels city from South to North over a distance of about 20 km. For sanitation purposes, the Zenne River was covered in the 19th century (1867–1871) in most of the stretch located in the Brussels region. In Brussels area, the Zenne River receives the sewage waters from two large WWTPs: the Brussels South WWTP (360,000 equivalent-inhabitants) in operation since the year 2000 and the Brussels North WWTP (1.2 million equivalent-inhabitants) in operation since 2007. The Zenne River also receives waters from two tributaries around the Brussels region the Zuunbeek, and the Woluwe (Fig. 1). Some small tributaries located in the Brussels area are diverted in the sewer collectors so that their waters reach Brussels WWTPs.

The two Brussels WWTPs function according to different technologies. The Brussels South WWTP treatment line includes a primary settling stage and a secondary biological treatment (activated sludge). At Brussels North WWTP, there are two treatment lines. The first one (biological line) includes a primary settling stage followed by a modern tertiary treatment technology (simultaneous removal of biodegradable organic carbon, nitrogen and phosphorus by an activated sludge process). The other treatment line (rain line) runs in parallel to the biological line when the discharge reaching the WWTP is too high in wet weather situations; this rain line uses only a coagulation-assisted high-rate lamellar settling process. On an annual basis, the volume treated in the biological line accounts for roughly 90% of the total volume reaching the WWTP.

Four sampling campaigns were performed in 2010 (January, April, July and October). Overall these sampling campaigns took place during marked dry weather periods; the discharge of the Zenne River just upstream from Brussels was always low during these campaigns:, 2.5 $m^3\ s^{-1}$ in January 2010, 1.4 $m^3\ s^{-1}$ in April 2010 and 2.0 $m^3\ s^{-1}$ in July 2010, 2.9 $m^3\ s^{-1}$ in October 2010. Comparatively, the April campaign was characterized by the driest conditions, both upstream and downstream Brussels city. Regarding October, there was actually a small rainfall occurring in Central Belgium during the sampling campaign but this did not even cause combined sewer overflow (CSO) discharge in the Zenne.

Seven stations were sampled along the Zenne River (Fig. 1) in the stretch located downstream from the confluence with its major right-bank tributary the Sennette. Accordingly, a kilometric scale along the longitudinal course of the river was defined; it is arbitrarily set at zero at Lembeek (station Z1) and increases from upstream to downstream. Stations Z1 (0 km) and Z3 (13 km) are located upstream from Brussels. Station Z5 (20 km) is located just downstream from the Brussels South WWTP effluents release. Stations Z7 (30 km) and Z9 (34 km) are located upstream and downstream from the Brussels North WWTP, respectively. Stations Z10 (39 km) and Z12 (51 km) are significantly downstream from the Brussels conurbation area. Besides, in each of the sampling campaigns, the effluents of both Brussels WWTPs were sampled (average daily samples collected with refrigerated automatic samplers) and analyzed.

Physico-chemical Analysis and Bacterial Enumeration

Temperature, pH, and electrical conductivity (EC) were measured directly on site using a portable WTW 340 multiprobe. Dissolved oxygen was measured on the spot with a WTW oxi 323 field probe. Total alkalinity was determined by titration with HCl. Inorganic nitrogen (NH_4, NO_2 and NO_3) and phosphorus (PO_4) concentrations were determined by automatic colorimetric methods using a QuAAtro (Seal, Analytical) segmented flow analyzer

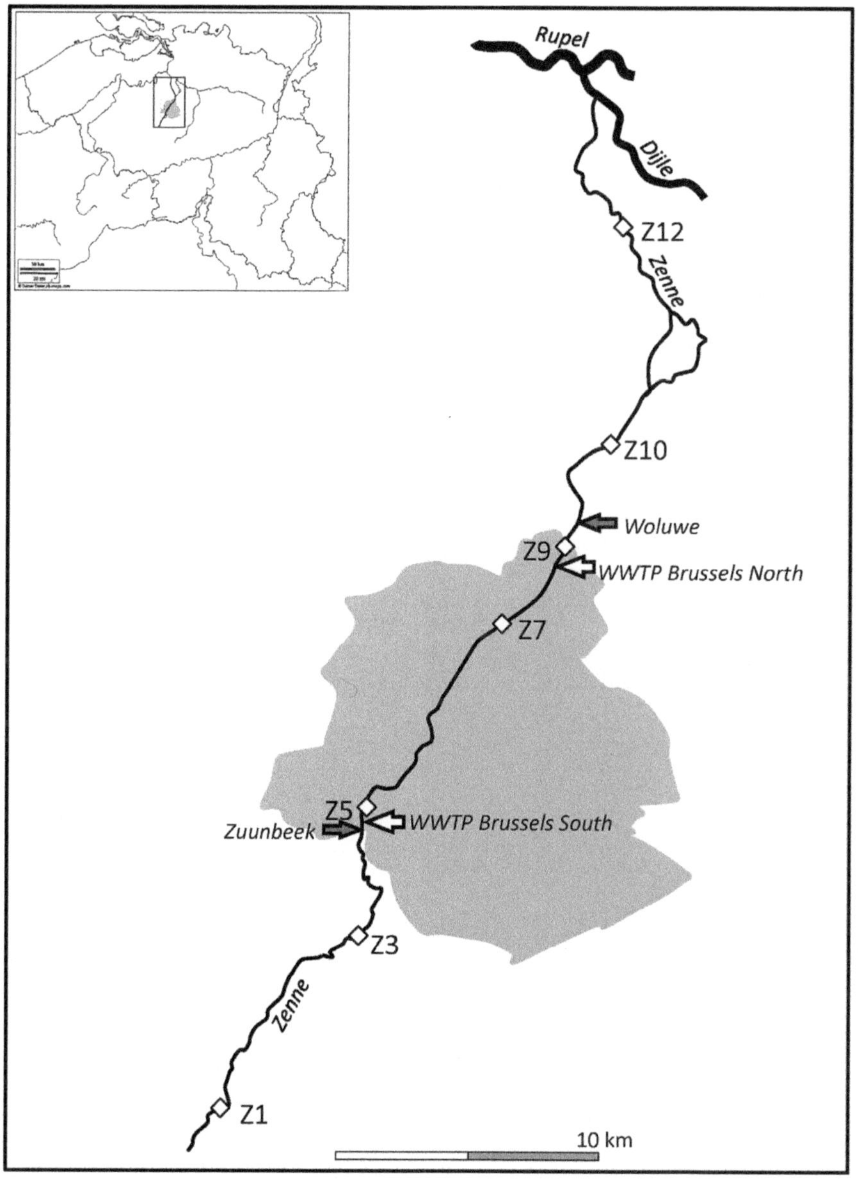

Figure 1. Map of the Zenne drainage network. Location of the sampling stations and the two Brussels WWTPs are represented.

system. Suspended particulate matter (SPM) was estimated as the weight of material retained on Whatman GF/F glass fiber filter (diameter 4.7 cm, particle retention size 0.7 µm) per volume unit after drying the filter at 105°C. For dissolved (DOC) and particulate organic carbon (POC) analysis, samples were filtered on GF/F glass fiber filters previously combusted at 550°C. DOC was measured in the filtrate while the material retained on the filter was used to estimate POC. Glassware receiving the water samples for DOC analysis was muffled at 550°C for 4 h after cleaning; samples for DOC analysis were preserved with H_3PO_4 (0.1% final concentration). DOC concentrations were measured with a total organic carbon analyzer in which inorganic carbon is eliminated by bubbling in the presence of phosphoric acid and organic carbon is oxidized by UV promoted persulfate oxidation and the produced CO_2 detected by infrared spectrometry (Hiper-TOC, Thermo). The GF/F filters with the particulate material were first acidified to remove inorganic carbon. POC was analyzed with an elemental analyzer (Flash EA 1112, Thermo) where organic material is oxidized by catalytic flash combustion at 1020°C and the produced CO_2 is measured by thermal conductivity. The DOC analyzer was calibrated using standard sucrose solutions, the POC analyzer was calibrated with acetanilide powder.

Bacterial abundance was determined by epifluorescence microscopy at 1000×magnification, after DAPI staining according to the procedure proposed by Porter and Feig [32].

DNA Extraction

Microbial biomass was collected from water fraction and concentrate by filtration. An aliquot (from 0.8 L to 5.0 L) of each sample was filtered in triplicate directly through 0.2 µm pore-size, 47-mm-diameter polycarbonate filters (Millipore, MA). All filters were placed in 3 mL lysis buffer (50 mM Tris-HCl, pH 8.0, 40 mM EDTA, 750 mM sucrose) and stored at −20°C until use.

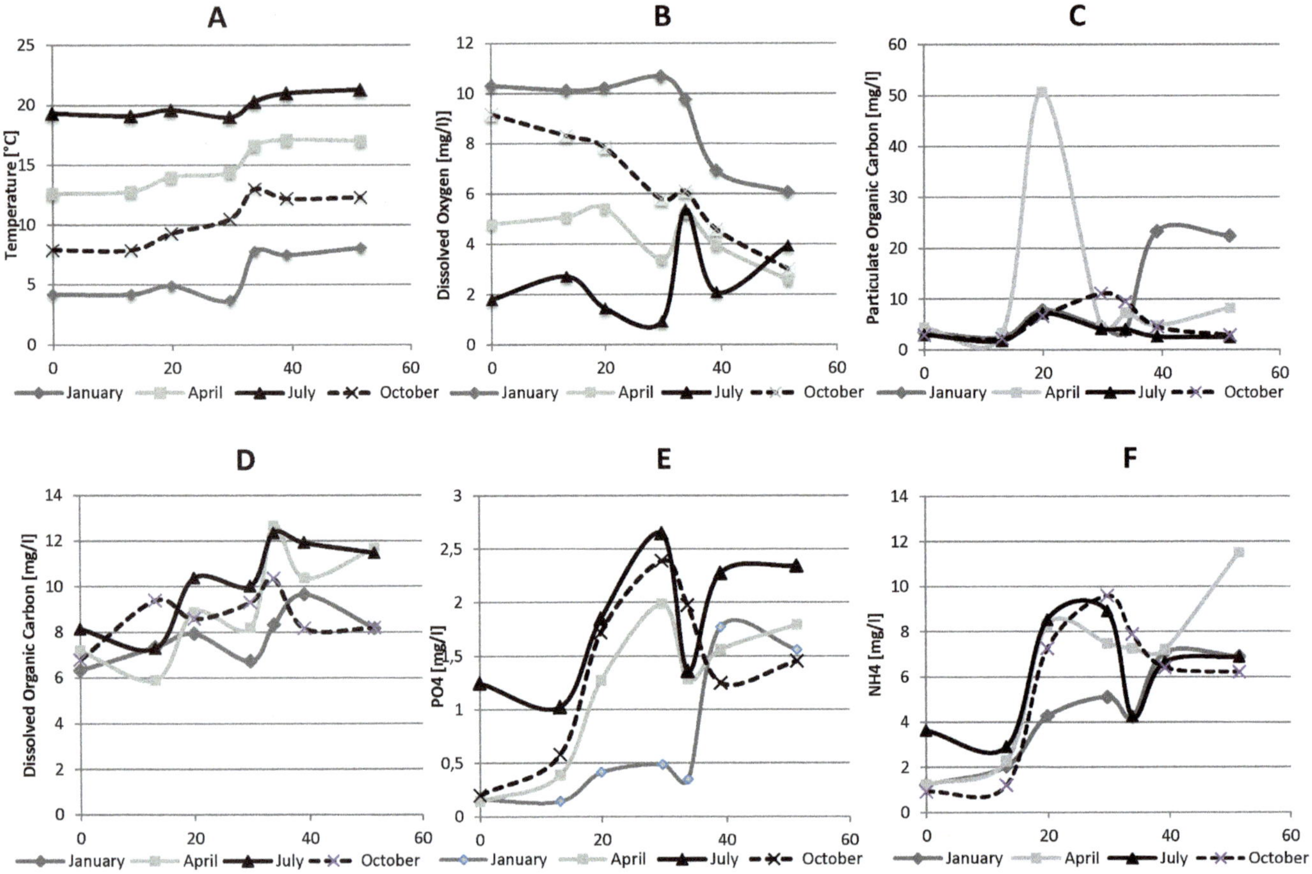

Figure 2. Longitudinal profiles of some physico-chemical parameters. (**A**) Temperature; (**B**) Dissolved Oxygen; (**C**) Particulate organic carbon (POC); (**D**) Dissolved organic carbon (DOC) (**E**) PO_4, (**F**) NH_4.

DNA was extracted using a combination of enzymatic cell lysis and cetyltrimethyl ammonium bromide (CTAB) extraction protocol as previously described [33]. Dry DNA pellets were finally rehydrated in 100 μL of 10 mM Tris-HCl buffer (pH 7.4) and further purified using Min Elute reaction cleanup kit (QIAGEN). DNA concentration and purity were then determined using a Nanodrop ND-2000 UV-Vis spectrophotometer (Nanodrop, DE). Purified DNA extracts were stored at −20°C until use.

Bacterial 16S rRNA Gene Tag-encoded FLX-titanium Amplicon Pyrosequencing

Bacterial tag-encoded FLX gene amplicon pyrosequencing (bTEFAP) analysis was carried out by means of a Roche 454 FLX instrument with titanium reagents. Titanium procedures were performed at the Research and Testing Laboratory (Lubbock, TX, USA) based upon RTL protocols (www.researchandtesting.com) as previously described [34,35]. DNA samples were diluted to a final concentration of 20 ng/μL prior to bTEFAP. One-step PCR with a mixture of Hot Start and HotStar high-fidelity Taq polymerases was used. The PCR primers for FLX amplicon pyrosequencing were chosen to span the variable regions V1–V3 in the 16S rRNA gene: 27F (5′-GAGTTT-GATCNTGGCTCAG-3′) and 519R (5′- GWNTTACNGC-GGCKGCTG-3′) for *Bacteria*. These selected primers cover about 78% of publicly available 16S rDNA sequences of *Bacteria* (TestPrime tool at SILVA webpage http://www.arb-silva.de/search/testprime/). All sequences generated in this study can be downloaded from NCBI Short Read Archive, accession numbers: (SRP021937).

16S rRNA Data Processing

The Research and Testing pipeline (Lubbock, TX, USA) performed the denoising were performed on each region. Briefly data were decompressed and sequencing errors were reduced by trimming flows (i.e., denoising) and sequences by applying the following criteria: amplicons shorter than 200 bp in length, - Afterwards, using USEARCH [36], reads were dereplicated meaning they were clustered together into groups such that each sequence was an exact match to portion of the seed sequence for the cluster. Then, the seed sequence from each cluster was sorted by abundance, largest cluster to smallest cluster, without any minimum size restrictions on the clusters. Afterwards, putative chimeras (checked by using USEARCH [36]) were removed from our data set. Then singletons were also removed. The denoised and chimera checked reads were condensed into a fasta file. Then failed sequences and sequences that have low quality tags, primers, or ends were removed.

The open-source, platform-independent, community-supported software program, Mothur (Mothur v.1.25.1; http://www.mothur.org) was used to analyze the clean sequence data. We trimmed the barcodes and primers from the resulting sequences.

Rarefaction curves based on identified OTU, Shannon diversity index and richness estimator Chao1 were generated using Mothur for each sample at 97% similarity. Sequences were classified using

Table 1. Observed bacterial richness and diversity estimates from analyzed Zenne River [28 samples] and WWTPs [8 samples] samples.

Samples	Cutoff	N	Richness	Chao	LCI95	UCI95	H'	varH
WWTPs*	0.03	22126	1816	2113	2051	2192	6.17	0.00012
WWTPs	0.05	22126	1366	1538	1494	1598	5.84	0.00012
Zenne River*	0.03	80188	4408	4551	4521	4590	6.72	0.00004
Zenne River	0.05	80188	3000	3073	3053	3101	6.23	0.00004

N:Number of sequences.
H': Shannon Index.
varH: Shannon index variance.
LCI:95% lower confidential interval.
UCI:95% upper confidential interval.
Cutoff 0.03- Species level, 0.05- Genus level.
*Only data at species level was further discussed in detail.

the greengenes database at 80% confidence threshold with Mothur. After taxonomic assignment of the sequences down to the phylum, class and genus level, relative abundance of a given taxonomic group was set as the number of sequences affiliated with that group divided by the total number of sequences per sample.

Statistical Analyses

Genera abundances were standardized, square root transformed, and assembled into a Bray Curtis similarity matrix to generate a multidimensional scaling (MDS) plot, using PRIMER 6 [37] and plotted to assess similarities among samples in two-dimensional space. MDS plots represent relative distances among samples in relation to the rank order of their relative similarities. The Spearman correlation test was used to analyze the correlations between relative abundance of genera and phylum/environmental parameters. The discrimination of bacterial assemblages based on water type was tested with one-way analysis of similarities (ANOSIM) using PRIMER 6 [37]. Similarity percentage (SIMPER) [38] was used to determine which genus contributed most to the dissimilarity between upstream stations (Z1, Z3), downstream stations (Z5, Z7, Z9, Z12 and WWTPs samples).

Rest of the analyses was conducted in R [39], using the vegan [40] and labdsv [41] packages. Procrustes analyses were done to determine the degree of concordance between MDS ordinations obtained at different taxonomic resolution levels (phylum, genera and $OTU_{0.03}$) on R [39]. Overall patterns in the microbiome were illustrated using heatmaps. When constructing the heatmaps, differences among sites were characterized using Bray-Curtis distances and differences among taxa were characterized using Euclidean distances. In addition, data within samples were standardized to facilitate comparisons. Species that occurred only once were removed for clustering analysis and only those having a relative abundance of more than 5% were shown in the heatmaps. We quantified the common and unique influences of environmental variables (T + pH + O2 + EC + SPM + NO2 + NO3 + NH4 + POC + DOC + PO4 + Alkalinity), distance and season on the bacterial community variation using variation partitioning analyses [42], using the varpart function in the R package 'vegan' package.

Results and Discussion

Physico-chemical and Biological Characterization of the Zenne River Samples

The four sampling campaigns were characterized by very contrasting temperatures (Fig. 2A), covering a range from 4°C (January) to 22°C (July). An increase of the temperature is systematically observed after the discharge of WWTPs effluents regardless of the season. This effect is particularly marked after Brussels-North WWTP considering its high volumes of effluent discharge into the river.

In fall and winter, the amount of dissolved oxygen (DO, Fig. 2B) in water was quite high at the entrance of the study zone and decreased in the Brussels area. In winter a rapid decrease was observed after the discharge of the Brussels North WWTP, whereas in autumn the decrease was continuous all along the river. In contrast, in summer, DO was already very low at the entrance of the study zone (around 2 mg L^{-1}) and dropped to 1 mg L^{-1} in the Brussels area values which are below the lethal concentration for fish [43,44]. The release of well aerated effluents from Brussels North WWTP obviously increased the oxygen level in the river that decreased rapidly further downstream.

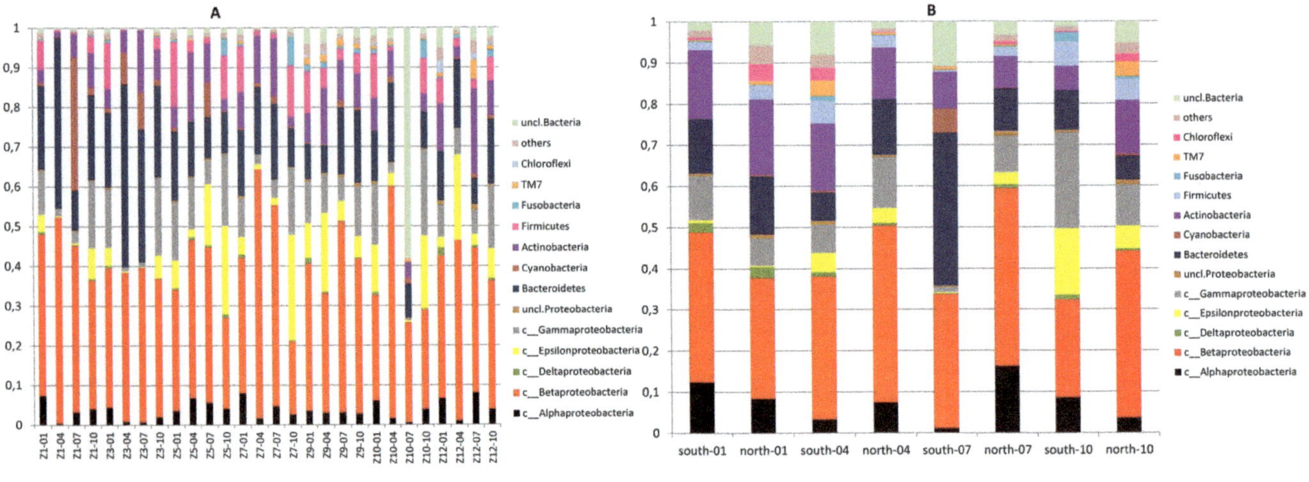

Figure 3. Taxonomic classification of bacterial reads at the phylum level. (**A**) River (28 samples of the Zenne River from all stations and seasons together) and (**B**) wastewater treatment plant effluents.

During our sampling campaigns, the particulate organic carbon (POC) content was detected in the range of 2–11 mg/L in all the stations, except three samples: Z5 in April, (51 mg/L, downstream from the Brussels South WWTP discharge) and Z10 and Z12 in January (23 mg/L, downstream from the Brussels North WWTP discharge). These abnormal increases in POC concentrations seem to be associated in April with the discharge of sludge from the activated sludge process at Brussels South WWTP. Such sludge release often occurs at Brussels South WWTP and was confirmed by microscopic observation for the studied April 2010 situation. In addition, in April, the driest studied period, the water mass at Z5 was very heterogeneous (due to the absence of a good mixing between WWTP South effluents and river water) so that particles were overrepresented in this sample. In January, the increase in POC downstream from Brussels area seems to be due to some untreated wastewater released into the river from station Z10. A significant increase in the ratio POC/SPM at Z10 and Z12 stations (25 to 30% C compared to 15% in Z9) was an indication of a high release of untreated sewage (with SPM rich in organic carbon) upstream from station Z10. On the other hand, DOC content was detected in the range of 5–13 mg/L in all the sampling stations. An increase in DOC concentration along the river was observed (Fig. 2D). Furthermore, in all the parameters measured, only NH_4 and PO_4 were positively correlated with each other, regardless of sampling season. These two parameters had a peak concentration downstream from the Brussels South WWTP (Z5) during each sampling campaign, indicating a major load of NH_4 and PO_4 from this WWTP effluent, due to the lack of tertiary treatment in Brussels South WWTP (Fig. 2E and 2F).

The total number of bacteria enumerated by epifluoresence microscopy ranged from 2.2×10^9 to 2.3×10^{10} cells/L (data not shown). These bacterial abundances are in the upper range of the values reported in freshwaters but abundances of bacteria higher than 10^{10} cells/L were already reported in rivers receiving high amounts of wastewaters [45]. Although ANOVA analysis revealed no significant difference between seasons, median values in January and April were higher than in July and October.

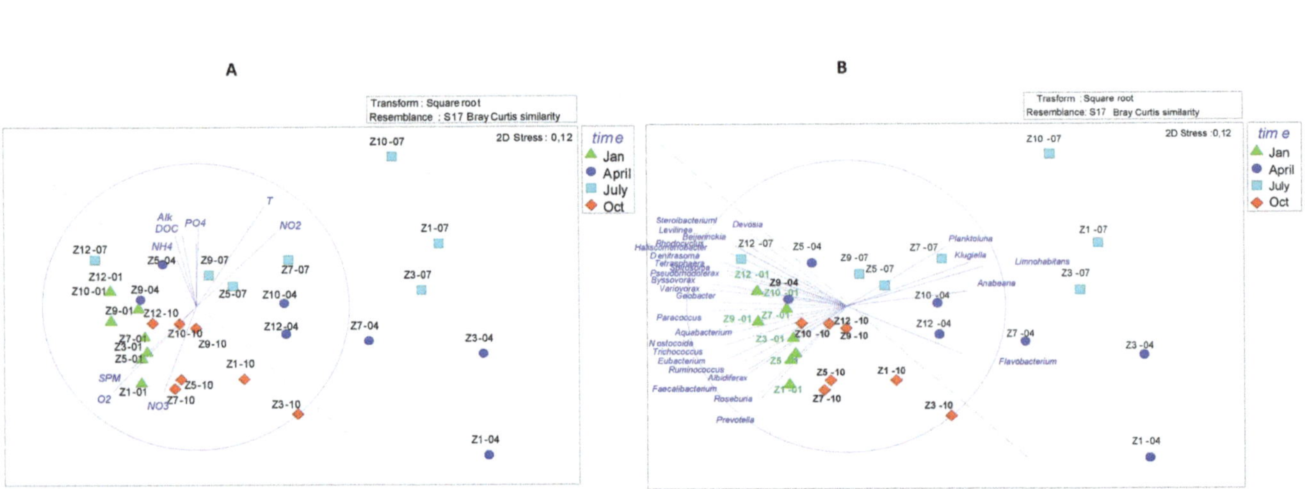

Figure 4. Nonmetric multidimensional scaling (MDS) plot. Environmental variables (**A**) and genera (**B**) are plotted as correlations with the river samples studied. Increasing distance between points equates to decreasing similarity in BCC. MDS plot are based on Bray-Curtis distances generated from square root transformed data. Analysis was conducted using Primer 6. Abbreviations: Jan-January, Oct-October.

Table 2. Pearson's correlation coefficients between environmental parameters and $OTUs_{0.03}$.

Environmental parameter	$OTU_{0.03}$	Assignation	coeffiecient
COD	Otu0032	p_Actinobacteria(100);c_Actinobacteria(100);o_Actinomycetales(100);	0.82
POC	Otu0032	p_Actinobacteria(100);c_Actinobacteria(100);o_Actinomycetales(100);	0.89
POC	Otu0053	p_Actinobacteria(100);c_Actinobacteria(100);o_Actinomycetales(100);f_Nocardiaceae(100); g_Rhodococcus(100);s_Rhodococcus coprophilus(100);	0.85
COD	Otu0119	p_Proteobacteria(100);c_Gammaproteobacteria(100);o_Pseudomonadales(100); f_Moraxellaceae(100);g_Acinetobacter(100);	0.84
POC	Otu0119	p_Proteobacteria(100);c_Gammaproteobacteria(100);o_Pseudomonadales(100); f_Moraxellaceae(100);g_Acinetobacter(100);	0.88
POC	Otu0156	p_Proteobacteria(100);c_Betaproteobacteria(100);o_Rhodocyclales(100); f_Rhodocyclaceae(100);g_Thauera(100);	0.81
SPM	Otu0183	k_Bacteria(100)	0.80
COD	Otu0206	p_Bacteroidetes(100);c_Flavobacteria(100);o_Flavobacteriales(100); f_Flavobacteriaceae(100);g_Flavobacterium(100)	0.83
POC	Otu0206	p_Bacteroidetes(100);c_Flavobacteria(100);o_Flavobacteriales(100); f_Flavobacteriaceae(100);g_Flavobacterium(100)	0.86
SPM	Otu0206	p_Bacteroidetes(100);c_Flavobacteria(100);o_Flavobacteriales(100); f_Flavobacteriaceae(100);g_Flavobacterium(100)	0.81
COD	Otu0303	p_Proteobacteria(100);c_Gammaproteobacteria(100);o_Pseudomonadales(100); f_Moraxellaceae(100);g_Acinetobacter(100)	0.81
POC	Otu0303	p_Proteobacteria(100);c_Gammaproteobacteria(100);o_Pseudomonadales(100); f_Moraxellaceae(100);g_Acinetobacter(100)	0.89
COD	Otu0410	k_Bacteria(100)	0.81
POC	Otu0410	k_Bacteria(100)	0.81
COD	Otu0436	p_Bacteroidetes(100);	0.86
POC	Otu0436	p_Bacteroidetes(100);	0.93
COD	Otu0532	k_Bacteria(100)	0.82
POC	Otu0532	k_Bacteria(100)	0.86
POC	Otu0596	k_Bacteria(100)	0.84
POC	Otu0827	p_Proteobacteria(100);c_Betaproteobacteria(100);	0.80
COD	Otu0926	p_Bacteroidetes(100)	0.83
POC	Otu0926	p_Bacteroidetes(100)	0.83
POC	Otu1079	p_Proteobacteria(100);c_Betaproteobacteria(100);o_Burkholderiales(100);	0.84
POC	Otu1130	p_Actinobacteria(100);c_Actinobacteria(100);o_Actinomycetales(100);	0.84

COD: Chemical Oxygen Demand, POC : Particulate Organic Carbon, SPM:Suspended Particulate Matter.

Bacterial Community Structure and Composition in the Zenne River and WWTPs Effluents Samples

After denoising and chimera checking, a total of 80109 sequences, with a mean length of 378 bp, were obtained for further analysis from the 28 samples of the Zenne River (all stations and seasons together) and 21114 from the 8 samples of treated wastewaters (4 from Brussels North WWTP and 4 from Brussels South WWTP).

First of all, the data from river samples were compared to those from wastewater effluents, regardless of sampling time and station. Richness was estimated to be 1786 and 4407 $OTUs_{0.03}$ for WWTPs effluents and the river, respectively, while Chao indices were found to be more than 2000 for WWTPs samples, and 4500 for the river samples (Table 1). The rarefaction curves at 0.03 (Fig. S1A) showed that the global diversity of the two series of samples was well represented with the number of sequences analyzed. A recent study done on different treated wastewater effluents reported that species richness might vary between 1500 and 4000 based on the type of the treatment applied [46]. Detection of lower bounds for species richness of WWTPs effluents in comparison to river samples might be due the massive enrichment of some species which are very well adapted to the conditions encountered in some wastewater treatment processes such as activated sludge. When the samples were analyzed individually, richness was estimated between 573 and 1073 in January (Fig. S1B). In April, samples reached to the saturation between 132 and 1271, increase in richness was observed after WWTPs discharges (Fig. S1C). In July, no sample except Z3 has reached to saturation (Fig. S1D). Richness was estimated between 349 and 774 in October (Fig. S1E). The highest richness in WWTPs effluents was detected in Brussels North in January (Fig. S1F).

Figure 3 shows the taxonomic assignations at phylum/class level for all the 28 river samples together (A), and the 8 treated wastewaters samples (B). In the river (Fig. 3A), the dominant phylum was the *Proteobacteria* with up to 73% of the sequences, followed by *Bacteroidetes* up to 46%, Cyanobacteria up to 33%, *Actinobacteria* up to 21% and *Firmicutes* up to16%. Among the *Proteobacteria*, the most abundant class was the *Betaproteobacteria* (up to 63%), followed by the *Epsilonproteobacteria* (up to 26%) *Gammaproteobacteria* (up to 22%), and *Alphaproteobacteria* (up to 8%). Previous studies also reported the dominance of *Proteobacteria* in rivers [7,28] and the class *Betaproteobacteria* is known to be a typical freshwater clade [47]. At genus level (data not shown), 18% of the sequences remained unclassified. The most abundant genus was: *Flavobacterium* (*Bacteriodetes*) that represented 13% of the sequences, followed by *Arcobacter* and *Malikia* (both 7%, *Epsilon-* and *Betaproteobacteria*, respectively) and *Rhodoferax* (4%, *Betaproteobacteria*). Sixteen percent of the total sequences corresponded to genera that represented less than 1% of the sequences from the river samples.

Concerning the treated wastewater effluents (Fig. 3B), at phylum level the global composition of the bacterial community was very similar to the one observed in the river samples. The dominant phylum was also the *Proteobacteria* with up to 74% of the sequences, followed by *Bacteroidetes* (up to 37%), *Actinobacteria* (up to 18%). Among the *Proteobacteria*, the most abundant class was the *Betaproteobacteria* (up to 43%), followed by the *Alpha- and Epsilonproteobacteria* (up to 16%), *Gammaproteobacteria* (up to 23%). Although community structures of river and wastewater effluents samples were pretty similar at phylum level, BCC of the treated effluents and the upstream samples were found significantly different at the genus level ($R = 0.472$, $p<0.05$). Therefore, after the release of the WWTPs effluent to the river, the BCC of the river might alter significantly. Recently, Drury et al. [48] studied BCC of rivers sediments and reported that WWTP effluents had impact on bacterial community structure and contributed to a biotic homogenization of rivers with different chemical and physical characteristics.

Multivariate Analysis of the Bacterial Community of All the River Samples

MDS analysis at $OTU_{0.03}$ and genus level was performed to see the BCC changes along the river at the different seasons (Fig. 4). Procrustes analyses showed that bacterial diversity patterns were strongly reproducible at both taxonomic levels (correlation coefficient = 0.75, $p = 0.001$) Despite the expected important influence of WWTPs discharge, the two major clusters based on the observed BCC corresponded to, on one hand, the spring and summer samples and, on the other hand, the autumn and winter samples. Previous studies on different river systems also showed large seasonal differences in BCC and bacterial production [49,50].

In addition, the samples from spring and summer are more scattered than those from autumn and winter suggesting more important variations along the course of the river in the warmer months.

The environmental parameters were shown to be significantly correlated with the samples distribution (Fig. 4A). Temperature was found to be the strongest parameter driving the BCC as already reported in previous studies [50]. Other important parameters were: oxygen concentration, DOC, the SPM content and inorganic nitrogen (NH_4, NO_2 and NO_3) concentrations. These results suggest that the marked seasonality of bacterial community could be associated with *in situ* bacterial activities (DOC and POC biodegradation associated with oxygen consumption and N transformation) which are higher in spring and summer. Indeed, BCC in April and July were positively correlated with PO_4, NH_4, total alkalinity and negatively with oxygen, showing active associated respiration processes, whereas opposite interactions were demonstrated in January and October. *In situ* bacterial activities in the river are lower in the colder situations; therefore the input of allochthonous bacteria by WWTPs effluents had a stronger impact on the BCC of the river in fall and winter.

Furthermore, clear association between the phylotypes and the samples were shown (Fig. 4B). An important number of taxa were associated with autumn and winter samples while a lower number were associated with spring and summer samples. Four out of the five genera which were significantly correlated with the spring and summer samples: *Planktoluna, Limnohabitans, Flavobacterium, Anabaena*, have been clearly identified as planktonic bacteria autochthonous from freshwaters [51]. Among the numerous genera present in winter and autumn samples, several belong to groups associated with sludge and wastewaters (*Faecalibacterium, Haliscomenobacter, Beijerinckia, Rhodocyclus, Prevotella, Roseburia, Nostocoida*).

This supports the hypothesis of a higher importance of *in situ* microbial processes in spring and summer, carried by taxa well adapted to the environmental conditions in the river, while allochthonous bacteria associated with wastewaters were found dominant during winter and autumn.

To estimate the relative importance of environmental, seasonal and spatial components in shaping bacterial community structure, we used a variation partitioning analysis. The percentages of the total variation in the bacterial matrix that can be attributed to the different components of variation (environmental, spatial, seasonal and unexplained) were based on the adjusted (unbiased) fractions. When considering the total bacterial diversity, the highest part of the variation (65%) was remained unexplained. The highest part of variation was explained by environmental factors (29%). Despite the important discharges of wastewaters along the river and the significant modification of the river BCC due to the allochthonous bacteria inputs, followed by the strict influence of the sampling location and season (3% each); Besides, correlation between $OTUs_{0.03}$ and environmental parameters were calculated ($rho>0.8$, $p<0.05$), and POC, COD and SPM showed significant correlations with some species. For instance, *Acinetobacter, Rhodococcus corprophilus, Thauera* were found to have a linear relationship with POC (Table 2). Such correlations were not previously shown at a lower taxonomic level. However, a significant correlation between *Bacteriodetes* and POC was also previously shown in a study conducted in the English Channel [52].

Spatio-temporal Variations of Bacterial Community Structure

Seasonal fluctuations of bacterial community. In this section, all the samples from each sampling campaign have been grouped and the four groups have been compared in order to evaluate the seasonal impact.

Figure 5 presents box plots of Chao and Shannon indices for the different seasons (all sampling stations of each campaign grouped). The median of both parameters was the lowest in April; and a gradually increasing trend was observed from April to January (Fig. 5A). The richest and most diverse samples were those from January. Besides, in January the Shannon index had very low standard deviation, while the highest deviation was observed in April (Fig. 5B).

The medians of Chao indices ranged from 600 (April) to 1300 (January), it means that more than twice different species are expected to be found in January with regards to April. The Shannon index also varied a lot ranging from 3.4 (April) to 5.2 (January). There are different seasonal studies done in the aquatic environments showing higher richness estimations in winter [53,54]. Due to the lower bacterial activity in colder situations, species are more evenly distributed in colder season than in the warmer seasons. However, temperature cannot be the only parameter explaining the difference in diversity indices. Combi-

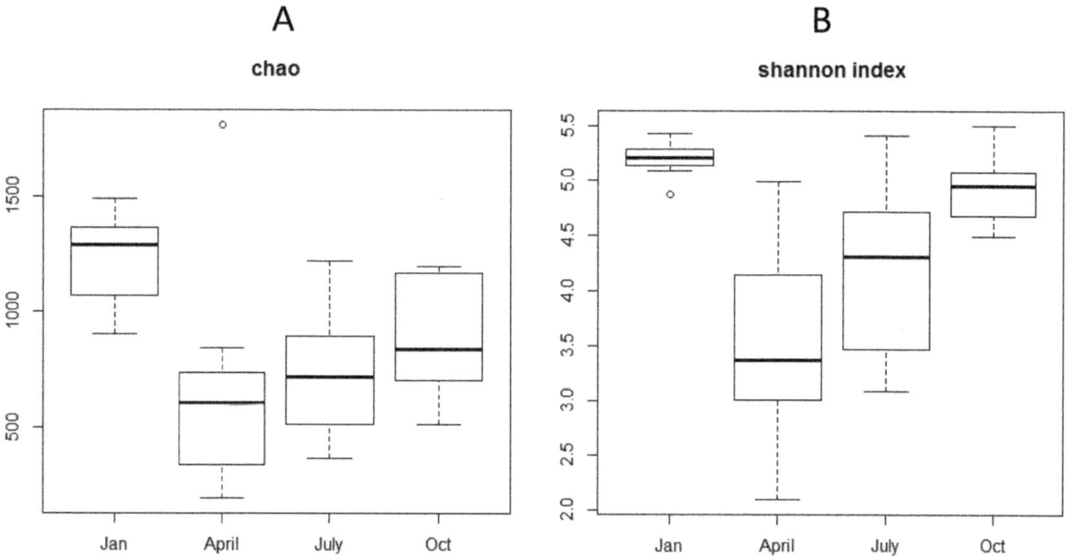

Figure 5. Box plot of richness and diversity estimates per season. (**A**) CHAO and (**B**) Shannon diversity index (All sampling stations of each campaign grouped). Abbreviations: Jan-January, Oct-October.

nation of different parameters such as oxygen, organic carbon and/or nitrogen components might have influenced the richness.

Accordingly, effect of seasonal variation was also observed on BCC. ANOSIM analysis (Table 3) revealed that the BCC of the samples from each season were significantly different from each other with the exception of BCCs in April and July which did not differ significantly. Besides, the SIMPER analysis was used to determine the percentage of similarity between the four sampling campaigns in the upstream stations (Z1 and Z3), in order to test if there was any seasonal impact on the BCC of upstream waters before receiving Brussels WWTPs effluents. SIMPER analysis also revealed which individuals contributed most to the dissimilarity between samples (Table S1). The data showed that BCC of the upstream samples in cold seasons (January and October) had the highest similarity with 62%, followed by similarity between April and July with 53%. On the contrary, the highest difference in the BCC was observed between January and April with 63%. Bloom of *Malikia* and *Flavobacterium* in April and relatively high abundant taxa such as *Dechloromonas*, *Lepthothrix*, *Pseudomonas* and *Arcobacter* in January explained the difference between the BBC of the two seasons.

Impact of WWTPs effluents on the river BCC. As already mentioned, WWTPs effluents have not the same characteristics in terms of BCC than the receiving water body; therefore, it is important to evaluate the impact of WWTPs effluents on the BCC of the river. First of all, we analyzed if the season has an impact on the BCC of the WWTPs' effluents. Based on the richness and diversity analysis, Brussels South WWTP had clearly the lowest richness and diversity index in July. Besides, BCC analysis revealed that Brussels South WWTP in July clustered apart from other WWTP effluents (Fig. S2). ANOSIM revealed no further clear separation between the WWTPs effluents based on the season or WWTP ($p>0.05$).

Figure S3 presents the longitudinal profiles of the Chao and Shannon indexes, each point representing the median of the four sampling seasons. The richness increased after the release of Brussels North WWTP, the standard deviations were large for all the sampling stations, reflecting the important seasonal variations. The diversity (Shannon index) slightly increased after the release of both WWTPs and these stations located just downstream from the WWTPs presented lower standard deviations (indicating lower seasonal variations) than the other sampling stations. To evaluate the impact of WWTPs discharge, three artificial groups of samples were compared: **(1)** the 8 samples of treated wastewaters; **(2)** the 8 river samples collected upstream from Brussels South WWTP discharge (Z1 and Z3 stations from the four sampling campaigns); **(3)** the 20 river samples collected downstream from Brussels South WWTP discharge (5 samples per sampling campaign). First of all, the richness estimation did not change significantly after the discharge, although significant effect of WWTPs discharges on the BCC in the water column was observed. In contrast, a recent study

Table 3. ANOSIM test for differences between time groups.

Groups	Global R	p
January/April	**0.433**	**0.008**
January/July	**0.489**	**0.003**
January/October	**0.452**	**0.003**
April/July	0.011	0.405
April/October	**0.374**	**0.01**
July./October	**0.459**	**0.003**

Table 4. ANOSIM test for differences between location groups.

Groups	Global R	p
wwtp/upstream	**0.448**	**0.005**
wwtp/downstream	0.119	11.7
upstream/downstream	**0.45**	**0.003**

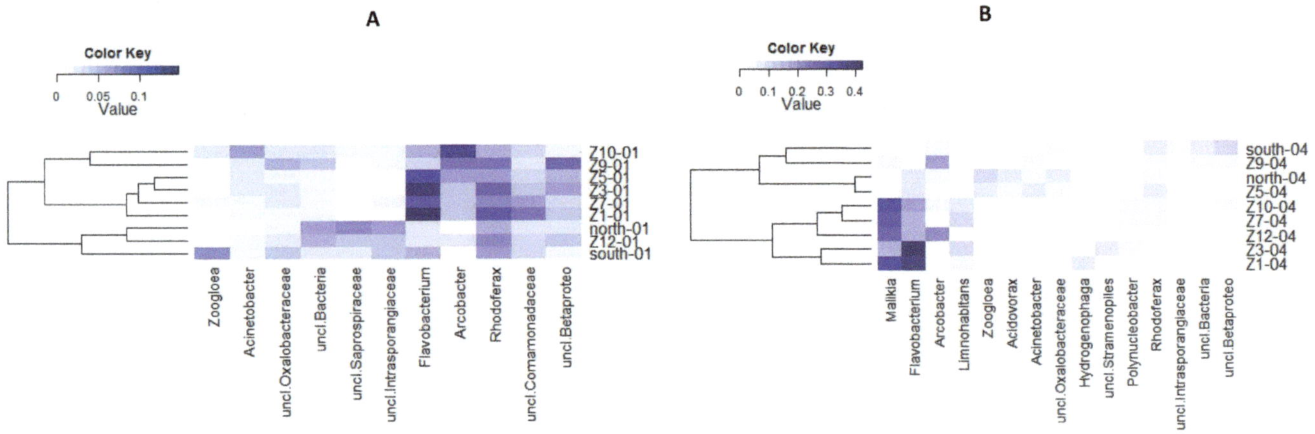

Figure 6. Double hierarchical dendrogram showing the bacterial distribution among the stations. In (**A**) January and (**B**) April. When constructing the heatmaps, differences among sites where characterized using Bray-Curtis distances and differences among taxa were characterized using Euclidean distances. The relative values for bacterial genera are depicted by color intensity.

showed that WWTP effluent significantly reduced bacterial diversity and richness in rivers sediments [48], which might indicate a different response of BCC to the WWTP discharges in the water column and the sediments. It has also been previously reported that the community structure of free living bacteria and particle-attached bacteria were different [16]. Due to the settling of the bacteria attached to particulates, the impact of WWTPs effluents on the bacterial richness in the water column and the sediments might be different.

Besides, ANOSIM analysis (Table 4) revealed that upstream samples were significantly different from downstream and WWTP effluents, while no significant differences were found between WWTPs effluents and the downstream samples. SIMPER analysis (Table S2) shows that no single genus can explain more than 5% of the differences. This means that the differences between the upstream and downstream samples are due to a shift after the discharge of very diverse bacterial community from the WWTPs. The impact of WWTPs on the BCC was studied further per season.

Longitudinal profiles of BCC in the river at different seasons. As January and April presented the most contrasting seasonal impact on the bacterial community, they are discussed in more detail below. Longitudinal profiles of BCC in July and October are further discussed in Results S1 and their dendrograms were illustrated in Figure S4.

In January, two major clusters were observed: river samples (Z1 to Z9), on one hand, and wastewater effluents and river sample Z12, in which a release of raw wastewater is suspected, on the other hand (Fig. 6A). River samples had three subclusters: samples collected upstream of the Brussels South WWTP (Z1, Z3); samples

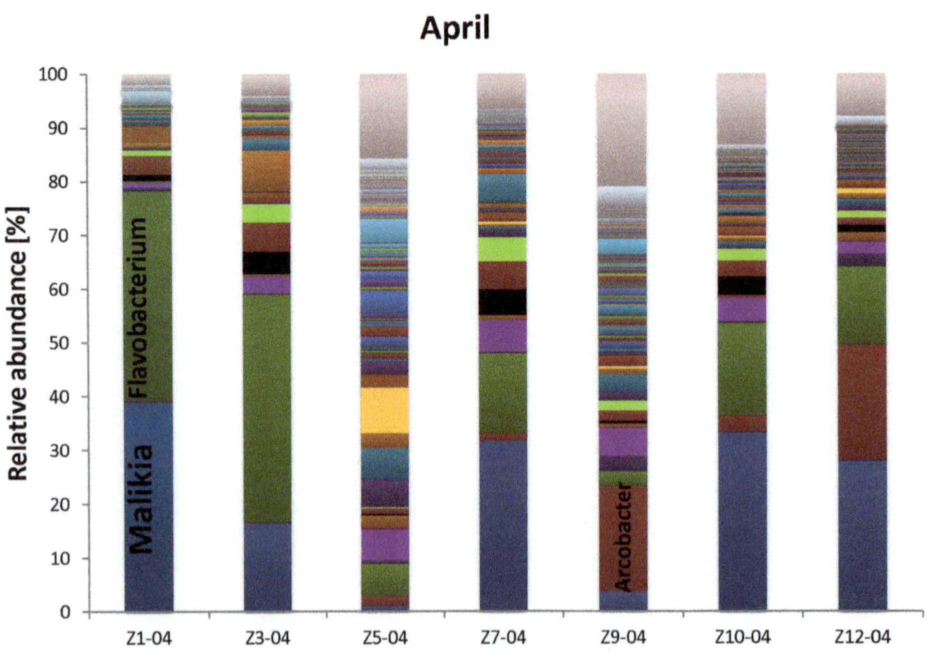

Figure 7. Taxonomic classification of bacterial reads at genus level along the river in April.

collected downstream from the Brussels South WWTP (Z5, Z7) and samples collected downstream from the Brussels North WWTP (Z9, Z10). It is interesting to note that the BCC of the samples taken just downstream from the WWTPs effluents discharge were not dominated immediately by the taxa brought by the effluents. The higher abundance of *Intrasporangiaceae* (closest match to *Tetrasphaera*), *Saprospiraceae* (closest match to *Haliscomenobacter*), and *Rhodoferax* originating from Brussels North WWTPs effluents were only observed at Z12 station. Gram-positive *Tetrasphaera*-related organisms (*Actinobacteria*) are putative polyphosphate-accumulating organisms that are abundant in many full-scale enhanced biological phosphorus removal plants [55] and *Haliscomenobacter* are one of the filamentous bacteria detected in wastewater treatment plants which are responsible for biomass bulking [56,57]. Their survival in the cold temperature was previously reported [57,58].

In contrast to January, a strong resilience of bacterial communities in the river was observed in April. Bacterial community was strongly modified just downstream from the WWTPs effluents discharges (Fig. 7) and in few kilometers the initial bacterial community composition was recovered and this phenomenon was observed after both WWTPs discharges. This difference between BCC in April and January was also shown by clustering. Two major clusters were observed in April, on one hand, the WWTPs samples and the river samples collected just downstream from the effluents and, on the other hand, the remaining river samples (Fig. 6B). The system seems to be resilient, since the initial situation is recovered already a few kilometers downstream from the discharges of WWTPs (disturbance). SIMPER analysis revealed the dissimilarity between WWTPs and upstream samples 71%. This was due to the bloom of *Malikia* and *Flavobacterium* in the upstream stations and the dominance of *Zooglea*, *Arcobacter*, *Acidovorax*, and *Acinetobacter* in the WWTPs' effluents. *Arcobacter* and *Zooglea* were previously isolated from sewage [59,60]. *Flavobacteria* are found in different environment such as soil and freshwater, some of whose species are known to cause disease in freshwater fishes [61] whereas hydrogen-oxidizing and polyhydroxyalkanoate accumulating *Malikia* are restricted to freshwater habitats [62]. We could not find any correlation between *Malikia* bloom and any environmental parameters measured in April. *Malikia* bloom might be due to the low level of river flow at this driest sampling time. Besides, in contrast to January, the impact of WWTPs effluents release has only been observable just downstream from the WWTPs discharges (Z5 and Z9 stations), while the BCC of samples collected further downstream from the WWTPs discharges were similar to the BCC in the upstream part of the river (Fig. 6B). In this case, the composition and structure of bacterial community seems to be dominated by *in situ* processes carried by bacterial groups well adapted to the riverine conditions, with an exception of the sampling points just downstream from the WWTPs discharges.

As it is described above, the impact of WWTPs effluents was season dependent. A recent study showed that sediments of two different rivers with different characteristics in their upstream watershed had almost indistinguishable BCC downstream from the WWTPs discharges, suggesting the potential of WWTPs to reduce the natural variability [48]. Our data suggest that the resilience of the BCC in the water column is season dependent. Studies done by Yannarell et al. [63] and Crump et al. [50] had also contrasting results based on the seasonal impacts on the variability of bacterial communities, which indicates the influence of environmental factors other than temperature like in our case.

Conclusions

Our data demonstrated that BCC was directly or indirectly influenced by complex seasonal shifts in environmental parameters. Our study showed that disturbance strength has the potential to change the composition of bacterial communities and confirms previous findings that bacterial communities are generally not resistant to disturbances [48,64,65]. Indeed, community composition remained altered after the perturbation by WWTPs discharges in three out of the four seasons. However in April, - the driest sampled season -, recovery to pre-disturbance community composition was observed just few kilometers downstream from the WWTPs (pulse disturbance). This is an evidence of the bacterial community resilience in a strongly polluted urban river. Studies dealing with in-situ evidences of freshwater bacterial community stability are scarce and essential for accurate modeling and management of water resources. Environmental and biological factors associated with resilience and resistance must be further studied to assess and prevent the impact of (treated or untreated) sewage pollution on the freshwater ecosystem functioning.

Supporting Information

Figure S1 Rarefaction analysis for individual and pooled samples of river and wastewater treatment plant at 97% cutoff level.

Figure S2 Longitudinal profiles of CHAO and Shannon index.

Figure S3 The community structure of bacterial genera in wastewater treatment plants. The complete linkage clustering of the samples based on the Bray-Curtis similarity metric demonstrated that, samples were grouped neither based on time nor based on wastewater treatment plant.

Figure S4 Double hierarchical dendrogram showing the bacterial distribution among the stations in [A] July and in [B] October. When constructing the heatmaps, differences among sites where characterized using Bray-Curtis distances and differences among taxa were characterized using Euclidean distances. In addition, data within samples were standardized to facilitate comparisons. The relative values for bacterial genera are depicted by color intensity.

Table S1 Percent of similarity (calculated through SIMPER analysis) between upstream and downstream samples per season.

Table S2 Results of SIMPER analyses indicating the contribution of specific genus to observed differences in community structure between upstream and downstream samples.

Results S1 Longitudinal profile of BCC in the river in July and October.

Acknowledgments

The authors thank Hydrologisch Informatie Centrum (HIC) for giving access to flow rates data.

Author Contributions

Conceived and designed the experiments: TGA PS MV NB NKO. Performed the experiments: TGA AA NB. Analyzed the data: TGA OI. Contributed reagents/materials/analysis tools: PS NB MV. Wrote the paper: TGA OI PS.

References

1. Abraham WR (2010) Megacities as sources for pathogenic bacteria in rivers and their fate downstream. Inter. J. Microbiol. 2011: 798282.
2. United Nations (UN) (2009) Revision of World Urbanization Prospects. www.unpopulation.org.
3. Kirchman DL (1994) The uptake of inorganic nutrients by heterotrophic bacteria. Microb. Ecol. 28: 255–271.
4. Kent AD, Yannarell AC, Rusak JA, Triplett EW, McMahon KD (2007) Synchrony in aquatic microbial community dynamics. ISME J 1: 38–47.
5. Murray AE, Preston CM, Massana R, Taylor LT, Blakis A, et al. (1998) Seasonal and spatial variability ofbacterial and archaeal assemblages in the coastal waters near Anvers Island, Antarctica. Appl Environ Microbiol 64: 2585–2595.
6. Lindstrom ES, Kamst-Van Agterveld MP, Zwart G (2005) Distribution of typical freshwater bacterial groups is associated with pH, temperature, and lake water retention time. Appl Environ Microbiol 71: 8201–8206.
7. Ibekwe MA, Leddy MB, Bold RM, Graves AK (2012) Bacterial community composition in low-flowing river water with different sources of pollutants. FEMS Microbiol Ecol 79: 155.
8. Araya R, Yamaguchi N, Tani K, Nasu M (2003) Change in the bacterial community of natural river biofilm during biodegradation of aniline-derived compounds determined by denaturing gradient gel electrophoresis Journal of Health Science 49(5): 379–385.
9. Echeveste P, Dachs J, Berrojalbiz N, Agusti S (2010) Decrease in the abundance and viability of oceanic phytoplankton due to trace levels of complex mixtures of organic pollutants Chemosphere, 81 (2): 161–168.
10. Caracciolo BA, Grenni P, Falconi F, Caputo MC, Ancona V, et al. (2011) Pharmaceutical waste disposal: assessment of its effects on bacterial communities in soil and groundwater. Chemistry and Ecology 27(1): 43–51.
11. Huerta B, Marti E, Gros M, López P, Pompêo M, et al. (2013) Exploring the links between antibiotic occurrence, antibiotic resistance, and bacterial communities in water supply reservoirs. Sci Total Environ. Epub 2013 Apr 13. doi:10.1016/j.scitotenv.2013.03.071.
12. Shanks OC, Newton RJ, Kelty CA, Huse SM, Sogin ML, et al. (2013). Comparison of the microbial community structures of untreated wastewaters from different geographic locales. Appl Environ Microbiol. 79(9): 2906–13.
13. Evison LM (1989) Comparative studies on the survival of indicator organisms and pathogens in fresh and seawater. Water Research 20: 309–315.
14. Garnier J, Billen G, Servais P (1992) Physiological characteristics and ecological rôle of small and large sized bacteria in a polluted river (Seine River, France). Archiv für Hydrobiologie-Ergebnisse Limnologie, 37: 83–94.
15. Okpookwasili GC, Akujobi TC (1996) Bacteriological indicators of tropical water quality. Environmental Toxicology and Water Quality 11: 77–81.
16. Crump BC, Armbrust EV, Baross JA (1999) Phylogenetic analysis of particle-attached and free-living bacterial communities in the Columbia River, its estuary, and the adjacent coastal ocean. Appl. Environ. Microbiol. 65: 3192–3204.
17. Sekiguchi H, Watanabe M, Nakahara T, Xu BH, Uchiyama H (2002).Succes-Succession of bacterial community structure along the Changjiang River determined by denaturing gradient gel electrophoresis and clone library analysis. Appl. Environ. Microbiol. 68: 5142–5150.
18. Beier S, Witzel KP, Marxsen J (2008) Bacterial community composition in central European running waters examined by temperature gradient gel electrophoresis and sequence analysis of 16S rRNA genes. Appl. Environ. Microbiol. 74: 188–199.
19. Shade A, Chiu CY, McMahon KD (2010) Seasonal and episodic lake mixing stimulate differential planktonic bacterial dynamics. Microb Ecol 59: 546–554.
20. Jaffe R (1991). Fate of hydrophobic organic pollutants in the aquatic environment: a review. Environ Pollut. 69(2–3): 237–57.
21. Murray K, Fisher L, Therrien J, George B, Gillespie J (2001). Assessment and use of indicator bacteria to determine sources of pollution to an urban river. J Great Lakes Res 27: 220–229.
22. Servais P, Garcia-Armisen T, George I, Billen G (2007) Fecal bacteria in the rivers of the Seine drainage network (France): Sources, fate and modelling. Science of the Total Environment, 375(1–3): 152–167.
23. Ouattara NK, Passerat J, Servais P (2011) Faecal contamination of water and sediment in the rivers of the Scheldt drainage network. Environ Monit Assess. 183(1–4): 243–57.
24. Kostanjsek R, Lapanje A, Drobne D, Perović S, Perović A, et al. (2005) Bacterial community structure analyses to assess pollution of water and sediments in the Lake Shkodra/Skadar, Balkan Peninsula. Environ Sci Pollut Res Int. 12(6): 361–8.
25. Li D, Yang M, Li Z, Qi R, He J, et al. (2008) Change of bacterial communities in sediments along Songhua River in Northeastern China after a nitrobenzene pollution event. FEMS Microbiol Ecol. 65(3): 494–503.
26. Wakelin SA, Colloff MJ, Kookana R (2008) Effect of Wastewater Treatment Plant Effluent on Microbial Function and Community Structure in the Sediment of a Freshwater Stream with Variable Seasonal Flow. Appl Environ Microbiol. 74(9): 2659–2668.
27. Zhu J, Zhang J, Li Q, Han T, Xie J, et al. (2013) Phylogenetic analysis of bacterial community composition in sediment contaminated with multiple heavy metals from the Xiangjiang River in China. Mar Pollut Bull. pii: S0025-326X(13)00079-9.
28. Kenzaka T, Yamaguchi N, Prapagdee B, Mikami E, Nasu M (2001) Bacterial community composition and activity in urban rivers in Thailand and Malaysia. J. Health Sci. 47: 353–361.
29. Winter C, Hein T, Kavka G, Mach RL, Farnleitner AH (2007) Longitudinal changes in the bacterial community composition of the Danube River: a whole-river approach. Appl. Environ. Microbiol. 73: 421–431.
30. Zhang M, Yu N, Chen L, Jiang C, Tao Y et al. (2012) Structure and seasonal dynamics of bacterial communities in three urban rivers in China. Aquat Sci 74: 113–120.
31. Brion N, Servais P, Bauwens W, Verbanck M (2012) Past and present chemical and microbiological quality of the Zenne River. In: Wynants and Nuytten (eds) "Bridge over troubled waters", Crosstalks, VUB press, 241–264.
32. Porter KG, Feig YS (1980) The use of DAPI for identifying and counting aquatic microflora. Limnol. Oceanog. 25: 943–948.
33. Llirós M, Casamayor EO, Borrego C (2008) High archaeal richness in the water column of a freshwater sulfurous karstic lake along an interannual study. FEMS Microbiol Ecol. 66: 331–342.
34. Callaway TR, Dowd SE, Wolcott RD, Sun Y, Mcreynolds JL, et al. (2009) Evaluation of the bacterial diversity in cecal contents of laying hens fed various molting diets by using bacterial tagencoded FLX amplicon pyrosequencing. Poult Sci.88: 298–302.
35. Smith DM, Snow DE, Rees E, Zischkau AM, Hanson JD, et al. (2010) Evaluation of the bacterial diversity of pressure ulcers using bTEFAP pyrosequencing. BMC Med Genomics 3: 41.
36. Edgar RC, Haas BJ, Clemente JC, Quince C, Knight R (2011) UCHIIME improves sensitivity and speed of chimera detection. Oxford Journal of Bioinformatics, 27(16): 2194–2200.
37. Clarke KR, Gorley RN (2006) PRIMER V6: User Manual/Tutorial. Plymouth, UK: PRIMER-E.
38. Clarke KR, Warwick RM (2001) Change in Marine Communities: An Approach to Statistical Analysis and Interpretation, PRIMERE Ldt, Plymouth, UK.
39. R Development Core Team (2011) R: A Language and Environment for Statistical Computing. R Foundation for Statistical Computing, Vienna, Austria. ISBN 3-900051-07-0.
40. Oksanen J, Blanchet FG, Kindt R, Legendre P, O'Hara RB, et al. (2011) vegan: Community Ecology Package. R package version 1.17–8.
41. David R (2010) labdsv: Ordination and Multivariate Analysis for Ecology. R package version 1.4–1.
42. Peres-Neto PR, Legendre P, Dray S, Borcard D (2006)Variation partitioning of species data matrices: estimation and comparison of fractions. Ecology 87: 2614–2625.
43. Weithman AS, Haas MA (1984) Effects of dissolved-oxygen depletion on the rainbow trout fishery in Lake Taneycomo, Missouri. Trans. Am. Fish. Soc. 113: 109–124.
44. Ademoroti CMA (1996) Mini water Development in Ibadan. Environmental Chemistry and Toxicology 1st Edition Foludex Press Ibadan, Nigeria. 70–121.
45. Garnier J, Servais P, Billen G (1992b) Bacterioplankton in the Seine River: impact of the Parisian urban effluents. Can. J. Microbiol. 38: 56–64.
46. Hu M, Wang X, Wen X, Xia Y (2012) Microbial community structures in different wastewater treatment plants as revealed by 454-pyrosequencing analysis. Bioresource Technology 117: 72–79.
47. Methé BA, Hiorns WD, Zehr JP (1998) Contrasts between marine and freshwater bacterial community composition: Analyses of communities in Lake George and six other Adirondack lakes. Limnol Oceanogr 43: 368–374.
48. Drury B, Rosi-Marshall E, Kelly JJ (2013) Wastewater treatment effluent reduces the abundance and diversity of benthic bacterial communities in urban and suburban rivers. Appl Environ Microbiol. 79(6): 1897–905.
49. Van der Gucht K, Sabbe K, De Meester L, Vloemans N, Zwart G, et al. (2001) Contrasting bacterioplankton community composition and seasonal dynamics in two neighbouring hypertrophic freshwater lakes. Environ. Microbiol. 3: 680–690.
50. Crump BC, Hobbie J (2005) Synchrony and seasonality in bacterioplankton communities of two temperate rivers. Limnol. Oceanogr. 50(6): 1718–1729.
51. Newton RJ,Jones SE, Eiler A, McMahon KD, Bertilsson S (2011) A guide to the natural history of freshwater lake bacteria. Microbiol Mol Biol Rev 75: 14–49.
52. Lamy D, Obernosterer I, Laghdass M, Artigas LF, Breton E, et al. (2009) Temporal changes of major bacterial groups and bacterial heterotrophic activity during a Phaeocystis globosa bloom in the eastern English Channel. Aquat. Microb. Ecol., 58: 95–107.

53. Gilbert JA, Field D, Swift P, Thomas S, Cummings D, et al. (2010) The taxonomic and functional diversity of microbes at a temperate coastal site: a 'multiomic' study of seasonal and diel temporal variation. PLoS One 5: e15545.
54. Ghiglione JF, Murray AE (2012) Pronounced summer to winter differences and higher winter time richness in coastal Antarctic marine bacterioplankton. Environmental Microbiology 14(3): 617–629.
55. Nielsen PH, Mielczarek AT, Kragelund C, Nielsen JL, Saunders AM, et al. (2010) A conceptual ecosystem model of microbial communities in enhanced biological phosphorus removal plants. Water Res 44: 5070–5088.
56. Eikelboom D (1975) Filementous Organisms observed in activated sludge. Water Research 9: 365–388.
57. Kotay SM, Datta T, Choi J, Goel R (2011) Biocontrol of biomass bulking caused by Haliscomenobacter hydrossis using a newly isolated lytic bacteriophage. Water Res. 45(2): 694–704.
58. Finneran KT, Johnsen CV, Lovley DR (2003) *Rhodoferax ferrireducens* sp. nov., a psychrotolerant, facultatively anaerobic bacterium that oxidizes acetate with the reduction of Fe(III). Int. J. Sys. Evol. Microbiol. 53: 669–673.
59. Belanchea L, Valdea JJ, Comas J, Rodab IR, Poch M (2000) Prediction of the bulking phenomenon in wastewater treatment plants. Artificial Intelligence in Engineering 14: 307–317.
60. Moreno Y, Botella S, Alonso JL, Ferrús MA, Hernández M, et al. (2003) Specific Detection of Arcobacter and Campylobacter Strains in Water and Sewage by PCR and Fluorescent *In situ* Hybridization. Appl Environ Microb, 69: 1181–1186.
61. Starliper C (2011) Bacterial coldwater diseased fished by *Flavobacterium psychrophilum*. J. Adv. Res. 2(2): 97–108.
62. Spring S, Wagner M, Schumann P, Kämpfer P (2005) Malikia granosa gen. nov., sp. nov., a novel polyhydroxyalkanoate- and polyphosphate-accumulating bacterium isolated from activated sludge, and reclassification of Pseudomonas spinosa as Malikia spinosa comb. nov. IJSEM 55(2): 621–629.
63. Yannarell AC, Kent AD, Lauster GH, Kratz TK, Triplett EW (2003) Temporal patterns in bacterial communities in three temperate lakes of different trophic status. Microb. Ecol. 46: 391–405.
64. Allison G (2004) The influence of species diversity and tress intensity on community resistance and resilience. Ecological Monographs 74: 117–134. doi:10.1890/02-068.
65. Berga M, Székely A, Langenheder S (2012) Effects of disturbance intensity and frequency on bacterial community composition and function. PLoS ONE 7(5): e36959. doi:10.1371/journal.pone.0036959.

4

Integrating Water Flow, Locomotor Performance and Respiration of Chinese Sturgeon during Multiple Fatigue-Recovery Cycles

Lu Cai[1], Lei Chen[2], David Johnson[3,4], Yong Gao[2], Prashant Mandal[3], Min Fang[1,3], Zhiying Tu[1,3], Yingping Huang[1,3]*

1 Collaborative Innovation Center for Geo-Hazards and Eco-Environment in Three Gorges Area, Hubei Province, Yichang, PR China, **2** Institute of Chinese Sturgeon Research, China Three Gorges Project Corporation, Yichang, PR China, **3** Engineering Research Center of Eco-environment in Three Gorges Reservoir Region, Ministry of Education, China Three Gorges University, Yichang, PR China, **4** School of Natural Sciences and Mathematics, Ferrum College, Ferrum, Virginia, United States of America

Abstract

The objective of this study is to provide information on metabolic changes occurring in Chinese sturgeon (an ecologically important endangered fish) subjected to repeated cycles of fatigue and recovery and the effect on swimming capability. Fatigue-recovery cycles likely occur when fish are moving through the fishways of large dams and the results of this investigation are important for fishway design and conservation of wild Chinese sturgeon populations. A series of four stepped velocity tests were carried out successively in a Steffensen-type swimming respirometer and the effects of repeated fatigue-recovery on swimming capability and metabolism were measured. Significant results include: (1) critical swimming speed decreased from 4.34 bl/s to 2.98 bl/s; (2) active oxygen consumption (i.e. the difference between total oxygen consumption and routine oxygen consumption) decreased from 1175 mgO_2/kg to 341 mgO_2/kg and was the primary reason for the decrease in U_{crit}; (3) excess post-exercise oxygen consumption decreased from 36 mgO_2/kg to 22 mgO_2/kg; (4) with repeated step tests, white muscle (anaerobic metabolism) began contributing to propulsion at lower swimming speeds. Therefore, Chinese sturgeon conserve energy by swimming efficiently and have high fatigue recovery capability. These results contribute to our understanding of the physiology of the Chinese sturgeon and support the conservation efforts of wild populations of this important species.

Editor: Josep V. Planas, Universitat de Barcelona, Spain

Funding: This work was supported by Major Science and Technology Program for Water Pollution Control and Treatment in the National Twelfth Five-Year Plan of China (Grant number: 2012ZX07104-002-04) and the National Nature Science Foundation of China (Grant number: 50979049, 51309140). The funders had no role in study design, data collection and analysis, decision to publish, or preparation of the manuscript.

Competing Interests: The authors have declared that no competing interests exist.

* E-mail: yingpinghuang@126.com

Introduction

Water flow is commonly believed to be of an important environmental factor to fish. Natural selection favors locomotor and respiration strategies appropriate for an existing flow [1]. Chinese sturgeon (*Acipenser sinensis*, Gray 1834) has been listed as Class I Endangered Species by the Chinese Government [2] and is also listed as Critically Endangered (CR) in the IUCN Red List of Threatened Species (version 2013.1). The species is anadromous and currently occurs only in the middle and lower reaches of the Yangtze River and near-shore in the Yellow and East China Seas [3]. Although captive breeding of Chinese sturgeon is now feasible [4,5], wild populations have declined significantly because of dam construction, overfishing and pollution [6–10]. Chinese sturgeon once migrated further than any other sturgeon, over 3,200 km in the Yangtze River. Restoration of the wild sturgeon populations is difficult due to long maturation time, infrequent spawning and low productivity [11–12]. Although the species has a life span of 50–60 years, the species breeds only three or four times and the hatchling survival rate is estimated at <1%.

Dam construction has blocked spawning migration [13] and the sturgeon populations have declined dramatically [10]. Augmenting wild populations by release of captive breed sturgeon is a great achievement for species conservation but not sufficient. Building fishways that allow migration is another important part of the effort to maintain wild populations [14,15]. Data on swimming capability and the physiological response to exercise and fatigue provide important design criteria for fishways [16–19].

Publications on Chinese sturgeon have been more frequent in recent years [20] but have focused on biochemistry and habitat assessment. Research on swimming capability and excess post-exercise oxygen consumption, for one repetition, has been reported (e.g., [21–24]. However, fish typically undergo multiple fatigue-recovery cycles while moving through fishways of large dams [25] and data on the effects of repeated fatigue-recovery is necessary for the design of effective fishways. In this study, the effect of repeated fatigue-recovery cycles on swimming capability and metabolism were investigated using a stepped velocity test in a Steffensen-type swimming respirometer. The study documents the change in respiration and swimming capability of fish undergoing repeated fatigue-recovery cycles. These results contribute to our understanding of the physiology of the Chinese sturgeon and support the conservation efforts of wild populations of this important species.

Materials and Methods

Ethics Statement

This study was conducted in strict accordance with the laws governing animal experimentation in China. The protocol was approved by the China Three Gorges University. All efforts were made to minimize suffering.

Test fish

Juvenile sturgeon were obtained from the institute of Chinese sturgeon research, China Three Gorges Project Corporation, in Yichang, China (30°56′ N; 111°15′ E) and were of the second filial generation of cultured Chinese sturgeon. The sturgeon (body length: 10.89±0.11 cm and body weight 8.38±0.27 g, mean ± S.E.) were maintained in cylindrical tanks (80 cm in diameter and 90 cm deep). They were fed to satiation each morning at 8:00 am with Tubificidae. Water temperature varied from 19.3°C to 20.8°C and dissolved oxygen (*DO*) maintained above 7.0 mg/L. Ammonia-N and nitrite-N were lower than 0.050 mg/L and 0.007 mg/L, respectively. Forty eight hours prior to the experiment, feeding was interrupted to prevent the elevated oxygen consumption rate associated with digestion [26].

Respirometer

The sturgeon were tested in a sealed 14 L Steffensen-type swimming respirometer submerged in a 55.7 L tank (84.0 cm long ×39.9 cm wide ×17.0 cm deep) to maintain temperature. A schematic diagram of the respirometer was included in an earlier publication [27]. Respirometer components include a motor with speed controller and propeller and a rectangular swimming chamber (4.5 L, 35.5 cm long ×11.0 cm wide ×11.5 cm deep) with a Lucite multi-aperture rectifier at the entrance to maintain laminar flow and a wire grid at the exit. Normal respirometer operating assumptions were made: (1) swimming speed (*U*) is equal to water flow speed as measured with an acoustic Doppler velocity meter (Nortek AS, Oslo, Norway), and (2) oxygen consumption can be calculated from the change in dissolved oxygen, as measured by an oxygen electrode (Hach HQ30d, Loveland, USA).

Protocol for stepped velocity test

Ten sturgeons were tested and the body length (bl), fork length, total length and body mass of each sturgeon were measured. Before testing, fish were acclimated overnight in the respirometer at a very low flow speed (0.3~0.5 bl/s) and then allowed to swim at 0.5 bl/s from 7:00–8:00 am the next morning [12,28]. When testing began, flow speed was increased by 0.5 bl/s every 20 min and dissolved oxygen was measured at 5 min intervals. When the sturgeon was exhausted (swimming ceased and the fish rested against the wire grid for 20 s), flow speed was adjusted to 0.5 bl/s during the 30 min recovery period [27]. This comprises a single stepped velocity test and each sturgeon tested was subjected to four tests in succession to simulate a repeated cycle of fatigue and recovery.

Data collection and analyses

All experimental data was expressed as mean ± S.E. and was analyzed and fitted using Origin (Version 8.1). The level of significance was determined using ANOVA Fisher LSD. Fit to mathematical models was evaluated following the Akaike information criterion *AIC* [29,30] using the model correcting for finite sample size and frequently applied in behavioral ecology [31];

$$AICc = n\log(RSS/n) + 2k + 2k(k+1)/(n-k-1) \quad (1)$$

where *RSS* is the residual sum of squares from the fitted model, *n* is sample size, and *k* is the total number of parameters. Its use helps prevent overfitting and offers a relative estimate of the information lost by a given model. The value of *AICc* is inversely related to the goodness of fit of a given model.

Based on data from the stepped velocity test, the critical swimming speed (U_{crit}) of the test fish is calculated as described by Brett [32]. Oxygen consumption rate (MO_2) is calculated using Equation 2;

$$MO_2 = (d(DO)/dt - d(DO)'/dt) \times V/m \quad (2)$$

where *V* (L) is the volume of the respirometer, *m* (kg) is the mass of the sturgeon, *d(DO)/dt* ($mgO_2/(L\times h)$), which is the slope of the linear regression of *DO* decreasing over time ($R^2>0.99$), is the rate of change in dissolved oxygen in the respirometer during the step test and *d(DO)'/dt* ($mgO_2/(L\times h)$) is the rate of change in dissolved oxygen in the control (respirometer with no fish). Routine oxygen consumption rate ($MO_{2,\ routine}$) of each sturgeon is the MO_2 at 0.5 bl/s (Cai et al., 2013b). Maximum oxygen consumption rate ($MO_{2,\ max}$) of each sturgeon is the maximum value of MO_2 during the stepped velocity test. The aerobic scope (*AS*) is the difference between $MO_{2,\ max}$ and $MO_{2,\ routine}$ and is related to the energy potentially available for swimming [33,34].

During the swimming period, variation of MO_2 with *U* was obtained by fitting the data to Equation 3 [35,36]:

$$MO_2 = a + bU^c \quad (3)$$

where *a*, *b* and *c* are constants. Active oxygen consumption (*AOC*) is the difference between aerobic oxygen consumption and routine oxygen consumption during the period and was obtained by computing the area under the MO_2 curve during the period. During the recovery period, variation of MO_2 over time (*t*) was obtained by fitting the data to Equations 4 and 5 [37,38]:

$$MO_2 = a + be^{ct} \quad (4)$$

$$MO_2 = e^{a+bt+ct^2} \quad (5)$$

where *a*, *b* and *c* are constants. Excess post-exercise oxygen consumption (*EPOC*) is the difference between aerobic oxygen consumption and routine oxygen consumption during recovery from exhaustion and was obtained by computing the area under the MO_2 curve during the period. Assuming that *EPOC* is the oxygen consumed to recover from anaerobic metabolism, total oxygen consumption was obtained by adding *EPOC* to aerobic oxygen consumption [32]. Total oxygen consumption was calculated by the iterative method of Lee et al. [37] until *EPOC* and the area between the MO_2 curve and the broken line (representing anaerobic metabolism) differed <0.5%. The broken line was then used to estimate *U* at the onset of anaerobic respiration (*U*a).

Results

The U_{crit} of test sturgeon decreased with successive recoveries from fatigue (Fig. 1), yielding values of 4.34±0.18, 3.77±0.20, 3.27±0.22, 2.98±0.26 bl/s, respectively, over four cycles. The relationship between the U_{crit} and the test number (N) was linear and the decrease was significant:

$$U_{crit}=4.76 \quad 0.47N(F=118,P=0.00836,R^2=0.975)$$

The $\Delta_{U crit}$ ($U_{crit,\ N+1}$-$U_{crit,\ N}$) values were 0.57, 0.50, 0.29 bl/s and the recovery ratios ($U_{crit,\ N+1}/U_{crit,\ N}$) were 86.9%, 86.7% and 91.1%.

Metabolic parameters were displayed in Table 1. Differences in $MO_{2,\ routine}$ among the four tests were not significant. And differences in $MO_{2,\ max}$ were also not significant; values for $MO_{2,\ routine}$ were similar but $MO_{2,\ max}$ and AS decreased with test number.

As shown in Fig. 2, U_{crit} varied exponentially with AS:

$$U_{crit}=2.78448+0.00114\times e^{0.0224AS}(F=1882,P=0.00163,$$
$$R^2=0.998)$$

Data for the variation of MO_2 with U was fitted to Equation 3 and the results are shown in Table 2. The speed coefficient (c), a measure of swimming efficiency, increased during tests 2 and 3 and then decreased in test 4.

Figure 3 illustrates the results of further analysis of the physiological changes that occurred during the four stepped velocity tests. The decreases in $EPOC$ indicates that anaerobic capacity decreases with successive cycles of fatigue and recovery; the decrease in the last test was significant larger than in the second and third tests. The onset of anaerobic metabolism occurred at lower swimming speeds (absolute U_a, 1.82, 1.79, 1.64, 1.57 bl/s, respectively, for tests 1~4). However, because the decrease in U_{crit} was more pronounced, relative U_a increased with successive testing (41.94%, 47.48%, 50.15% and 52.68%).

During the recovery period, MO_2 decreased rapidly over time; the relationship fitted to Equations 4 and 5 is shown in Table 3.

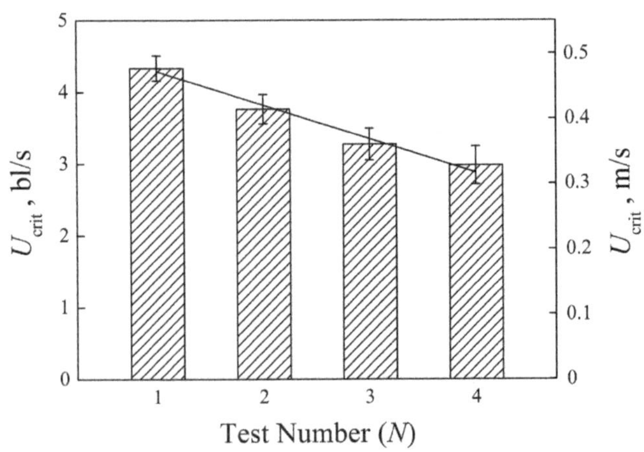

Figure 1. Variation of critical swimming speed (U_{crit}, mean±S.E.) over four cycles of fatigue-recovery.

Table 1. Respiratory metabolic parameters*.

Test	$MO_{2,\ routine}$	$MO_{2,\ max}$	Mean AS	AOC	$EPOC$
	$mgO_2/(kg\times h)$	$mgO_2/(kg\times h)$	$mgO_2/(kg\times h)$	mgO_2/kg	mgO_2/kg
1	266.60±30.94	598.80±40.55	332.20	1,175.48	35.70
2	231.51±33.10	532.51±35.06	301.00	625.67	31.51
3	241.40±20.70	512.65±26.07	271.25	458.38	29.99
4	263.81±22.50	492.37±68.45	228.56	341.34	21.57

*Routine oxygen consumption rate ($MO_{2,\ routine}$, mean ± S.E.) is the oxygen consumption rate (MO_2) at 0.3~0.5 BL/s; maximum oxygen consumption rate ($MO_{2,\ max}$, mean ± S.E.) is the maximum value of MO_2 during the stepped velocity test; the aerobic scope (AS) is the difference between $MO_{2,\ max}$ and $MO_{2,\ routine}$; active oxygen consumption (AOC) is the difference between aerobic oxygen consumption and routine oxygen consumption during swimming period; excess post-exercise oxygen consumption ($EPOC$) is the difference between aerobic oxygen consumption and routine oxygen consumption during recovery from exhaustion. Differences in $MO_{2,\ routine}$ and $MO_{2,\ max}$ not significant among successive tests: $MO_{2,\ routine}$, $F=0.392$, $P=0.759$; $MO_{2,\ max}$, $F=1.050$, $P=0.390$.

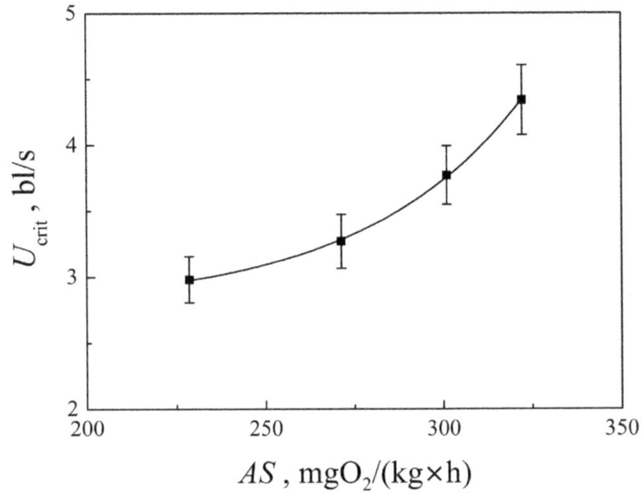

Figure 2. Variation of critical swimming speed (U_{crit}, mean±S.E.) with aerobic scope (AS). The AS is the difference between maximum oxygen consumption rate and routine oxygen consumption rate.

Table 2. Relationship between oxygen consumption rate and swimming speed*.

Test	Equation	F	P	R^2
1	$MO_2=241.8+60.9U^{1.20}$	1.6×10^6	<0.00001	0.964
2	$MO_2=232.7+30.5U^{1.70}$	0.8×10^6	<0.00001	0.959
3	$MO_2=233.7+32.3U^{1.71}$	6.3×10^6	<0.00001	0.997
4	$MO_2=247.1+50.0U^{1.42}$	8.1×10^6	<0.00001	0.993

*Oxygen consumption rate, MO_2, mgO2/(kg×h); swimming speed, U, bl/s.

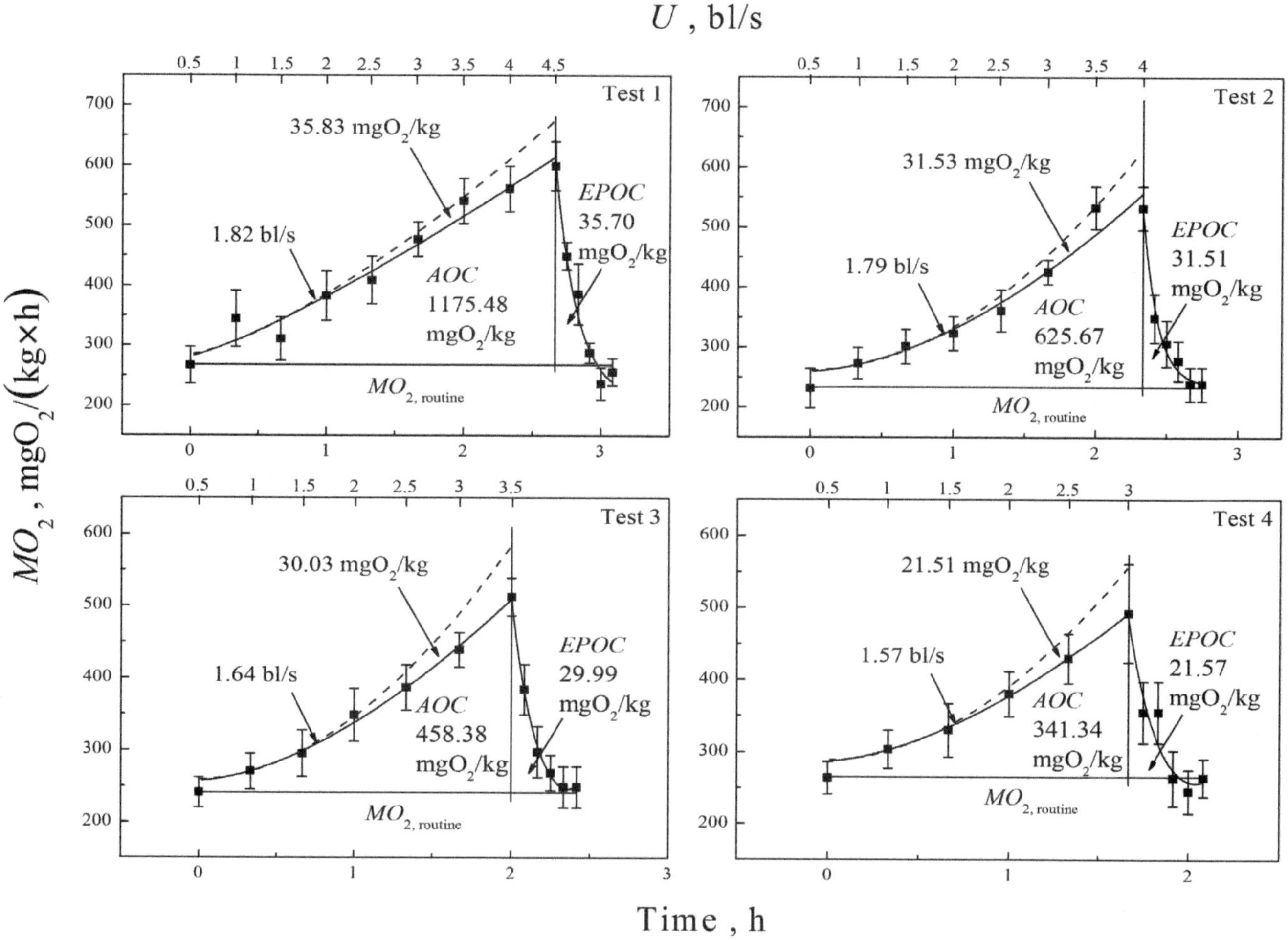

Figure 3. Variation of oxygen consumption rate (MO_2, mean±S.E.) over time. Active oxygen consumption (*AOC*) is the difference between aerobic oxygen consumption and routine oxygen consumption during swimming period; excess post-exercise oxygen consumption (*EPOC*) is the difference between aerobic oxygen consumption and routine oxygen consumption during recovery from exhaustion.

Discussion

Critical swimming speed, U_{crit}

Juvenile sturgeon, such as shortnose sturgeon *Acipenser brevirostrum* [39] and Amur sturgeon (*Acipenser schrenckii*) [27], generally display U_{crit} levels of 2 bl/s~3 bl/s. However, the U_{crit} of Chinese sturgeon was found to be above that range, indicating that Chinese sturgeon have better swimming capability. In addition, there was a linear decrease ($R^2 = 0.975$) in U_{crit} of the Chinese sturgeon with repeated fatigue-recovery cycles (Fig. 1). However, the difference in U_{crit} declined with repeated testing (Δ_{Ucrit} 0.57, 0.50, 0.29 bl/s) and the recovery ratio increased from 86.9% to 91.1%. A better fit could have been obtained using a higher order equation but would have resulted in overfitting according to AIC analysis. The recovery ratio for Chinese sturgeon is comparable to sockeye salmon (*Oncorhynchus nerka*), with a recovery ratio of approximately 85% after a resting time of 1 hour [40]. Because exertion decreases U_{crit}, the slot velocities of upstream fishways should be lower than those in downstream fishways.

Table 3. Variation of oxygen consumption rate over time during the recovery period*.

Test	Equation	*F*	*P*	R^2
1	$MO_2 = 211.6 + 8.8 \times 10^9 e^{-6.35t}$	0.5×10^6	<0.00001	0.968
2	$MO_2 = 235.0 + 1.6 \times 10^{12} e^{-9.61t}$	1.1×10^6	<0.00001	0.979
3	$MO_2 = e^{36.1 - 25.9t + 5.5t^2}$	8.5×10^6	<0.00001	0.998
4	$MO_2 = e^{24.8 - 18.9t + 4.6t^2}$	0.4×10^6	<0.00001	0.837

*Oxygen consumption rate, MO_2, mgO2/(kg×h); time, *t*, h.

$MO_{2, routine}$ and $MO_{2, max}$

Differences in $MO_{2, routine}$ over the four test cycles were small and erratic (Table 1), indicating little, if any, influence of fatigue on the routine metabolic rate. While differences in $MO_{2, max}$ were not significant, the changes were more pronounced and decreased with each test cycle. This leads to a decrease in *AS* and is consistent with the observed decrease in the U_{crit}. The relationship between *AS* and U_{crit} is shown in Fig. 2, indicating an exponential increase in U_{crit} with *AS*.

Speed exponent, *c*

The exponent *c* (Equation 3) is inversely related to swimming efficiency [34,38]. In the first test, *c* was 1.20, similar to those of the white sturgeon, lake sturgeon and Amur sturgeon [27,41]; not surprising as the four *Acipenser* species are closely related [42]. In the three successive tests, the values of *c* did not change consistently ($c = 1.70$, 1.71, 1.42, respectively, for Tests 2, 3, and 4). There are numerous reports on the effect of fish morphology on swimming efficiency [43,44]. In this investigation morphology was consistent among the tests and results indicate that repeated fatigue-recovery cycles affect physiological function. It has been reported that different fish use different swimming strategies [45,46], but it remains unclear as to whether the sturgeon are adjusting swimming strategy in response to repeated fatigue-recovery cycles. Swimming efficiency was highest in Test 1, lower in Tests 2 and 3, with some recovery displayed in Test 4.

Excess post-exercise oxygen consumption, EPOC

During fatigue recovery the variation of oxygen consumption rate with time is usually described by Equation 4 [27,37]. The data fit for Test 1 and Test 2 was good (Table 3) but data for Test 3 and Test 4 could not be fit to Equation 4 because the fit did not converge. After fitting the data to several other functions Equation 5 was selected, based on fit and *AIC* analysis, to estimate the onset of anaerobic respiration and its contribution to U_{crit}.

Information on *EPOC* is helpful for establishing design criteria for the resting pools of fishways [27,47]. The *EPOC* is inversely related to the recovery capability from exhaustion [37]. The *EPOC* in the first test was 35.7 mgO_2/kg, much lower than for salmon but similar to Amur sturgeon [27,37], indicating high recovery capability for Chinese sturgeon. Although *EPOC* decreased with repeated recovery from fatigue, it does not necessarily follow that the recovery capability increased because U_{crit} and *AS* also decreased and *AOC* decreased by 70% from Test 1 to Test 4 (Table 1).

The *EPOC* is directly related to anaerobic respiration [37] and the decrease in *EPOC* indicates that repeated recovery from fatigue lowers capacity for anaerobic metabolism. With each fatigue-recovery cycle anaerobic respiration begins at a lower swimming speed but contributes less to achieving U_{crit}. However, it can be seen in Fig. 3 and Table 1 that oxygen consumption is primarily from aerobic respiration and the decrease in aerobic respiration is the primary cause for the decrease in U_{crit}. It was reported that salmon pass fishways using primarily aerobic respiration [25], consistent with our results.

Swimming speed at onset of anaerobic respiration, U_a

Aerobic metabolism with red muscle fiber provides propulsion in general, but white muscle fiber contributes to achieving high speeds [36,48,49]. It was reported that anaerobic swimming in cyprinids begins to augment propulsion when swimming speed reaches 30~50% U_{crit} [50]. However, aerobic power is different in different fish species [37,50,51]. In this study, the relative U_a was between 41.94% and 52.68% U_{crit} (absolute U_a, 1.82, 1.79, 1.64, 1.57 bl/s, respectively, for tests 1~4) similar to the higher range of cyprinids. So, with repeated step tests, white muscle (anaerobic metabolism) began contributing to propulsion at lower absolute swimming speeds.

Conclusions

Repeated cycles of fatigue and recovery affect the swimming capability of Chinese sturgeon; U_{crit}, *AOC* and absolute U_a decrease with decreasing swimming capacity. These results contribute to our understanding of the physiology of the Chinese sturgeon and support the conservation efforts of wild populations of this important species.

Acknowledgments

We thank Miss Tian Suyi, Mr. Guo Baifu and Mr. Yuan Tao who also contributed to the work providing assistance with writing and aquaculture operations. This work was supported by Major Science and Technology Program for Water Pollution Control and Treatment in the National Twelfth Five-Year Plan of China (Grant number: 2012ZX07104-002-04) and the National Nature Science Foundation of China (Grant number: 50979049, 51309140).

Author Contributions

Conceived and designed the experiments: L. Cai. Performed the experiments: L. Cai. Analyzed the data: L. Cai DJ. Contributed reagents/materials/analysis tools: L. Chen YG. Wrote the paper: L. Cai. Contributed substantially to revisions: YH DJ PM MF ZT.

References

1. Langerhans RB (2009) Predictability of phenotypic differentiation across flow regimes in fishes. Integr Comp Biol 48: 750–768. doi:10.1093/icb/icn092.
2. Wei Q, Ke F, Zhang J, Zhuang P, Luo J, et al. (1997) Biology, fisheries, and conservation of sturgeons and paddlefish in China. Environ Biol Fish 48: 421–255. doi:10.1023/A:1007395612241.
3. Wei Q, He J, Yang D, Zheng W, Li L (2004) Status of sturgeon aquaculture and sturgeon trade in China: a review based on two recent nationwide surveys. J Appl Ichthyol 20: 321–332. doi:10.1111/j.1439-0426.2004.00593.x.
4. Guo B, Chang J, Xiao H, Zhu B, Wan J, et al. (2011) The reproductive biology of first filial generation of *Acipenser sinensis* growing up in the freshwater environment. Acta Hydrob Sinica 35: 940–945. doi:10.3724/SP.J.1035.2011.00940.
5. Wei Q, Li L, Du H, Zhang X, Xiong W, et al. (2013) Research on technology for controlled propagation of cultured Chinese sturgeon (*Acipenser sinensis*). *J Fish Sci China* 20:1–11. doi:10.3724/SP.J.1118.2013.00001.
6. Xie P (2003) Three-Gorges Dam: risk to ancient fish. Science 302: 1149–1151.
7. Xie S, Li Z, Liu J, Xie S, Wang H, et al. (2007) Fisheries of the Yangtze River show immediate impacts of the Three Gorges Dam. Fisheries 32: 343–344.
8. Hu J, Zhang Z, Wei Q, Zhen H, Zhao Y, et al. (2009) Malformations of the endangered Chinese sturgeon Acipenser sinensis, and its causal agent. P Natl Acad Sci USA 106: 9339–9344. doi:10.1073/iti2309106.
9. Turvey ST, Barrett LA, Hao Y, Zhang L, Zhang X, et al. (2010) Rapidly shifting baselines in Yangtze fishing communities and local memory of extinct species. Conserv Biol 24: 778–797. doi:10.1111/j.1365-2761.2004.00590.x.
10. Wang C, Kynard B, Wei Q, Du H, Zhang H (2013) Spatial distribution and habit suitability indices for non-spawning and spawning adult Chinese sturgeons below Gezhouba Dam, Yangtze River: Effects of river alterations. J Appl Ichthyol 29: 31–40. doi:10.1111/jai.12094.
11. Pikitch KE, Doukakis P, Lauck L, Chakrabarty P, Erickson LD (2005) Status, trends and management of sturgeon and paddlefish fisheries. Fish Fish 6: 233–265. doi:10.1111/j.1467-2979.2005.00190.x.
12. He X, Lu S, Liao M, Zhu X, Zhuang M, et al. (2013) Effects of age and size on critical swimming speed of juvenile Chinese sturgeon *Acipenser sinensis* at seasonal temperatures. J Fish Biol 82: 1047–1056. doi:10.1111/j.1095-8649.2012.12015.x.

13. Barthem RB, Ribeiro M, Petrere M (1991) Life strategies of some long-distance migratory catfish in relation to hydroelectric dams in the Amazon Basin. Biol Conserv 56: 339–345.
14. Cheong ST, Kavvas LM, Anderson KE (2006) Evaluation of adult white sturgeon swimming capabilities and applications to fishway design. Environ Biol Fish 77: 197–208. doi:10.1007/s10641-006-9071-y.
15. Rodriguez TT, Agudo PJ, Mosquera PL, Gonzalez PE (2006) Evaluating vertical-slot fishway design in terms of fish swimming capabilities. Ecol Eng 27: 37–48. doi:10.1016/j.ecoleng.2005.09.015.
16. Cooke SJ, Hinch SG, Farrell AP, Patterson DA, Miller-Saunders K, et al. (2008) Developing a mechanistic understanding of fish migrations by linking telemetry with physiology, behavior, genomics and experimental Biology: an interdisciplinary case study on adult Fraser river sockeye salmon. Fisheries 33: 321–339. doi:10.1577/1548-8446-33.7.321.
17. Yagci O (2010) Hydraulic aspects of pool-weir fishways as ecologically friendly water structure. Ecol Eng 36: 36–46. doi:10.1016/j.ecoleng.2009.09.007.
18. Van Leeuwen TE, Rosenfeld JS, Richards JG (2011) Adaptive trade-offs in juvenile salmonids metabolism associated with habitat partitioning between coho salmon and steelhead trout in coastal streams. J Anim Ecol 80: 1012–1023. doi:10.1111/j.1365-2656.2011.01841.x.
19. Fu SJ, Cao ZD, Yan GJ, Fu C, Pang X (2013) Integrating environmental variation, predation pressure, phenotypic plasticity and locomotor performance. Oecologia 173: 343–354. doi:10.1007/s00442-013-2626-7.
20. Jaric I, Gessner J (2012) Analysis of publications on sturgeon research between 1996 and 2010. Scientometrics 90: 715–735. doi:10.1007/s11192-011-0583-7.
21. Farrell AP, Gamperl AK, Birtwell IK (1998) Prolonged swimming, recovery and repeat swimming performance of mature Sockeye Salmon *Oncorhynchus nerka* exposed to moderate hypoxia and pentachlorophenol. J Exp Biol 201: 2183–2193.
22. Jain KE, Farrell AP (2003) Influence of seasonal temperature on the repeat swimming performance of rainbow trout *Oncorhynchus mykiss*. J Exp Biol 206: 3569–3579. doi:10.1242/jeb.00588.
23. Macnutt JM, Hinch SG, Farrell AP, Stopp S (2004) The effect of temperature and acclimation period on repeat swimming performance in cutthroat trout. J Fish Biol 65: 342–353. doi:10.1111/j.0022-1112.2004.00453.x
24. Tiemey KB, Farrell AP (2004) The relationships between fish health, metabolic rate, swimming performance and recovery in return-run sockeye salmon, Oncorhynchus nerka (Walbaum). J Fish Dis 27: 663–671. doi:10.1111/j.1365-2761.2004.00590.x.
25. Pon LB, Hinch SG, Cooke SJ, Patterson DA, Farrell AP (2009) Physiological, energetic and behavioural correlates of successful fishway passage of adult sockeye salmon (*Oncorhynchus nerka*) in the Seton River, British Columbia. J Fish Biol 74: 1323–1336. doi:10.1111/j.1095-8649.2009.02213.x.
26. Herrmann JP, Enders EC (2000) Effect of body size on the standard metabolism of horse mackerel. J Fish Biol 57: 746–760. doi:10.1111/j.1095-8649.2000.tb00272.x.
27. Cai L, Taupier R, Johnson D, Tu Z, Liu G, et al. (2013) Swimming capability and swimming behavior of juvenile *Acipenser schrenckii*. J Exp Zool Part A 319: 149–155. doi:10.1002/jez.1780.
28. Pang X, Yuan XZ, Cao ZD, Fu SJ (2013) The effects of temperature and exercise training on swimming performance in juvenile qingbo (*Spinibabus sinensis*). J Comp Physiol B 183: 99–108. doi:10.1007/s00360-012-0690-7.
29. Akaike H (1987) Factor analysis and AIC. Psychometrik 52: 317–332. doi:10.1007/BF02294359.
30. Hurvich CM, Tsai CL (1989) Regression and time series model selection in small samples. Biometrika 76: 297–307. doi:10.1093/biomet/76.2.297.
31. Burnham KP, Anderson DR, Huyvaert KP (2011) AIC model selection and multimodel inference in behavioral ecology: some background, observation, and comparisons. Behav Ecol Sociobiol 65: 23–35. doi:10.1007/s00265-010-1029-6.
32. Brett JR (1964) The respiratory metabolism and swimming performance of young sockeye salmon. J Fish Res Board Can 21: 1183–1226. doi:10.1139/f64-103.
33. Killen SS, Marras S, Steffensen JF, McKenzie DJ (2012) Aerobic capacity influences the spatial position of individuals within fish schools. Proc R Soc Lond B 279: 357–364. doi:10.1098/rspb.2011.1006.
34. Tu Z, Li L, Yuan X, Huang Y, Johnson D (2012) Aerobic swimming performance of juvenile Largemouth bronze gudgeon (*Coreius guichenoti*) in the Yangtze River. J Exp Zool Part A 317: 294–302. doi:10.1002/jez.1723.
35. Behrens JW, Prabel K, Steffensen JF (2006) Swimming energetics of the Barents Sea capelin (*Mallotus villosus*) during the spawning migration period. J Exp Mar Biol Ecol 331: 208–216. doi:10.1016/j.jembe.2005.10.012.
36. Tu Z, Yuan X, Han J, Shi X, Huang Y, et al. (2011) Aerobic swimming performance of juvenile Schizothorax chongi (Pisces, Cyprinidae) in the Yalong River, southwestern China. Hydrobiologia 675: 119–127. doil10.1007/s10750-011-0809-y.
37. Lee CG, Farrell AP, Lotto A, Hinch SG, Healey MC (2003) Excess post-exercise oxygen consumption in adult sockeye (*Oncorhynchus nerka*) and coho (*O. kisutch*) salmon following critical speed swimming. J Exp Biol 206: 3253–3260. doi:10.1242/jeb.00548.
38. Cai L, Liu G, Taupier R, Fang M, Johnson D, et al. (2013) Effect of temperature on swimming performance of juvenile *Schizothorax prenanti*. *Fish* Physiol Biochem. doi:10.1007/s10695-013-9860-0.
39. Deslauriers D, Kieffer JD (2012) Swimming performance and behabiour of young-of-the-year shortnose sturgeon (*Acipenser brevirostrum*) under fixed and increased velocity swimming tests. Can J Zool 90: 345–351. doi:10.1139/z2012-004
40. Jain KE, Birtwell IK, Farrell AP (1998) Repeat swimming performance of mature sockeye salmon following a brief recovery period: a proposed measure of fish health and water quality. Can J Zool 76: 1488–1496.
41. Peake S (2005) Swimming and respiration. Sturgeons and Paddlefish of North America 27: 147–166. doi:10.1007/1-4020-2833-4_8.
42. Billard R, Lecointre G (2000) Biology and conservation of sturgeon and paddlefish. Rev Fish Biol Fisher 10: 355–392. doi:10.1023/A:1012231526151.
43. Pettersson LB, Hedenstrom A (2000) Energetics, cost reduction and functional consequences of fish morphology. Proc R Soc Lond B 267: 759–764. doi:10.1098/rspb.2000.1068.
44. Ohlberger J, Staaks G, Holker F (2006) Swimming efficiency and the influence of morphology on swimming costs in fishes. J Comp Physiol B 176: 17–25. doi:10.1007/s00360-005-0024-0.
45. Castro-Santos T (2005) Optimal swim speeds for traversing velocity barriers: an analysis of volitional high-speed swimming behavior of migratory fishes. J Exp Biol 208: 421–432. doi:10.1242/jeb.01380.
46. Castro-Santos T, Sanz-Ronda FJ, Ruiz-Legazpi J (2013) Breaking the speed limit - comparative sprinting performance of brook trout (*Salvelinus fontinalis*) and brown trout (*Salmo trutta*). Can J Fish Aquat Sci 70: 280–293. doi:10.1139/cjfas-2012-0186.
47. Webber JD, Chun SN, Maccoll TR, Mirise LT, Kawabata A, et al. (2007) Upstream swimming performance of adult White Sturgeon: effects of partial baffles and a ramp. T Am Fish Soc 136: 402–408. doi:10.1577/T06-064.1.
48. Rome LC, Funke RP, Alexander RM (1990) The influence of temperature on muscle velocity and sustained performance in swimming carp. J Exp Biol 154: 163–178.
49. Marras S, Killen SS, Domenici P, Claireaux G, Mckenzie DJ (2013) Relationships among traits of aerobic and anaerobic swimming performance in individual European Sea bass *Dicentrarchus labrax*. PLOS ONE 8: e72815. doi:10.1371/journal.pone.0072815.
50. Jones DR (1982) Anaerobic exercise in teleost fish. Can J Zool 60: 1131–1134. doi:10.1139/z82-157.
51. Svendsen JC, Tudorache C, Jordan AD, Steffensen JF, Aarestrup K, et al. (2010) Partition of aerobic and anaerobic swimming costs related to gait transitions in labriform swimmer. J Exp Biol 213: 2177–2183. doi:10.1242/jeb.041368.

5

Coral Reef Disturbance and Recovery Dynamics Differ across Gradients of Localized Stressors in the Mariana Islands

Peter Houk[1,2]*, David Benavente[3], John Iguel[3], Steven Johnson[3], Ryan Okano[3]

1 University of Guam Marine Laboratory, UOG Station, Mangilao, Guam, **2** Pacific Marine Resources Institute, Saipan, Northern Mariana Islands, **3** CNMI Bureau of Environmental and Coastal Quality, Saipan, Northern Mariana Islands

Abstract

The individual contribution of natural disturbances, localized stressors, and environmental regimes upon longer-term reef dynamics remains poorly resolved for many locales despite its significance for management. This study examined coral reefs in the Commonwealth of the Northern Mariana Islands across a 12-year period that included elevated Crown-of-Thorns Starfish densities (COTS) and tropical storms that were drivers of spatially-inconsistent disturbance and recovery patterns. At the island scale, disturbance impacts were highest on Saipan with reduced fish sizes, grazing urchins, and water quality, despite having a more favorable geological foundation for coral growth compared with Rota. However, individual drivers of reef dynamics were better quantified through site-level investigations that built upon island generalizations. While COTS densities were the strongest predictors of coral decline as expected, interactive terms that included wave exposure and size of the overall fish assemblages improved models (R^2 and AIC values). Both wave exposure and fish size diminished disturbance impacts and had negative associations with COTS. However, contrasting findings emerged when examining net ecological change across the 12-year period. Wave exposure had a ubiquitous, positive influence upon the net change in favorable benthic substrates (i.e. corals and other heavily calcifying substrates, $R^2 = 0.17$ for all reeftypes grouped), yet including interactive terms for herbivore size and grazing urchin densities, as well as stratifying by major reeftypes, substantially improved models ($R^2 = 0.21$ to 0.89, lower AIC scores). Net changes in coral assemblages (i.e., coral ordination scores) were more sensitive to herbivore size or the water quality proxy acting independently ($R^2 = 0.28$ to 0.44). We conclude that COTS densities were the strongest drivers of coral decline, however, net ecological change was most influenced by localized stressors, especially herbivore sizes and grazing urchin densities. Interestingly, fish size, rather than biomass, was consistently a better predictor, supporting allometric, size-and-function relationships of fish assemblages. Management implications are discussed.

Editor: Benjamin Ruttenberg, California Polytechnic State University, United States of America

Funding: This study was funded by the National Oceanic and Atmospheric Administration (NOAA) coral-reef monitoring grants administered to the CNMI Coastal Resources Management Office, and USEPA Region IX funding administered to the CNMI Division of Environmental Quality. Financial support for PH was provided through a NOAA domestic coral reef grant award (NA11-NOS4820015). The funders had no role in study design, data collection and analysis, decision to publish, or preparation of the manuscript.

Competing Interests: The authors have declared that no competing interests exist.

* Email: houkp@uguam.uog.edu

Introduction

A vast array of acute disturbances and chronic stressors threaten coral reefs [1–3]. Yet, our knowledge of the role that individual disturbances and stressors play in determining reef dynamics through time remains limited for most locales [4]. One cause of this uncertainty stems from the rarity of ecological datasets that span across sufficient timeframes to capture disturbance and recovery periods, which can help partition the variance associated with population dynamics, and attribute cause, proportionally, to individual stressors. For coral reefs, two putative, localized stressors of primary concern are unsustainable fishing and pollution that may act independently or in combination with disturbance cycles to diminish the growth of reef accreting organisms such as corals [5,6]. Many studies have compared reefs where high and low human influences existed to generalize how water quality and herbivory have impacted coral growth and resulted in macroalgal replacement [7,8]. However, less attention has been given to understanding their individual contributions towards growth dynamics outside of laboratory or manipulative settings [9–11], and thus, our ability to upscale evidence from manipulative studies to entire reefscapes remains limited. In turn, context-dependent roles of localized stressors continue to be the focus of much research, and varying findings provide support for many ideologies [12–14]. When used out of context, or when limited information exists to develop an appropriate context, uncertainty and improperly informed decision making can result.

Acute disturbances such as typhoons, *Acanthaster planci* outbreaks, and climate-induced bleaching are well appreciated for their role in driving coral population dynamics. Disturbance and recovery cycles have traditionally been investigated using coral cover trends, integrated across both local and regional scales, and through time [15,16]. Yet, coral cover can be an inconsistent metric of recovery due to varying natural environmental regimes, such as wave exposure, that dicate coral growth capacity [17–19], and thus the time needed for recovery. Improved assessments of reef condition and calcification potential have emerged from benthic-substrate datasets by simultaneously considering the abundances of macroalgae and other less-calcifying organisms in comparison to corals and other heavily-calcifying organisms [9,20,21]. Such integrated metrics can better account for the inherent environmental variation that drives coral cover trajectories, and represents one useful metric of overall recovery potential when comparing across reefs that is furthered within the present study.

In addition to coverage estimates, shifting species abundances [22] and colony-size distributions [23,24] have also proven to be sensitive indicators of disturbance-and-recovery cycles for coral populations. Studies show that faster growing corals (*Acropora*, *Pocillopora*, *Stylophora*) have lower tolerances to disturbance events, while others (*Porites* and numerous faviids) have a prolonged ability to deal with both acute disturbance and chronic stress [5,25–27]. However, many of the same faster-growing corals may be more resilient (or adaptable) to repeated climate-induced disturbance in the absence of chronic stress [28,29]. Thus, size distributions and abundance patterns of corals that have low thresholds, fast recovery, and the potential for adaptability provides an additional means of evaluating coral population dynamics [20,30].

The premise for the present study is that disturbance and recovery are not necessarily co-dependent processes on coral reefs, and by examining them independently with the aid of refined metrics described above, an improved understanding of causation and predicted resilience can emerge [3,31]. While many studies have described the nature of acute disturbances to reefs from a variety of agents, relatively few have followed recovery dynamics with respect to individual factors. In the Caribbean, Mumby and Harborne [10] reported significantly higher recovery of coral coverage and colony sizes in response to a no-take fishery closure over a 2.5 year period. Similarly, the recovery of coral colony sizes [22], but not recruitment [32], were found to be most heavily dependent upon fishing pressure in Kenya. In American Samoa, Houk et al. [9] found differential recovery of calcifying benthic substrates and coral species evenness to be interactively driven by herbivore biomass and water quality, with their hierarchical influence shifting based upon geological reef settings (i.e., a potential proxy for connectivity with groundwater discharge). In contrast, a meta-analysis of coral reef recovery across all major oceanic basins provided some evidence for unintuitive, reduced recovery rates within fisheries closures following disturbance events [3]. These findings were a perceived artifact of higher pre-disturbance coral coverage within no-take closures, and not attributed to management status. Interestingly, the causes behind recovery trajectories were not consistent across major geographic regions. Clearly recovery is context-dependent with respect to physical settings as well as management regimes [33–37]. Considering that predicting reef futures is becoming more of a priority for both local and global management efforts, further study is needed to interpret when and if generalizations may exist, and at what spatial scales they might operate at.

We examined 12-year trends in coral-reef assemblages across fore-reefs in the Commonwealth of the Northern Mariana Islands (CNMI) during a time period when significant disturbance impacts and differential recovery occurred. We first described the nature of *Acanthaster planci* (i.e., Crown-of-Thorn Starfish, COTS) population dynamics during the study period. COTS population trends were augmented with typhoon records to define pre-, during, and post-disturbance time periods across our 12-year study. Coral cover and colony sizes were then evaluated for two study islands that differed in human population, development, and geological setting. Island-scale coral decline and recovery trajectories provided an initial framework for interpreting the nature of coral growth cycles across the study period, and along with previous studies, highlighted primary factors influencing site-specific reef dynamics. Ensuing site-level regressions were performed between coral decline and COTS densities (i.e., the hypothesized, primary disturbance agent), while including key factors that quantified localized stressors and environmental regimes as interactive covariates. Similar site-level regressions were also performed to examine net ecological change across the study period. Combined, our study provided a descriptive framework for disentangling the differences between 'disturbance impacts' and 'net ecological change', and attributed cause, proportionally, to a suite of factors that help to inform management.

Methods

Study location

The Commonwealth of the Northern Mariana Islands (CNMI) represents a series of active volcanic and inactive raised limestone islands located in the Western Pacific Ocean (Figure 1). The present study focused upon the southernmost, inactive limestone islands where the majority of the human population resides. From north to south, the study islands were Saipan (48,220 people, capital island), Tinian (3,136 people), Aguijan (no inhabitants but the nearshore resources are accessed from Saipan and Tinian), and Rota (2,527 people) (Census 2010 statistics, http://commerce.gov.mp/divisions/central-statistics/) (Figure 1).

CNMI's local monitoring program has been collecting standardized benthic, coral, macroinvertebrate, and fish assemblage data since 2000 [38,39]. Monitoring designs were stratified based upon geological reeftypes, management status, and watershed development. Geological reeftypes have previously been described with respect to wave exposure and submarine groundwater discharge through karst watersheds that were attributed to specific coral assemblages and reef growth through time [18]. Cumulatively, Houk and van Woesik [18] described four distinct geological reeftypes in the CNMI: 1) optimal spur-and-groove structures, 2) constructional, high-relief interstitial framework, 3) low-relief framework with limited Holocene deposition found only on Rota, and 4) incipient coral assemblages with little to no deposition. Coral coverage, diversity, and evenness peak on the first two, with spur-and-groove structures being the most optimal settings for modern coral assemblage growth.

The most significant acute disturbance since the inception of monitoring efforts has been high *Acanthaster planci* populations (COTS herein) evident between mid-2003 and 2006. In complement to COTS disturbances, several tropical storms passed nearby the study islands during a similar timeframe, with the strongest being Pongsona (passing 40 km south of Rota, 95 knots maximum sustained windspeed) and Chaba (passing over Rota, 118 knot sustained windspeed) in December 2002 and August 2004, respectively. While large wave events were recorded, observations by the authors before and after these storms suggested negligible direct impacts compared with the onset of COTS, examined further within. COTS events became widespread throughout the North Pacific during this timeframe [40]. However, no information exists to understand any potential linkages between the storm events and the subsequent emergence of COTS. Initial examinations were undertaken by coupling COTS abundance and tropical storm histories to define three time periods within the present study *a priori*: pre-, during, and post-disturbance. Subsequently,

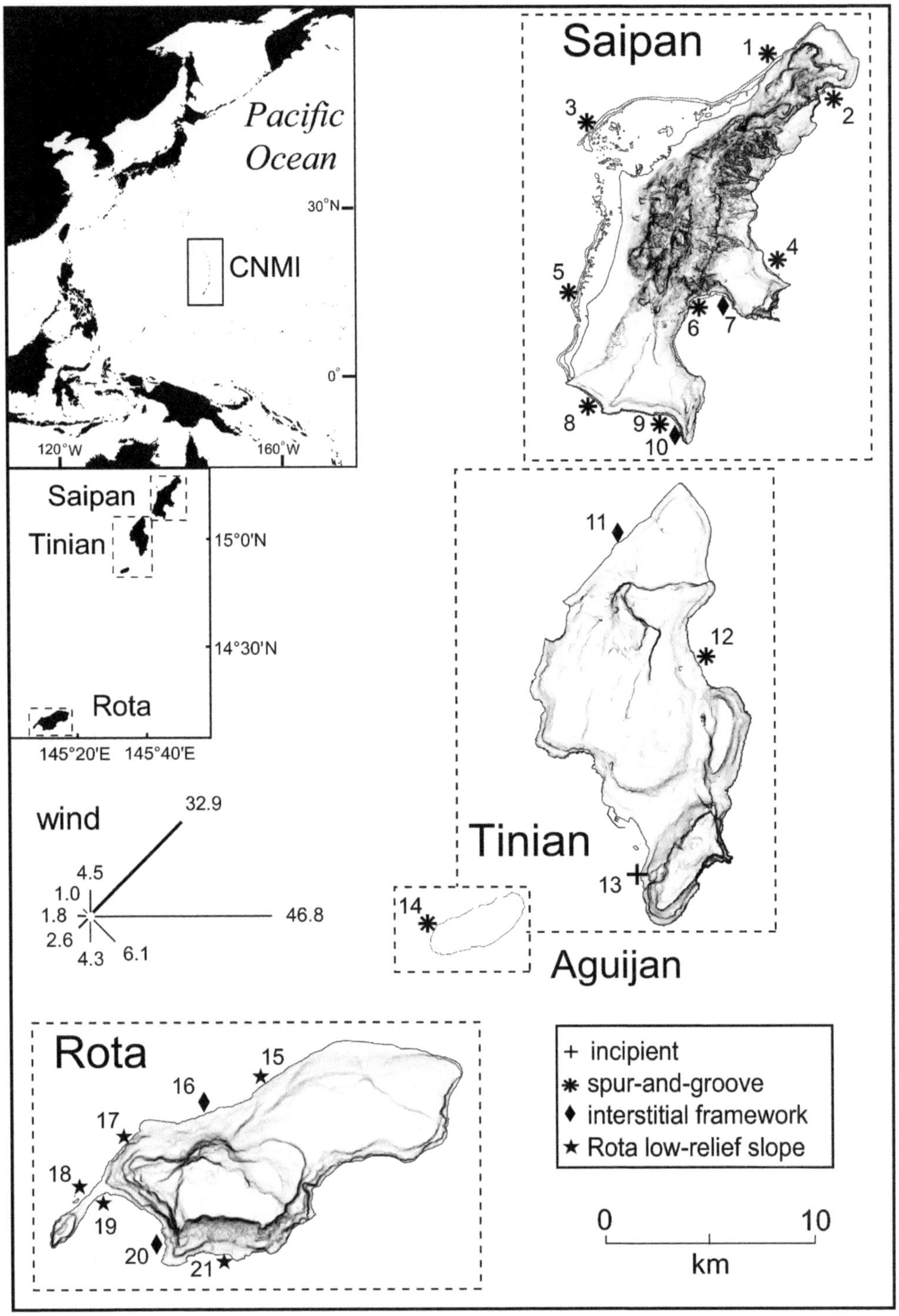

Figure 1. A map of the Western Pacific Ocean and the Commonwealth of the Northern Mariana Islands study islands. Distance between islands is not drawn to scale as dashed boxes indicate individual island entities. Wind vectors show the percent of time that winds originated from each of 8 quadrants (length and corresponding number) as well as the mean annual intensity (thickness). Reeftypes are indicated by symbols referred to in the legend. Topographic lines infer the steepness and size of watersheds.

we quantified how much of the site-level coral decline during the 'disturbance' period could be accounted for by COTS.

Ecological data collection

Data were collected as part of the CNMI Division of Environmental Quality and the CNMI Coastal Resources Management Office coral-reef monitoring program. These programs have the legislative authority to conduct monitoring activities, and given the non-invasive nature of this research, no further permits were required.

Long-term monitoring data have been collected at 1 to 3 year intervals for 25 monitoring stations across the CNMI in most cases, however, longer intervals between repeated site-visits existed in a few instances (Appendix 1). The prerequisite for inclusion in the present study was that coral, benthic, and/or macroinvertebrate data were available for all three timeframes: before, during, and after the disturbance period. Among the 25 monitoring sites, data from 21 met these criteria (Table S1, Figure 1).

Sites were identified by global positioning system coordinates coupled with directional bearings to indicate transect placement. During each survey event, five, 50 m transects were placed at the 8 m depth contour to guide fieldwork. In three instances, homogeneous substrates were not consistently available, and three, 50 m transects were used instead (sites 2, 16, and 18, Figure 1). Benthic substrate abundances were estimated from photographs of 0.5×0.5 m quadrats. In 2000 and 2001, 25 photographs were taken along each transect line, and the substrate under each of 16 data points was identified and recorded. Since 2002, 50 photographs were taken from each transect line, and the substrate under each of 5 data points was recorded. Methods were shifted in order to improve detection limits for temporal change [38]. With respect to the present study, the changes in benthic substrate abundances associated with COTS disturbance were a magnitude of order higher than the expected detection limits based upon 2002 datasets [38]. In all instances, benthic categories chosen for analysis were corals (to genus level), turf algae (non-identifiable turfs typically less than 2 cm), macroalgae (readily identifiable alga typically greater than 2 cm, to genus level if abundant), calcareous encrusting algae known to actively shed epithelial layers or inhibit the survival of juvenile corals (*Peyssonnelia, Pneophyllum*) [41–44], crustose coralline algae (CCA) known to promote reef accretion and juvenile coral settlement, sand, and other invertebrates (genus level if abundant). Using these substrate categories, we defined a benthic substrate ratio by the percent cover of heavily calcifying (corals and CCA) versus non-or-less calcifying (turf, shedding-calcareous, macroalgae) substrates.

Coral assemblage data have been collected since 2003 by the same observer using a standard quadrat-based technique. During each survey, 16 replicate 0.5×0.5 m quadrats were haphazardly tossed at equal intervals along the transect lines. All corals whose centerpoint resided within the quadrat boundary were identified, and the maximum diameter, and the diameter perpendicular to the maximum, were recorded. Surface area was calculated from these measurements assuming colonies were elliptical in nature. For three-dimensional corals, measurements were extended along the colony surface area. Coral taxonomy followed Veron [45]. Species-level data were used for analyses of richness and evenness, while species were grouped by 'sub-genus' (i.e., genus and growth form, digitate *Acropora*, massive *Porites*) to examine multivariate trends and subsequent species weightings described below.

Macroinvertebrate densities were estimated along individual transect lines (noted above) using 50×4m observation belts. All sea cucumbers, sea urchins, shellfish, and other conspicuous macroinvertebrates were identified to species level and recorded. Count data derived from macroinvertebrate belt transects were used to examine COTS and grazing urchin populations through time.

Fish census protocols were recently added into CNMI long-term monitoring program. The present study used data collected between 2011 and 2012 by a single observer. Fish assemblages were estimated from 12 stationary-point-counts (SPC's) conducted at equal intervals along the transect lines, following a modified version of [46]. During each SPC, the observer recorded the name (species) and size of all food-fish within a 5 m circular radius for a period of 3-minutes. Food fish were defined by acanthurids, scarids, serranids, carangids, labrids, lethrinids, lutjanids, balistids, kyphosids, mullids, and holocentrids that are a known to be harvested. Sharks were also included. Size data were binned into 5 cm categories (i.e., 12.5–17.5 cm = 15 cm, inferring the size estimates of 13, 14, 15, 16, an 17 were considered as 15 cm), and converted to biomass using coefficients reported in fishbase (www.fishbase.org). Species data were binned into functional categories based upon maximum adult sizes and family level taxonomy for some graphical interpretations and analyses, such as large/small bodied parrotfishes and surgeonfishes, and large/small bodied snappers and groupers. We defined large-and-small-bodied species based upon estimated mean reproductive sizes greater or less than 30 cm, respectively, based upon fishbase records, or the life-history wizard. Planktivores were excluded from the present analyses due to their low abundance in comparison to other trophic groups, and low sample sizes for defining confidence intervals. Last, fish recruits less than 7.5 cm (i.e., 8–12 cm size class bin) were omitted from all analyses to avoid potential bias from differential recruitment with unknown post-settlement mortality dynamics. Recruits comprised less than 3% of the surveyed population.

Environmental data collection

A proxy for water quality was developed from geographic information system (GIS) layers pertaining to topography, landuse, and human population. Digital elevation models (i.e., topographic data) were used to define watershed boundaries. Landuse data were then overlaid upon the watersheds, and a measure of disturbed land was calculated by combining the coverage of barren land, urbanized vegetation, and developed infrastructure within each watershed (United States Forest Service, http://www.fs.usda.gov/r5). Equal weighting was given to each category because variation in pollution contribution was expected within each. For example, urbanized lands were associated with both septic systems and sewer collection systems that differ with respect to their pollution contribution. Human population density in the watershed adjacent to each site was derived from the 2010 CNMI census. Landuse and human population data were also standardized to weight them equally, and averaged to establish an overall proxy to watershed pollution that represents the simplest assumptions of equal contributions from two known sources of watershed-based pollution. In order to match the low-to-high scale of other localized stressors (i.e., herbivore size), the inverse of the proxy was used in regression modeling.

Wave energy was derived from long-term wind datasets and estimates of fetch [47]. For each site, fetch (i.e., distance of unobstructed open water) was first calculated for 16 radiating lines equally distributed between 20 to 360 degrees. Fully develop sea conditions were considered if unobstructed exposure existed for 20 km or greater. Ten-year windspeed averages were calculated from Saipan airport data (http://www7.ncdc.noaa.gov/), and used as inputs to calculate wave height following Ekebom et al. [47]. Mean height was calculated by:

$$Hm = 0.018 \cdot U^{1.1} \cdot F^{0.45} \tag{1}$$

Where Hm is the wave height (m) for each quadrant, U is the windspeed at an elevation of 10 m, and F is the fetch (km). Windspeed corrections for varying elevations were made following Ekebom et al. [47]. Last, wave height was converted to energy following:

$$E = (1/8) \cdot \rho \cdot g \cdot H^2 \quad (2)$$

Where ρ is the water density (kg/m^3), g is the acceleration due to gravity (9.81 m/s^2), and H is the wave height (m).

Data Analyses

Island comparisons

We first generalized disturbance and recovery trajectories for two islands that differed with respect to human population, development, and geological setting, Rota and Saipan. These islands were selected because they both had sufficient sampling effort and spatial coverage over the study period to provide a generalization of temporal reef dynamics (Figure 1, Table S1). Initial, island-scale examinations were conducted for coral cover and colony sizes. Due to varying geological settings, coral growth capacity is inherently lower on Rota compared with Saipan [18], and thus we focused investigations upon the nature and relative rate of decline and recovery, and not absolute values of change. Coral cover data were site-averaged within each of the study time periods, and repeat measures ANOVA tests with post-hoc Tukey's pairwise tests were conducted between the study periods for each island. Repeat measures ANOVA tests were similarly used to examine coral population density across the disturbance time-frames. Last, differences in coral colony-size distributions were examined using Kolmogorov-Smirnov (K-S) tests that provided a P-value that is generated from the distances between cumulative frequency plots [48].

We next examined sea cucumber and grazing urchin densities across the study periods. Macroinvertebrate count data had zero-inflated, negative binomial distributions that best conformed to zero-inflated, repeat-measures ANOVA models [49]. Zero-inflated models have two distinct parts, one that describes the probability of obtaining a zero count, and one that describes the expected density given a non-zero count. Tests of significance took both simultaneously into account (i.e., hurdle models).

Fish biomass and size were compared between Rota and Saipan within each trophic category noted above: herbivores/detritivores, secondary consumers, and tertiary consumers. Standard comparative tests were used to examine differences in fish size and biomass between the two islands (t-tests, Mann-Whitnet U-test if normality assumptions were not met).

Site-level analyses

Deeper investigations of disturbance and recovery trajectories were conducted at the site-level, in order to match the spatial scale that localized stressors and key environmental regimes such as wave exposure operate at. A stepwise regression modeling process was performed to determine the likelihoods and magnitudes of influence for COTS, pertinent localized stressors, and wave exposure. These factors were considered based upon both initial island-scale analyses and previous studies describing their driving influence upon CNMI's reefs [18], with more details provided below to describe localized stressors. Stepwise modeling to understand the impacts during the disturbance time frame utilized coral cover as the primary, sensitive dependent variable. Modeling to understand net ecological change across the 12-year period used two dependent variables that better described coral reef 'condition' and relied less upon coral cover alone, which is known to differ across species and environments [18]. These were: 1) the net change in the benthic substrate ratio, and 2) net change in multivariate coral ordination scores (Figure 2). Benthic and coral ordination metrics were weakly correlated ($r<0.4$, all comparisons). Coral ordination metrics were calculated based upon two-dimensional principle component ordination plots. Coral species abundance data were log-transformed to produce a Bray-Curtis similarity matrix that quantified multivariate similarities between all sites and years. Similarity matrices were projected in two-dimensional space using principle component ordination (PCO) plots that depicted multivariate differences using axes (i.e., eigenvectors) that hierarchically account for the variation in multi-species datasets (PCO) [50]. Corals that were the strongest drivers of multivariate dissimilarities were overlaid on the plot to indicate the nature of shifting assemblages over the years. The percent net change for coral assemblages represented the two-dimensional PCO-movement (i.e., vector magnitude) away from *Montipora*, *Acropora*, *Stylophora*, and *Pocillopora* assemblages to faviid and *Porites* assemblages, and the magnitude of return proportional to its starting position (Figure 2). These corals cumulatively accounted for the top 30% of the variation in the PCO plot. Percent change values were calculated because they are comparable across sites, whereas absolute values would not be. Second, we used mean values of the benthic substrate ratio defined above. Percent net change was again defined by differences between the post- and pre-disturbance period, divided by the initial pre-disturbance period value (Figure 2). Cumulatively, larger reductions in percent change values indicated a shift to reefs with non-calcifying substrates and smaller, tolerant coral assemblages.

A forward, stepwise regression modeling process was conducted to evaluate the likelihood and magnitude of independent factors in determining the observed biological changes. This process began by searching for factors with a ubiquitous influence upon biological change across all reeftypes. These included COTS in the disturbance timeframe models and wave exposure in the benthic net-change models. No ubiquitous factor was revealed for models describing the net change in coral ordinations, so only individual terms were used. Forward steps were only taken when/if interactive models improved the fit (R^2-values) and likelihood (AIC-values, described below) of predictions. Interactive combinations were considered because independent variables were all scaled in a consistent, low-to-high manner describing a gradient of weak to strong influences, and are known to have contextual effects that are not independent of one another [9,33]. If stepwise terms did not improve the model fit and likelihood they were dropped, not presented in the results, and not further considered. This process continued until all interaction terms were examined. We first examined regression models including sites from all reeftypes grouped together, and subsequently examined subsets of the most favorable reef settings for coral growth to determine if herbivory, water quality, or wave exposure might have context-dependent roles that differed in accordance with reeftype. Comparisons of the explanatory power and likelihood of independent variables were only made across models examining the same reeftypes. Our goal was not to suggest that a single "best-fit" model existed, but rather to highlight that several plausible models existed, and provide the details of each. The influence of each factor was assessed based upon: 1) the (added) model fit, 2) overall presence across the suite of models, and 3) the likelihood scores and influence of outliers.

Regression modeling was performed using R [51]. Independent variables were standardized to provide equal scaling, and a constant value was added to make all numbers positive. Positive numbers were required when log-transformations were needed to ensure residual normality. Models were examined for normality using Shapiro-Wilk tests, and ranked based upon their explanatory

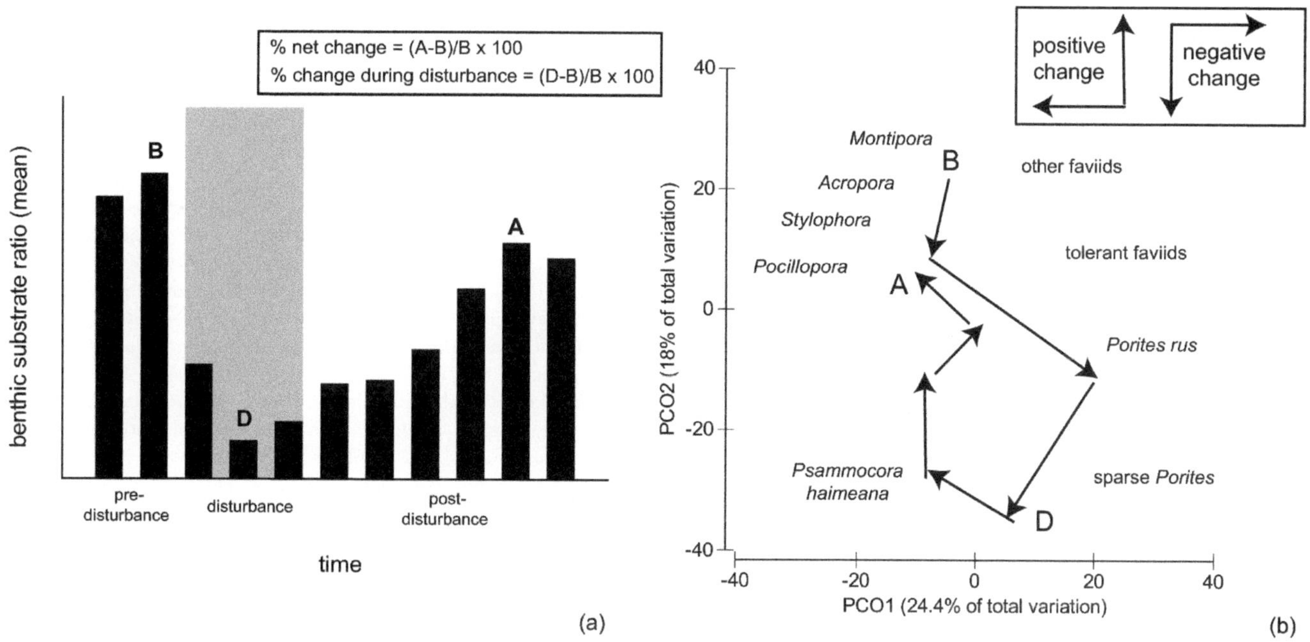

Figure 2. Methods used to calculate the percent decline and net change in (a) the benthic substrate ratio and (b) multivariate measures of the coral assemblages. B, D, A – before, during, and after the disturbance period, respectively. Percent declines were calculated by taking the difference between the minimum and maximum values during and before the disturbance period, and dividing by the pre-disturbance values. Net change values were calculated by taking the difference between maximum values before and after the disturbance period. Values for coral assemblages were calculated based upon their vector magnitude from the origin (0,0), with positive values given for PCO movement towards *Acropora*, *Montipora*, and *Pocillopora* assemblages, and negative values for PCO movement towards tolerant faviids and sparse *Porites*. See methods and Figure 6 for a better description of the corals depicted on the PCO plot.

power (R2-value) and likelihood as measured by Akaike's Information Criterion (AIC). Lower AIC scores indicated a better fit, based upon the least number of parameters and greatest residual normality that together maximize the probability of given outcomes based upon independent predictors. In all instances, the independent predictors used in interactive regression models were not highly correlated ($r<0.30$, $P>0.05$).

Independent variables were selected to represent factors that have a strong influence on CNMI's coral-reef assemblages and dictate disturbance and recovery dynamics in general. These were COTS densities (i.e., as a disturbance agent), wave exposure, the water quality proxy, mean fish size and biomass, and grazing urchin density (defined by the sum of *Echinothrix*, *Echinometra*, and *Diadema* urchins). When examining models dealing with net ecological change, we used mean herbivore/detritivore fish size and biomass. When examining models dealing with coral cover decline during the disturbance period, we used overall fish size and biomass. This was done in accordance with: 1) the literature describing herbivores as key ecosystem engineers facilitating net recovery dynamics [6,22], and 2) existing/emerging relationships between enhanced overall fish size/biomass and reduced COTS impact [12,30,52].

Last, the present study placed a reliance upon 2010/2011 fish data to predict disturbance and recovery trajectories over the past decade. We do not assume that fish populations were static across the disturbance period, as long-term studies across disturbance events elsewhere clearly define fish assemblage dynamics [53,54,55]. However, this approach does assume that site-level relationships were preserved, and relative differences with respect to environmental regimes and/or human exploitation maintained. This was supported by several lines of evidence. Fishery-dependent studies conducted in the CNMI over the past 20 years that showed consistent trends in fish sizes across numerous geographic sectors in the CNMI based upon wave exposure and proximity to human population centers [56,57,58], despite evidence for an overall shift (i.e., decline) in several measures of the fishery resource. These studies also highlighted a shift towards increased herbivore dominance and smaller herbivores over the years, suggesting the relevance of herbivore assemblage metrics. In addition, a recent coral reef resilience assessment conducted for 35 sites around Saipan found that wave exposure and MPA-status together were highly correlated with a fishing pressure metric, and fish abundance trends in the present dataset, which were derived from an independent opinion survey taken by CNMI's resource managers [59]. Beyond CNMI, long-term studies show responses of fish assemblages to disturbances whereby a decline in overall fish diversity, and a decline in the abundance of coral-associated species are most often noted with coral loss [53,54,60]. In response to algal growth following disturbances, these studies also show increases in some acanthurids and/or scarids that can respond rapidly to algal substrate availability, with their abundances gradually decreasing to pre-disturbance levels after several, typically 4 to 5, years. Thus, the present regression models were based upon: 1) environmental gradients and management factors known to persist over longer time periods and more extensive spatial gradients than the present study, and 2) fish abundance datasets that were collected 4 to 5 years following the disturbance period when any pulses of herbivores were expected to diminish, and 3) utilized metrics of food-fish size and biomass that were not influenced by smaller, coral-associated species.

Results

High COTS densities were evident across the CNMI between 2003 and 2006 (Figure 3a, Table S2), concomitant with two tropical storms that passed by the study islands. Together, these findings formed the basis for determining 'before', 'during', and 'after' study periods. Transect-based densities of COTS were largest during the disturbance years on Saipan, however, increased densities were noted across all islands. Following disturbance years, COTS densities declined to pre-disturbance levels for Saipan and Rota, while one unique site on Tinian had an anomalous high density become emergent during 2011 (i.e., site 11 was driving the trend in Figure 3a for Tinian in 2011, Table S2). Cumulatively, COTS densities agreed with 2003 to 2006 as peak disturbance years, but also suggest that Rota and Saipan had similar temporal dynamics, with higher absolute abundances being found on Saipan.

Island comparisons

The disturbance period had a negative impact on coral cover throughout the CNMI, however rates of change were not uniform across the study islands. The largest decline was evident on Saipan, where high human populations and development existed, and smallest on Rota, where human presence and geological foundations for optimal reef growth were lower (Figure 3b). Coral cover declined from 34% to 23% on Saipan (32% decline, F-Statistic = 3.7, P = 0.05, repeat measures ANOVA, pairwise Tukey's post-hoc q-statistic = 3.6, P = 0.05), with no significant recovery since the disturbance years. Conversely, on Rota where coral cover is naturally lower than Saipan as a consequence of island geology, cover had a non-significant decrease from 11.5% to 9.7%, with recovery back to 12%. Yet, Rota coral cover trends were heavily influenced by a single locality where the coral *Porites rus* is abundant and dominant (Site 20 coral cover 30% whereas all other sites have 10%, Figure 1, Table S2). Re-analysis of the coral cover trends with this site omitted indicated that coral cover had a more substantial decline (8.5% to 4.6%) and recovery back to 7.7% (F-Statistic = 4.0, P = 0.06, repeat measures ANOVA). The impact of disturbances to coral-colony sizes and population densities were also markedly different (Figure 4). Mean coral colony-size on Saipan declined from 7.5 to 5.1cm (P<0.001, K-S test), with no change in the years after disturbance (Figure 3). Demographic changes were complemented by significant increases in coral population density during the disturbance years (F-Statistic = 5.96, P = 0.01, repeat measures ANOVA, pairwise Tukey's post-hoc q-statistic = 4.8, P<0.05), mainly due to small faviids and *Porites*. In contrast, mean colony size on Rota had a non-significant decline and recovery (4.3 to 3.7cm, followed by a recovery to 4.1cm). Population density comparisons were also non-significant.

Grazing urchin densities declined across the disturbance period on both islands, yet the magnitude of decline was much larger for Rota, where *Echinothrix* densities were reduced from over 6 individuals per 100 m^2 to less than 2 during the onset of disturbances (Figure 3c, z-statistics = 4.6 and1.8, respectively, for the increased probability of obtaining zero counts and reduction in densities where non-zero counts existed, P<0.01 for both, repeat-measure, zero-inflated hurdle comparisons before and during). Urchin densities were initially lower on Saipan and had a gradual decrease across the study period that was most pronounced in the years following disturbance (z-statistics = 2.8 and1.4, P<0.05, for greater zero counts and reduced densities, respectively, zero-inflated comparisons during and after), representing a timeframe when more abundant COTS existed compared to Rota (Figure 3a). Models indicated that declines were attributed to both higher probabilities of sites with no urchins being present (30% and 38% decline in significant modeled estimates, respectively for Rota and Saipan), and sites with urchins present at lower densities (48% and 58% decline, respectively). No significant differences in sea cucumber densities existed across the study periods.

Fish biomass and density were consistently greater for large-bodied primary and secondary consumers on Rota compared with Saipan (i.e., species that attain larger reproductive sizes, Figure 5). In contrast, smaller-bodied counterparts were higher in biomass and density on Saipan. These trends were most pronounced for large-bodied groupers, snappers, and parrotfish on Rota, and small-bodied acanthurids and parrotfish on Saipan (P<0.05, comparative t-tests and Mann-Whitney U-tests). The findings support that fewer, large-bodied fish with varying functional roles existed on Saipan, with more numerous, small-bodied species comprising a majority of the biomass.

Stepwise regression modeling at the site level

Focusing investigations at the site level revealed that disturbance and recovery cycles varied markedly, with differences attributable, in part, to varying reeftypes and islands that dictated site geomorphology (Figures 6 and 7, Table S3). Percent decline in coral cover during the disturbance period was primarily driven by the observed, transect-based COTS densities regardless of reeftype or island (Table 1). When considering all reeftypes grouped together, only 18% of the variance in coral decline was attributable to COTS (P = 0.06). Yet, the amount of variance explained increased to over 50% when examining the subset of reeftypes with the highest coral growth capacity (52% for interstitial and spur-and-groove reefs combined, and 67% for spur-and-groove reefs alone, P<0.01 for both). The addition of an interaction term describing the mean size of the fish assemblages enhanced the amount of variance accounted for (R^2 = 0.60 and 0.86, P = 0.009 and 0.005, for interstitial and spur-and-groove reefs combined, and spur-and-groove reefs alone, respectively), while wave exposure improve the fit for spur-and-groove reefs alone (R^2 = 0.83, P = 0.003, Table 1). Fish size had a negative association with COTS densities for coral-dominated reefs (r = 0.28, Pearson correlation coefficient, interstitial and spur-and-groove reefs combined), while an interactive term of *fish size×wave exposure* had a stronger association with COTS (r = 0.48), highlighting the inter-dependence among these three factors. In sum, COTS were primary predictors of coral decline (i.e., the disturbance period), while COTS densities and coral impacts were diminished with higher wave exposure and larger fish assemblages, especially on reefs with a high capacity for coral growth.

The general response of coral assemblages to disturbance was a decline in *Acropora*, *Montipora*, *Stylophora*, and *Pocillopora* corals, with faviid and *Porites* assemblages becoming emergent, and differential rates of return in the post-disturbance years. Two-dimensional PCO-plots accounted for 40 to 50% of the variance in coral assemblage similarities, and provided a quantitative basis to assess disturbance and net change across individual sites (Figures 2 and 6). Similarly, photo-quadrat data revealed that benthic substrate ratios were sensitive to disturbance (i.e., the ratio of heavily calcifying corals and crustose coralline algae divided by less-calcifying turf, fleshy-coralline, and macroalgae, Figure 7), with differential levels of decline and net change.

In contrast to regression models describing spatial patterns in coral cover declines, COTS densities did not emerge as a significant predictor of net ecological change. This was the case when considering COTS densities during the disturbance time-

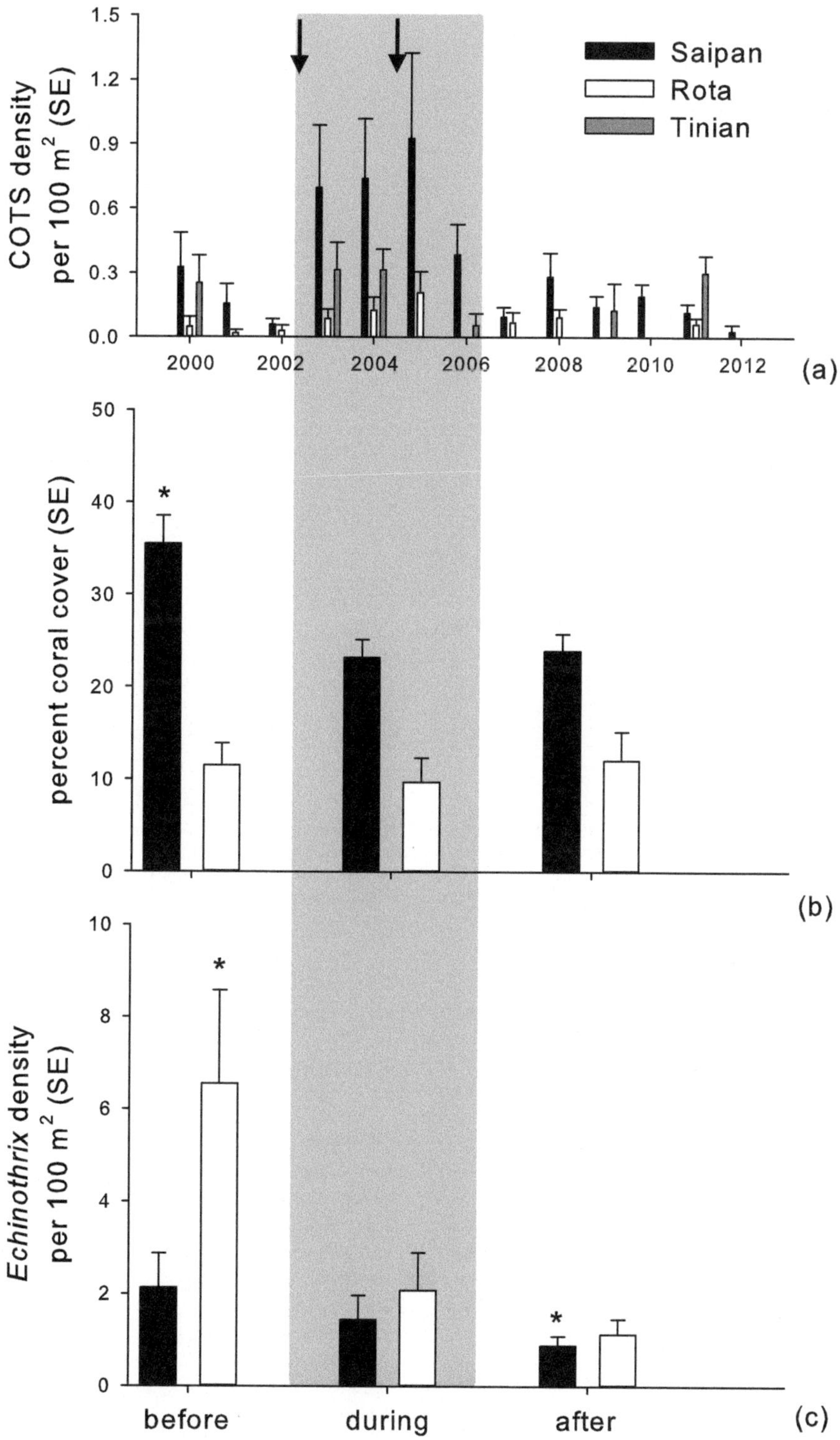

Figure 3. Density of Crown-of-Thorn starfish (COTS) based upon belt-transect data across Saipan, Tinian, and Rota from 2000 to 2012. (a). Densities represent island-based averages that diminish the highest and lowest observations in order to establish patterns across study years. Disturbance reduced coral cover on all islands, but recover trajectories, or the net rate of change, differed by island (b). Coral cover declined on Saipan with no significant recovery (*indicates $P<0.05$, repeat measures ANOVA and post-hoc tests), while a non-significant decline and recovery was noted on Rota. *Echinothrix* urchins also declined (c) in density during the disturbance period, with a further decline in the years after disturbance (*indicates $P<0.05$, zero-inflated hurdle models). Black arrows indicate tropical storms that passed by the study islands during the disturbance timeframe (grey rectangle box indicates the disturbance timeframe).

frame only, or integrated across the entire study period. Stepwise regression modeling for the net change in favorable benthic substrates reported that wave exposure was the strongest individual factor across all reeftypes, accounting for 20% of the

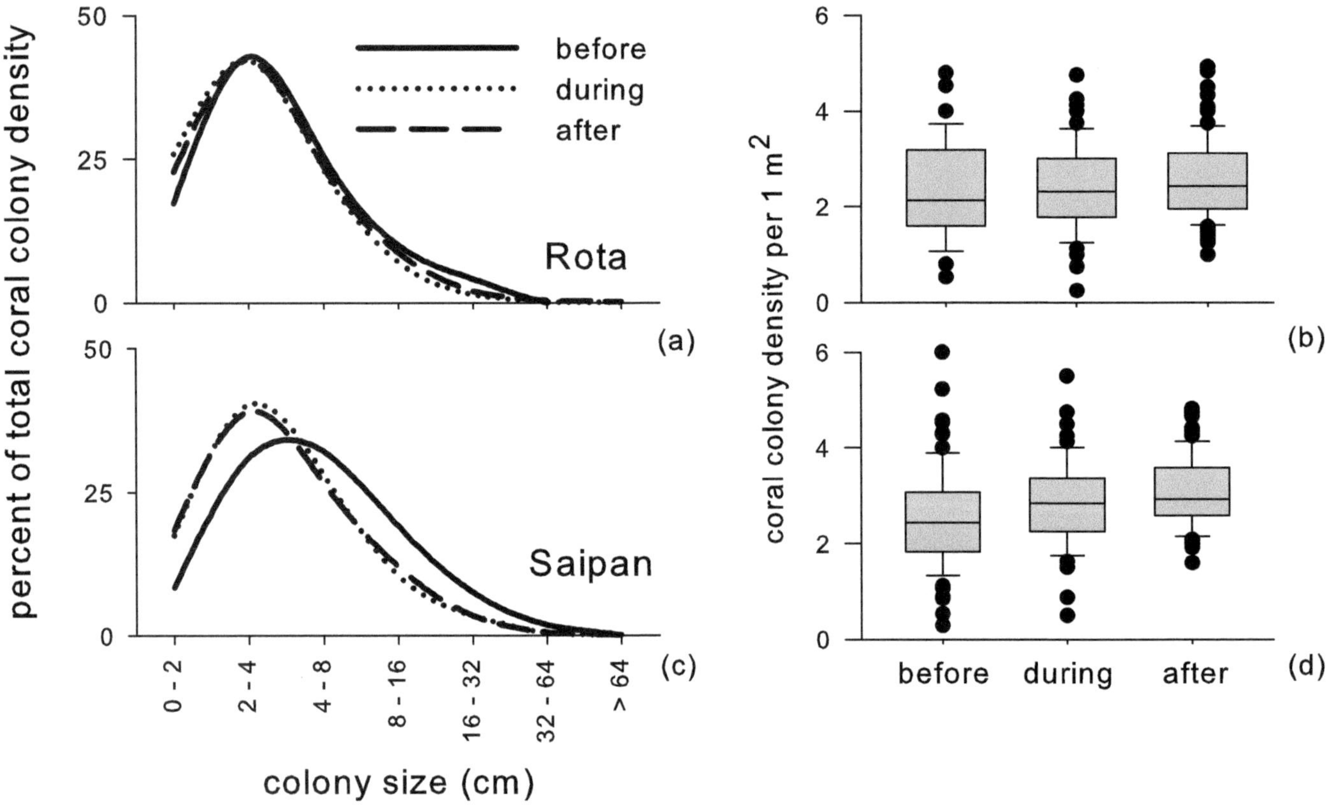

Figure 4. Dynamics of coral colony-size distributions and population densities across the disturbance periods. Rota had a non-significant decline and recovery in colony-size across the study periods (a), as well as a non-significant, sequential increase in population density (b). Reductions in colony-size were evident for Saipan (c) during the COTS period, accompanied by increases in population densities (d), attributed to the emergence of numerous small faviid and *Porites* corals (*indicates $P<0.05$, repeat measures ANOVA with post-hoc tests, see also Fig. 6).

variance ($P = 0.03$, Table 2). The influence of wave exposure grew when considering the subset of reefs with highest coral growth capacity ($R^2 = 0.22$, interstitial and spur-and-groove combined, $R^2 = 0.5$, spur-and-groove only). Yet, improved model fitting consistently required the inclusion of mean herbivore size and/or grazing urchin density that both increased the explanatory power and likelihood of resultant models. Explanatory power increased by 3%, 16%, and 14% when including herbivore size for all reeftypes, interstitial and spur-and-groove reefs combined, and spur-and-groove reefs alone, respectively (Table 2). Explanatory power increased by 0%, 20%, and 33% when including grazing urchin densities for these same reeftypes, respectively. Last, for coral-dominated reeftypes, including the interaction between herbivore size and grazing urchin density increased model fit by 33%. In all instances, interaction models lowered AIC scores suggesting their greater likelihoods.

No single factor consistently emerged as the primary driver of net change in the coral assemblages. Net change was predicted individually and not interactively by both herbivore size and the water quality proxy. When examining sites across all reeftypes, the water quality proxy had a slightly greater explanatory power ($R^2 = 0.28$ versus 0.38, herbivore size and water quality proxy, respectively), however the AIC-based likelihood was lower for water quality due to the relatively strong influence of 1 to 2 sites that diminished the normality of residuals (i.e., residuals still met the requirements of normality, but less so as compared with the herbivore size model, reducing the AIC score). The only other notable models emerged when examining spur-and-groove reefs, whereby herbivore size ($R^2 = 0.44$, $AIC = 18.9$) was a slightly better predictor of net coral assemblage change as compared to the water quality proxy ($R^2 = 0.39$, $AIC = 27.1$, Table 1).

Discussion

Significant coral loss occurred in the CNMI between 2003 and 2006 concurrent with high COTS densities and several typhoons that passed through CNMI. Given that tropical storm paths during these years were consistently in closer proximity to Rota, passing between 60 and 100 km from Saipan, we purport that COTS activity was the primary driver of disturbance that simultaneously impacted the study islands during the mid-2000's. In support, COTS densities were ubiquitous in models explaining the spatial patterns in coral decline during the disturbance years. However, potential synergies and/or linkages between storm activity and COTS were not approached, and deeper mechanisms may have existed.

While starfish densities were similarly elevated during the disturbance years for two islands that differed in human presence and geology, Rota and Saipan, abundances were highest and most persistent on Saipan. These findings help to explain the greater impacts to coral decline on Saipan. Yet, COTS abundances provided no significant explanation of net recovery patterns to the favorable benthic substrates or coral assemblage dynamics across the entire 12 year study period. We hypothesized the island-scale differences in resistance and recovery were attributable to a suite of factors. First, due to varying island geomorphology, Rota naturally had less coral to begin with (i.e., less prey), including a reduction in preferential prey, *Acropora* and *Montipora* [18]. This

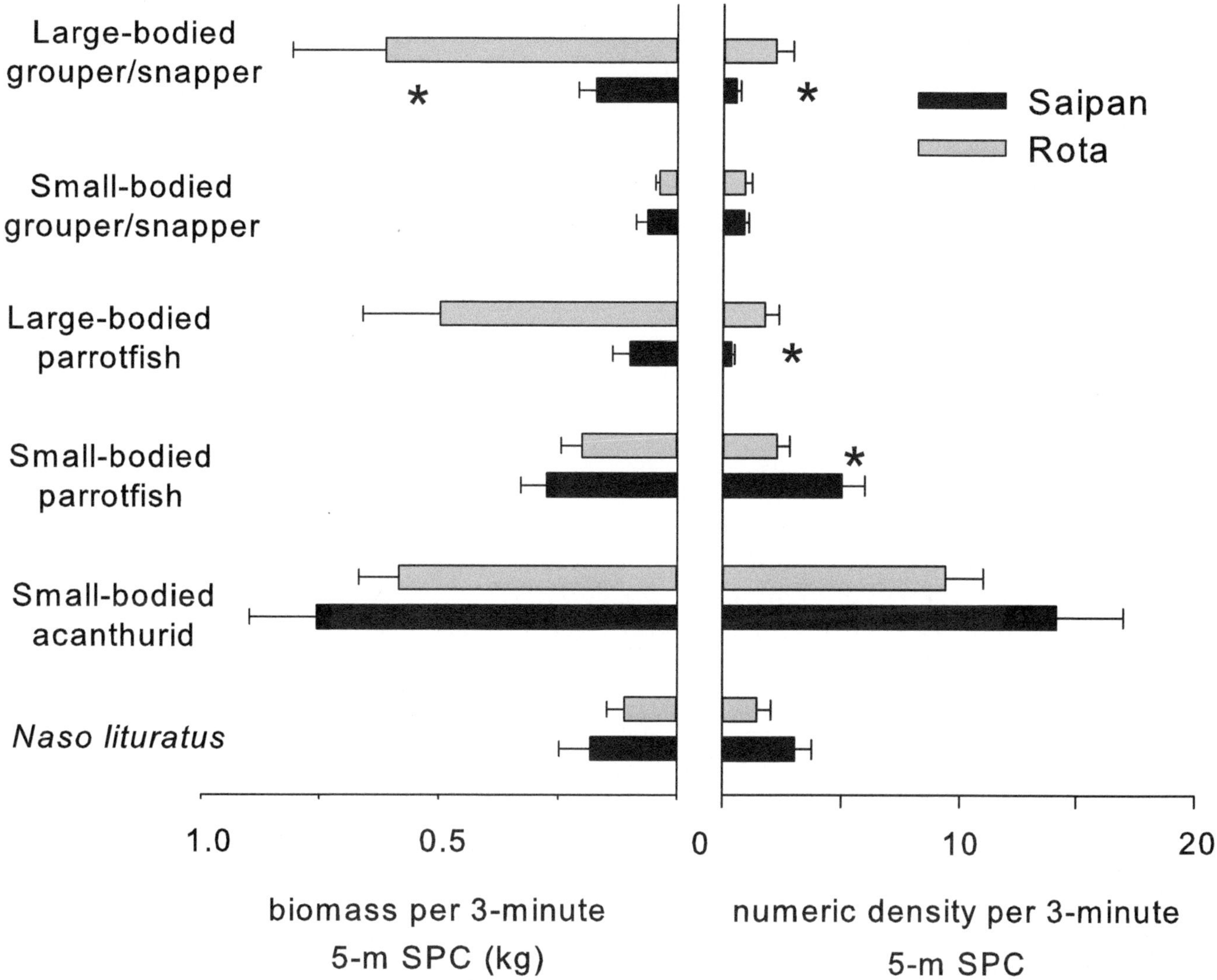

Figure 5. Comparisons of fish biomass and numeric density for several functional fish groups on Saipan and Rota. (*indicates $P<0.05$, comparative tests described in methods).

situation seems most relevant when interpreting why COTS persistence and coral impacts were diminished on Rota. Second, numerous studies continue to support that fish assemblages comprised of larger individuals across all trophic levels are associated with reduced impacts from COTS disturbances [12,30,51], and may help to explain COTS persistence and coral recovery dynamics [10,22]. Larger biomass and body-size of fish assemblages on Rota compared with Saipan supported this notion. Third, nutrient enrichment from watershed runoff is known to contribute to persistent, localized COTS populations [61,62]. CNMI water quality reports have consistently found better water quality on Rota over the past decade, with a 50% reduction in bacteria violations on Rota compared with Saipan in 2012 [63]. When considering where persistent populations existed (site 16 on Rota, sites 5, 6, 7, 8, and 10 on Saipan all had densities of 0.2 individuals per 100 m^2 across several post-disturbance years, Table S2), support for these combined hypotheses grows. Sites 7, 16, and 10 were associated with interstitial framework reefs, posited to have high connections with the karst aquifers compared to others [18] (Figure 1). These sites, in addition to others (5, 6, and 8), all had the smallest overall fish sizes. We synthesize that inherent geological difference as well as localized stressors were influential in describing disturbance dynamics across Rota and Saipan, and utilized site-based analyses to better approach the individual and interactive roles of a suite of factors.

Site-level drivers of change on CNMI's reefs

Wave exposure has long been considered to shape modern coral assemblages and reef growth through time in CNMI [64], whereby full exposure to prevailing northeast trade winds has selected against geological reef development through time (i.e., reeftype 4 noted in the methods). Yet, beside the incipient reef development that exists along much of CNMI's eastern shoreline, significant variation in wave exposure remained among the subset of reeftypes where high coral growth capacity existed, despite having a lower overall magnitude. In fact, this secondary gradient in wave exposure (i.e, low to moderate levels) was the most influential, positive determinant of net benthic substrate change in all reef settings. This may be a result of greater flushing, nutrient transfer rates, and/or the removal of detrital build-up with wave energy [65]. After accounting for wave exposure, the process of grazing, as represented by herbivore/detritivore size and grazing

Table 1. Stepwise regression models predicting coral decline.

Independent variables	Slope	SE	Intercept	R^2	P-Value	AIC
All reeftypes (n = 16)						
$COTS^{-1}$	**6.57**	**3.20**	**0.68**	**0.18**	**0.06**	**46.2**
Interstitial and spur-and-groove reefs (n = 12)						
$COTS^{-1}$	7.84	2.18	0.47	0.52	0.005	25.7
$COTS^{-1}$ × fish_size	1.55	0.43	1.62	0.60	0.009	18.4
Spur-and-groove reefs (n = 7)						
$COTS^{-1}$	8.88	2.43	0.23	0.67	0.01	16.8
$COTS^{-1}$ × log(exposure)	8.21	1.50	0.44	0.83	0.003	12.2
$COTS^{-1}$ × fish_size	2.17	0.38	1.30	0.86	0.005	8.6

Summary of forward, stepwise regression models that examined the drivers of coral decline during disturbance years. Methods describe the suite of independent variables examined and the basis for their selection. Significant independent variables presented below include mean *Acanthaster planci* densities during disturbance years (COTS), wave exposure, and mean fish size for all trophic groups combined. COTS densities were inversely scaled for consistency with other localized stressors (i.e., low-bad/high-good). AIC-scores were used to indicate the relative likelihood of models being able to predict outcomes, and are only comparable within each reeftype grouping.

urchin densities, was the strongest and most reliable predictor of favorable benthic substrates based upon their presence across the suite of models, the added variance accounted for, and the improved model likelihood scores (AIC values). These findings were amplified when stratifying by reeftype, and including only reefs with the most favorable geological foundations for coral growth. Water quality emerged as a significant predictor of change for coral assemblages ordinations along with mean herbivore size, however, we purport contextual roles of water quality that are furthered below.

Synthesizing the findings reveals that localized stressors were most influential to reefs with low to moderate wave exposure, and high inherent capacity for coral growth. Given the distribution of geological reef settings in the CNMI (Figure 1), this means that

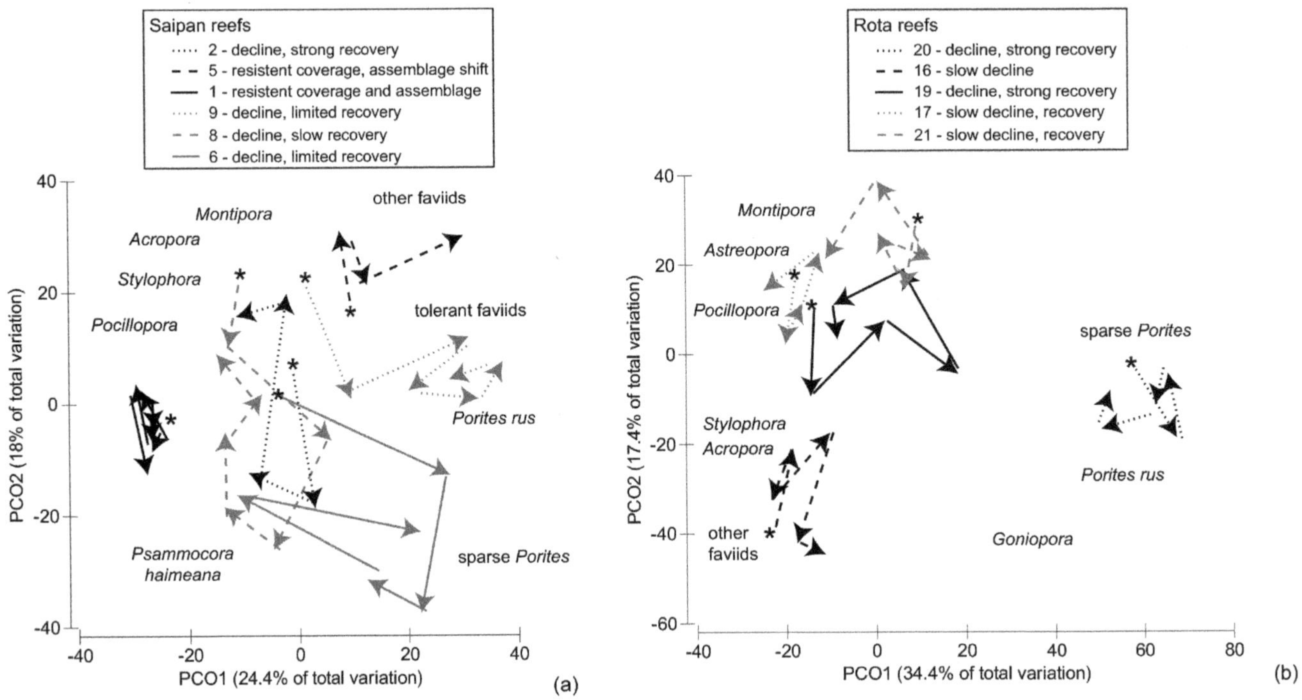

Figure 6. Principle components ordination of coral assemblages for six representative monitoring sites around Saipan (a) and Rota (b). See Figure 1 for site identification and Table S3 for summary statistics. Pre-disturbance assemblages are indicated with an asterisk (*), while vectors depict directional change through time. Sparse *Porites* refers to a dominance of *P. lichen*, *P. vaughani*, and small colonies of other massive species. Tolerant faviids consisted of *Leptastrea purpurea*, *Goniastrea retiformis*, *G. edwardsi*, *Favia matthaii*, *F. pallida*, and *F. favus*. Other faviids consisted of *Favia stelligera*, *Platygyra* spp., *Cyphastrea* spp., and *Favites abdita*.

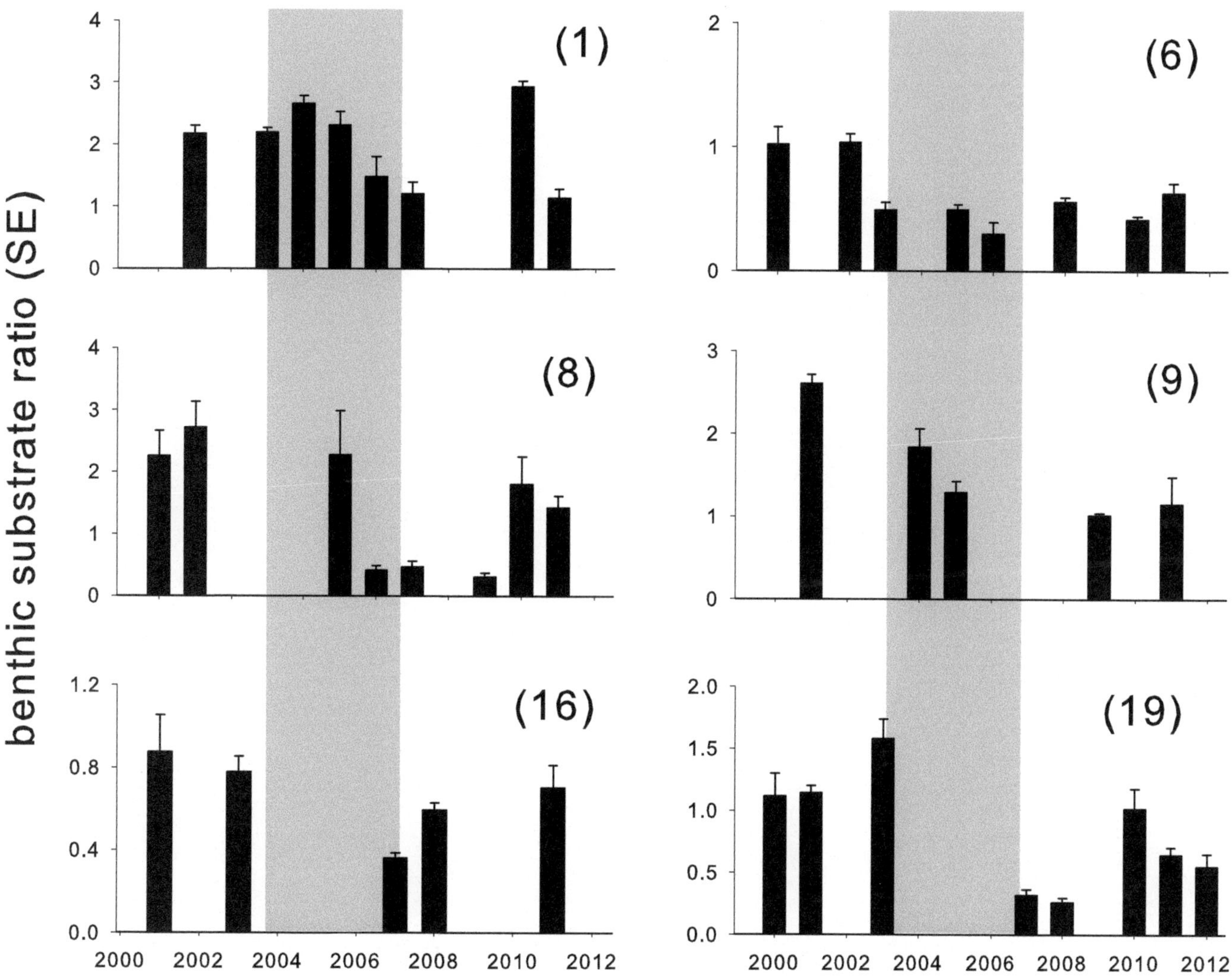

Figure 7. Benthic substrate ratio dynamics for representative monitoring sites around Saipan (sites 1, 9, 8, and 6) and Rota (sites 19 and 16). See Figure 1 for site identification and Table S3 for trends from all sites. Grey bars indicate the disturbance period. Benthic substrate ratios indicate the proportion of heavily-calcifying versus less-or-non-calcifying substrates (*see methods*).

Saipan reefs were weighted disproportionally within the subset of regression models associated with localized stressors, representing 63% of the sites with favorable geological settings (i.e., interstitial or spur-and-groove reefs). Hence, localized stressors were most influential for Saipan, the most populated island where reef-based tourism is centralized, and constitutes a key component of the economy [66].

The present results further an interesting and emerging association between high COTS impacts, smaller-bodied fish assemblages, and low wave exposure. While mechanisms remain unclear and of interest, similar patterns describing diminished COTS impacts with higher fish abundances in successful, no-take marine protected areas have been observed elsewhere [12,30,51]. Fish sizes (i.e., herbivores) were also influential to net change metrics, and the collective findings pertaining to the mean sizes of the fish assemblages rather than their biomass resonated well with power laws that describe relationships between body-size, physiology, and function in ecology [67]. Specific to the present study, power laws have been shown to govern numerous physiological traits such as grazing efficiency [68] and reproductive potential [69], whereby a doubling in fish size equates to an exponential increase in function. Thus, even if similar biomass exists, fish assemblages comprised of mainly small-bodied species are expected to have a reduced ecological function within coral-reef food webs [70,71]. Rasher et al. [72] described that a subset of larger, functionally-dissimilar herbivores play a disproportional role in macroalgal grazing (key constituents include *Chlorurus* spp., *Siganus* spp., and two *Naso* spp., *N. lituratus* and *N. unicornis*), while a suite of other species were more reliant upon generalized detritus and turf grazing from the reef substrate. Species within these functional groups represent highly desirable food fish in CNMI. Market studies have revealed their declining sizes and abundance over the past two decades [55,56,57], and also reported a disproportionally small reef-area-per-person, and mean fish size-at-capture, compared to other Micronesia jurisdictions [57]. We reconcile that both size and functional diversity appear to be key attributes of CNMI herbivore assemblages that are sensitive to harvesting pressure, and influential to coral reef recovery patterns, making their improved management desirable for reef futures.

The reduction in *Echinothrix* urchin densities concomitant with COTS disturbances was a novel association to our knowledge.

Table 2. Stepwise regression models predicting net ecological change.

Benthic substrate ratio net change						
All reeftypes (n = 18)						
Independent variables	*Slope*	*SE*	*Intercept*	R^2	*P-Value*	*AIC*
log(exposure)	3.81	1.64	−0.76	0.20	0.03	50.8
log(exposure) × herb_size	0.62	0.27	1.33	0.23	0.05	44.6
Interstitial and spur-and-groove reefs (n = 12)						
log(exposure)	5.46	2.71	−2.26	0.22	0.07	35.3
log(exposure) × urchin	0.84	0.28	0.74	0.42	0.01	31.7
log(exposure) × herb_size	0.87	0.36	0.85	0.38	0.04	27.4
log(exposure) × herb_size × urchin	0.19	0.06	1.58	0.57	0.01	24.2
Spur-and-groove reefs (n = 7)						
log(exposure)	9.23	3.50	−6.06	0.50	0.05	21.8
log(exposure) × herb_size	1.13	0.35	0.40	0.64	0.03	17.6
log(exposure) × urchin	1.11	0.20	0.08	0.83	0.003	14.4
Coral assemblage net change						
All reeftypes (n = 12)						
herb_size	0.66	0.32	1.37	0.28	0.08	25.0
$poll\_proxy^{-1}$	3.89	1.4	7.14	0.38	0.02	32.2
Spur-and-groove reefs (n = 7)						
$poll\_proxy^{-1}$	4.29	1.71	7.5	0.39	0.04	27.2
herb_size	0.93	0.41	0.64	0.44	0.08	18.9

Summary of forward, stepwise regression models that examined the drivers of net change in the benthic substrate ratio and coral ordination scores across the study period. Methods describe the suite of predictor variables examined and the basis for their selection. Significant independent variables presented below include wave exposure, mean herbivore/detritivore size, mean grazing urchin density, and the water quality proxy. The water quality proxy was inversely scaled (i.e., low-bad/high-good) for consistence with other localized stressors. AIC-scores were used to indicate the relative likelihood of models being able to predict outcomes, and are only comparable within each reeftype grouping.

Field observations and photographs provided anecdotal evidence of competition for refuge within the reef matrix, but clearly these relationships remain speculative. In the event that disturbances not only diminished coral cover but structural complexity as well, (i.e., fewer large *Acropora*, *Pocillopora*, *Stylophora* colonies), a decrease in urchin densities was the anticipated ecological response following disturbance [73]. Yet, urchin declines were concomitant with the onset of disturbance, and continued to decline throughout the study period. Given these trends, deeper investigations into the cause(s) of urchin declines seem warranted.

Last, the water quality proxy was an influential driver of net coral assemblage dynamics across the 12 year study period as well, despite its diminished presence across the suite of regression models (i.e., diminished presence in Tables 1 and 2). These findings support previous relationships between water quality proxies and coral species richness in CNMI [18,20], and reinforce that coral species composition may be a sensitive metric of water quality. It is beyond the purview of this study to formally discuss the linkages between diversity and ecosystem function, however, diversity is well known to facilitate functional redundancy in ecological systems, thereby providing for enhanced resistance and recovery to disturbance. Pollution contribution appeared to be influential to reefs where high human development existed, however, extensive human development was not common within the majority of CNMI's coastal watersheds. In addition, anchor points in the water quality regression models were driven by the presence of interstitial reefs associated with higher groundwater connectivity (Figure 1, Table 2, and Table S3). We conclude that watershed restoration strategies aimed at improving reef condition might obviously focus upon the largest urban centers, but less obviously, focus upon karst watersheds adjacent to high-value reef assemblages. We note that a major watershed restoration project addressing Laolao Bay (eastern Saipan) remains ongoing, and if successful would serve to address one of the key anchor points.

Conclusions

Over the past 12 years in CNMI, a period of high COTS densities led to significant coral declines. Yet, the failure of some reefs to recovery was best attributed to localized stressors, which transformed the substrates opened up by coral loss into persistent stands of turf and macroalgae, less conducive for coral replenishment and recovery. Declining trends were strongest for reefs that have favorable geomorphology (i.e., a gently sloping reef foundation), which disproportionally occur on Saipan alongside lower wave exposure. These same reefs represent centers for reef-based tourism that constitutes a major part of CNMI's economy, highlighting a need to improve upon compromised fish assemblages, grazing urchin populations, and specific localities where water quality concerns exist.

Supporting Information

Table S1 Monitoring site frequencies. Monitoring frequency for each of the long-term sites incorporated into the present study (*see Fig. 1*). Lowercase letters indicate the type of survey conducted in each year: (b) benthic substrate, (i) macroinvertebrate, (c) coral, and (f) fish.

Table S2 Site-based coral coverage and *Acanthaster* density data. Coral coverage and *Acanthaster planci* density summary statistics for each of the long-term monitoring sites incorporated into the present study. Site-based data formed the basis for regression modeling (Table 1). Reeftypes follow: "sg" - optimal spur-and-groove structures, "int" - high-relief, interstitial framework, "rot" - low relief Holocene framework found on Rota only, and "pl" - incipient coral assemblages residing upon a Pleistocene basement (*see methods*).

Table S3 Site-based summary statistics for regression models. Summary statistics for each of the long-term monitoring sites incorporated into the present study. Site-based data formed the basis for regression modeling. Dependent variables included the net change in the benthic substrate ratio and coral assemblages, noted as the sum of the percent decline (−) and subsequent recovery (+) of these ecological metrics (*see methods*). Reeftypes follow: "sg" - optimal spur-and-groove structures, "int" - high-relief, interstitial framework, "rot" - low relief Holocene framework found on Rota only, and "pl" - incipient coral assemblages residing upon a Pleistocene basement (*see methods*).

Acknowledgments

We are grateful for continuous discussions and support from several local and federal resource management agencies, including the United States Environmental Protection Agency Region IX, National Oceanic and Atmospheric Administration Coral Reef Conservation Program, and the Commonwealth of the Northern Mariana Islands (CNMI) Bureau of Environmental and Coastal Quality. We especially thank CNMI agency directors Frank Rabauliman and Frances Castro for logistical support throughout our study. Finally, several reviewers and the topic editor provided constructive criticism that greatly benefitted our manuscript.

Author Contributions

Conceived and designed the experiments: PH. Performed the experiments: PH DB JI SJ RO. Analyzed the data: PH DB JI SJ RO. Contributed reagents/materials/analysis tools: PH. Wrote the paper: PH DB JI SJ RO.

References

1. Hughes TP, Rodrigues MJ, Bellwood DR, Ceccarelli D, Hoegh-Guldberg O, et al. (2007) Phase shifts, herbivory, and the resilience of coral reefs to climate change. Curr Biol 17: 360–365.
2. Baker AC, Glynn PW, Riegl B (2008) Climate change and coral reef bleaching: An ecological assessment of long-term impacts, recovery trends and future outlook. Estuar Coastal Shelf S 80: 435–471.
3. Graham NAJ, Nash KL, Kool JT (2011) Coral reef recovery dynamics in a changing world. Coral Reefs 30: 283–294.
4. Hughes TP, Graham NAJ, Jackson JBC, Mumby PJ, Steneck RS (2010) Rising to the challenge of sustaining coral reef resilience. Trends Ecol Evol 25: 633–642.
5. Fabricius K, De'ath G, McCook L, Turak E, Williams DM (2005) Changes in algal, coral and fish assemblages along water quality gradients on the inshore Great Barrier Reef. Mar Pollut Bull 51: 384–398.
6. Mumby PJ, Dahlgren CP, Harborne AR, Kappel CV, Micheli F, et al. (2006) Fishing, trophic cascades, and the process of grazing on coral reefs. Science 311: 98–101.
7. Costa OS, Leao Z, Nimmo M, Attrill MJ (2000) Nutrification impacts on coral reefs from northern Bahia, Brazil. Hydrobiologia 440: 307–315.
8. Lapointe BE, Barile PJ, Matzie WR (2004) Anthropogenic nutrient enrichment of seagrass and coral reef communities in the Lower Florida Keys: discrimination of local versus regional nitrogen sources. J Exp Mar Biol Ecol 308: 23–58.
9. Houk P, Musburger C, Wiles P (2010) Water Quality and Herbivory Interactively Drive Coral-Reef Recovery Patterns in American Samoa. Plos One 5: e13913.
10. Mumby PJ, Harborne AR (2010) Marine Reserves Enhance the Recovery of Corals on Caribbean Reefs. Plos One 5: e8657.
11. Walsh SM (2010) Ecosystem-scale effects of nutrients and fishing on coral reefs. J Mar Biol doi:10.1155/2011/187248.
12. Dulvy NK, Freckleton RP, Polunin NV (2004) Coral reef cascades and the indirect effects of predator removal by exploitation. Ecol Lett 7: 410–416.
13. Aronson RB, Precht WF (2006) Conservation, precaution, and Caribbean reefs. Coral Reefs 25: 441–450.
14. De'ath G, Fabricius K (2010) Water quality as a regional driver of coral biodiversity and macroalgae on the Great Barrier Reef. Ecol App 20: 840–850.
15. Connell JH, Hughes TP, Wallace CC (1997) A 30-year study of coral abundance, recruitment, and disturbance at several scales in space and time. Ecol Monogr 67: 461–488.
16. Gardner TA, Côté IM, Gill JA, Grant A, Watkinson AR (2005) Hurricanes and Caribbean coral reefs: impacts, recovery patterns, and role in long-term decline. Ecology 86: 174–184.
17. Done TJ (1999) Coral community adaptability to environmental change at the scales of regions, reefs and reef zones. Am Zool 39: 66–79.
18. Houk P, van Woesik R (2010) Coral assemblages and reef growth in the Commonwealth of the Northern Mariana Islands (Western Pacific Ocean). Mar Ecol 31: 318–329.
19. Chollett I, Mumby PJ, Müller-Karger FE, Hu C (2012) Physical environments of the Caribbean Sea. Lim Oceanogr 57: 1233.
20. Cooper TF, Gilmour JP, Fabricius KE (2009) Bioindicators of changes in water quality on coral reefs: review and recommendations for monitoring programmes. Coral Reefs 28: 589–606.
21. Perry CT, Murphy GN, Kench PS, Smithers SG, Edinger EN, et al. (2013) Caribbean-wide decline in carbonate production threatens coral reef growth. Nat Commun 4: 1402.
22. McClanahan TR (2008) Response of the coral reef benthos and herbivory to fishery closure management and the 1998 ENSO disturbance. Oecologia 155: 169–177.
23. Bak RPM, Meesters EH (1998) Coral population structure: the hidden information of colony size-frequency distributions. Mar Ecol Prog Ser 162: 301–306.
24. Bak RPM, Meesters EH (1999) Population structure as a response of coral communities to global change. Am Zool 39: 56–65.
25. Loya Y, Sakai K, Yamazato K, Nakano Y, Sambali H, et al. (2001) Coral bleaching: the winners and the losers. Ecol Lett 4: 122–131.
26. Brown BE, Clarke KR, Warwick RM (2002) Serial patterns of biodiversity change in corals across shallow reef flats in Ko Phuket, Thailand, due to the effects of local (sedimentation) and regional (climatic) perturbations. Mar Biol 141: 21–29.
27. Pratchett MS, Trapon M, Berumen ML, Chong-Seng K (2011) Recent disturbances augment community shifts in coral assemblages in Moorea, French Polynesia. Coral Reefs 30: 183–193.
28. van Woesik R, Sakai K, Ganase A, Loya Y (2011) Revisiting the winners and the losers a decade after coral bleaching. Mar Ecol Prog Ser 434: 67–76.
29. Guest JR, Baird AH, Maynard JA, Muttaqin E, Edwards AJ, et al. (2012) Contrasting Patterns of Coral Bleaching Susceptibility in 2010 Suggest an Adaptive Response to Thermal Stress. Plos One 7: e33353.
30. Houk P, Benavente D, Fread V (2012a) Characterization and evaluation of coral reefs around Yap Proper, Federated States of Micronesia. Biodivers Conserv 21: 2045–2059.
31. Blackwood JC, Hastings A, Mumby PJ (2011) A model-based approach to determine the long-term effects of multiple interacting stressors on coral reefs. Ecol App 21: 2722–2733.
32. McClanahan TR, Maina J, Starger CJ, Herron-Perez P, Dusek E (2005) Detriments to post-bleaching recovery of corals. Coral Reefs 24: 230–246.
33. Burkepile DE, Hay ME (2006) Herbivore vs. nutrient control of marine primary producers: Context-dependent effects. Ecology 87: 3128–3139.
34. Banse K (2007) Do we live in a largely top-down regulated world? J Bioscience 32: 791–796.

35. Mork E, Sjoo GL, Kautsky N, McClanahan TR (2009) Top-down and bottom-up regulation of macroalgal community structure on a Kenyan reef. Estuar Coastal Shelf S 84: 331–336.
36. Sjoo GL, Mork E, Andersson S, Melander I (2011) Differences in top-down and bottom-up regulation of macroalgal communities between a reef crest and back reef habitat in Zanzibar. Estuar Coastal Shelf S 91: 511–518.
37. Wilson SK, Graham NAJ, Fisher R, Robinson J, Nash K, et al. (2012) Effect of Macroalgal Expansion and Marine Protected Areas on Coral Recovery Following a Climatic Disturbance. Conserv Biol 26: 995–1004.
38. Houk P, Van Woesik R (2006) Coral reef benthic video surveys facilitate long-term monitoring in the commonwealth of the Northern Mariana Islands: Toward an optimal sampling strategy. Pac Sci 60: 177–189.
39. Starmer J, Houk P (2008) Marine and Water Quality Monitoring Plan for the Commonwealth of the Northern Mariana Islands. CNMI Division of Environmental Quality, Saipan, MP.
40. Houk P, Bograd S, van Woesik R (2007) The transition zone chlorophyll front can trigger Acanthaster planci outbreaks in the Pacific Ocean: Historical confirmation. J Oceanogr 63: 149–154.
41. Keats DW, Chamberlain YM, Baba M (1997) Pneophyllum conicum (Dawson) comb nov (Rhodophyta, Corallinaceae), a widespread Indo-Pacific non-geniculate coralline alga that overgrows and kills live coral. Bot Mar 40: 263–279.
42. Antonius A (1999) Metapeyssonnelia corallepida, a new coral-killing red alga on Caribbean Reefs. Coral Reefs 18: 301–301.
43. Antonius A (2001) Pneophyllum conicum, a coralline red alga causing coral reef-death in Mauritius. Coral Reefs 19: 418–418.
44. O'Leary JK, Potts DC, Braga JC, McClanahan TR (2012) Indirect consequences of fishing: reduction of coralline algae suppresses juvenile coral abundance. Coral Reefs 31: 547–559.
45. Veron JEN (2000) Corals of the World. Stafford-Smith, Townsville.
46. Bohnsack JA, Bannerot SP (1986) A stationary visual census technique for quantatively assessing community structure of coral reef fishes. National Oceanic and Atmospheric Administration Technical Report NMFS 41.
47. Ekebom J, Laihonen P, Suominen T (2003) A GIS-based step-wise procedure for assessing physical exposure in fragmented archipelagos. Estuar Coastal Shelf S 57: 887–898.
48. Zar J (1999) Biostatistics. Prentice Hall, New Jersey.
49. Zuur AF, Ieno EN, Walker NJ, Saveliev AA, Smith GM (2009) Mixed effects models and extensions in ecology with R. Springer, New York.
50. Anderson M, Gorley R, Clarke K (2008) PERMANOVA+ for PRIMER: Guide to software and statistical methods. PRIMER-E, Plymouth, UK.
51. R Development Core Team RDC (2008) R: A language and environment for statistical computing. R foundation for statistical computing, Vienna, Austria. Available: http://wwwR-projectorg.
52. McCook LJ, Ayling T, Cappo M, Choat JH, Evans RD, et al. (2010) Adaptive management of the Great Barrier Reef: A globally significant demonstration of the benefits of networks of marine reserves. Proc Nat Acad Sci USA 107: 18278–18285.
53. Gilmour JP, Smith LD, Heyward AJ, Baird AH, Pratchett MS (2013) Recovery of an isolated coral reef system following severe disturbance. Science 340: 69–71.
54. Adams TC, Schmitt RJ, Holbrook SJ, Brooks AJ, Edmunds PJ, et al. (2011) Herbivory, connectivity, and ecosystem resilience: response of a coral reef to a large-scale perturbation. PLoS ONE 5: e23717.
55. Glynn PW, Enochs IC, Afflerbach JA, Brandtneris VW, Serafy JE (2014) Eastern Pacific reef fish responses to coral recovery following El Nino disturbances. Mar Ecol Prog Ser 495: 233–247.
56. Graham T (1994) Biological analysis of the nearshore reef fish fishery of Saipan and Tinian. CNMI Division of Fish and Wildlife Technical Report 94–02, Saipan, MP.
57. Trianni MS (1998) Summary and Further Analysis of the Nearshore Reef Fishery of the Northern Mariana Islands. Technical report submitted to the CNMI Division of Fish and Wildlife, Saipan, MP.
58. Houk P, Rhodes K, Cuetos-Bueno J, Lindfield S, Fread V, et al. (2012b) Commercial coral-reef fisheries across Micronesia: A need for improving management. Coral Reefs 31: 13–26.
59. Maynard J, McKagan S, Johnson S, Houk P, Ahmadia G, et al. (2012) Coral reef resilience to climate change in Saipan, CNMI; field-based assessments and implications for vulnerability and future management. Technical report submitted to the CNMI Division of Environmental Quality, Saipan, MP.
60. Halford A, Cheal AJ, Ryan D, Williams DM (2004) Resilience to large-scale disturbance in coral and fish assemblages on the Great Barrier Reef. 85: 1892–1905.
61. Brodie J, Fabricius K, De'ath G, Okaji K (2005) Are increased nutrient inputs responsible for more outbreaks of crown-of-thorns starfish? An appraisal of the evidence. Mar Pollut Bull 51: 266–278.
62. Fabricius K, Okaji K, De'ath G (2010) Three lines of evidence to link outbreaks of the crown-of-thorns seastar Acanthaster planci to the release of larval food limitation. Coral Reefs 29: 593–605.
63. Bearden C, Chambers D, Okano R (2012) Commonwealth of the Northern Mariana Islands Integrated 305(b) and 303(d) Water Quality Assessment Report. CNMI Division of Environmental Quality Saipan, MP.
64. Cloud PE (1959) Geology of Saipan, Mariana Islands, Part 4. Submarine topography and shoal-water ecology. Geological Survey Professional Paper 280-K, Washington, DC.
65. Crossman DJ, Choat JH, Clements KD, Hardy T, McConochie J (2001) Detritus as food for grazing fishes on coral reefs. Limnol Oceanogr 46: 1596–1605.
66. van Beukering P, Haider W, Wolfs E, Liu Y, van der Leeuw K, et al. (2006) The economic value of the coral reefs of Saipan, Commonwealth of the Northern Mariana Islands. Report prepared by Cesar Environmental Consulting for the National Oceanic and Atmospheric Administration, Silver Springs, MD.
67. Peters RH (1983) The ecological implications of body size. Cambridge University Press.
68. Lokrantz J, Nyström M, Thyresson M, Johansson C (2008) The non-linear relationship between body size and function in parrotfishes. Coral Reefs 27: 967–974.
69. Birkeland C, Dayton PK (2005) The importance in fishery management of leaving the big ones. Trends Ecol Evol 20: 356–358.
70. Houk P, Musburger C (2013) Trophic interactions and ecological stability across coral reefs in the Marshall Islands. Mar Ecol Prog Ser 488: 23–34.
71. Bellwood DR, Hoey AS, Hughes TP (2012) Human activity selectively impacts the ecosystem roles of parrotfishes on coral reefs. Proc R Soc B 279: 1621–1629.
72. Rasher DB, Hoey AS, Hay ME (2013) Consumer diversity interacts with prey defenses to drive ecosystem function. Ecology 94: 1347–1358.
73. Graham NAJ, Nash KL (2013) The importance of structural complexity in coral reef ecosystems. Coral Reefs 32: 315–326.

6

Modeling the Effect of Hypoxia on Macrobenthos Production in the Lower Rappahannock River, Chesapeake Bay, USA

Samuel Kersey Sturdivant[1]*, Mark J. Brush[2], Robert J. Diaz[2]

1 Division of Marine Science and Conservation, Nicholas School of the Environment, Duke University, Beaufort, North Carolina, United States of America, **2** Virginia Institute of Marine Science, College of William & Mary, Gloucester Point, Virginia, United States of America

Abstract

Hypoxia in Chesapeake Bay has substantially increased in recent decades, with detrimental effects on macrobenthic production; the production of these fauna link energy transfer from primary consumers to epibenthic and demersal predators. As such, the development of accurate predictive models that determine the impact of hypoxia on macrobenthic production is important. A continuous-time, biomass-based model was developed for the lower Rappahannock River, a Bay tributary prone to seasonal hypoxia. Phytoplankton, zooplankton, and macrobenthic state variables were modeled, with a focus on quantitatively constraining the effect of hypoxia on macrobenthic biomass. This was accomplished through regression with *Z′*: a sigmoidal function between macrobenthic biomass and dissolved oxygen concentration, derived using macrobenthic data collected from the Rappahannock River during the summers of 2007 and 2008, and applied to compute hypoxia-induced mortality as a rate process. The model was verified using independent monitoring data collected by the Chesapeake Bay Program. Simulations showed that macrobenthic biomass was strongly linked to dissolved oxygen concentrations, with fluctuations in biomass related to the duration and severity of hypoxia. Our model demonstrated that hypoxia negatively affected macrobenthic biomass, as longer durations of hypoxia and greater hypoxic severity resulted in an increasing loss in biomass. This exercise represents an important contribution to modeling anthropogenically impacted coastal ecosystems, by providing an empirically constrained relationship between hypoxia and macrobenthic biomass, and applying that empirical relationship in a mechanistic model to quantify the effect of the severity, duration, and frequency of hypoxia on benthic biomass dynamics.

Editor: Ruth H. Carmichael, Dauphin Island Sea Lab; University of South Alabama, United States of America

Funding: Virginia DEQ funded the benthic monitoring program in the lower Bay. Support for this work was provided by National Oceanic and Atmospheric Administration grant NA05NOS4781202, a National Science Foundation funded Hall-Bonner Fellowship, and a National Oceanic and Atmospheric Administration Graduate Sciences Program fellowship. The funders had no role in study design, data collection and analysis, decision to publish, or preparation of the manuscript. No additional external funding received for this study.

Competing Interests: The authors have declared that no competing interests exist.

* E-mail: kersey.sturdivant@duke.edu

Introduction

Macrobenthic organisms (retained on a sieve size >500 μm) are of importance to ecological processes in estuarine ecosystems like Chesapeake Bay [1], regulating or modifying most physical, biological, chemical, and geological processes [2]. Macrobenthos influence these sediment geochemical and physical properties [3] through bioturbation, the biological reworking of sediments [4]. In the estuarine environment, macrobenthos are a primary pathway through which organic carbon is cycled in the sediments [1]. However, the role of macrobenthos is limited to normoxic conditions; anaerobic metabolism becomes an increasingly important means of recycling organic carbon during hypoxia [5]. Macrobenthos also serve as a major energetic link between primary producers and demersal fish and epibenthic predators [6]; however, the sessile nature of macrobenthos makes them susceptible to natural and anthropogenic perturbations such as hypoxia [7], a significant concern given the documented importance of estuarine macrobenthic communities [1].

Since colonial times, the number of humans in Chesapeake Bay watershed has grown exponentially, with a 3-fold increase during the last 100 years [8]. Human activity adversely affects land topography, chemistry of the Earth's atmosphere and water, rates and balance of biogeochemical processes, and biodiversity [9]; Chesapeake Bay estuary is no different. Anthropogenic disturbance has greatly increased the flux of nitrogen and phosphorous compounds through land clearing, application of fertilizer, discharge of human waste, animal production, and combustion of fossil fuels, leading to eutrophication of the Bay [10]. Hypoxia, dissolved oxygen (DO) concentrations ≤ 2 mg O_2 l^{-1} [11], is closely associated with eutrophication, an increase in the rate of supply of organic matter to a system [12]. Low DO concentrations have been documented in mainstem Chesapeake Bay since the early 1930s [13] and in the Potomac since the 1910s [14]. Presently, seasonal hypoxia forms in early to late spring and lasts approximately 120 days, with the most severe low DO events occurring during mid-summer in mainstem Chesapeake Bay [8]. From the 1950s to the present, hypoxic volume in the Bay has increased substantially, from approximately 3 km^3 to 10 km^3 [15].

This increase is of concern given documentation of low DO impairing growth and reproduction and stressing living resources, increasing faunal susceptibility to disease and other environmental stressors [11,16–25].

As hypoxia continues to increase in the Bay and many other coastal systems worldwide [15,26], the development of accurate predictive models that quantify the ecological impacts of hypoxia becoming increasingly important. Recent models have begun to include functions relating hypoxia to increased mortality and/or reduced filtration or ingestion of macrobenthos, fish, and shellfish (e.g. [27–30]), but typically these functions are hypothetical and have yet to be constrained by empirical data. Models generally take two forms; detailed complex models that attempt to replicate as much of the natural environment as possible [28,31], and more simplistic models that only incorporate what is functionally necessary [32]. The focus of the current study was to model the effect of observed hypoxia, and various scenarios of hypoxia, on the biomass of macrobenthos; therefore our approach was to force DO concentrations into a simple model to predict macrobenthic responses, rather than to first simulate hypoxia with a highly resolved, linked hydrodynamic-water quality model. The approach we developed here can be incorporated into more complex models as they continue to add additional mechanistic detail with respect to ecological responses to hypoxia. We began by empirically constraining a formulation for hypoxia-induced mortality, incorporated this function into a simplified version of the Chesapeake Bay Eutrophication Model [28], and used the resulting model to predict the potential effects of hypoxia on macrobenthos in the Rappahannock River, a tributary of Chesapeake Bay that experiences seasonal hypoxia.

Materials and Methods

2.1 Study Area

Seasonal hypoxia occurs throughout Chesapeake Bay and some of its tributaries during the summer months [8], but in the lower Chesapeake Bay, the Rappahannock River is the only major tributary with the hydrography that allows for the development of sustained seasonal hypoxia [33]. In the Rappahannock, a combination of tidal mixing and advection of undersaturated mainstem waters into the tributary controls the seasonal hypoxia, which develops in late May and abates in early September [34,35]. Macrobenthic taxa in the Rappahannock River exhibit a variety of life history traits, such as resiliency and recruitment, which affects their response to seasonal hypoxia [1]. These responses allow some taxa to rapidly recruit to areas post-hypoxia, while other species struggle to re-establish, influencing variation in macrobenthic biomass; information on the functional characteristics of the dominant macrobenthos taxa in the lower Rappahannock are displayed in Table 1. No permits were required for the described study, which complied with all relevant regulations.

2.2 Field Collection

Macrobenthos data from a previous study were used to develop a formulation for hypoxia-induced mortality in our benthic model [36]. During the summers of 2007 and 2008, two random sites were chosen in each year in the lower Rappahannock for continuous monitoring of water quality, and biweekly sampling of macrobenthos from May to October. Based on water quality data from the Chesapeake Bay Long-Term Benthic Monitoring Program (www.baybenthos.versar.com) one normoxic site and one site known to experience seasonal hypoxia were chosen in each year. Data from a hypoxic and normoxic site in both 2007 and 2008, four sites total, were used for model construction.

At each monitored location a Hach DS500X Hydrolab datasonde was deployed approximately 0.5 m above the sediment surface attached to a small tripod. DO concentration, salinity, temperature, and depth were recorded every 20 minutes. Approximately every two weeks the datasonde was replaced with another Hydrolab datasonde, and a sediment sample was collected with a Young grab (440 cm^2 to a depth of 10 cm) for benthic community analysis. Grabs were sieved in the field through a 0.5 mm screen, and organisms and detritus retained on the screen transferred into labeled jars, preserved in a 10% formaldehyde solution, and stained with Rose Bengal. Samples were processed to identify and enumerate each species present as described in Dauer and Llansó [37]. Ash-free dry weight (AFDW) biomass was measured for each species by drying to a constant weight at 60°C and ashing in a muffle furnace at 500°C for four hours.

2.3 Model Construction

A continuous-time, biomass-based model was constructed using STELLA Modeling and Simulation Software®. The model was based on the benthic sub-model in the 2002 Chesapeake Bay Eutrophication Model [28,38], and contained three governing equations. All formulations and parameter values from the parent model [28,38] were used in the current project with the exception of the modifications described in sections 2.3, 2.4 and 2.5. Phytoplankton biomass was modeled as:

$$\frac{\delta}{\delta t}P=[(G-R-Wa)*P-PR] \tag{1}$$

where:
P = phytoplankton biomass (g C m^{-3})
G = growth rate of phytoplankton (d^{-1})
R = respiration rate of phytoplankton (d^{-1})
Wa = phytoplankton settling loss (d^{-1})
PR = predation on phytoplankton (g C m^{-3} d^{-1}).

Zooplankton was modeled as the combined biomass of micro- and mesozooplankton for simplicity as:

$$\frac{\delta}{\delta t}M=[(Gz-BMz-Mz)*M-PRz] \tag{2}$$

where:M = zooplankton biomass (g C m^{-3})
Gz = growth rate of zooplankton (d^{-1})
BMz = basal metabolic rate of zooplankton (d^{-1})
Mz = hypoxic mortality (d^{-1})
PRz = predation on zooplankton (g C m^{-3} d^{-1}).

Macrobenthos were modeled as the combined biomass of deposit and suspension feeders, as:

$$\frac{\delta}{\delta t}B=\left[\alpha*\left(\frac{I_0}{m_2*10^6}\right)*(POC+PM)*k_{mn1}*B\right] - \left[\left(r-\beta^2-m\right)*B\right] \tag{3}$$

where:
B = macrobenthos biomass (g C m^{-2})
α = assimilation efficiency for carbonI_0 = ingestion rate of macrobenthos (g sediment g C^{-1} biomass^{-1} d^{-1})
m_2 = sediment solids concentration (kg l^{-1})POC = sediment particulate organic carbon concentration (g C m^{-3})
PM = phytoplankton and zooplankton biomass (g C m^{-3})k_{mn1} = Michaelis-Menton growth limitation term for carbon$r$ = respiration rate of macrobenthos (d^{-1})
β = predation rate (m^2 g C d^{-1})

Table 1. The major macrobenthic taxa of the lower Rappahannock River.

Taxon	Feeding Type	Mobility	Sediment Reworking Rate (mg dry wt. of indiv)	Hypoxia LT_{50} (h)	Reproduction	Notes	Source(s)
Acteocina canaliculata (G)	C	M	2080 (displacement)				Myers 1977; Schaffner 1987
Capitella spp. (A)	HD	LM		168–312	Spawn (throughout the year) Larval development completely benthonic (~1 mo.)		Rosenberg 1972; Warren 1976; Kravitz 1983
Heteromastus filiformis (A)	HD	LM	262.5–700	312	Spawn (Spring)	Eggs deposited in the sediment wrapped in a cocoon. Hatched larvae may start their burrowing mode of life immediately.	Rosenberg 1972; Cadee 1979; Kravitz 1983; Hines and Comtois 1985
Mediomastus ambiseta (A)	HD	LM	262.5–700		Spawn (Apr-Sept)		Kravitz 1983; Schaffner 1987
Nereis succinea (A)	SD,O	LM	3.9–103.7	62–84	Spawn (Mar-Oct)	Planktonic larvae remain in the water column until they possess 9–12 segments, at which time they settle to the benthos	Cammen et al. 1978; Fauchald and Jumars 1979; Kravitz 1983; Hines and Comtois 1985; Hardedge et al. 1990; Fong 1991; Sagasti etal. 2001; Tiffany et al. 2002
Paraprionospio pinnata (A)	SD,SF	LM	39.6–51.9 (juveniles) 90.9–232.8 (adults)		Spawn (summer) Recruitment (June-Dec) Clutch Size (6000 eggs)	Dominant during hypoxia	Dauer et al. 1981; Kravitz 1983; Schaffner 1987; Luckenbach et al. 1988,;Mayfield 1988
Streblospio benedicti (A)	SD,SF	LM	11.7–18.7	27–43	Spawn (spring and summer) Clutch Size (15–70 planktotrophy or 70–460 lecithotrophy)	Larvae to Juvenile (3 days – 2 weeks) 4 annual cohorts	Dauer et al. 1981; Hines and Comtois 1985; Kravitz 1983; Levin and Hugget 1990; Llansó 1991; Levin and Bridges 1994
Tubificoides spp. (A)	HD	LM	20.7–31.9	720			Tevesz et al. 1980; Hines and Comtois 1985; Giere et al. 1999

Compilation of the functional characteristics of major macrobenthos taxa in the lower Rappahannock River. Feeding type: C- carnivore, HD- head-down, SD- surface deposit, SF- suspension feeder, O- omnivore. Mobility: M- mobile, LM- limited mobility.

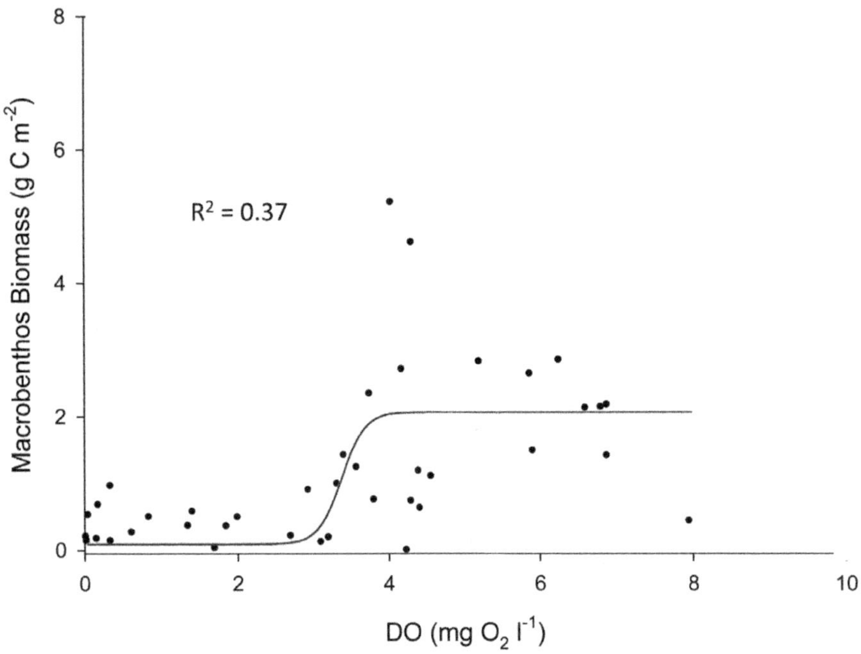

Figure 1. Relationship between DO concentration and macrobenthos biomass. Comparison of macrobenthos biomass and DO concentration from the four Rappahannock River sites monitored bi-weekly from May to October. Trendline is a sigmoid curve, where equation $Z' = \frac{1.97}{1+e^{\left(\frac{x-3.35}{0.15}\right)}}$.

m = hypoxia mortality rate (d^{-1}).

Phytoplankton and zooplankton groups were included in the model given the tight benthic-pelagic (B-P) coupling that exists in estuarine and shallow coastal systems [1,39–41], and the importance of both groups as a source of food for macrobenthos [41]. The recycling of nutrients is another major feature of B-P coupling [42], but nutrient recycle terms are not included in the model to maintain simplicity. Our model excluded the state equation for suspension feeders in the Chesapeake Bay Eutrophication Model, and it combined macrobenthic suspension and deposit feeders into a single state equation. The benthic suspension feeder equation was not included in our model because its construction was based on large bivalve suspension feeders [28] that are generally rare in the lower Rappahannock River [43]. Based on the community composition of macrobenthos collected during the summers of 2007 and 2008 [36], a single governing equation for macrobenthos was assumed to be sufficient to model macrobenthic response to DO concentrations in the lower Rappahannock. Many of the macrobenthos in our samples demonstrate both suspension and deposit feeding traits (Table 1), and no definable difference in response to hypoxia was observed from either group in terms of a change in biomass [36]. The combination of these two groups also maintained our goal of keeping the model as simple as functionally possible.

Water quality variables (DIN, POC, and DO) were obtained from daily interpolations of Chesapeake Bay Water Quality Monitoring Program data from 1985 to 2001 [44], with the exception of photosynthetically active radiation (PAR) and water temperature. Daily PAR and water temperature were forced using equations derived by Wetzel and Neckles [45] for lower Chesapeake Bay.

2.4 Adaptations to the Original Model

Some specific changes were made to the original governing equations of the Chesapeake Bay Eutrophication Model (addressed below). The formulation that represented the response of zooplankton mortality to hypoxia (*Mz*) was altered; in our model, if DO concentration was less than 2 mg DO l^{-1} then:

$$Mz = MZERO_Z * \left(1 - \frac{DOREF}{DOCRIT_Z}\right) \quad (4)$$

in which:

Mz = hypoxic mortality of zooplankton group Z (d^{-1})

MZEROz = mortality at zero dissolved oxygen concentration (d^{-1})

DOREF = dissolved oxygen concentration when DO < $DOCRIT_z$, otherwise 2 (mg DO l^{-1})

DOCRITz = threshold below which dissolved-oxygen-induced mortality occurs, this value equals 2 (mg DO l^{-1}).

In the original equation *DOCRITz* was always 2 mg O_2 l^{-1}, and *DOREF* was the dissolved oxygen concentration when DO < *DOCRITz*, otherwise it was zero. However, during model simulation this resulted in a linear increase in modeled zooplankton population through time, therefore, the formulation was amended to equation 4, where the *DOREF* was the dissolved oxygen concentration when DO < *DOCRITz*, otherwise it was 2 mg DO l^{-1}.

The parent Chesapeake Bay Eutrophication model simulates three fractions of sediment organic carbon, a labile, semi-labile, and refractory pool. In the original version of the macrobenthos model, the following portion of Eq. 3, $\left[\alpha * \left(\frac{I_0}{m_2 * 10^6}\right) *\right.$

$POC * k_{mn1} * B$], was computed twice, once for the labile and once for the semi-labile carbon pool. Since our model was not coupled to a larger eutrophication model, in the interest of maintaining simplicity we computed this term in Eq. 3 once using total sediment POC from field measurements.

2.5 Rappahannock Function Relating Biomass to Hypoxia

In the original Eutrophication model, the impact of DO concentration on macrobenthos respiration (r), ingestion (I_0), and mortality (m) was represented by the sigmoid equation Z, where:

$$Z = \frac{1}{1 + e^{1.1 * \left(\frac{DO_{gx} - DO}{DO_{gx} - DO_{qx}}\right)}} \quad (5)$$

where:

DO_{gx} = DO at which macrobenthos function is 50% of maximum.

DO_{qx} = DO at which macrobenthos function is 25% of maximum.

In the case of hypoxia-induced mortality, Z is then used in Eq. 5 above. The sigmoid equation that represents Z was not supported by any data. Our analysis of Chesapeake Bay field data and results from Seitz et al. [24] suggests a sigmoid relationship between DO and macrobenthic biomass, as mortality occurs below rather sharply defined oxygen concentrations. Therefore, a sigmoid function was empirically constrained using macrobenthos data collected from the Rappahannock River during the summers of 2007 and 2008 (Fig. 1), and a function (Z') was fit to the data to model the impact of DO concentration on macrobenthicbiomass:

$$Z' = \frac{1.97}{1 + e^{\left(\frac{x - 3.35}{0.15}\right)}} \quad (6)$$

It is important to note that these relationships are quantitative representations of the actual various underlying causal factors between DO and macrobenthic biomass. In our model, Z is still used to model the impact of DO concentration on r and I_0 (see [38]), but Z' is used to more accurately model the impact of DO on macrobenthic biomass, replacing Z in Eq. 5 and providing an empirically constrained method for estimating hypoxia-induced mortality. Equation 6 was normalized (0 to 1, dimensionless) by replacing the numerator with 1, such that Equation 7 was the version applied in our model:

$$Z' = \frac{1}{1 + e^{\left(-\frac{DO - 3.35}{0.15}\right)}} \quad (7)$$

where:

DO = dissolved oxygen concentration (mg l^{-1}).

2.6 Model Verification and Simulation Analysis

A single model run encompassed a period of 365 days with a time step of one calculation per day. Model output for the three governing equations were verified using Chesapeake Bay Benthic, Water Quality, and Plankton Monitoring Program data from stations LE3.2, LE3.4, and LE3.6 which are located approximately 25 km, 19 km, and 2 km from the mouth of the Rappahannock River, respectively. While the stations used for verification differ spatially, the physical dynamics over this relatively small spatial scale of the lower Rappahannock River are not significantly different [44]. The macrobenthic output was verified with data from stations LE3.2 and LE3.4, and data from station LE3.6 were used to verify the phytoplankton and zooplankton equations; three stations were used due to data availability. Model output was visually compared to observations to determine validity [46], and by assessment of mean percent error (MPE) and root mean square deviation (RMSD) [32].

Predicted phytoplankton biomass was verified against the mean annual cycle of chlorophyll-a concentrations computed using Chesapeake Bay Program data from 1985 to 2001. To reflect mean water column conditions, this annual cycle was based on averages of surface and bottom concentrations; differences between the surface and bottom were usually less than 10 mg m^{-3}. Predicted zooplankton biomass was verified using computed average annual cycles of combined micro- and mesozooplankton biomass using Chesapeake Bay Program data from 1985 to 2001. Depth-integrated zooplankton counts from the Plankton Monitoring Program were converted to biomass using species-specific carbon contents [47]. Predicted macrobenthic biomass was verified using Bay Program data from 1992, using site LE 3.4 for verification under normoxia and site LE3.2 for verification under hypoxia. Year 1992 was chosen at random from years 1985–2001. The approach of using a randomly selected year for macrobenthos verification instead of averaged annual cycles was due to the stochastic nature of benthic data collection through time; data were collected at variable dates each year by the Chesapeake Bay Program. Following verification, sensitivity analyses were conducted for the phytoplankton, zooplankton, and macrobenthic state variables by adjusting selected parameters that directly impacted growth or loss (i.e. consumption or predation parameters). Maximum photosynthetic rate ($P^b m$), phytoplankton settling velocity (Wa), and predation rate on algae ($Phtl$) were tested for the phytoplankton state variable; predator biomass and clearance rate ($PHTlz$) for the zooplankton state variable; and assimilation efficiency for carbon (α) and ingestion limitation (K_I) for the macrobenthic state variable. All parameters tested in sensitivity analysis were adjusted at an increment of ±20% and the relative percent difference from the standard run was calculated for each. Parameters with percentage differences greater than 10% were deemed to be sensitive parameters [48].

A set of simulations analyses were then conducted, adjusting DO concentration to model the effect of the severity and duration of hypoxia on the three modeled state equations (Table 2). DO concentrations were gradually adjusted during a period of 3 days to avoid artificial rapid changes in the model. Simulations S1 to S3 focused on the sustained duration of hypoxia. S4 simulated the development of hypoxia during the neap-spring tidal cycle with DO cycling for 14 days. Verified model output from the state equations (i.e. the base model results during normoxia after model verification and sensitivity analyses; Fig. 2, 3, and 4) was used as a baseline for comparison against simulations S1 to S4. This was conduct through visual observation and comparison of the means and standard deviations between the verified model output and the simulations (e.g. phytoplankton, macrobenthos, and zooplankton output in S1 was compared to the models in Fig. 2, 3, and 4, respectively). Simulations S5 to S9 modeled the severity of hypoxia from 0.0 to 2.0 mg O_2 l^{-1} with a 0.5 mg O_2 l^{-1} step and constant duration of 60 days. Comparison of means and standard deviations were used to justify "significant" differences. DO concentration was forced for each model simulation, with consideration of historical data of DO dynamics in the Rappahannock River [44].

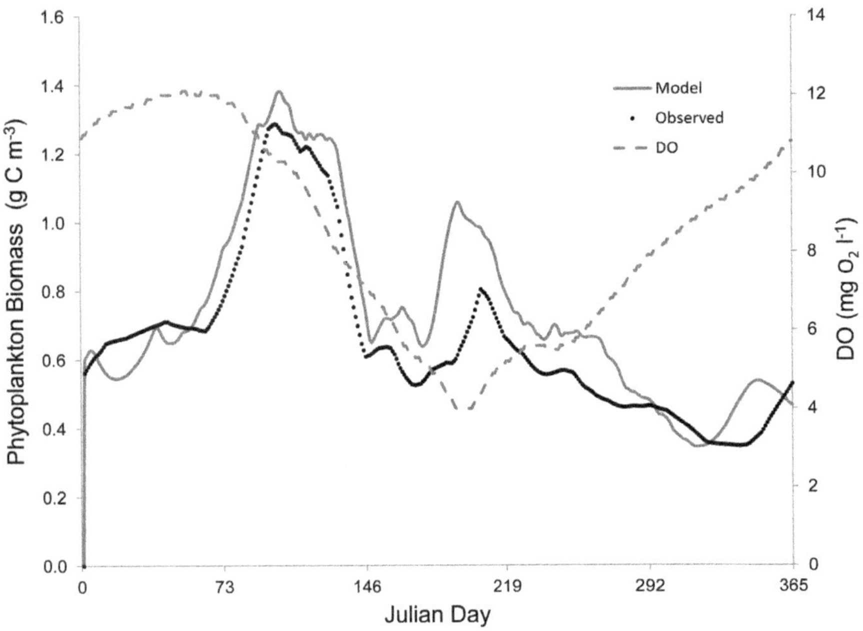

Figure 2. Verification of the phytoplankton state variable. Left y-axis: the gray solid line represents modeled phytoplankton biomass and the black dots denote mean observed phytoplankton biomass from site LE3.6 of the Chesapeake Bay Water Quality Monitoring Program from 1985–2001. Right y-axis: the grey dashed line indicates the dissolved oxygen concentration. Modeled phytoplankton biomass matched the trend and magnitude of observed phytoplankton biomass in 1992.

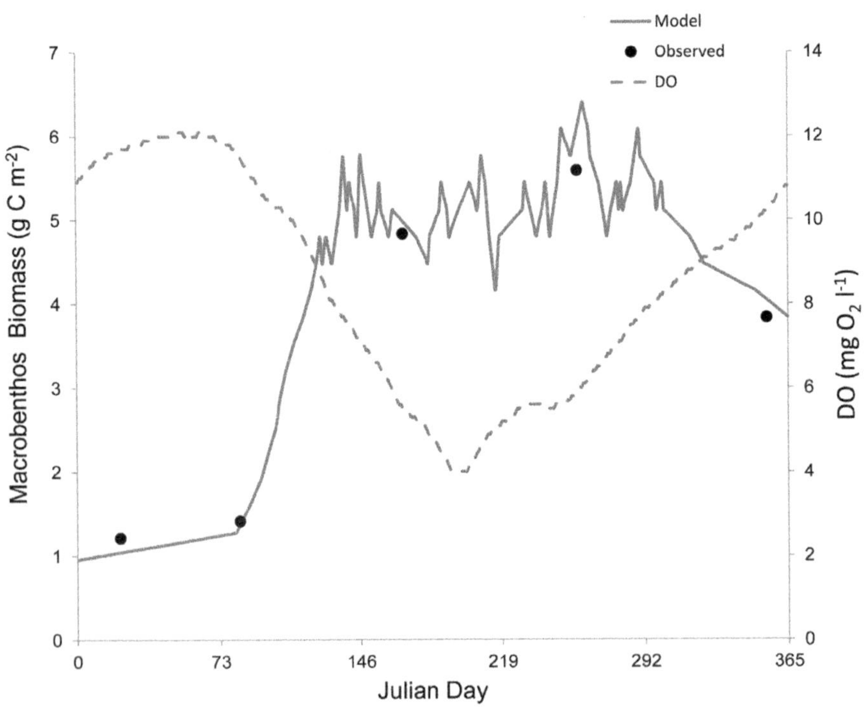

Figure 3. Verification of the macrobenthos state variable during normoxia. Left y-axis: the gray solid line represents modeled macrobenthos biomass and the black dots denote observed macrobenthos biomass from site LE3.4 of the Chesapeake Bay Benthic Monitoring Program in 1992. Right y-axis: the grey dashed line indicates the dissolved oxygen concentration. Modeled macrobenthos biomass matched the trends and magnitude of observed macrobenthos biomass in 1992.

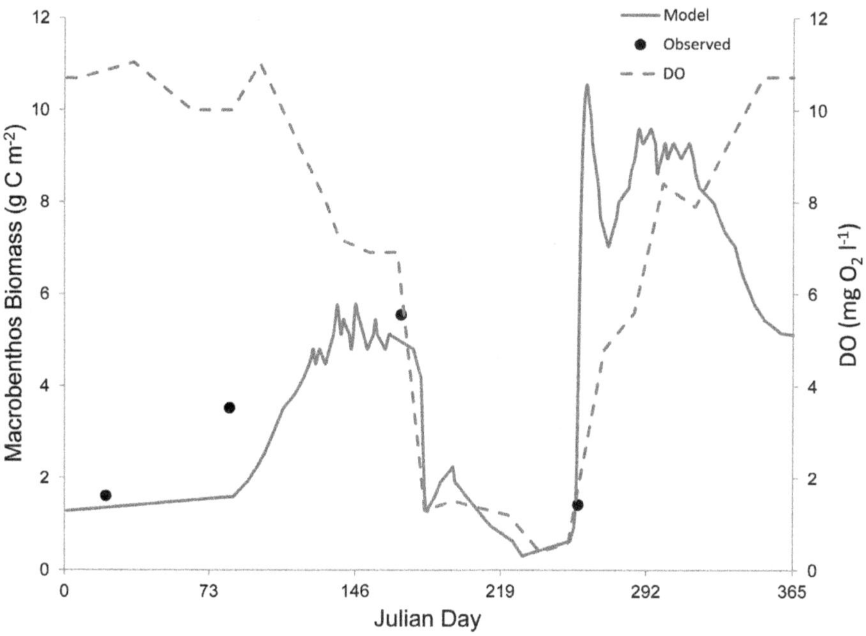

Figure 4. Verification of the zooplankton state variable. Left y-axis: the gray solid line represents modeled zooplankton biomass and the black dots denote mean observed zooplankton biomass from site LE3.6 of the Chesapeake Bay Water Quality Monitoring Program from 1985–2001. Right y-axis: the grey dashed line indicates the dissolved oxygen concentration. Modeled zooplankton biomass approximated the observed annual cycle and matched the magnitude of observed zooplankton biomass in 1992.

Results

Modeled phytoplankton biomass (*P*) matched the general trends in observed biomass in the lower Rappahannock River, with blooms and declines consistent with the data from 1992 (Fig. 2). The model also reproduced the approximate magnitude of phytoplankton biomass with MPE of 13.3% and RMSD of 0.18 g C m^{-3}. Modeled macrobenthic biomass (*B*) also matched the patterns in observed biomass in 1992 at both a normoxic site (Fig. 3) and a hypoxic site (Fig. 5). The model accurately portrayed the temporal dynamics in macrobenthic biomass with respect to DO concentration, as well as the magnitude of biomass with MPE of 12.1% and RMSD of 0.56 during normoxia and MPE of 15.7% and RMSD of 0.22 g C m^{-2} during hypoxia. Modeled zooplankton biomass matched the magnitude of observed biomass in 1992, but did not follow the trends in the observed data (Fig. 4), with MPE of 4.5% and RMSD of 0.38 g C m^{-2}. Modeled zooplankton dynamics appeared to be delayed when compared to observed increases and decreases in zooplankton biomass. This discrepancy may be due to the combination of micro- and mesozooplankton into a single state variable. Additionally, there were noted calibration difficulties for zooplankton in the parent eutrophication model [49], which may have extended into our adaptation of that model.

Discrepancies between modeled and observed zooplankton biomass could be expected to have detrimental effects on modeled

Table 2. Model simulations.

Simulations	Hypoxia Duration (d)	Julian Day	Ordinal Date	DO (mg O_2 l^{-1})
S1	120	148–267	May 28– Sept 24	0.5
S2	60	178–237	Jun 26– Aug 26	0.5
S3	30	191–222	Jul 11– Aug 11	0.5
S4	14 d intervals	162–176, 190–204, 218–232, 246–260	Jun 11–25, Jul 9–23, Aug 6–20, Sept 3–17	0.5
S5	60	178–237	Jun 26– Aug 26	2.0
S6	60	178–237	Jun 26– Aug 26	1.5
S7	60	178–237	Jun 26– Aug 26	1.0
S8	60	178–237	Jun 26– Aug 26	0.5
S9	60	178–237	Jun 26– Aug 26	0.0

Ecosystem model simulations with varying hypoxic duration and severity. S1 to S4 (Figures 6 and 7) modeled hypoxic duration at a constant concentration of 0.5 mg O_2 l^{-1}; S4 modeled intermittent hypoxia (hypoxia occurring every 14 days on a neap/spring tidal cycle); S5 to S9 modeled the affect of hypoxic severity at a constant duration of 60 days.

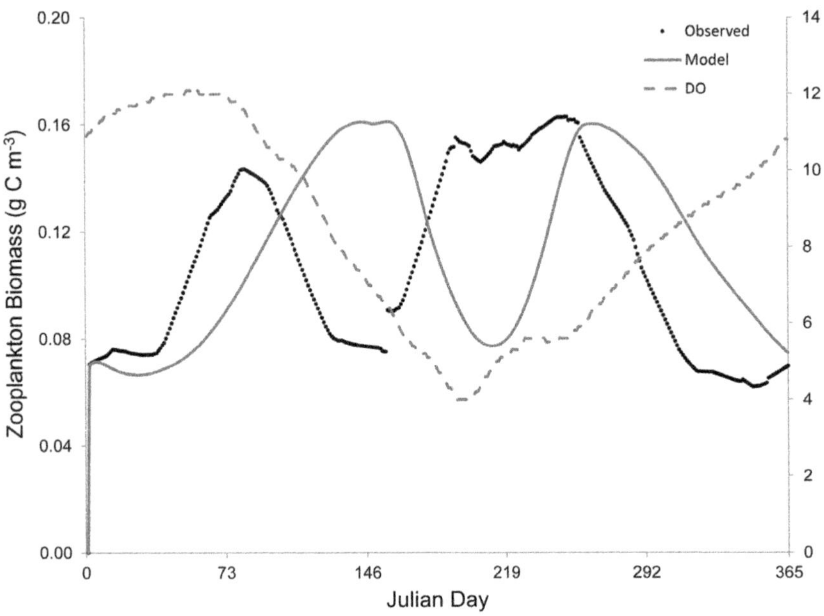

Figure 5. Verification of the macrobenthos state variable during hypoxia. Left y-axis: the gray solid line represents modeled macrobenthos biomass and the black dots denote observed macrobenthos biomass from site LE3.2 of the Chesapeake Bay Benthic Monitoring Program in 1992. Right y-axis: the grey dashed line indicates the dissolved oxygen concentration. Modeled macrobenthos biomass matched the trends and magnitude of observed macrobenthos biomass in 1992.

phytoplankton biomass, given the significant influence of zooplankton predation rate on algae (see Table 3 below). However, modeled phytoplankton biomass successfully reproduced the observations (Fig. 2). Additionally, phytoplankton biomass in the lower Rappahannock River is 3.3–14.3 times greater (mean = 6.5) than zooplankton biomass (Fig. 2 and 4), which indicates that phytoplankton will have the dominant effect on modeled macrobenthic biomass. The successful simulation of phytoplankton biomass, simulation of zooplankton biomass in the correct range (Fig. 4), the overall much greater biomass of phytoplankton compared to zooplankton, and the accurate simulation of macrobenthic biomass under both normoxia (Fig. 3) and hypoxia (Fig. 5), confirm the ability of the model to predict the dynamics of macrobenthic biomass in response to hypoxia in further simulation analyses.

Sensitivity analyses were conducted on model parameters for each state variable (Table 3). The model was sensitive to a majority of tested parameters, with the phytoplankton state variable sensitive to all tested parameters. The zooplankton state variable was found to be sensitive to an increase and decrease in predator biomass and clearance rate. The macrobenthic state variable was sensitive to assimilation efficiency for carbon and insensitive to ingestion limitation. The sensitivity of these parameters is a problem in the parent model that we did not attempt to address. The goal of this manuscript was not to resolve nuances in the original model, but to take a simple approach and constrain the effect of DO concentration on macrobenthic biomass, within the context of the more complex Chesapeake Bay Eutrophication Model.

Simulations were run assessing the impact of hypoxic duration on phytoplankton, zooplankton, and macrobenthic biomass. Macrobenthic biomass began to decrease as scenario S1, which simulated a hypoxic duration of 120 days, approached hypoxia (Fig. 6A). At the start of hypoxia in S1 the steady decrease in

Table 3. Sensitivity analysis.

State variable	Parameter	−20%	+20%	Average RMS	% Diff -20%	% Diff +20%
Phytoplankton	$P^b m$	0.032	0.041	0.037	12.7*	11.8*
	Wa	0.041	0.032	0.037	11.5*	12.2*
	Phtl	0.044	0.030	0.037	18.8*	18.6*
Zooplankton	*PHTlz*	0.001	0.003	0.002	11.8*	13.1*
Macrobenthos	α	0.118	0.369	0.192	38.8*	91.8*
	K_I	0.191	0.194	0.192	0.8	0.5

Results of sensitivity analysis for phytoplankton, zooplankton, and macrobenthic state variables. The root mean square (RMS) was the average variance in the base state equation over all time steps of a single year (n = 365). RMS values are shown for ±20% variation for each state variable by parameter. The model was deemed to be sensitive when % difference exceeded 10% (Ragone-Calvo et al. 2001). Parameters: Maximum photosynthetic rate ($P^b m$), phytoplankton settling velocity (*Wa*), predation rate on algae (*Phtl*), predator biomass and clearance rate (*PHTlz*), assimilation efficiency for carbon (α), and ingestion limitation (K_I).
***Denotes model sensitivity.**

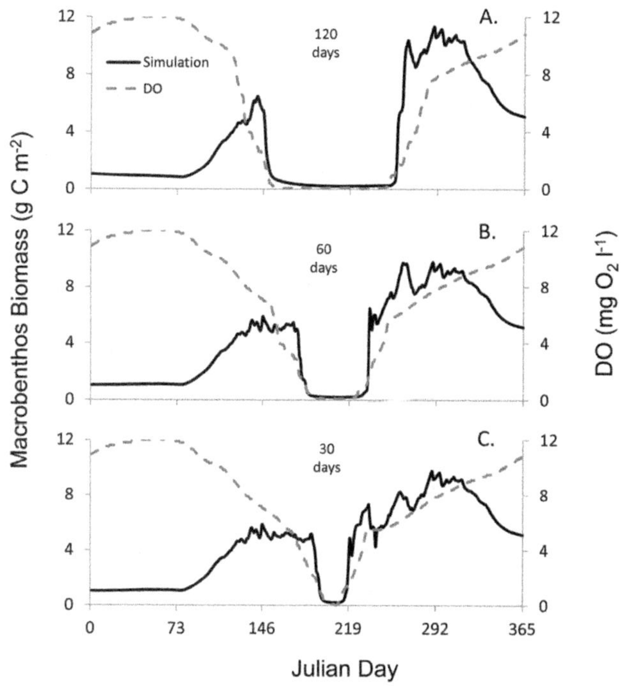

Figure 6. Simulations of hypoxic duration and macrobenthic response. Simulated macrobenthos biomass (*B*) under hypoxic durations of (A) 120, (B) 60, and (C) 30 days; S1, S2, and S3, respectively. Left y-axis: the gray solid line represents modeled macrobenthos biomass and the black dots denote observed macrobenthos biomass from site LE3.4 of the Chesapeake Bay Benthic Monitoring Program in 1992. Right y-axis: the grey dashed line indicates the dissolved oxygen concentration. *B* was influenced by the duration of hypoxia.

macrobenthic biomass accelerated to an immediate collapse in biomass that lasted the duration of the hypoxic event, with a temporal trend in macrobenthic biomass different from that modeled under normoxic conditions. Macrobenthic biomass began to respond and increase before DO concentrations in S1 became normoxic. A few days after hypoxia ended, macrobenthic biomass had increased to above pre-hypoxia levels. Similar trends were observed in S2 and S3 involving hypoxic durations of 60 and 30 days (Fig. 6B and C), with the length of the collapse the main difference in macrobenthic biomass. Simulations with shorter durations of hypoxia resulted in less time with macrobenthic biomass near 0 g C m^{-2}. Macrobenthic biomass of S2, with a mean of 4.1 g C m^{-2} (SD = 3.1), and S3, with a mean of 4.4 g C m^{-2} (SD = 2.8), were different than the macrobenthic biomass model under normoxia, with a mean of 3.5 g C m^{-2} (SD = 1.5).

In simulation S4, with intermittent hypoxia, macrobenthic biomass decreased at the onset of hypoxia and remained near zero for the duration of hypoxia (Fig. 7). During the 14-day intervals when DO concentration was normoxic, macrobenthic biomass began to increase but decreased back near zero with the onset of hypoxia. Macrobenthic biomass fluctuated through this pattern throughout the hypoxic simulation. A few days after hypoxia abated permanently, macrobenthic biomass began to increase to biomass levels greater than those observed pre-hypoxia.

Modeled phyto- and zooplankton biomass responded oppositely during hypoxic simulations (Fig. 8). As DO concentrations began to decline, phytoplankton biomass initially decreased, however, with the onset of hypoxia, phytoplankton biomass increased. The length of increased phytoplankton biomass was dependent on the duration of hypoxia, with a longer duration of hypoxia resulting in higher overall phytoplankton biomass, and to some extent an even greater magnitude of phytoplankton biomass. Hypoxia had the reverse effect on zooplankton biomass. As DO concentration decreased to hypoxic levels, zooplankton biomass initially increased and then declined to near 0 g C m^{-3}. The length of time that zooplankton biomass stayed near 0 g C m^{-3} was dependent on the duration of hypoxia; lengthy durations of hypoxia coincided with longer durations of reduced zooplankton biomass. Modeled zooplankton biomass did not respond to increased DO concentration until days to weeks after hypoxia ended.

The effect of hypoxic severity on macrobenthic biomass was tested by adjusting DO concentration between 2.0 and 0.0 mg O_2 l^{-1} in increments of 0.5 mg O_2 l^{-1}. Macrobenthic biomass was not substantially different between simulations when compared during the full year (Table 4). The mean biomass of these four simulations was approximately equal, with large standard deviations. However, these simulations had similar macrobenthic biomass during normoxia, as no parameters were changed; macrobenthic biomass did not differ between simulations until DO concentrations became hypoxic. Therefore, simulations S5 to S9 were analyzed starting at the onset of hypoxia on day 178 through the end of hypoxia on day 237 (Table 5). S5 had appreciably higher macrobenthic biomass than S6 to S9. Mean biomass in S5, which depicted 60 days of hypoxia at 2.0 mg O_2 l^{-1}, was >5 times higher than simulations that modeled hypoxia at DO concentrations of 0.0–1.0 mg O_2 l^{-1}, and 3 times higher than S6. In S6 when the DO was 1.5 mg O_2 l^{-1} macrobenthic biomass was lower than S5 and >2 times higher than S7 to S9. There were negligible differences in biomass at DO concentrations of 1.0 mg O_2 l^{-1} and lower. The mean biomass in S7, S8, and S9 was approximately similar with large standard deviations.

Discussion

The sigmoid function ($\mathcal{Z}$) applied to this ecosystem model resulted in clear changes in simulated macrobenthic biomass (*B*) during various hypoxic scenarios. In our model, hypoxic duration resulted in prolonged reductions of macrobenthic biomass relative to the length of hypoxia, with the model suggesting near defaunation (macrobenthic biomass equal to 0 g C m^{-2}) during the 120, 60, and 30 day hypoxic scenarios (simulations 1, 2, and 3, Fig. 6) at a DO concentration of 0.5 mg O_2 l^{-1}. This is in good agreement with laboratory and field studies that show duration and severity of hypoxia to impact benthic communities (see [11,26]). Periods of prolonged hypoxia have been observed previously in Chesapeake Bay and elsewhere [8,50]. The deep trough of the mainstem Bay experiences sustained seasonal hypoxia year after year [15]. While some species have documented resistance to hypoxia [7], during extended periods of hypoxic exposure (~40 days) even the most tolerant of species experience total mortality [51].

After DO levels in our model returned to normoxia, macrobenthic biomass recovered greater than pre-hypoxic levels. This was unexpected as conditions in Chesapeake Bay that stimulate ecological production in the pre-hypoxic spring differ in the post-hypoxic fall [8]. Increased nutrient run-off from the spring freshet promotes plankton production; the particulate organic matter from these blooms eventually settles to the bottom promoting benthic growth [52]. Large plankton blooms seen in the spring are noticeably absent in the fall, and with less primary production one would expect the rate of recovery of macrobenthic biomass to be less in the fall than in the spring. However, data from the

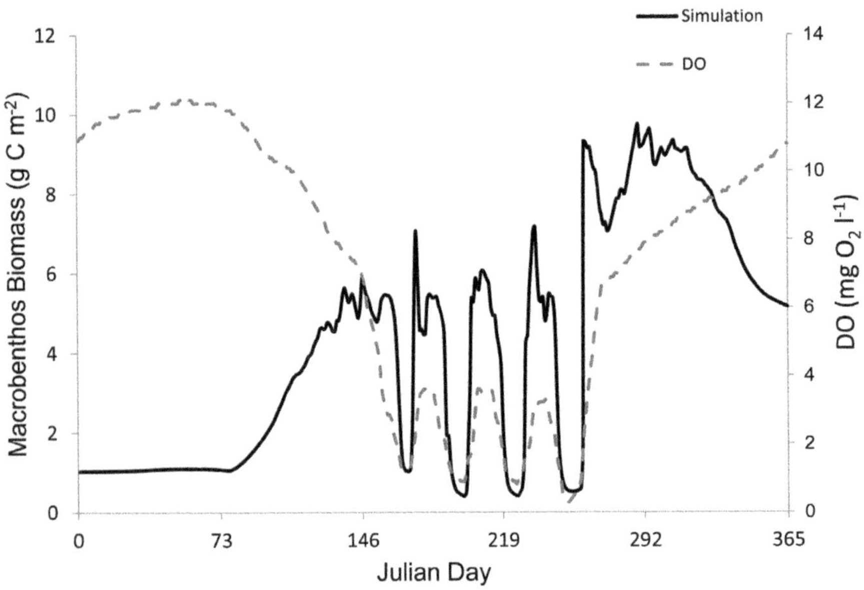

Figure 7. Simulations of intermittent hypoxic duration and macrobenthic response. Simulated macrobenthos biomass (*B*) under intermittent hypoxia (14 days hypoxic and 14 days normoxic) representing effect of neap/spring tide cycle. Left y-axis: the gray solid line represents modeled macrobenthos biomass and the black dots denote observed macrobenthos biomass from site LE3.4 of the Chesapeake Bay Benthic Monitoring Program in 1992. Right y-axis: the grey dashed line indicates the dissolved oxygen concentration on the right y-axis. *B* responded to modeled intermittent hypoxia.

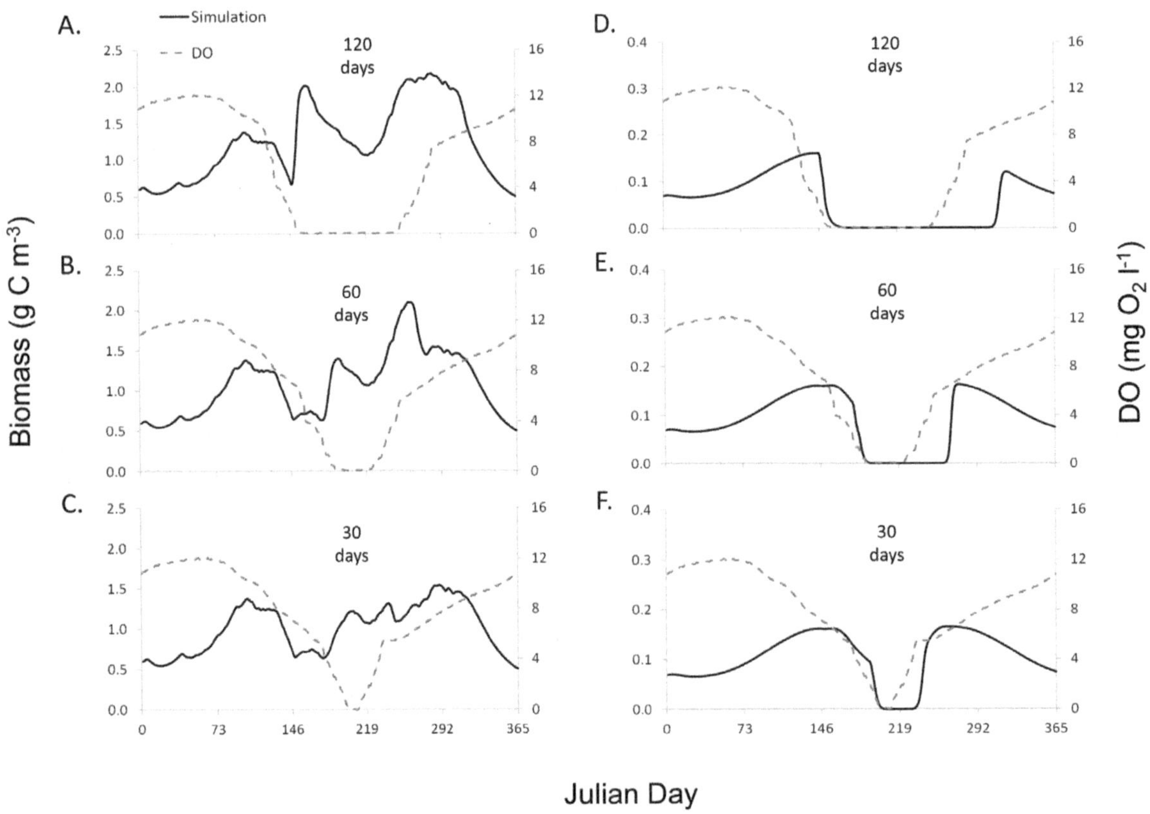

Figure 8. Simulations of hypoxic duration and phyto- and zooplankton response. Simulated phytoplankton biomass under hypoxic durations of (A) 120, (B) 60, and (C) 30 day, and zooplankton biomass under hypoxic durations of (D) 120, (E) 60, and (F) 30 days. Left y-axis: the gray solid line represents modeled biomass and the black dots denote mean observed biomass from site LE3.6 of the Chesapeake Bay Water Quality Monitoring Program from 1985–2001. Right y-axis: the grey dashed line indicates the dissolved oxygen concentration on the right y-axis. Modeled phytoplankton and zooplankton biomass responded inversely to hypoxic simulations.

Table 4. Response of macrobenthic biomass to hypoxic severity (365 d).

Simulation	DO (mg O_2 l^{-1})	Julian Day	Mean Biomass (g)
S5	2.0	1–365	3.20 (1.84)
S6	1.5	1–365	3.04 (2.07)
S7	1.0	1–365	2.88 (2.07)
S8	0.5	1–365	2.88 (2.07)
S9	0.0	1–365	2.88 (2.07)

Comparison of macrobenthos biomass to hypoxic severity over a full year. Macrobenthos biomass was not measurably different between simulations over a full year.

Table 5. Response of macrobenthic biomass to hypoxic severity (60 days).

Simulation	DO (mg O_2 l^{-1})	Julian Day	Mean Biomass (g)
S5	2.0	178–237	4.07 (0.46)
S6	1.5	178–237	1.77 (0.69)
S7	1.0	178–237	0.76 (1.15)
S8	0.5	178–237	0.62 (1.15)
S9	0.0	178–237	0.62 (1.15)

Comparison of macrobenthos biomass to hypoxic severity over a partial year, covering the 60 day time-frame of simulated hypoxia. Macrobenthos biomass was measurably variable during the 60 day assessment. Mean biomass is shown with ±1 SD in parentheses.

continuously monitored sites used to derive *Z'* indicated macrobenthic production can increase to pre-hypoxic levels a few weeks post hypoxia, suggesting the macrobenthic biomass increases we observed post-hypoxia may be reasonable [36]. A decrease in predation pressure may also have contributed to the observed increase in macrobenthic biomass. In an ecological context, decreases in temperature during the fall and winter months post-hypoxia lower metabolic demand, and thus the need of predators to obtain food [53]. In our model this was accounted for through Arrhenius temperature dependencies in the predation formulation, which is a simple but accurate formula for the temperature dependence of reaction rates [54]. This however, suggests that during summer hypoxia, the loss of macrobenthic biomass and its impacts to higher consumers such as epibenthic predators and demersal fish is not compensated for by the increase in macrobenthic biomass post-hypoxia due to a change in metabolic demand and activity.

Modeled intermittent hypoxia (S4, Fig. 7) resulted in reduced macrobenthic biomass during hypoxic periods but some recovery during normoxia. This cycling continued throughout the intermittent series of hypoxic events. Given the severity at which the DO concentration was set for this simulation (0.5 mg O_2 l^{-1}), it is not surprising that macrobenthic biomass decreased to the observed level. Neubauer [55] found *in situ* decreases in instantaneous macrobenthic production coincided with a hypoxic event, however observed no cyclical pattern between macrobenthic production and hypoxia. Mean DO concentration during Neubauer's [55] hypoxic event were ≈2.1–2.5 mg O_2 l^{-1}, which would be considered mild hypoxia for macrobenthos [56]. Further, Neubauer [55] experienced large recruitment events for some macrobenthic species during hypoxia at the time of his study. In contrast, a cyclical pattern between hypoxia and macrobenthic production was observed in data from the continuously monitored sites used to derive *Z'*. DO concentrations at our *in situ* sites were as low as ~0.02 mg O_2 l^{-1} on two separate occasions, and the similarities in hypoxic severity between our model and our *in situ* observations may account for analogous results [36].

As with simulations S1, S2, and S3, the growth of macrobenthic biomass during simulation S4 was very rapid with the return of normoxia, an observation that was also documented *in situ* [36]. A number of the dominant macrobenthic species in the lower Rappahannock spawn and recruit throughout the summer [57–63]. The larvae of these species are found distributed throughout the water column, and this pattern does not change in response to low DO [64]. When favorable conditions return, the planktonic larvae of these benthic species are available to settle in an organic-rich environment with few competitors, which may explain why quick benthic recruitment post-hypoxia is observed. Modeled macrobenthic biomass also began recovering ~2–3 days before hypoxia abated, which represents the time frame that DO was increasing from 0.5 to 2.0 mg O_2 l^{-1}, indicating a sensitivity to hypoxic severity. Macrobenthos have been previously documented to recruit before the DO concentration is above 2.0 mg O_2 l^{-1}, although this low level of DO does delay settlement of some hypoxia sensitive taxa [64]. In our intermittent simulation, once macrobenthic biomass reached a level where it appeared sustainable, hypoxia returned and biomass was reduced to near zero again.

Hypoxic severity had a measureable impact on macrobenthic biomass with higher biomass in simulations with less severe hypoxia. Studies have shown the severity of hypoxia to affect the response of benthic communities; the more severe the hypoxia, the greater the impact on the benthos, directly and indirectly [7]. Directly, benthic species vary in their tolerances to low DO concentrations [11], and as the severity of hypoxia increases towards anoxia, sensitive species die-off, usually due to asphyxiation, decreasing the diversity of the affected area and overall biomass [7]. Additionally, the toxic compound hydrogen sulfide is present in severely hypoxic sediments, and has been suggested as a mechanism contributing to macrobenthic mortality during hypoxia [65]. Indirectly, DO concentrations can positively and negatively affect benthic predation; Nestlerode and Diaz [66] showed that benthos may actually have a refuge from predation under mild hypoxic conditions, and Brante and Hughes [67] demonstrated that hypoxia reduced *Carcinus maenas* predation on mussels. If macrobenthos are able to avoid mortality via asphyxiation, such actions during prolonged hypoxic events could also indirectly lead to starvation [7].

During model simulations, there was no increase in macrobenthic biomass as DO concentrations declined towards hypoxia. In our model β:

$$\beta = \beta' \frac{DO}{DO + K_{DO}} \tag{8}$$

in which: β' = the predation rate before considering hypoxic effects K_{DO} = predation DO half-saturation.

accounts for the predation rate on macrobenthos and denotes predation rate as a function of temperature and DO concentration, where predation increases with increasing DO. During mild hypoxia predators may not effectively prey upon benthos, and hypoxia tolerant benthos would survive and maintain their biomass. The model excludes the effect of increasing predation during hypoxia, an artifact of the parent model. Seitz et al. [68]

and Long and Seitz [21] showed that epibenthic predators and demersal fish can at times capitalize on stressed benthos during mild hypoxic events. As oxygen concentrations become lethal, stressed macrobenthos extend their appendages and bodies out of the sediment in an attempt to escape severely hypoxic conditions below the sediment-water interface [56]. Opportunistic mobile predators have been shown to re-enter hypoxic areas and prey on exposed macrobenthos during mild hypoxia [69,70].

Zooplankton biomass (*M*) was negatively impacted by hypoxia directly, causing zooplankton biomass to be drastically reduced. Marcus et al. [71] considered the effect of reduced DO concentration on the survival and population dynamics of zooplankton, demonstrating the deleterious effect hypoxia has on zooplankton population and community dynamics. In our model, phytoplankton biomass was sensitive to predation by zooplankton, with a 20% increase and 20% decrease in algal predation resulting in an 18% difference in phytoplankton biomass (Table 3). As a result, phytoplankton biomass (*P*) in our model was indirectly positively influenced by the onset of hypoxia, due to the release of phytoplankton from zooplankton grazing pressure.

Conclusions

Macrobenthos data from the lower Rappahannock River were used to derive Z', a sigmoid relationship, to model the effect of DO concentration on macrobenthic biomass (*B*). Z' was then used to empirically constrain a formulation for hypoxia-induced mortality in a biomass-based ecosystem model and used to assess the impact of hypoxia on macrobenthic biomass, while including the important biological interactions that occur through benthic-pelagic coupling. Z' is a useful tool in that it can be applied to existing models to simulate the impact of hypoxia on the macrobenthos, and the methods used to derive Z' can be applied to other systems to develop site specific and species specific parameterizations of Z'. Further, in terms of application, the simplicity of our model makes it easy to implement, interpret, and update, while at the same time the approach developed here can be readily incorporated into more complex models.

From our modeling efforts we found that the duration and severity of hypoxia negatively affected macrobenthic biomass; longer durations and greater hypoxic severity resulted in less biomass. Further, our model suggests that post hypoxia, macrobenthic biomass can return to pre-hypoxic levels, which implies a level of resiliency in the macrobenthic community to hypoxia. However, it is important to remember that the loss of macrobenthic biomass occurs at a critical time when energy demands of epibenthos and demersal fish, predators of macrobenthic organisms, are at their highest. It is unknown if the rebound in macrobenthic biomass observed in our model post-hypoxia would be enough to compensate for the loss in biomass during hypoxia.

The ecological importance of macrobenthos to estuarine systems underlies the significance in understanding processes that positively and negatively impact this group. The ability to accurately model the influence of low DO and strength of interactions between ecosystem components will improve our understanding of the impacts of hypoxia and provide a holistic view of a major anthropogenic stressor on ecosystem functioning.

Acknowledgments

We thank the staff at Old Dominion University for data from the lower Chesapeake Bay Program Benthic Monitoring Program.

Author Contributions

Conceived and designed the experiments: SKS. Performed the experiments: SKS. Analyzed the data: SKS MJB. Contributed reagents/materials/analysis tools: SKS MJB RJD. Wrote the paper: SKS MJB RJD.

References

1. Diaz RJ, Schaffner LC (1990) The functional role of estuarine benthos. In: Haire, M, Krome, E.C., editors. Perspectives on the Chesapeake Bay: advances in estuarine sciences. Virginia: Chesapeake Research Consortium; 25–56.
2. Day JW, Hall CAS, Kemp WM, Yanez-Arancibia A (1989) Estuarine ecology. New York: John Wiley & Sons. 558 p.
3. Rhoads DC, Boyer LF (1982) Effects of marine benthos on physical properties of sediments: a successional perspective. In: McCall PL, Tevesz MJS, editors. Animal-Sediment Relations. Plenum Press, New York. 3–51.
4. Meysman FJR, Middelburg JJ, Heip CHR (2006) Bioturbation: a fresh look at Darwin's last idea. Trends Ecol Evol 21: 688–695.
5. Roden EE, Wetzel RG (1996) Organic carbon oxidation and suppression of methane production by microbial Fe(III) oxide reduction in vegetated and unvegetated freshwater wetland sediments. Limnol Oceanogr 41: 1733–1748.
6. Nilsen M, Pedersen T, Nilssen EM (2006) Macrobenthic biomass, productivity (P/B) and production in a high-latitude ecosystem, North Norway. Mar Ecol Prog Ser 321: 67–77.
7. Diaz RJ, Rosenberg R (1995) Marine benthic hypoxia: a review of its ecological effects and the behavioural responses of benthic macrofauna. Oceanogr Mar Biol 33: 245–303.
8. Kemp WM, Boynton WR, Adolf JE, Boesch DF, Boicourt WC, et al. (2005) Eutrophication of Chesapeake Bay: historical trends and ecological interactions. Mar Ecol Prog Ser 303: 1–29.
9. Vitousek PM, Mooney HA, Lubchenco J, Melillo JM (1997) Human domination of Earth's ecosystems. Science 277: 494–499.
10. Cloern JE (2001) Our evolving conceptual model of the coastal eutrophication problem. Mar Ecol Prog Ser 210: 223–253.
11. Vaquer-Sunyer R, Duarte CM (2008) Thresholds of hypoxia for marine biodiversity. Proc Natl Acad Sci 105: 15452–15457.
12. Nixon SW (1995) Coastal marine eutrophication: A definition, social causes, and future concerns. Ophelia 41: 199–219.
13. Newcombe CL, Horne WA, Shepherd BB (1939) Studies on the physics and chemistry of estuarine waters in Chesapeake Bay. J Mar Res 2: 87–116.
14. Sale JW, Skinner WW (1917) The vertical distribution of dissolved oxygen and the precipitation of salt water in certain tidal areas. Franklin Institute J 184: 837–848.
15. Hagy JD, Boynton WR, Keefe CW, Wood KV (2004) Hypoxia in Chesapeake Bay, 1950–2001: Long-term change in relation to nutrient loading and river flow. Estuaries 37: 634–658.
16. Dauer DM, Rodi AJ, Ranasinghe JA (1992) Effects of low dissolved oxygen events on the macrobenthos of the lower Chesapeake Bay. Estuaries 15: 384–391.
17. Diaz RJ, Neubauer RJ, Schaffner LC, Pihl L, Baden SP (1992) Continuous monitoring of dissolved oxygen in an estuary experiencing periodic hypoxia and the effect of hypoxia on macrobenthos and fish. Sci Total Environ Supplement 1: 1055–1068.
18. Jørgensen BB (1980) Seasonal oxygen depletion in the bottom waters of a Danish fjord and its effect on the benthic community. Oikos 34: 68–76.
19. Llansó RJ (1992) Effects of hypoxia on estuarine benthos: the lower Rappahannock River (Chesapeake Bay), a case study. Estuar Coast Shelf Sci 35: 491–515.
20. Tallqvist M (2001) Burrowing behaviour of the Baltic clam *Macoma balthica*: effects of sediment type, hypoxia, and predator presence. Mar Ecol Prog Ser 212: 183–191.
21. Long WC, Seitz RD (2008) Trophic interactions under stress: hypoxia enhances foraging in an estuarine food web. Mar Ecol Prog Ser 362: 59–68.
22. Riedel B, Zuschin M, Haselmair A, Stachowitsch M (2008) Oxygen depletion under glass: Behavioural responses of benthic macrofauna to induced anoxia in the Northern Adriatic. J Exp Mar Biol Ecol 367: 17–27.
23. Rosenberg R, Agrenius S, Hellman B, Nilsson HC, Norling K (2002) Recovery of marine benthic habitats and fauna in a Swedish fjord following improved oxygen conditions. Mar Ecol Prog Ser 234: 43–53.
24. Seitz RD, Dauer DM, Llansó RM, Long WC (2009) Broad-scale effects of hypoxia on benthic community structure in Chesapeake Bay, USA. J Exp Mar Biol Ecol 381: 4–12.
25. van Colen C, Montserrat F, Vincx M, Herman PMJ, Ysebaert T, et al. (2010) Long-term divergent tidal flat benthic community recovery following hypoxia-induced mortality. Mar Poll Bull 60: 178–186.
26. Diaz RJ, Rosenberg R (2008) Spreading dead zones and consequences for marine ecosystems. Science 321: 926–929.

27. Luo J, Hartman KJ, Brandt SB, Cerco CF, Rippetoe TH (2001) A spatially-explicit approach for estimating carrying capacity: an application for the Atlantic menhaden (*Brevoortia tyrannus*) in Chesapeake Bay. Estuaries 24: 545–556.
28. Cerco CF, Noel MR (2004) The 2002 Chesapeake Bay eutrophication model. Annapolis MD: E.P.A. 903–R–04–004, Chesapeake Bay Program, U.S. Environmental Protection Agency, Annapolis, MD, 349 p.
29. Cerco CF, Noel MR (2007) Can oyster restoration reverse cultural eutrophication in Chesapeake Bay? Estuar Coast 30: 331–343.
30. Fulford RS, Breitburg DL, Newell RIE, Kemp WM, Luckenbach M (2007) Effects of oyster population restoration strategies on phytoplankton biomass in Chesapeake Bay: a flexible modeling approach. Mar Ecol Prog Ser 336: 43–61.
31. Bonnet MP, Wessen K (2000) ELMO, a 3-D water quality model for nutrients and chlorophyll: first application on a lacustrine ecosystem. Ecol Model 141: 19–33.
32. Brush MJ, Nixon SW (2010) Modeling the role of macroalgae in a shallow sub-estuary of Narragansett Bay, RI (USA). Ecol Model 221: 1065–1079.
33. Kuo AY, Neilson BJ (1987) Hypoxia and salinity in Virginia estuaries. Estuaries 10: 277–283.
34. Kuo AY, Park K, Moustafa MZ (1991) Spatial and temporal variabilities of hypoxia in the Rappahannock River, Virginia. Estuaries 14: 113–121.
35. Park K, Kuo AY, Neilson J (1996) A numerical model study of hypoxia in the tidal Rappahannock River of Chesapeake Bay. Est Coast Shelf S 42: 563–581.
36. Sturdivant SK, Seitz RD, Diaz RJ (2013) Effects of seasonal hypoxia on macrobenthic production and function in the Rappahannock River, Virginia, USA. Mar Eol Prog Ser 490: 53–68.
37. Dauer DM, Llansó RJ (2003) Spatial scales and probability based sampling in determining levels of benthic community degradation in the Chesapeake Bay. Environ Monitor Assess 81: 175–186.
38. HydroQual (2000) Development of a suspension feeding and deposit feeding benthos model for Chesapeake Bay. Final report, project no. USCE0410, U.S. Army Engineer Research and Development Center, Vicksburg MS, 117 p.
39. Haven DS, Morales-Alamo R (1972) Biodeposition as a factor in sedimentation of fine suspended solids in estuaries. Geol S Am Mem 133: 121–130.
40. Pryor WA (1975) Biogenic sedimentation and alteration of argillaceous sediments in shallow marine environments. Geol Soc Am Bull 86: 1244–1254.
41. Frithsen JB, Doering PH (1986) Active enhancement of particle removal from the water column by tentaculate benthic polychaetes. Ophelia 25: 169–182.
42. Garber JH (1987) Benthic-pelagic coupling in the Chesapeake Bay. In Perspectives in the Chesapeake Bay: advances in estuarine sciences. Eds Lynch MP, Krome EC, Gloucester Point, Virginia: Chesapeake Research Consortium.
43. Schaffner LC, Friedrichs CT, Dauer DM (2002) Review of the benthic processes model with recommendations for future modeling efforts. Benthic Process Model Review Team, Environmental Protection Agency Chesapeake Bay Program, Annapolis MD, 1–38.
44. Chesapeake Bay Program, Chesapeake Bay Benthic Monitoring Program (2010) Versar Inc. May-Aug 2010. Available: http://www.baybenthos.versar.com/.
45. Wetzel RL, Neckles HA (1986) A model of *Zostera marina* L. photosynthesis and growth: simulated effects of selected physical-chemical variables and biological interactions. Aquat Bot 26: 307–323.
46. Kremer JN, Nixon SW (1978) A Coastal Marine Ecosystem. Springer-Verlag, New York, Berlin, Heidelberg.
47. Steinberg DK, Pilskaln CH, Silver MW (1998) Contribution of zooplankton associated with detritus to sediment trap "swimmer" carbon in Monterey Bay, CA. Mar Ecol Prog Ser 164, 157–166.
48. Ragone-Calvo LM, Wetzel RL, Burreson EM (2001). Development and verification of a model for the population dynamics of the protistan parasite, Perkinsus marinus, within its host, the eastern oyster, Crassostrea virginica, in Chesapeake Bay. J Shellfish Res 20: 231–241.
49. Cerco CF, Cole T (1993) Three-dimensional eutrophication model of Chesapeake Bay. J Environ Eng 119: 1006–1025.
50. Officer CB, Biggs RB, Taft JL, Cronin LE, Tyler MA, et al. (1984) Chesapeake Bay anoxia: origin, development, and significance. Science 223: 22–27.
51. Rosenberg R, Hellman B, Johansson B (1991) Hypoxic tolerance of marine benthic fauna. Mar Ecol Prog Ser 79: 127–131.
52. Rabalais NN (2004) Eutrophication. In The Global Coastal Ocean Multiscale Interdisciplinary Processes, The Sea. Vol 13, eds Robinson AR, McCarthy J, Rothschild BJ, 819–865. Boston, Massachusetts: Harvard University Press.
53. Brown JH, Gillooly JF, Allen AP, Savage VM, West GB (2004) Toward a metabolic theory of ecology. Ecology 85: 1771–1789.
54. Laidler JK (1987) Chemical kinetics, Third edition. Harper and Row, New York.
55. Neubauer JR (1993) The relationship between dominant macrobenthos and cyclical hypoxia in the lower York River. Dissertation. College of William & Mary, School of Marine Science.
56. Tyson RV, Pearson TH (1991) Modern and ancient continental shelf anoxia: an overview. Geol Soc Spec Pub 55: 1–24.
57. Warren LM (1976). A population study of the polychaete *Capitella capitata* at plymouth. Mar Biol 38: 209–216.
58. Mayfield SM (1988) Aspects of the life history and reproductive biology of the worm *Paraprionospio pinnata*. Thesis, Texas A&M University, College Station, Texas. 102 p.
59. Hardege JD, Bartels-Hardege HD, Zeeck E, Grimm FT (1990) Induction of swarming in *Nereis succinea*. Mar Biol 104: 291–295.
60. Levin LA, Huggett DV (1990) Implication of alternative reproductive modes for seasonality and demography in an estuarine polychaete. Ecology 71: 2191–2208.
61. Fong PP (1991) The effects of salinity, temperature, and photoperiod on epitokal metamorphosis in *Neanthes succinea* (Frey et Leuckart) from San Francisco Bay. J Exp Mar Biol Ecol 149: 177–190.
62. Levin LA, Bridges TS (1994) Control and consequences of alternative development modes in a poecilogonous polychaete. Am Zool 34: 323–332.
63. Tiffany MA, Swan BK, Watts JM, Hurlbert SH (2002) Metazooplankton dynamics in the Salton Sea, California, 1997–1999. Hydrobiologia 473: 103–120.
64. Powers SP, Harper Jr DE, Rabalais NN (2001) Effect of hypoxia/anoxia on the supply and settlement of benthic invertebrate larvae. Coast Estuar Stud 58: 185–210.
65. Vaquer-Sunyer R, Duarte CM (2010) Sulfide exposure accelerates hypoxia-driven mortality. Limnol Oceanogr 55: 1075–1082.
66. Nestlerode JA, Diaz RJ (1998) Effects of periodic environmental hypoxia on predation of a tethered polychaete, *Glycera americana*: implication for trophic dynamics. Mar Ecol Prog Ser 172: 185–195.
67. Brante A, Hughes RN (2001) Effect of hypoxia on the prey-handling behavior of *Carcinus maenas* feeding on *Mytilus edulis*. Mar Ecol Prog Ser 209: 301–305.
68. Seitz RD, Marshall Jr LS, Hines AH, Clark KL (2003) Effects of hypoxia on predator-prey dynamics of the blue crab *Callinectes sapidus* and the Baltic clam *Macoma balthica* in Chesapeake Bay. Mar Ecol Prog Ser 257: 179–188.
69. Pihl L, Baden SP, Diaz RJ (1991) Effects of periodic hypoxia on distribution of demersal fish and crustaceans. Mar Biol 108: 349–360.
70. De Robertis A, Eiane MK, Rau G (2001) Eat and run: anoxic feeding and subsequent aerobic recovery by *Orchomene obtusus* in Saanich Inlet, British Columbia, Canada. Mar Ecol Prog Ser 219: 221–227.
71. Marcus NH, Richmond C, Sedlacek C, Miller GA, Oppert C (2004) Impact of hypoxia on the survival, egg production and population dynamics of *Acartia tonsa* Dana. J Exp Mar Biol Ecol 301: 111–128.

7

Heterogeneous Occupancy and Density Estimates of the Pathogenic Fungus *Batrachochytrium dendrobatidis* in Waters of North America

Tara Chestnut[1,2]*, Chauncey Anderson[2], Radu Popa[3], Andrew R. Blaustein[4], Mary Voytek[5], Deanna H. Olson[6], Julie Kirshtein[7]

1 Oregon State University, Environmental Science Graduate Program, Corvallis, Oregon, United States of America, 2 US Geological Survey, Oregon Water Science Center, Portland, Oregon, United States of America, 3 Biological Sciences, University of Southern California, Los Angeles, California, United States of America, 4 Oregon State University, Department of Integrative Biology, Corvallis, Oregon, United States of America, 5 Astrobiology Program, National Aeronautics and Space Administration Headquarters, Washington DC, United States of America, 6 US Forest Service, Pacific Northwest Research Station, Corvallis, Oregon, United States of America, 7 US Geological Survey, National Research Program, Reston, Virginia, United States of America

Abstract

Biodiversity losses are occurring worldwide due to a combination of stressors. For example, by one estimate, 40% of amphibian species are vulnerable to extinction, and disease is one threat to amphibian populations. The emerging infectious disease chytridiomycosis, caused by the aquatic fungus *Batrachochytrium dendrobatidis* (*Bd*), is a contributor to amphibian declines worldwide. *Bd* research has focused on the dynamics of the pathogen in its amphibian hosts, with little emphasis on investigating the dynamics of free-living *Bd*. Therefore, we investigated patterns of *Bd* occupancy and density in amphibian habitats using occupancy models, powerful tools for estimating site occupancy and detection probability. Occupancy models have been used to investigate diseases where the focus was on pathogen occurrence in the host. We applied occupancy models to investigate free-living *Bd* in North American surface waters to determine *Bd* seasonality, relationships between *Bd* site occupancy and habitat attributes, and probability of detection from water samples as a function of the number of samples, sample volume, and water quality. We also report on the temporal patterns of *Bd* density from a 4-year case study of a *Bd*-positive wetland. We provide evidence that *Bd* occurs in the environment year-round. *Bd* exhibited temporal and spatial heterogeneity in density, but did not exhibit seasonality in occupancy. *Bd* was detected in all months, typically at less than 100 zoospores L^{-1}. The highest density observed was ~3 million zoospores L^{-1}. We detected *Bd* in 47% of sites sampled, but estimated that *Bd* occupied 61% of sites, highlighting the importance of accounting for imperfect detection. When *Bd* was present, there was a 95% chance of detecting it with four samples of 600 ml of water or five samples of 60 mL. Our findings provide important baseline information to advance the study of *Bd* disease ecology, and advance our understanding of amphibian exposure to free-living *Bd* in aquatic habitats over time.

Editor: Benedikt R. Schmidt, Universität Zurich, Switzerland

Funding: Support was provided by the National Science Foundation GK-12 Graduate STEM Fellows in K-12 Education Program (Award #0948041) to TC, US Forest Service - Pacific Northwest Research Station and US Geological Survey Amphibian Research and Monitoring Initiative (ARMI). The funders had no role in study design, data collection and analysis, decision to publish, or preparation of the manuscript.

Competing Interests: The authors have declared that no competing interests exist.

* Email: chestnut@usgs.gov

Introduction

Loss of biodiversity in terrestrial, freshwater, and marine systems is occurring at a global scale [1–5]. The causes of losses are often complex and include synergistic effects of natural and human-induced stressors, such as habitat loss and fragmentation [6,7], urbanization [8], invasive species [7,9,10], contaminants [11–13], global climate change [14–16], and emerging infectious diseases [17,18]. In the last 35 years, one estimate suggests that the amphibian extinction rate is at least 105 times higher than the expected rate [19], with 32.5 to 41% of amphibian species threatened [7,20]. In the USA, the number of sites occupied by amphibians is declining by an estimated 3.7% per year [20,21].

Among the many threats to amphibians, the role of disease in population declines has been recognized increasingly over the last two decades. Numerous amphibian diseases have been identified, with mass mortality events attributed to water molds (*Saprolegnia* spp.) [22–24], *Aeromonas* bacterial infections [25], iridoviruses [26–29], alveolate infections [30,31], and malformations caused by trematodes [32,33]. Chytridiomycosis, the emerging infectious disease caused by the amphibian chytrid fungus, *Batrachochytrium dendrobatidis* (*Bd*), is implicated as a causal agent in many recent global amphibian population declines and extinctions [34–36].

Of the over 3,000 fungal species described from aquatic habitats [37], chytrid fungi are the earliest of extant fungi to diverge phylogenetically, and now have a global distribution [38,39]. *Bd* is one of over 1,200 chytrid species described from freshwater, marine, and terrestrial systems occurring across temperate, tropical, and tundra environments [40–42]. Although chytrids function primarily as plant saprobes and parasites [37], some also parasitize animals [43,44]. *Bd* is one of only two chytrids known to infect vertebrate hosts [44,45].

To understand the pathology and conservation implications of *Bd*, a better understanding of its distribution, ecology, and the life history of *Bd* in the wild is needed. The life cycle of *Bd* includes forms that are free-living in the aquatic environment [46,47]. *Bd* has been detected by filtering water samples to capture free-living zoospores and zoosporangia, then performing a genetic analysis on the filtered particulates [48–50]. *Bd* has not been reliably detected from sediments [48]. Laboratory experiments have demonstrated that *Bd* has survived on sterilized moist sand for up to three months and remained infective in lake water for up to seven weeks [51]. *Bd* cultures can be maintained under laboratory conditions for several years (TC, ARB personal observation), which suggests that *Bd* can survive in the environment without a host as long as conditions are favorable. In laboratory settings, *Bd* growth and reproduction depended on temperature (4–25°C, ideal 17–23°C) and pH (4–10, 6–7 ideal) [52], and differences in both generation time and fecundity of *Bd* in response to different thermal regimes are observed in at least one *Bd* strain [53].

A greater understanding of *Bd* distribution and spread is of paramount importance to more fully understand its threat to amphibians [36]. Although it has an impact on the persistence of selected amphibian populations around the world [21,35,54], the full scope of the effects of chytridiomycosis on global amphibian population declines is not well understood. Most *Bd* research efforts have focused on *Bd* in amphibian hosts *per se*, with little focus on the ecology of free-living *Bd* outside of the amphibian host [55–57]. In amphibian hosts, *Bd* exhibits sensitivity to a number of environmental variables including temperature [36,58–61]and elevation [59,62], suggesting a tendency to associate high-elevation areas or regions with cool temperatures with increased risk for *Bd*-related declines and extinctions [63]. From a large global data set of amphibians sampled for *Bd* infections, it was found that *Bd* occurrences were inversely correlated with the temperature range at sites [36]. Seasonal patterns of amphibians infected with *Bd* have emerged in several studies [58,64–67], further implicating microclimatic associations and potential differences between strains [68]. In wild amphibian populations from temperate areas, *Bd* infections appear to follow predictable patterns, with the highest prevalence and intensity in the cooler spring months, and decreasing prevalence in the warmer summer and autumn months, sometimes to non-detectable limits [66,67]. We describe spatial and temporal patterns in occupancy and density of free-living *Bd* in aquatic habitats of the United States, outside of the amphibian host, to better understand the geographic distribution and ecology of this pathogen, and to further inform pathogen dynamics research.

Occupancy modeling is a powerful tool to estimate site occupancy and detection probability when detection is imperfect [69]. Occupancy models have been used to investigate disease ecology in amphibians, fishes, and birds, where the focus was on pathogen occurrence in the host [70–72]. We used an occupancy modeling approach to investigate free-living *Bd* in North American surface waters, from water samples collected across several sites in North America (Figure 1) to assess: 1) the seasonality, if any, of *Bd* detection; 2) the relationships between *Bd* site occupancy and habitat attributes including water quality (temperature, pH, specific conductance), geography (latitude, longitude, elevation), climate metrics (minimum/maximum/range in precipitation and temperature, dew point), and time (Julian day); and 3) the probability of *Bd* detection from water samples as a function of the number of samples taken, sample volume, and water quality (temperature, pH, specific conductance). We predicted that free-living *Bd* occupancy and density would follow the same trends as observed in *Bd*-infected amphibians, with the highest detections in the North American spring season and the lowest in the late summer/autumn [64,66,71]. Based on reported *Bd*-amphibian host relationships and laboratory studies of *Bd*, we expected to see corresponding temperature and pH associations in our analyses. We also predicted that detection probability would increase as the number of samples and volume of water per sample increased, and would increase as temperature and pH approached the ideal range for *Bd* growth [52]. Finally, we report on the temporal patterns of *Bd* density from a four-year case study of a *Bd*-positive wetland, and the relationship between water quality and *Bd* density.

Our findings address important baseline information to advance the study of *Bd* disease ecology in temperate-zone systems, and specifically to advance our understanding of the likelihood of amphibian exposure to free-living *Bd* in aquatic habitats over time. Our results are applicable to *Bd* risk assessment, sampling protocols, policy development, and regulatory decision-making processes.

Methods

Ethics Statement

Field studies involved collecting water samples. No animals were sampled as part of this study. Sampling did not involve endangered or protected species, although they may have occurred at or near the sites sampled. Permission to access National Park Service lands was obtained by Bill Commins, NPS Research Permit Coordinator, Washington, DC (bill_commins@nps.gov). Private lands were accessed with landowner permission; TC may be contacted for future permissions. Public right-of-ways did not require permission/permits to sample. Given the sensitive nature of some sites, GPS coordinates are available upon request.

Field Methods

All Sites. We collected *Bd* water samples from 41 amphibian breeding sites along a latitudinal gradient spanning the North American continent between 45 and 48 degrees North, and a longitudinal gradient focused on, but not exclusive to, the western United States west of 105 degrees West (Figure 1). Site selection occurred opportunistically at locations where on-going amphibian research was being conducted; hence our scope of inference is restricted to the sampled sites. Sites were located in temperate ecoregions in a variety of land-use settings ranging from urban/suburban to designated wilderness, and at elevations ranging from near sea level to ~3500 m—39% were low-elevation, 54% were mid-elevation, and 7% were high-elevation. Three replicate water samples were collected, separated by >50 m, during a single survey occasion that occurred between May and September, 2007 to 2010. Most sites (63%) were sampled in June and July. We measured water quality (i.e., temperature, pH, specific conductance) at 22 of 41 sites, herein WQ sites (Figure 1).

Case Study Site (Beaverton, Oregon). We also conducted a four-year case study in Beaverton, Oregon (45.3108, −122.4752, Datum: WGS84), which we monitored monthly between 2007 and 2011 to assess temporal variation in *Bd* occurrence and density. The site was a palustrine wetland in a suburban setting, dominated by reed canary grass (*Phalaris arundinacea*). It was approximately 1.2 hectares in area, bisected by a road, and connected to larger wetland complexes 6.4 hectares in area associated with Johnson

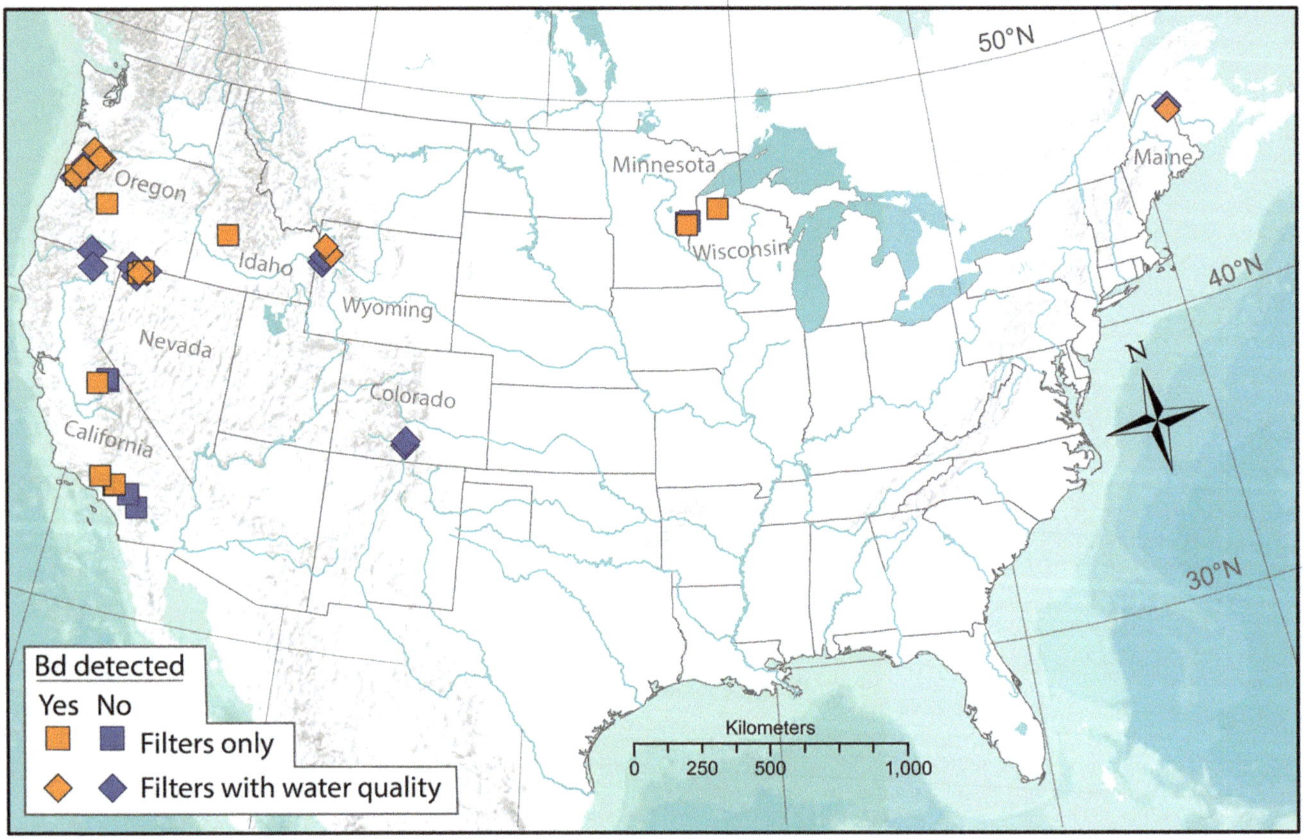

Figure 1. Sampling locations for *Batrachochytrium dendrobatidis* (*Bd*) in temperate North America with results of *Bd* detections from water samples.

Creek, both upstream and downstream of the site. Surrounding land use was historically agriculture (row crops and horse pasture), and is now primarily residential. In the 1990s the site was converted to a wetland mitigation site and has undergone recent efforts to reestablish a forested wetland.

Water samples were collected using the protocol described in [48], with modifications to improve *Bd* DNA recovery and sampling efficiency, which are described below. At each site, three spatial replicate water samples were collected for *Bd* testing, also at least 50 m apart in the pond or wetland. Each sample was collected from shallow water, between 5 and 20 cm below the water surface. A general overview of sample collection is as follows: using a sterile 60-mL syringe rinsed 3 times with native water, we filtered water drawn from below the surface through a Sterivex 0.22-μm capsule filter (luer-loc with male end) until the filter was nearly clogged. The volume was recorded (range: between 20 mL and 2.4 L, mean 350 mL) and the filter was flushed using a new sterile syringe with 50 mL of DNAase/RNAse-free 0.01 Mol Phosphate Buffered Saline (PBS) to remove excess dissolved organic carbon which can inhibit the PCR reaction. The filter was purged of water by removing it from the syringe, drawing air into the syringe, reattaching the filter and pushing air through the filter and repeating if necessary. The outflow end of the capsule filter was sealed with Hemato-Seal clay (we found that other types dried and cracked in the field, compromising the seal), then we injected the filter with 0.9 mL cell lysis buffer solution, and sealed it with a luer-loc cap. Each sample was placed in an individual Whirl-pak bag and kept cool. Upon returning from the field, samples were refrigerated until DNA extraction occurred, within three months of sample collection. Water quality was measured with a single YSI multiparameter data sonde, calibrated before each sampling event using standardized U.S. Geological Survey (USGS) protocols [73]. The YSI was cleaned between each use.

The method we used to collect *Bd* samples involves water filtration and molecular techniques but differs from what is usually termed environmental DNA (eDNA) sampling [74,75]. Typical eDNA methods sample the environment for DNA from parts of an organism (e.g., shed skin, tissue fragments), whereas the method we used is a standard method for characterizing microbial communities, where whole microbes are captured on the filter and dissolved DNA is passed through the filter. This ensured we were sampling only *Bd* that occupied the site at the time the sample was collected. However, it is possible that if we filtered recently shed amphibian skin infected with *Bd*, it may have been captured by our water sampling process.

Laboratory Methods

DNA was extracted from the cell lysis solution using the Gentra Puregene Tissue Kit [48]. We added proteinase K (0.1 mg mL^{-1}) directly to the filter by opening the luer-loc cap, injecting the solution and resealing the capsule. We placed each filter in a sterile 50-mL Falcon tube, and incubated it at 55°C for 60 min in a continuously rotating rack inside the incubator to ensure the filter was completely bathed in the lysis/proteinase K solution. Following this step, we drew the solution out of the filter with a

Table 1. Covariates hypothesized to relate to the probability of detection (p) or occurrence (Ψ) of *Batrachochytrium dendrobatidis* (*Bd*) in amphibian habitats, including data source and model fit.

Covariate abbreviation and description	Probability of occurrence (Ψ)	Detection probability (p)	Data source	Included in best fitting models
vol = volume of water filtered from an amphibian habitat to the nearest mL	-	X	Field measurement	-
temp = temperature in degrees Celsius to the nearest tenth at the location a sample was collected	-	X	Field measurement	-
ph = pH at the location a sample was collected	-	X	Field measurement	X
sp_cond = Specific conductance in microsiemens (uS), standardized to 25°C, at the location a sample was collected	-	X	Field Measurement	-
lat = latitude in decimal degrees, datum WSG84	X	-	field measurement, GIS verified	-
long = longitude in decimal degrees, datum WGS84	X	-	field measurement, GIS verified	-
elev = elevation in m to the nearest whole number	X	-	field measurement, GIS verified	X
precip = precipitation in mm in the 24 hours prior to sampling	X	-	PRISM data (Daly et al. 2008)	x
tmin = minimum temperature in degrees C in the 24 hours prior to sampling	X	-	PRISM data (Daly et al. 2008)	-
tmax = maximum temperature in degrees C in the 24 hours prior to sampling	X	-	PRISM data (Daly et al. 2008)	-
Trange = temperature range in degrees C in the 24 hours prior to sampling	X	-	PRISM data (Daly et al. 2008)	x
day = day of year	X	-	calculated value	-

In our exploratory analysis, we built models for all possible combinations of the a priori covariates and eliminated covariates that had low predictive power. X indicates the best fitting model, based on AIC value and weight; x indicates a covariate identified in a model with a delta AIC value of <2 but less support as indicated by AIC weight.

sterile syringe, and continued the extraction following the manufacturer's protocol. Quantitative PCR assays were conducted using the Qiagen QuantiTect SYBR Green PCR kit following standard protocols [48]. To quantify the *Bd* zoospore equivalents, we calculated a conversion factor from a dilution series of known zoospore numbers (i.e., 169), which is within the range of 60 to 220 for rRNA fungal gene locus copy numbers [76]. *Bd* genomic equivalents are therefore 169 times greater than the numbers we report. A site was considered *Bd*-positive if at least one of three replicate filters from a site returned a positive qPCR result.

Statistical Methods

Occupancy models require multiple observations per site. We treated each water sample as an independent observation, for a total of 3 observations per site. We used a single-season occupancy model to simultaneously estimate the site-level probability of *Bd* occurrence in amphibian habitats (Ψ) while accounting for the observation-level probability of detecting *Bd* in amphibian habitats when it is present (p), and relating both parameters to our hypothesized covariates. We investigated eight covariates that we hypothesized were related to the probability of occurrence (Ψ) and five covariates we hypothesized were related to detection probability (p) in amphibian habitats, and built models for all possible combinations of the a priori covariates (Table 1). Temporal (day of year) and geographic (latitude, longitude, and elevation) predictors of occupancy were recorded at the time of sampling. Other environmental predictors of occupancy (precipitation, minimum temperature, maximum temperature, temperature range) were derived from PRISM (Parameter-elevation Regressions on Independent Slopes Model) climate mapping

Table 2. Model selection statistics for a priori models relating to occupancy of *Batrachochytrium dendrobatidis* (*Bd*) from 41 amphibian habitats with environmental and geographic covariates in temperate North America, where detection probability is constant p(.), as differences in detection probability were negligible when the pH range was considered.

Model	AIC	deltaAIC	AIC wgt	Model likelihood	# of parameters	.-2*LogLike
psi(elev),p(.)	139.89	0	0.2284	1	3	133.89
psi(elev+precip),p(.)	141.3	1.41	0.1128	0.4941	4	133.3
psi(elev+Trange),p(.)	141.61	1.72	0.0966	0.4232	4	133.61
psi(elev+lat),p(.)	141.82	1.93	0.087	0.381	4	133.82

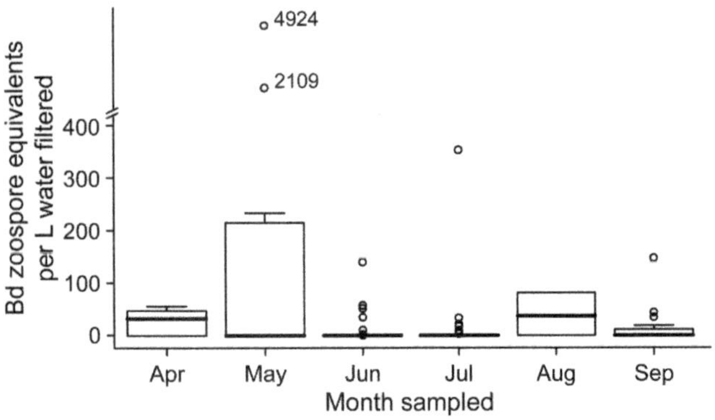

Figure 2. *Batrachochytrium dendrobatidis* (*Bd*) zoospore density from 41 amphibian survey sites measured between April and September, 2007 to 2010.

system data at a spatial resolution of 1 km^2 [77]. Two covariates of p were recorded from all sites (p as a function of volume, and p as constant across surveys), and three covariates of water quality (temperature, pH, specific conductance) were recorded from the subset of 22 WQ sites. Continuous covariates were standardized to mean = 0 and standard deviation = 1. We evaluated the degree of support for each competing model based on Akaike's Information Criteria (AIC) and the resulting Akaike's weights. We considered that models within 2 delta AIC received similar degree of support. Using the maximum likelihood estimates from the top-performing model for all 41 sites, we also estimated the detection probability given the number of samples collected and the volume of the water filtered per sample. In the subset of WQ sites, we further investigated three water quality covariates (temperature, pH, and specific conductance) that potentially related to the probability of *Bd* detection, using the top-performing detection probability models (Table 2). In the Oregon case study, we examined associations between *Bd* detection and four water quality metrics (temperature, pH, specific conductance, and turbidity) using Pearson's correlation coefficient. To include this site in our occupancy analysis, we randomly selected one sampling event that occurred within the sampling period of the other sites, (between May and September, and between the years 2007 and 2010). Data were summarized using the base packages available in R 2.14.0 [78]. To examine the relationships between *Bd* occupancy of amphibian habitats and our covariates of interest, we developed models using the software Presence 4.4 (http://www.mbr-pwrc.usgs.gov/software/presence.html). *Bd* data are archived in the USGS Western Ecological Research Center Multitaxa Database and water quality data are archived in the USGS National Water Information System.

Results

Bd exhibited temporal and spatial heterogeneity in detection and density in amphibian habitats. In both our landscape-scale study of North America and the Beaverton, Oregon case study we detected *Bd* in all months surveyed, but not in all samples. *Bd* density was highest in spring (Figures 2 & 3), and at two Nevada sites sampled in May, *Bd* densities were more than 50 times the densities that we observed at other sites (2,109 and 4,924 zoospores L^{-1}). At all sites, mean zoospore densities were highest in April and August (Figure 2, overall mean: 40 zoospores L^{-1}; median: 15 zoospores L^{-1}; range: 0–4,924 zoospores L^{-1}.). In some cases, we observed marked heterogeneity between the individual samples, e.g., at the two Nevada sites, one sample recovered 2,109 zoospores L^{-1} from the wetland and the other two filters recovered an order of magnitude less, 192 and 227 zoospores L^{-1}. At the site where 4,924 zoospores L^{-1} were recovered, *Bd* was not detected from the other two filter samples. Our Oregon case study had a higher mean density than observed in our landscape-scale study (i.e., 100 zoospores L^{-1}) and the highest *Bd* density observed from a field collected sample, 3,159,289 zoospores L^{-1} (Figure 3). There was a 95% chance of detecting *Bd* at a site when it was present with 4 samples of 600 mL of water or 5 samples of 60 mL of water (Figure 4), estimated by calculating 1–(1-p)^#samples.

All Sites

We detected *Bd* in 19 of 41 (47%) of sites sampled, however our occupancy estimate was much higher, suggesting 61% of amphibian habitats were occupied by *Bd* (SE = 11%, CI = 39%–80%), underestimating *Bd* in our study by 14%. Models with elevation were the best fit to predict *Bd* occupancy (Table 2); as elevation decreased, *Bd* occupancy increased (Figure 5). At 100 m

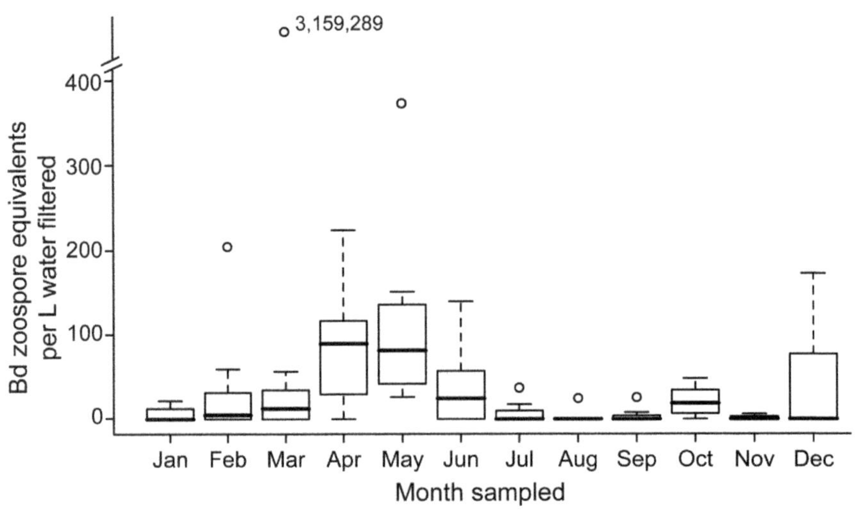

Figure 3. *Batrachochytrium dendrobatidis* (*Bd*) zoospore density from a four year case study site in Oregon, July 2007-March 2011.

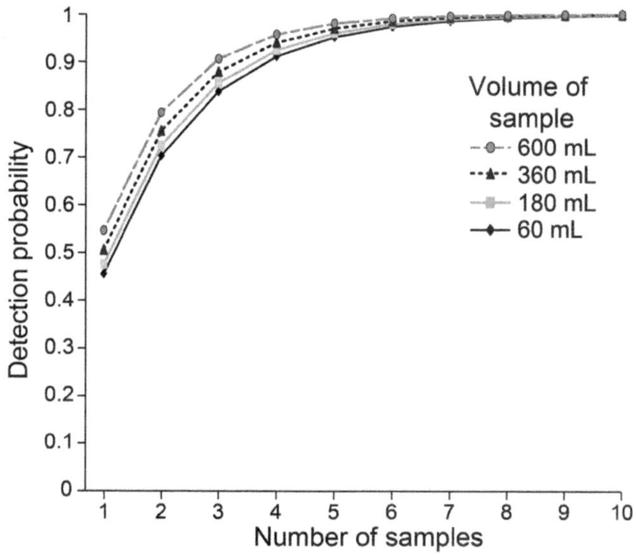

Figure 4. Variation in detection probability of *Batrachochytrium dendrobatidis* (*Bd*) in water collected from amphibian habitats (vol = 60–600 mL/sample) as a function of (i) the number and (ii) the volume of water samples collected.

elevation, the probability of detecting *Bd* when it was present was between 92–99%. At 500 m elevation, the probability of detection decreased slightly to 81–96%. However, at higher elevations, detection probability decreased sharply, 36–59% at 1500 m, and only 10–20% at 2500 m. The best-supported model structure for parameter 'p' was p(pH); the probability of detecting *Bd*, when present, increased as pH increased, however differences in detection probability were negligible when the pH range was considered (Figure 6). Observed pH ranged from 6.5 to 10.4 (mean = 7.8, median = 7.6). *Bd* was detected in samples with a pH between 6.5 and 8.5 (mean = 7.1, median = 7.1, n = 14 samples from 8 sites). Because there was a negligible difference between the p(ph) model and the p(.) model ($\Delta AIC = 1.1$), we proceeded with the most parsimonious model for estimates of Ψ (Table 2).

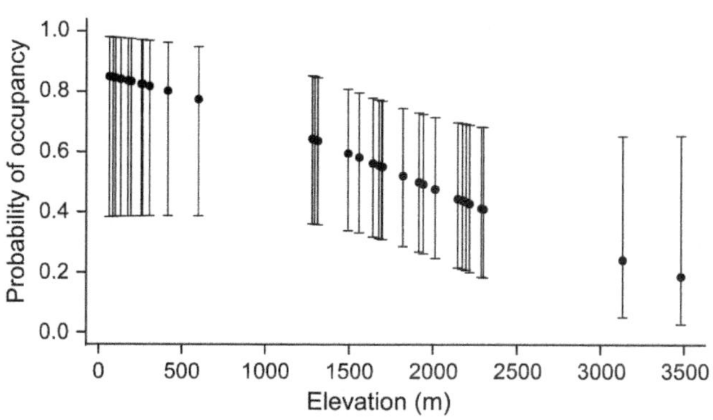

Figure 5. Variation in detection probability of *Batrachochytrium dendrobatidis* (*Bd*) in water collected from amphibian habitats *Batrachochytrium dendrobatidis* (*Bd*) as a function of elevation in meters.

Case Study Site (Beaverton, Oregon)

Across all years, we detected *Bd* at the Oregon site at least once per month, although not necessarily in every month in a given year (e.g., *Bd* was detected in December during 2 of 4 years sampled). As predicted, *Bd* zoospore density was highest in the spring, March to June, and contrary to our landscape-scale analysis, lowest in summer (Figure 3). Mean *Bd* density was similar in March and October, and an unexpected peak was observed in December in 2 of 4 years (2007 and 2008). From one sampling event in March 2011, we recovered 3,159,289 zoospores L^{-1} from one filter, approximately four orders of magnitude larger than any previous observation in our study (e.g., Nevada sites mentioned above). The *Bd* densities from the other two filters from this March 2011 sampling event were 0 and 4 zoospores L^{-1}. *Bd* detection and density at the Oregon site were not correlated with any of the water quality variables that we measured (temperature $R^2 = 0.01$, pH $R^2 = 0.02$, specific conductance $R^2 = 0.02$ and turbidity $R^2 = 0.02$).

Discussion

Bd is emerging as an extremely challenging pathogen to fully comprehend, and our baseline information of the temporal and spatial patterns of *Bd* occurrence in USA waters, free-living, apart from amphibian hosts, adds a new dimension to understanding the dynamics of the pathogen. Host-pathogen systems are more complex when a pathogen can infect multiple host species rather than a single species [79]. Unlike most chytrid parasites, *Bd* is a generalist, infecting at least 42% of the species of amphibians examined [36]. Globally, *Bd* occurrence in amphibians increased as species richness increased, although in North America when species richness was <10, as was the case in our study area, the odds of *Bd* detection were constant [36]. *Bd* detection probability in water also increased as amphibian density at a site increased [80]. This may be due to an increase in the number of competent hosts, direct skin contact with infected individuals, or exposure to free-living *Bd*. Also, *Bd*-positive amphibians arriving at breeding sites from upland overwintering sites may release *Bd* into the aquatic environment, causing an increase in *Bd* density in the water. Alternatively (or concurrently), spring-season conditions (e.g., increasing temperature and light) and amphibian arrival at a breeding site may stimulate free-living *Bd* to grow in the aquatic environment. Adding to the complexity is context-dependency, where different species, life stages, and populations of amphibians may respond to a pathogen such as *Bd* differently, depending on

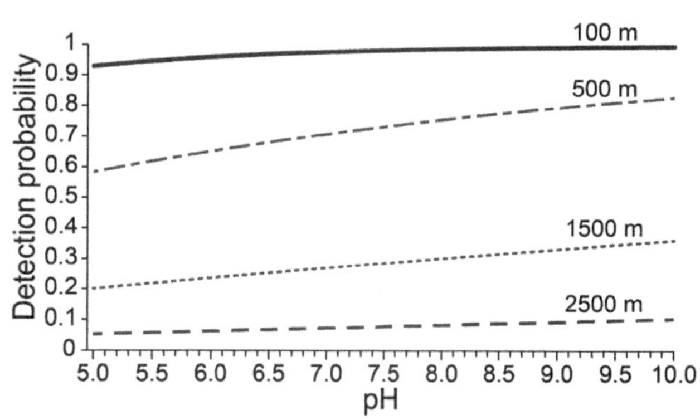

Figure 6. The probability of detecting *Batrachochytrium dendrobatidis* (*Bd*) in water samples at four elevations within a range of pH.

biotic and abiotic factors acting in concert with each other, including strain differences and variation in organismal responses between sites [81–83].

Bd did not exhibit seasonality in occupancy (e.g., presence) in our study. We detected *Bd* in all months, which is evidence that *Bd* persists in the environment year-round in sites in the U.S. In amphibians within the geographic scope of our study, *Bd* has been estimated to occur in 53% of amphibian populations sampled, and in 66% of non-native bullfrog (*Lithobates catesbeianus*) populations sampled [71], estimates that are similar to our occupancy estimates for *Bd* in amphibian habitats (61%). Areas for future research include analyses of: (i) the substrates upon which *Bd* lives in the environment in the cold season when amphibian densities at the ponds are low or zero; (ii) spatial heterogeneity within a wetland or pond; (iii) alternative hosts, vectors, and reservoirs in different seasons; (iv) *Bd* strain diversity, especially in areas with introduced species; and (v) pathogenicity, or infectious potential, of *Bd* when amphibians are absent for months or years.

The occurrence of a species and its abundance are influenced by habitat factors at different temporal and spatial scales, from microhabitat to landscape [84–89]. In temperate areas, there is seasonal variation in *Bd* infection prevalence (i.e., the proportion of a population infected) in amphibian populations, with the highest prevalence in the early spring and lowest in the autumn [66,90]. Our study calls attention to the heterogeneity that may exist within a site and across the landscape over space and time. We observed a clear signal of increased *Bd* density in the spring. Our results were consistent with our prediction that *Bd* density in water would be high in the spring, and contrary to our prediction that *Bd* density would be low in the summer. The effect of season may be driven by various factors; the temperatures during the spring months are within the range for optimal *Bd* growth [52], and *Bd*-infected animals are returning to breeding sites, shedding zoospores into the water and transmitting the infection to each other as they congregate. Alternatively or additionally, springtime increases in host abundance may elicit a chemotactic response from free-living *Bd* [91], which could contribute to an observed increase in the zoospore density in water.

Adding complexity to the effort of understanding *Bd*'s temporal dynamics in prevalence, alternative *Bd* hosts and reservoirs need to be ascertained, and their potential interactive role in amphibian-*Bd* systems is only beginning to be explored. Several documented and suggested alternative *Bd* hosts and reservoirs (e.g., crayfish [92]; nematodes [93]; birds [94,95]) occur within the same water bodies as amphibians, and some potential host/reservoir species (e.g., zooplankton) consume chytrid spores, including *Bd* [43,96], although their post-consumption fate is unknown. When consumed, up to 50% of parasitic fungal spores may survive passage through the gut, which may further fuel disease epidemics [97]. However, there is experimental evidence that consumption by zooplankton can negatively affect free-living *Bd* such that infections in tadpoles are reduced [98], which suggests the aquatic-community context is an important consideration with regard to disease dynamics. Furthermore, *Bd* can survive in pond water for months without a host [51], and occurrence in benthic sediments appears to be likely, although detection is inhibited, perhaps due to high levels of organic carbon inherent in sediment samples from these habitats [48]. These combined factors make interpreting the *Bd*-amphibian disease system in a field-based setting particularly challenging, and highlight the need for a more focused effort to study *Bd* life history in the aquatic environment.

In our study, *Bd* exhibited seasonality in density, but the pattern depended on the spatial scale considered (North American landscape-scale or single-site case study). At the landscape level, the mean densities observed did not follow all of the patterns we predicted. While we did observe high densities in spring months, we also observed high densities in at least one of the summer months, August. This may have been a limitation of our study design; if sites had been visited more than once, we could have made comparisons of relative densities between visits. At our Oregon case-study site, we observed the predicted temporal patterns, with exceptions. We repeatedly observed similar mean densities in late spring and late autumn (March and October), as well as a secondary peak in winter (December), when we expected *Bd* density to be low. There were no major storm or flood events in the days prior to our sampling that would have flushed or resuspended *Bd* or potential physical reservoirs such as sediments into the wetland. Our observations lead to more questions about the role of potential secondary host species and reservoirs in regulating *Bd* densities in amphibian habitats. Crayfish are identified as a potential secondary host [92], but we did not observe them at the Oregon case-study site in our four-year monitoring period. Although it was a single observation, our sample of over 3 million zoospore equivalents from a liter of water is notable because it was several orders of magnitude higher than any other sample collected (the sample was run multiple times to ensure no lab errors). The sample was collected in March 2011, during the predicted spring peak, and it is possible that we captured a large free-floating colony of zoosporangia. Such colonies might be expected to be patchy in distribution, and our sample may have haphazardly collected such a localized high-density cluster. This observation strongly suggests that *Bd* zoospores can have a patchy/aggregated distribution and highlights the importance of collecting several spatial replicates from a pond or wetland during water sampling. The observed temporal variation in our study is intriguing, and suggests an interplay of abiotic and biotic factors, highlighting the relevance of finer-scale assessments. Our results demonstrate that we can correct for this clustering and reliably detect *Bd* by adjusting the number of samples and volume of water collected. Investigations of the patterns and processes influencing *Bd* occupancy in amphibian habitats may be most appropriate at an ecoregional or local scale [21,36]. Factors influencing detection probability may be site-specific and context-dependent [66,71]. Importantly, such studies can be informed by both landscape approaches (like ours), and insights from laboratory experiments (e.g., investigating primary and secondary hosts, reservoir species, and amphibian species and life stage) that can shed light on local, regional, and global amphibian declines [83,92,98].

Elevation as a predictor of occupancy

The fact that *Bd* occupancy within our study area was highest at low-elevation sites suggests that distribution patterns and geographic range may be influenced by human-induced landscape change, or differences in host-species composition. The probability of detecting *Bd* when it is present increases as the Human Footprint, an index human activities, increases (i.e., human population size, density of secondary roads, proportion of agricultural lands) [71]. This generalized index doesn't identify specific human activities, but provides a direction for future studies of ecophysiology, or the links between environmental stress, endocrine-immune interactions, and disease [99]. Bullfrogs are a common non-native species implicated as disease vectors throughout our study area at lower elevations that are absent from higher elevations, although they demonstrate differential responses to North American versus Globally Distributed *Bd* strains [68]. The proportion of sites occupied by *Bd* is higher when non-native bullfrogs are present (66%, versus 53% when bullfrogs were absent), which may explain why we observed higher occupancy as elevation decreased, though bullfrog presence alone does not explain widespread *Bd* occupancy of amphibian habitats [71,100].

Although temperature was not a predictor of *Bd* occupancy, the temperature at high elevations often has a greater range than at lower elevations. Temperature influences both generation time and fecundity of *Bd*, and differences in long-term responses to different thermal regimes are observed in at least one *Bd* strain [53,57]. Cold-adapted *Bd* develops faster than warm-adapted *Bd*, however, overall developmental time is longer; it may take 4 times longer to encyst, and take 6 times longer to mature and release zoospores [53]. These factors may explain differences in occupancy between our high- and low-elevation sites, and they could influence detection probability if *Bd* density is lower as a result of decreased generation time or fecundity.

At a global scale, there are family-level differences in the odds of *Bd* infection related to elevation. Elevation was found to be (i) positively correlated with Bd detection/infection in toads (Bufonidae); (ii) negatively correlated with *Bd* detection in true frogs (Ranidae); and (iii) not a predictor for *Bd* detection in treefrogs (Hylidae) [36]. However, at the North American scale, elevation was not a predictor for *Bd* detection, rather species richness, temperature, and biome were [36]. This lends further support that regional and local spatial scales may be more appropriate levels of investigation to assess risk to amphibian populations and species. Species representing these three amphibian families occur throughout our study area, and the effect of the amphibian community assemblage on *Bd* occupancy of their breeding habitats warrants further investigation.

Water quality as a predictor of detection probability

Our study revealed a potential relationship between pH and *Bd* detection probability. Other water quality covariates that we had predicted could be related to *Bd* occupancy and detection showed no relationships, perhaps because of the scale of our measurements. Specific conductance (conductivity in microSiemens, $\mu S\ cm^{-1}$, standardized to 25°C) and turbidity are coarse measurements that indicate the relative amount of dissolved ions and optical properties of the water but do not identify the specific constituents in the water that may be important to *Bd* occupancy and detection (e.g., Ca^{+2}, Mg^{+2}, Na^{+1}, K^{+1}, Cl^{-1}, nitrogen and phosphorus species, organic carbon, turbidity source, sediment loading, phytoplankton community). The observed temperature and pH ranges at our sites were always within the range that *Bd* grows in the laboratory, and both metrics were often within the range for optimal growth. Our results did not support our prediction that detection probability would be highest when pH was within the range for ideal *Bd* growth [52]. Although the best models included pH and suggested detection probability increased as pH increased, the differences in detection probability were negligible when the pH range was considered (Figure 6). While there may be a narrow pH range that is optimal for *Bd* growth (pH 6–7) in a laboratory setting [52], this narrow range may not be ecologically relevant given the fluctuations in pH at different time scales, such as diel, episodic, and seasonal [101,102].

pH is influenced by abiotic (e.g., acid-neutralizing capacity) and biotic (e.g., organic carbon, aquatic plant community) characteristics of a system [103–105], and it is not clear if pH itself is a mechanism influencing *Bd* detection probability within our study area or if pH is instead reflective of other processes of the aquatic system that are important to *Bd* ecology. Lower pH can inhibit microbial metabolism [106], which could decrease detection probability. Changes in pH are related to the acid-neutralizing capacity in a system, which is strongly tied to the amount of organic carbon present [104,105], which is in turn an important nutrient for aquatic fungi [107]. Higher pH may indicate higher organic carbon [105], which could increase detection probability. Generally, chytrids decompose organic particulate matter [107], and it is not known if *Bd* uses non-animal organic carbon sources. Diel variation of up to pH 1.8 is not uncommon in mesotrophic and eutrophic aquatic systems, which may be driven by the plant communities present [101,102]. pH increases with the addition of algae and vascular plants [103]. If *Bd* is associating with plants, perhaps as a secondary host, pH as the predictor of detection probability would be indirect. This result provides direction for future research into how pH influences *Bd* site occupancy and detection probability.

Management Implications

Our results show that water filtration sampling is a reliable way to assess the occurrence of *Bd* in amphibian habitats when 4–5 samples of a small volume, i.e., 60–600 mL of water, are filtered directly from ponds and wetlands, especially at sites less than 1500 m elevation. *Bd* densities recovered were consistently above the detection limit of the qPCR for a quantitative result (10 genomic equivalents), which affirms water filtration as a viable method for both *Bd* detection and quantification in amphibian habitats. Our results are context-dependent, as found in other studies from the southwestern United States where more samples were necessary to achieve 95% confidence that *Bd* was detected when it was present [80]. Our results suggest that a single visit collecting multiple samples in the spring or summer months may be sufficient (and perhaps more effective) for *Bd* detection from water samples, rather than multiple visits collecting one sample throughout the season, however spatial replicates within the same site visit are important [50]. The number of samples required to detect *Bd* is representative of our area of inference and is not intended to serve as a recommendation for a standard protocol, as regional and site-level variation in *Bd* occupancy may require fewer or additional samples [80].

There are few, if any systems, where species detection is perfect [69]. Imperfect detection (false negatives) may occur because a species is rare or cryptic [108,109], or alternatively, a false positive report or misidentification may occur [110–112]. We underestimated *Bd* occupancy of amphibian habitats within our study area by 14%, which is greater than other observations from a different geographic region that used the same filtration method [80]. These results stress the importance of using methods that account for imperfect detection when evaluating the potential occurrence and distribution of organisms when detection probability is less than 1. The probability of detecting *Bd* in amphibians varies based on life stage, and is highest in adults at low elevations [71]. In temperate areas, lentic-breeding amphibians are often explosive breeders [113,114], and the likelihood of encountering many adult amphibians is limited to the narrow window of time (hours to days) they are present at a site. While larval amphibians are present at a site for longer time periods, the probability of detecting *Bd* when it is present in them is less than 20% [71]. If the objective of a study is to determine the *Bd* status of a site, sampling the habitat rather than the animals may be the most cost-effective approach in terms of time and resources. Sampling the environment also resolves problems with detection due to differences in capture probability between infected and uninfected individuals related to behavioral changes, however the results are limited to *Bd* occupancy of the aquatic habitat sampled, and inferences to the disease status of an amphibian population, or potential secondary hosts or vectors, cannot be made except by sampling them directly.

By focusing research efforts on understanding the ecology of this fungal pathogen outside of the host, conservation efforts can be more informed and focused to meet the management goals and objectives for species at-risk and common species alike. Patterns in amphibian response to *Bd* are different based on biome [36]. The strategy for assessing *Bd* status at a site may be different according to the ecoregion, and may include sampling the habitat, other species that are present, as well as the amphibians that occur at a site. When water sampling is coupled with amphibian sampling, scientists can begin to understand the relationships between *Bd* occupancy and density in the environment, and the occurrence of disease in amphibian populations.

Acknowledgments

Larissa Bailey, Erin Muths, and two anonymous reviewers, provided comments that improved the manuscript. Mike Adams, Chris Pearl, Bill Battaglin, Jeremy Buck, Elizabeth Materna, Steve Corn, Blake Hossack, Erin Muths, Gary Fellers, Pat Kleeman, Sophie St. Hilaire, Peter Murphy, Perry Jones, and Mike Lico collected samples. We thank Kathryn Ronnenberg for assistance with figures. This manuscript is contribution no. 478 of the US Geological Survey Amphibian Research and Monitoring Initiative (ARMI).

Author Contributions

Conceived and designed the experiments: TC CWA MAV JK. Performed the experiments: TC CWA MAV JK. Analyzed the data: TC. Contributed reagents/materials/analysis tools: TC CWA RP DHO MAV ARB JK. Wrote the paper: TC CWA RP DHO MAV ARB JK.

References

1. Vaughn CC (2010) Biodiversity losses and ecosystem function in freshwaters: emerging conclusions and research directions. BioScience 60: 25–35.
2. Pereira HM, Leadley PW, Proença V, Alkemade R, Scharlemann JPW, et al. (2010) Scenarios for global biodiversity in the 21st century. Science 330: 1496–1501. doi:10.1126/science.1196624
3. Jones JP, Collen BEN, Atkinson G, Baxter PW, Bubb P, et al. (2011) The why, what, and how of global biodiversity indicators beyond the 2010 target. Conserv Biol 25: 450–457.
4. Mora C, Sale PF (2011) Ongoing global biodiversity loss and the need to move beyond protected areas: a review of the technical and practical shortcomings of protected areas on land and sea. Mar Ecol Prog Ser 434: 251–266. doi:10.3354/meps09214
5. Hooper DU, Adair EC, Cardinale BJ, Byrnes JE, Hungate BA, et al. (2012) A global synthesis reveals biodiversity loss as a major driver of ecosystem change. Nature 486: 105–108.
6. Krauss J, Bommarco R, Guardiola M, Heikkinen RK, Helm A, et al. (2010) Habitat fragmentation causes immediate and time-delayed biodiversity loss at different trophic levels. Ecol Lett 13: 597–605. doi:10.1111/j.1461-0248.2010.01457.x
7. Hoffmann M, Hilton-Taylor C, Angulo A, Böhm M, Brooks TM, et al. (2010) The impact of conservation on the status of the world's vertebrates. Science 330: 1503–1509. doi:10.1126/science.1194442
8. Seto KC, Güneralp B, Hutyra LR (2012) Global forecasts of urban expansion to 2030 and direct impacts on biodiversity and carbon pools. P Natl Acad Sci USA 109: 16083–16088. doi:10.1073/pnas.1211658109
9. Chapin Iii FS, Zavaleta ES, Eviner VT, Naylor RL, Vitousek PM, et al. (2000) Consequences of changing biodiversity. Nature 405: 234–242. doi:10.1038/35012241
10. Pimentel D, Zuniga R, Morrison D (2005) Update on the environmental and economic costs associated with alien-invasive species in the United States. Ecol Econ 52: 273–288. doi:10.1016/j.ecolecon.2004.10.002
11. Relyea RA (2005) The impact of insecticides and herbicides on the biodiversity and productivity of aquatic communities. Ecol Appl 15: 618–627.
12. Relyea RA (2009) A cocktail of contaminants: how mixtures of pesticides at low concentrations affect aquatic communities. Oecologia 159: 363–376.
13. Clements WH, Rohr JR (2009) Community responses to contaminants: Using basic ecological principles to predict ecotoxicological effects. Environ Toxicol Chem 28: 1789–1800. doi:10.1897/09-140.1
14. Lawler JJ, Shafer SL, White D, Kareiva P, Maurer EP, et al. (2009) Projected climate-induced faunal change in the Western Hemisphere. Ecology 90: 588–597.
15. Dawson TP, Jackson ST, House JI, Prentice IC, Mace GM (2011) Beyond predictions: biodiversity conservation in a changing climate. Science 332: 53–58. doi:10.1126/science.1200303
16. Mantyka-pringle CS, Martin TG, Rhodes JR (2012) Interactions between climate and habitat loss effects on biodiversity: a systematic review and meta-analysis. Glob Change Biol 18: 1239–1252. doi:10.1111/j.1365-2486.2011.02593.x
17. Daszak P, Cunningham AA, Hyatt AD (2000) Emerging infectious diseases of wildlife–threats to biodiversity and human health. Science 287: 443–449.
18. Jones KE, Patel NG, Levy MA, Storeygard A, Balk D, et al. (2008) Global trends in emerging infectious diseases. Nature 451: 990–993. doi:10.1038/nature06536
19. McCallum ML (2007) Amphibian decline or extinction? Current declines dwarf background extinction rate. J Herpetol 41: 483–491.
20. Stuart SN, Chanson JS, Cox NA, Young BE, Rodrigues ASL, et al. (2004) Status and trends of amphibian declines and Extinctions worldwide. Science 306: 1783–1786. doi:10.1126/science.1103538
21. Adams MJ, Miller DAW, Muths E, Corn PS, Grant EHC, et al. (2013) Trends in amphibian occupancy in the United States. PLoS ONE 8: e64347. doi:10.1371/journal.pone.0064347
22. Blaustein AR, Grant Hokit D, O'Hara RK, Holt RA (1994) Pathogenic fungus contributes to amphibian losses in the Pacific Northwest. Biol Conserv 67: 251–254.
23. Romansic JM, Johnson PTJ, Searle CL, Johnson JE, Tunstall TS, et al. (2011) Individual and combined effects of multiple pathogens on Pacific treefrogs. Oecologia 166: 1029–1041. doi:10.1007/s00442-011-1932-1
24. Ault KK, Johnson JE, Pinkart HC, Wagner RS (2012) Genetic comparison of water molds from embryos of amphibians *Rana cascadae*, *Bufo boreas* and *Pseudacris regilla*. Dis Aquat Org 99: 127–137. doi:10.3354/dao02456
25. Bradford DF (1991) Mass mortality and extinction in a high-elevation population of *Rana muscosa*. J Herpetol 25: 174–177. doi:10.2307/1564645
26. Mao J, Green D., Fellers G, Chinchar V (1999) Molecular characterization of iridoviruses isolated from sympatric amphibians and fish. Virus Res 63: 45–52. doi:10.1016/S0168-1702(99)00057-X
27. Chinchar VG (2002) Ranaviruses (family Iridoviridae): emerging cold-blooded killers. Arch Virol 147: 447–470. doi:10.1007/s007050200000
28. Jancovich JK, Davidson EW, Parameswaran N, Mao J, Chinchar VG, et al. (2005) Evidence for emergence of an amphibian iridoviral disease because of human-enhanced spread. Mol Ecol 14: 213–224. doi:10.1111/j.1365-294X.2004.02387.x
29. Chinchar VG, Robert J, Storfer AT (2011) Ecology of viruses infecting ectothermic vertebrates—the impact of ranavirus infections on amphibians. In: Hurst CJ, editor, Studies in viral ecology: animal host systems, volume 2. Hoboken, NJ: John Wiley & Sons. pp. 231–259.
30. Davis AK, Yabsley MJ, Keel MK, Maerz JC (2007) Discovery of a novel alveolate pathogen affecting southern leopard frogs in Georgia: description of the disease and host effects. EcoHealth 4: 310–317. doi:10.1007/s10393-007-0115-3
31. Jones MEB, Armin AG, Rothermel BB, Pessier AP (2012) Granulomatous myositis associated with a novel alveolate pathogen in an adult southern leopard frog (*Lithobates sphenocephalus*). Dis Aquat Org 102: 163–167. doi:10.3354/dao02539
32. Johnson PT, Lunde KB, Thurman EM, Ritchie EG, Wray SN, et al. (2002) Parasite (*Ribeiroia ondatrae*) infection linked to amphibian malformations in the western United States. Ecol Monogr 72: 151–168.
33. Blaustein AR, Johnson PTJ (2003) The complexity of deformed amphibians. Front Ecol Environ 1: 87–94.
34. Rohr JR, Raffel TR, Romansic JM, McCallum H, Hudson PJ (2008) Evaluating the links between climate, disease spread, and amphibian declines. P Natl Acad Sci USA 105: 17436–17441.
35. Vredenburg VT, Knapp RA, Tunstall TS, Briggs CJ (2010) Dynamics of an emerging disease drive large-scale amphibian population extinctions. P Natl Acad Sci USA 107: 9689–9694.
36. Olson DH, Aanensen DM, Ronnenberg KL, Powell CI, Walker SF, et al. (2013) Mapping the global emergence of *Batrachochytrium dendrobatidis*, the amphibian chytrid fungus. PLoS ONE 8: e56802.
37. Shearer C, Descals E, Kohlmeyer B, Kohlmeyer J, Marvanová L, et al. (2007) Fungal biodiversity in aquatic habitats. Biodiv Conserv 16: 49–67. doi:10.1007/s10531-006-9120-z
38. Adl SM, Simpson AGB, Lane CE, Lukes J, Bass D, et al. (2012) The revised classification of Eukaryotes. J Eukaryot Microbiol 59: 429–493. doi:10.1111/j.1550-7408.2012.00644.x
39. Hibbett DS, Binder M, Bischoff JF, Blackwell M, Cannon PF, et al. (2007) A higher-level phylogenetic classification of the Fungi. Mycol Res 111: 509–547. doi:10.1016/j.mycres.2007.03.004
40. Gleason FH, Küpper FC, Amon JP, Picard K, Gachon CM, et al. (2011) Zoosporic true fungi in marine ecosystems: a review. Mar Freshwater Res 62: 383–393.
41. Ibelings B, De Bruin A, Kagami M, Van Donk E (2007) Diatom blooms, chytrid epidemics and the evolutionary. J Phycol 43: 2–2.

42. James TY, Kauff F, Schoch CL, Matheny PB, Hofstetter V, et al. (2006) Reconstructing the early evolution of Fungi using a six-gene phylogeny. Nature 443: 818–822. doi:10.1038/nature05110
43. Kagami M, Van Donk E, de Bruin A, Rijkeboer M, Ibelings BW (2004) Daphnia can protect diatoms from fungal parasitism. Limnol Oceanogr 49: 680–685.
44. Longcore JE, Pessier AP, Nichols DK (1999) *Batrachochytrium dendrobatidis* gen et sp nov, a chytrid pathogenic to amphibians. Mycologia 91: 219–227. doi:10.2307/3761366
45. Martel A, Sluijs AS der, Blooi M, Bert W, Ducatelle R, et al. (2013) *Batrachochytrium salamandrivorans* sp. nov. causes lethal chytridiomycosis in amphibians. PNAS: 201307356. doi:10.1073/pnas.1307356110
46. Berger L, Hyatt AD, Speare R, Longcore JE (2005) Life cycle stages of the amphibian chytrid *Batrachochytrium dendrobatidis*. Dis Aquat Org 68: 51–63. doi:10.3354/dao068051
47. Rosenblum EB, Voyles J, Poorten TJ, Stajich JE (2010) The deadly chytrid fungus: a story of an emerging pathogen. PLoS Pathog 6: e1000550. doi:10.1371/journal.ppat.1000550
48. Kirshtein JD, Anderson CW, Wood JS, Longcore JE, Voytek MA (2007) Quantitative PCR detection of *Batrachochytrium dendrobatidis* DNA from sediments and water. Dis Aquat Org 77: 11.
49. Walker SF, Baldi Salas M, Jenkins D, Garner TW, Cunningham AA, et al. (2007) Environmental detection of *Batrachochytrium dendrobatidis* in a temperate climate. Dis Aquat Org 77: 105.
50. Hyman OJ, Collins JP (2011) Evaluation of a filtration-based method for detecting *Batrachochytrium dendrobatidis* in natural bodies of water. Dis Aquat Org 97: 185–195. doi:http://dx.doi.org.ezproxy.proxy.library.oregonstate.edu/10.3354/dao02423
51. Johnson ML, Speare R (2003) Survival of *Batrachochytrium dendrobatidis* in water: quarantine and disease control implications. Emerg Infect Dis 9: 922.
52. Piotrowski JS, Annis SL, Longcore JE (2004) Physiology of *Batrachochytrium dendrobatidis*, a chytrid pathogen of amphibians. Mycologia 96: 9–15. doi:10.2307/3761981
53. Voyles J, Johnson LR, Briggs CJ, Cashins SD, Alford RA, et al. (2012) Temperature alters reproductive life history patterns in *Batrachochytrium dendrobatidis*, a lethal pathogen associated with the global loss of amphibians. Ecol Evol 2: 2241–2249. doi:10.1002/ece3.334
54. Cheng TL, Rovito SM, Wake DB, Vredenburg VT (2011) From the cover: Coincident mass extirpation of neotropical amphibians with the emergence of the infectious fungal pathogen *Batrachochytrium dendrobatidis*. P Natl Acad Sci USA 108: 9502–9507. doi:10.1073/pnas.1105538108
55. Mitchell KM, Churcher TS, Garner TW., Fisher MC (2008) Persistence of the emerging pathogen *Batrachochytrium dendrobatidis* outside the amphibian host greatly increases the probability of host extinction. P Roy Soc B-Biol Sci 275: 329–334. doi:10.1098/rspb.2007.1356
56. Schmidt BR (2010) Estimating the impact of disease in species threatened by amphibian chytrid fungus: comment on Murray et al. Conserv Biol 24: 897–899. doi:10.1111/j.1523-1739.2010.01507.x
57. Woodhams DC, Alford RA, Briggs CJ, Johnson M, Rollins-Smith LA (2008) Life-history trade-offs influence disease in changing climates: strategies of an amphibian pathogen. Ecology 89: 1627–1639.
58. Berger L, Speare R, Hines HB, Marantelli G, Hyatt AD, et al. (2004) Effect of season and temperature on mortality in amphibians due to chytridiomycosis. Aust Vet J 82: 434–439.
59. Drew A, Allen EJ, Allen LJS (2006) Analysis of climatic and geographic factors affecting the presence of chytridiomycosis in Australia. Dis Aquat Org 68: 245.
60. Raffel TR, Michel PJ, Sites EW, Rohr JR (2010) What drives chytrid infections in newt populations? Associations with substrate, temperature, and shade. EcoHealth 7: 526–536. doi:http://dx.doi.org.ezproxy.proxy.library.oregonstate.edu/10.1007/s10393-010-0358-2
61. Rohr JR, Raffel TR (2010) Linking global climate and temperature variability to widespread amphibian declines putatively caused by disease. P Natl Acad Sci USA 107: 8269–8274.
62. Seimon TA, Seimon A, Daszak P, Halloy SR p., Schloegel LM, et al. (2007) Upward range extension of Andean anurans and chytridiomycosis to extreme elevations in response to tropical deglaciation. Glob Change Biol 13: 288–299. doi:10.1111/j.1365-2486.2006.01278.x
63. Fisher MC, Garner TW, Walker SF (2009) Global emergence of *Batrachochytrium dendrobatidis* and amphibian chytridiomycosis in space, time, and host. Annu Rev Microbiol 63: 291–310.
64. Retallick RW, McCallum H, Speare R (2004) Endemic infection of the amphibian chytrid fungus in a frog community post-decline. PLoS Biol 2: e351.
65. Kriger KM, Hero J-M (2006) Large-scale seasonal variation in the prevalence and severity of chytridiomycosis. J Zool: 060905012106004–??? doi:10.1111/j.1469-7998.2006.00220.x
66. Pearl CA, Bull EL, Green DE, Bowerman J, Adams MJ, et al. (2007) Occurrence of the amphibian pathogen *Batrachochytrium dendrobatidis* in the Pacific Northwest. J Herpetol 41: 145–149.
67. Kinney VC, Heemeyer JL, Pessier AP, Lannoo MJ (2011) Seasonal pattern of *Batrachochytrium dendrobatidis* infection and mortality in *Lithobates areolatus*: affirmation of Vredenburg's "10,000 zoospore rule." PLoS ONE 6: e16708.
68. Gervasi SS, Urbina J, Hua J, Chestnut T, Relyea RA, et al. (2013) Experimental evidence for American Bullfrog (*Lithobates catesbeianus*) susceptibility to chytrid fungus (*Batrachochytrium dendrobatidis*). EcoHealth: 1–6. doi:10.1007/s10393-013-0832-8
69. MacKenzie DI, Nichols JD, Lachman GB, Droege S, Andrew Royle J, et al. (2002) Estimating site occupancy rates when detection probabilities are less than one. Ecology 83: 2248–2255.
70. Thompson KG (2007) Use of site occupancy models to estimate prevalence of *Myxobolus cerebralis* infection in trout. J Aquat Anim Health 19: 8–13. doi:10.1577/H06-016.1
71. Adams MJ, Chelgren ND, Reinitz D, Cole RA, Rachowicz LJ, et al. (2010) Using occupancy models to understand the distribution of an amphibian pathogen, *Batrachochytrium dendrobatidis*. Ecol Appl 20: 289–302.
72. McClintock BT, Nichols JD, Bailey LL, MacKenzie DI, Kendall WL, et al. (2010) Seeking a second opinion: uncertainty in disease ecology. Ecol Lett 13: 659–674. doi:10.1111/j.1461-0248.2010.01472.x
73. U.S. Geological Survey (2010) National field manual for the collection of water-quality data: U.S. Geological Survey Techniques of Water-Resources Investigations, book 9, chaps. A1-A9. Available: http://pubs.water.usgs.gov/twri9A US Department of the Interior, US Geological Survey.
74. Ficetola GF, Miaud C, Pompanon F, Taberlet P (2008) Species detection using environmental DNA from water samples. Biol Lett 4: 423–425.
75. Goldberg CS, Pilliod DS, Arkle RS, Waits LP (2011) Molecular detection of vertebrates in stream water: a demonstration using Rocky Mountain Tailed Frogs and Idaho Giant Salamanders. PLoS ONE 6: e22746. doi:10.1371/journal.pone.0022746
76. Simon D, Moline J, Helms G, Friedl T, Bhattacharya D (2005) Divergent histories of rDNA group I introns in the lichen family Physciaceae. J Mol Evol 60: 434–446. doi:10.1007/s00239-004-0152-2
77. Daly C, Halbleib M, Smith JI, Gibson WP, Doggett MK, et al. (2008) Physiographically sensitive mapping of climatological temperature and precipitation across the conterminous United States. Int J Climatol 28: 2031–2064.
78. R Core Team (2012) R: A language and environment for statistical computing. R Foundation for Statistical Computing, Vienna, Austria. Available: http://www.R-project.org/
79. Dobson A (2004) Population dynamics of pathogens with multiple host species. Am Nat 164: S64–S78.
80. Schmidt BR, Kéry M, Ursenbacher S, Hyman OJ, Collins JP (2013) Site occupancy models in the analysis of environmental DNA presence/absence surveys: a case study of an emerging amphibian pathogen. Method Ecol Evol 4: 646–653. doi:10.1111/2041-210X.12052
81. Blaustein AR, Kiesecker JM (2002) Complexity in conservation: lessons from the global decline of amphibian populations. Ecol Lett 5: 597–608.
82. Bancroft BA, Han BA, Searle CL, Biga LM, Olson DH, et al. (2011) Species-level correlates of susceptibility to the pathogenic amphibian fungus *Batrachochytrium dendrobatidis* in the United States. Biodivers Conserv 20: 1911–1920. doi:10.1007/s10531-011-0066-4
83. Gervasi S, Gondhalekar C, Olson DH, Blaustein AR (2013) Host identity matters in the amphibian-*Batrachochytrium dendrobatidis* system: fine-scale patterns of variation in responses to a multi-host pathogen. PLoS ONE 8. doi:10.1371/journal.pone.0054490
84. Welsh HH, Lind AJ (2002) Multiscale habitat relationships of stream amphibians in the Klamath-Siskiyou region of California and Oregon. J Wildlife Manage 66: 581. doi:10.2307/3803126
85. Euliss NH, LaBaugh JW, Fredrickson LH, Mushet DM, Laubhan MK, et al. (2004) The wetland continuum: A conceptual framework for interpreting biological studies. Wetlands 24: 448–458. doi:10.1672/0277-5212(2004)024[0448:TWCACF]2.0.CO;2
86. Browne CL (2009) The relationship of amphibian abundance to habitat features across spatial scales in the Boreal Plains. Ecoscience: 209–223. doi:10.2980/16-2-3220
87. Murray KA, Retallick RWR, Puschendorf R, Skerratt LF, Rosauer D, et al. (2011) Assessing spatial patterns of disease risk to biodiversity: implications for the management of the amphibian pathogen, *Batrachochytrium dendrobatidis*. J Appl Ecol 48: 163–173. doi:10.1111/j.1365-2664.2010.01890.x
88. Lannoo MJ, Petersen C, Lovich RE, Nanjappa P, Phillips C, et al. (2011) Do frogs get their kicks on Route 66? Continental U.S. transect reveals spatial and temporal patterns of *Batrachochytrium dendrobatidis* infection. PLoS ONE 6: e22211. doi:10.1371/journal.pone.0022211
89. Gsell AS, Domis LNDS, Naus-Wiezer SMH, Helmsing NR, Van Donk E, et al. (2013) Spatiotemporal variation in the distribution of chytrid parasites in diatom host populations. Freshw Biol 58: 523–537. doi:10.1111/j.1365-2427.2012.02786.x
90. Retallick RW, McCallum H, Speare R (2004) Endemic infection of the amphibian chytrid fungus in a frog community post-decline. PLoS Biol 2: e351.
91. Moss AS, Reddy NS, Dortaj IM, San Francisco MJ (2008) Chemotaxis of the amphibian pathogen *Batrachochytrium dendrobatidis* and its response to a variety of attractants. Mycologia 100: 1–5.
92. McMahon TA, Brannelly LA, Chatfield MW, Johnson PT, Joseph MB, et al. (2013) Chytrid fungus *Batrachochytrium dendrobatidis* has nonamphibian hosts and releases chemicals that cause pathology in the absence of infection. P Natl Acad Sci USA 110: 210–215.
93. Shapard EJ, Moss AS, San Francisco MJ (2012) *Batrachochytrium dendrobatidis* can infect and cause mortality in the nematode *Caenorhabditis elegans*. Mycopathologia 173: 121–126.

94. Johnson ML, Speare R (2005) Possible modes of dissemination of the amphibian chytrid *Batrachochytrium dendrobatidis* in the environment. Dis Aquat Org 65: 181–186. doi:10.3354/dao065181
95. Garmyn A, Van Rooij P, Pasmans F, Hellebuyck T, Van Den Broeck W, et al. (2012) Waterfowl: potential environmental reservoirs of the chytrid fungus *Batrachochytrium dendrobatidis*. PLoS ONE 7: e35038. doi:10.1371/journal.pone.0035038
96. Buck JC, Truong L, Blaustein AR (2011) Predation by zooplankton on *Batrachochytrium dendrobatidis*: biological control of the deadly amphibian chytrid fungus? Biodiv Conserv 20: 3549–3553. doi:10.1007/s10531-011-0147-4
97. Duffy MA (2009) Staying alive: The post-consumption fate of parasite spores and its implications for disease dynamics. Limnol Oceanogr 54: 770–773. doi:10.4319/lo.2009.54.3.0770
98. Searle CL, Mendelson JR, Green LE, Duffy MA (2013) *Daphnia* predation on the amphibian chytrid fungus and its impacts on disease risk in tadpoles. Ecol Evol 3: 4129–4138. doi:10.1002/ece3.777
99. Blaustein AR, Gervasi SS, Johnson PT, Hoverman JT, Belden LK, et al. (2012) Ecophysiology meets conservation: understanding the role of disease in amphibian population declines. Philos T Roy Soc B 367: 1688–1707.
100. Hayes MP, Jennings MR (1986) Decline of ranid frog species in western North America: are Bullfrogs (*Rana catesbeiana*) responsible? J Herpetol 20: 490. doi:10.2307/1564246
101. Reddy KR (1981) Diel variations of certain physico-chemical parameters of water in selected aquatic systems. Hydrobiologia 85: 201–207. doi:10.1007/BF00017610
102. Maberly SC. (1996) Diel, episodic and seasonal changes in pH and concentrations of inorganic carbon in a productive lake. Freshwater Biol 35: 579–598. doi:10.1111/j.1365-2427.1996.tb01770.x
103. Halstead BG, Tash JC (1982) Unusual diel pHs in water as related to aquatic vegetation. Hydrobiologia 96: 217–224. doi:10.1007/BF00010613
104. Litaor IM, Thurman EM (1988) Acid neutralizing processes in an alpine watershed front range, Colorado, U.S.A.—1: Buffering capacity of dissolved organic carbon in soil solutions. Appl Geochem 3: 645–652. doi:10.1016/0883-2927(88)90096-0
105. Herlihy AT, Kaufmann PR, Mitch ME (1991) Stream chemistry in the eastern United States: 2. Current sources of acidity in acidic and low acid-neutralizing capacity streams. Water Resour Res 27: 629–642. doi:10.1029/90WR02768
106. Chamier A-C (1987) Effect of pH on microbial degradation of leaf litter in seven streams of the English Lake District. Oecologia 71: 491–500. doi:10.1007/BF00379287
107. Gleason FH, Kagami M, Lefevre E, Sime-Ngando T (2008) The ecology of chytrids in aquatic ecosystems: roles in food web dynamics. Fungal Biol Rev 22: 17–25. doi:10.1016/j.fbr.2008.02.001
108. MacKenzie DI, Royle JA, Brown JA, Nichols JD (2004) Occupancy estimation and modeling for rare and elusive populations. In: Thompson WL, editor. Sampling rare or elusive species. Washington, DC: Island Press. pp. 149–172.
109. Hines JE, Nichols JD, Royle JA, MacKenzie DI, Gopalaswamy AM, et al. (2010) Tigers on trails: occupancy modeling for cluster sampling. Ecol Appl 20: 1456–1466.
110. Royle JA, Link WA (2006) Generalized site occupancy models allowing for false positive and false negative errors. Ecology 87: 835–841.
111. Miller DA, Nichols JD, McClintock BT, Grant EHC, Bailey LL, et al. (2011) Improving occupancy estimation when two types of observational error occur: non-detection and species misidentification. Ecology 92: 1422–1428. doi:10.1890/10-1396.1
112. Miller DAW, Nichols JD, Gude JA, Rich LN, Podruzny KM, et al. (2013) Determining occurrence dynamics when false positives occur: estimating the range dynamics of wolves from public survey data. PLoS ONE 8: e65808. doi:10.1371/journal.pone.0065808
113. Berven KA (1990) Factors affecting population fluctuations in larval and adult stages of the wood frog (*Rana sylvatica*). Ecology 71: 1599–1608.
114. Blaustein AR, Han B, Fasy B, Romansic J, Scheessele EA, et al. (2004) Variable breeding phenology affects the exposure of amphibian embryos to ultraviolet radiation and optical characteristics of natural waters protect amphibians from UV-B in the US Pacific Northwest: Comment. Ecology 85: 1747–1754.

8

Laurentian Great Lakes Phytoplankton and Their Water Quality Characteristics, Including a Diatom-Based Model for Paleoreconstruction of Phosphorus

Euan D. Reavie[1]*, Adam J. Heathcote[2], Victoria L. Shaw Chraïbi[3]

1 Center for Water and the Environment, Natural Resources Research Institute, University of Minnesota Duluth, Duluth, Minnesota, United States of America, **2** Groupe de Recherche Interuniversitaire en Limnologie, Département des Sciences Biologiques, Université du Québec à Montréal, Montréal, Québec, Canada, **3** Department of Earth and Atmospheric Sciences, University of Nebraska-Lincoln, Lincoln, Nebraska, United States of America

Abstract

Recent shifts in water quality and food web characteristics driven by anthropogenic impacts on the Laurentian Great Lakes warranted an examination of pelagic primary producers as tracers of environmental change. The distributions of the 263 common phytoplankton taxa were related to water quality variables to determine taxon-specific responses that may be useful in indicator models. A detailed checklist of taxa and their environmental optima are provided. Multivariate analyses indicated a strong relationship between total phosphorus (TP) and patterns in the diatom assemblages across the Great Lakes. Of the 118 common diatom taxa, 90 (76%) had a directional response along the TP gradient. We further evaluated a diatom-based transfer function for TP based on the weighted-average abundance of taxa, assuming unimodal distributions along the TP gradient. The r^2 between observed and inferred TP in the training dataset was 0.79. Substantial spatial and environmental autocorrelation within the training set of samples justified the need for further model validation. A randomization procedure indicated that the actual transfer function consistently performed better than functions based on reshuffled environmental data. Further, TP was minimally confounded by other environmental variables, as indicated by the relatively large amount of unique variance in the diatoms explained by TP. We demonstrated the effectiveness of the transfer function by hindcasting TP concentrations using fossil diatom assemblages in a Lake Superior sediment core. Passive, multivariate analysis of the fossil samples against the training set indicated that phosphorus was a strong determinant of historical diatom assemblages, verifying that the transfer function was suited to reconstruct past TP in Lake Superior. Collectively, these results showed that phytoplankton coefficients for water quality can be robust indicators of Great Lakes pelagic condition. The diatom-based transfer function can be used in lake management when retrospective data are needed for tracking long-term degradation, remediation and trajectories.

Editor: Francois G. Schmitt, CNRS, France

Funding: This research was largely supported by grants to E. Reavie from the U.S. Environmental Protection Agency under Cooperative Agreements GL-00E23101 and GL-00E00790-2. This document has not been subjected to the Agency's required peer and policy review and therefore does not necessarily reflect the view of the Agency, and no official endorsement should be inferred. Sedimentary assemblages from Lake Superior were collected and assessed under research sponsored by the Minnesota Sea Grant College Program supported by the NOAA office of Sea Grant, United States Department of Commerce, under grant No. R/CC-01-10. The funders had no role in study design, data collection and analysis, decision to publish, or preparation of the manuscript.

Competing Interests: The authors have declared that no competing interests exist.

* Email: ereavie@d.umn.edu

Introduction

Recent trends from monitoring data demonstrate rapid changes in the Laurentian Great Lakes system associated with anthropogenic drivers. Most notably, changes in the water quality [1], phytoplankton (unpublished data), zooplankton [2] and benthos [3] collectively confirm or suggest detriments associated with invasive species, nutrient imbalances and climate. While algal abundances have been inordinately low in some regions (e.g., Lake Huron; unpublished data), new eutrophication problems have also manifested as substantial algal blooms (e.g., [4]). Such quantitative assessments of environmental problems have been valuable to support lake management [5] because they provide information on the abundance of various organisms, traits that are critical to understanding and managing food webs. However, taxon-specific evaluations of pelagic indicators in the Great Lakes are lacking. In addition to these more intuitive evaluations of organism numbers and biomass, taxon-specific ecological data can provide additional tools to inform management [6].

Developing effective indicators of ecological condition requires that they be calibrated to identify their responses to important environmental stressors [7]. The main goals of calibration are to identify environmental characteristics of potential indicator taxa so that they may be subsequently used to infer condition. Assemblages of algae, which are physiologically subject to water quality, have the potential to provide time-integrated inferences of limnological conditions. Such bioindicators are particularly needed to monitor the impacts of human activities that are increasing nutrient supplies to water bodies, introducing non-native species, and changing climate. Great Lakes coastal algae, particularly diatoms, have been shown to provide a more temporally integrated assessment of water quality conditions than discrete water quality measurements [8], but similar indicators have not been developed for the pelagic ecosystem. Diatoms are particularly advantageous as indicators in paleolimnological studies because their diagnostic cell walls (frustules) remain archived in lake sedimentary records [9]. Hence, the fossil diatom assemblages may be used to reconstruct historical degradation and recovery and estimate future trajectories of condition.

The characteristics of the Great Lakes phytoplankton community as indicators of the prevailing eutrophication that was noted in the 1960s and 1970s was well described by Stoermer [10]. At that time little taxonomic treatment of the phytoplankton had been accomplished, so Stoermer advocated the indicator potential that would result from better taxonomic assessment of the flora and species autecology. Significant collections of phytoplankton have occurred since that time, such as the United States Environmental Protection Agency's (USEPA) biological monitoring program that has been active since 1983 [11], among other monitoring data (e.g., [12]). We aimed to provide the tools to develop species- and assemblage-based indicators from Great Lakes phytoplankton data.

Modern datasets (also known as training sets) provide the basis for developing indicator transfer functions by relating contemporary assemblages with their corresponding environmental measurements [13]. Algal assemblages, in particular, are proven robust indicators of stressors such as nutrients ([14],[15]), water clarity [16] and acidification [17], as well as a suite of other water quality problems in freshwater ecosystems [18]. Training sets using diatoms have been developed for lakes, rivers and the coastlines of the Great Lakes [8]. Algae are known to have definable optima along gradients of environmental conditions. Species tend to be taxonomically distinct and abundant, and they respond rapidly to changing conditions. Hence, by using training set data to calibrate the environmental characteristics of algae, researchers can use changes in community composition to classify and quantify long-term environmental changes that result from anthropogenic activities. This is particularly needed because few pelagic monitoring data were collected prior to the 1970s; existing data provide a spatially and temporally incomplete picture of environmental conditions in the lakes. Species-environmental coefficients should allow reconstructions of past conditions based on sedimentary assemblages.

A transfer function is derived by relating taxa assemblages (usually diatoms) in a training set of samples (in this case, pelagic samples collected throughout the Great Lakes) to an environmental variable of interest [19]. The transfer function consists of taxa coefficients (environmental optima and, optionally, tolerances) that can be used to infer quantitative information about the variable of interest based on the abundance of each taxon in a sample assemblage. These assemblages are usually characterized from recently accumulated surface sediments [20]; to our knowledge, a transfer function based on diatom phytoplankton samples has not been attempted.

For the current assessment, modeling approaches were applied to characterize the environmental coefficients of the Great Lakes phytoplankton taxa. A checklist of the common taxa and their ecological indicator values for a suite of water quality variables is presented. To illustrate one way these coefficients may be used, a diatom-based transfer function was developed for paleolimnological applications. Transfer function evaluation and testing typically involve the comparison of algal-inferred water quality to measured water quality to evaluate function robustness, which is usually characterized by a coefficient of determination (r^2) and a prediction error. It is well-known that adjacent phytoplankton samples (e.g. 5–100 km apart) from the Great Lakes tend to have similar species assemblages and environmental conditions, and this lack of independence among sites can violate statistical assumptions. Such autocorrelation may result in misleading estimates of transfer function performance [21], and with new analyses revealing previously undocumented weaknesses with diatom-based nutrient models [22], it is imperative that transfer functions undergo thorough testing. So, we evaluated how (1) autocorrelation, (2) relationships between taxon abundance and phosphorus and (3) redundancies among environmental variables affected our training set and its predictive power. We further tested the model by applying it to fossil diatom assemblages from Lake Superior to determine whether pelagic diatom-inferred phosphorus serves as an effective paleoecological tool in the Great Lakes.

Materials and Methods

We evaluated phytoplankton data collected as part of the USEPA's biological monitoring program for the Great Lakes. The standard operating procedure for phytoplankton collection and analysis is described in detail in the published procedures [11], but abbreviated details were as follows. The EPA data were based on twice-annual synoptic sampling ("spring" = typically the month of April, "summer" = typically the month of August) from standard stations throughout the Great Lakes basin. Our analyses focused on phytoplankton and water quality samples collected from 2007 through 2011, a total of 717 unique sampling events. No specific permissions were required for these locations or activities; namely, sample locations in the pelagic Great Lakes. Field studies did not involve endangered or protected species.

Whole water samples were collected from the rosette sampler on-board the Research Vessel Lake Guardian. Phytoplankton samples were composites of water sampled at discrete depths from the euphotic zone of the water column. For isothermal spring samples, the sample integrated equal volumes of water from 1, 5, 10 and 20 m. In shallower locations in Lake Erie the 20-m sample was replaced by an above-bottom collection. If the total depth was less than 15 m, equal volumes were integrated from surface, mid-depth and above-bottom samples. For the stratified (summer) water column, equal volumes were taken from 1 m, 5 m, 10 m and the lower epilimnion and integrated. If the epilimnion was very shallow, equal volumes were integrated from a maximum of four and a minimum of two sampling depths. Samples were split and analyzed separately for the whole phytoplankton assemblage (i.e., "soft" algae) and diatoms. Analysis of soft algae used the quantitative Utermöhl method of counting preserved specimens in a settling chamber on an inverted microscope [23]). During soft algal analyses, diatoms containing cytoplasm were identified as centric or pennate forms. The second split sample was digested in nitric acid and subsequently in peroxide to isolate the diatom valves which were then plated on slides and counted using oil immersion (1000× or higher) to identify taxa. All counting included measurements of cell dimensions so that algal biovolumes could be calculated. Ultimately, analyses afforded detailed taxonomic resolution, and data were available in quantitative data formats (cell density [cells/ml] and biovolume [μm^3/ml]).

A 30-cm sediment core from eastern Lake Superior was used as a test subject for a diatom-based phosphorus transfer function. Details of sediment coring and sample preparation are provided by Shaw Chraïbi et al. [24]. The stratigraphic record of diatom assemblages dating back to the early 1700s was the subject of testing in the present assessment.

Algal associations with 11 environmental variables (Table 1), collected simultaneously with phytoplankton, were explored. Additional variables were considered but were redundant (specific conductivity with chloride, beam attenuation with turbidity) or poorly represented in the dataset (dissolved oxygen, irradiance, total nitrogen). Geophysical variables (e.g., depth, latitude, longitude) were not used in multivariate analyses in order to better describe relationships with water quality, although the importance of "lake" as a nominal variable was explored in terms of species specificity. Collection and analysis of environmental data is described in detail in the USEPA (2010) standard operating procedures. Three additional molar ratio variables (N:P, N:Si, P:Si) were calculated from other variables in the list so that they may be used in the evaluation of algal responses across nutrient ratio gradients. Because of substantial correlation among these variables (Figure 1), multivariate approaches were deemed necessary to determine species-environmental relationships.

All analyses were performed using the R statistical software [25] and associated packages. All scripts and datasets are available from the authors on request. Phytoplankton and water quality data are archived in the Great Lakes Environmental Database (http://www.epa.gov/greatlakes/monitoring/data_proj/glenda), and the paleoecological data are provided electronically in Shaw Chraïbi [26].

Ordination

Multivariate ordinations were performed using the R package "vegan" [27]. An initial principal components analysis (PCA) was used to characterize the major environmental gradients in the water quality data. All environmental variables were scaled to zero mean and unit variance. The first two principal components (axes), which are a linear combination of the greatest variation in environmental condition across sites, were used as new environmental variables (Figure 1) in the exploration of algal species autecologies.

For diatom-environmental analyses, redundancy analysis (RDA), the constrained form of PCA, was performed to condense the complex dataset into summary variables (axes) that capture the majority of environmental variation (Juggins and Birks [28]). An RDA with a Monte Carlo permutation test was used to identify the environmental variables with significant ($P<0.05$) relationships with the diatom data. The subset of significant variables was then selected to generate a new RDA to identify species-environmental relationships and orthogonal gradients that capture variation in the diatom data. Explained variation from this RDA was compared to a partial RDA that partitioned variation in the diatom data by characterizing the total and unique contributions

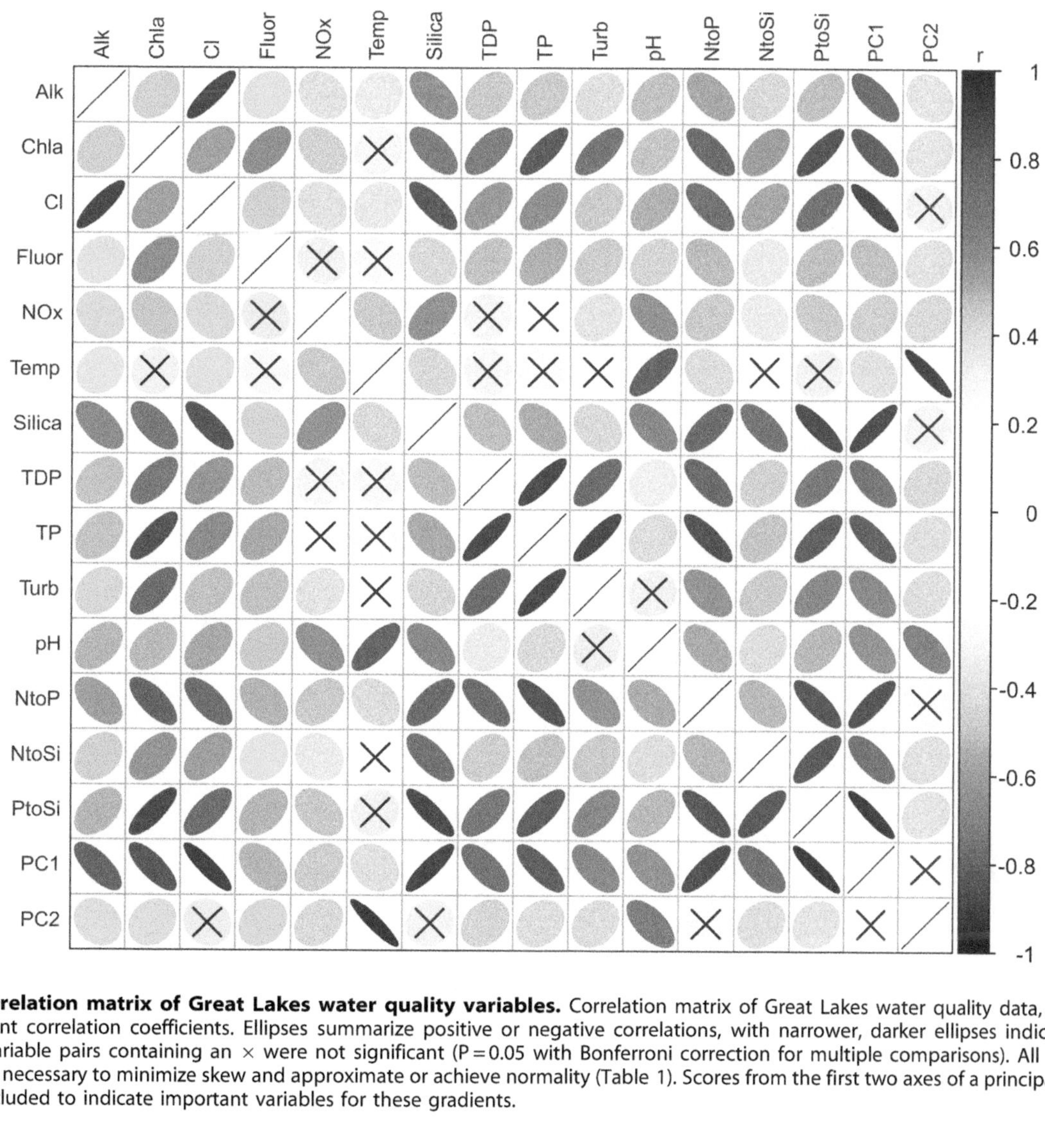

Figure 1. Correlation matrix of Great Lakes water quality variables. Correlation matrix of Great Lakes water quality data, using Pearson product-moment correlation coefficients. Ellipses summarize positive or negative correlations, with narrower, darker ellipses indicating stronger correlations. Variable pairs containing an × were not significant ($P=0.05$ with Bonferroni correction for multiple comparisons). All variables were transformed as necessary to minimize skew and approximate or achieve normality (Table 1). Scores from the first two axes of a principal components analysis are included to indicate important variables for these gradients.

Table 1. The environmental variables used in this study.

Variable	Abbreviation	Transformation	Minimum	5th percentile	Median	Mean	95th percentile	Maximum
Alkalinity (mg/L)	Alk	sqrt	41.90	43.00	86.00	78.40	109.00	124.00
Chlorophyll a (µg/L)	Chla	log	0.00	0.44	1.03	2.54	10.12	35.39
Chloride (mg/L)	Cl	sqrt	1.20	1.29	11.57	10.39	22.69	29.27
Fluorescence (µg/L)	Fluor	log	0.00	0.11	1.39	2.41	7.53	50.96
Nitrates + Nitrites (mg/L)	NOx	sqrt	0.00	0.08	0.32	0.30	0.43	1.63
Temperature (°C)	Temp	sqrt	0.11	1.02	7.19	11.23	24.99	26.81
Silica (mg/L)	Silica	-	0.00	0.02	0.82	0.69	1.11	1.63
Dissolved Phosphorus (µg/L)	TDP	log	0.45	0.85	1.90	2.75	6.40	69.70
Total Phosphorus (µg/L)	TP	log	0.72	1.54	2.96	6.26	18.08	226.78
Turbidity (NTU)	Turb	log	0.11	0.16	0.30	1.64	5.23	178.00
pH	pH	-	7.58	7.89	8.31	8.33	8.80	9.20
N:P (molar ratio)	NtoP	sqrt	0.00	0.01	0.10	0.10	0.22	0.37
N:Si (molar ratio)	NtoSi	log	0.00	0.29	0.37	1.49	5.74	120.95
P:Si (molar ratio)	PtoSi	log	0.98	1.58	3.37	84.48	414.08	3702.76
Principal Component 1	PC1	-	−0.91	−0.68	0.12	0.00	0.46	0.58
Principal Component 2	PC2	-	−0.63	−0.52	0.04	0.00	0.50	2.29

Provided are untransformed ranges of the variables and the transformations applied to remove skew. A principal components analysis was performed to derive the summary axes (PCs).

that can be attributed to TP. This was further applied to the variables alkalinity (Alk), chloride (Cl) and nitrates + nitrites (NOx) to examine similar characteristics for variables that may be confounded with TP (e.g., Alk, NOx) or have strong independent controls on diatom autecology (e.g, Cl).

Species-environmental relationships

Algae-based inference models typically assume the taxa respond unimodally across the environmental gradient of interest [29]; however, based on many possible factors (e.g., narrow gradients, multivariate controls, inadequate sampling or unique ecological distributions) taxa may vary in their responses, or have no apparent response. It is important to characterize the goodness of fit of species to the environmental variable, as well as to consider linear, Gaussian or skewed responses along gradients in order to validate the assumption that diatom species responses actually reflect the environmental variable of interest.

Responses of algae taxa along environmental gradients were evaluated using linear, Gaussian and weighted-average (WA) assumptions. First, environmental variables were transformed as necessary to minimize skew and achieve normality in their frequency distribution, if possible. Taxa that were observed in more than five samples were evaluated. Responses were evaluated using four measures of algal abundance: cell densities (cells/ml), biovolume (μm^3/ml), relative cell density and relative biovolume.

Using the "stats" (version 3.0.0) package [25] the function "lm" was used to generate a linear model, then the function "cor.test" was used to determine linear model relationships between taxon abundance and each environmental variable. Significance ($P = 0.05$) was tested based on Pearson's product moment correlation coefficient.

Using the "gam" (version 1.08) package [30], the "gam" function was used to fit Gaussian model relationships between taxon abundance and each environmental variable. Based on a comparison of residuals, a paired t-test ($P = 0.05$) was applied to determine whether a Gaussian fit was better than a linear fit. The point on the environmental gradient where a Gaussian fit was maximized was considered a taxon's optimum. The environmental weighted-average optimum for a taxon was calculated based on the average of environmental values where the taxon occurred, weighted by taxon abundance. Taxon optima based on Gaussian and WA calculations were rescaled to a range of 0 to 10, which was based on the range of the transformed (if applied) environmental variable. This rescaling resulted in more user-friendly coefficients (i.e. "environmental indicator values"), but coefficients based on actual variable units are available from the authors.

To support the utility of taxa included in the diatom-based transfer function, generalized additive models (GAMs [31]) were used to test the significance of the response of each diatom taxon to the measured TP gradient using the R package "mgcv" [32]. The additive models test several response forms (linear, unimodal, skewed) across the selected gradient. We further tested the significance of the unique explanatory power of TP once the effect of alkalinity (another important water quality variable in the Great lakes) was removed. GAMs were tested on the relative density data for taxa with 10 or more occurrences in samples and significance of fitted models was assessed using 199 Monte Carlo permutations of the deviance explained by TP for each species response GAM. (Although a less stringent criterion for occurrence [five samples] was used in the development of indicator values, here we use 10 samples to ensure sufficient abundance to identify responses along the TP gradient.) Data were evaluated based on relative density and relative biovolume as these are the data types that may be used in diatom-based paleoecological studies.

Diatom-based transfer functions

Transfer functions for Great Lakes total phosphorus (TP) were derived using the diatom components of the algal assemblages. As such a function is most likely to be used for down-core analyses in paleolimnological assessments [29], the soft algae that are less likely to preserve in the sedimentary record were not used. Transfer functions were developed using weighted averaging (WA) regression with inverse de-shrinking with log-normal taxa transformation ("rioja" package [33]). Diatom-inferred estimates of TP (DI-TP) for each sample were calculated by taking the optimum of each taxon to that variable, weighting it by its abundance in that sample, and calculating the average of the combined weighted taxa optima ([34], [35]). The apparent strengths of the transfer functions were evaluated by calculating the squared correlation coefficient (r^2) and the root mean square error of prediction (RMSEP) between measured TP and transfer function estimates of those values for all samples. The predictive error was estimated by jackknifing [36] which corrects for under-estimation of root-mean squared error due to predicting values only from samples that were included in the model. As above, variations of these functions were tested using two possible representations of the algal data, relative cell density and relative biovolume. Developing such functions based on absolute abundances was not deemed useful for future applications because species abundances in sedimentary assemblages (e.g., cells per g dry sediment) cannot be assumed to be comparable to prevailing phytoplankton concentrations (e.g., cells per mL in the water column).

Transfer function testing

We evaluated redundancy in the training set using the modern analog technique (MAT; [37]) with five analogs. Instead of using the weighted abundance of taxa, a reconstructed value was based on the five most taxonomically similar samples in the training set. The effect of autocorrelation on transfer function performance was evaluated using the "rne" function in the R package "palaeoSig" [38]. The rne function compares the performance (r^2) of the transfer function during cross-validation as sites are deleted (1) randomly, (2) that are geographically close to the test site, and (3) that are environmentally similar to the test site [21]. Reducing the number of sites will worsen the performance as the number of potential analogues decreases [39]. In the case of autocorrelation, deleting the nearest sites would remove the best analogues, and should be more detrimental to performance than random deletion. Such degradation would also be expected when sites in close environmental proximity (e.g., with similar TP) are selectively deleted. If performance deteriorates more by deletion of close sites than by deletion of environmental neighbors, the performance loss relative to the random deletion must in part be due to spatial autocorrelation between diatom assemblages.

We further checked the diatom-TP transfer function performance against several randomized simulations using multiple iterations of the "multi.mat" function in palaeoSig [21]. This method simulates the environmental variable across sites using the same autocorrelation structure as the measured data and recalculates performance (r^2). Creating simulated environmental data requires many geographical considerations, including development of an empirical variogram to determine the spatial structure of TP in the Great Lakes, fitting a theoretical variogram model, and a simulation to generate a spatially structured random variable with the same spatial structure as the original data. We employed the gstat package [40] in R to support these calculations, and the method is detailed by Telford and Birks [21]. A transfer function is considered to have statistically significant predictive power ($p<0.05$) if the real value of r^2 for the transfer function exceeds 95% of the simulations.

Application to paleolimnology

Reconstructed estimates of past TP concentrations were obtained from sedimentary samples collected as part of a recent paleolimnological analysis of eastern Lake Superior (Shaw Chraïbi et al. [24]). Despite the training set samples representing seasonal snapshots of modern diatom assemblages as opposed to sedimentary assemblages containing year-round integrations of settling diatoms, paleo-assemblages had several good modern analogues in the training set, thus providing preliminary evidence that the training set may be applicable (Shaw Chraïbi [26]). Diatom-inferred TP from the uppermost sediment samples matched well with contemporary TP monitoring data (Shaw Chraïbi et al. [24]). Downcore TP estimates were calculated using WA with inverse de-shrinking (Juggins and Birks [28]). To track the historical trajectory in relation to historical diatom conditions, sedimentary assemblages were projected on the RDA containing the suite of training set diatom data and significant environmental variables.

We then calculated the ratio of two values: (1) λ_R, the variance in sedimentary diatom assemblages captured by the first axis of a RDA constrained to DI-TP; (2) λ_P, the variance explained by the first axis of an unconstrained PCA of the diatom data. The λ_R/λ_P ratio expresses the proportion of variation explained by DI-TP as a fraction of the maximum explainable variance in the sedimentary samples. We also calculated the correlation coefficient for past DI-TP versus corresponding axis 1 sample scores from the unconstrained PCA of the sedimentary diatom assemblages. We would expect both of these values to be high (i.e. close to a proportion of 1.0) if downcore assemblages were strongly related to changes in TP [22].

Results and Discussion

Taxon-specific results

263 taxa were sufficiently abundant (i.e. in more than five samples) to be considered for evaluation along environmental gradients. The common taxa comprised centric diatoms (42 taxa), pennate diatoms (103 taxa), chrysophytes (37 taxa), green algae (45 taxa), cryptomonads (8 taxa), blue-green algae (17 taxa), euglenoids (2 taxa), dinoflagellates (6 taxa) and some unknown entities (3 taxonomic categories). Some were unknown species or genera not identifiable to the species level (a genus followed by "spp.") and others were members of known divisions but with few diagnostic characteristics (e.g., "unidentifiable chrysophyte ovoid"). Although these less specific taxonomic categories are difficult to understand ecologically due to the likelihood of multiple species comprising a category, we present them as possible inference tools because several of them had significant relationships with environmental variables.

Table 2 presents environmental characteristics for select taxa relative to nine variables, as well as their lake and seasonal specificity, based on observed algal biovolumes and relative cell densities in phytoplankton samples. It is beyond the scope of this article to characterize the numerous species-specific mechanisms for these environmental relationships, but some of these taxa are recognized as important components of the Great Lakes phytoplankton community. The full suite of species coefficients are provided in supplementary appendices that detail species coefficients based on cell densities (Table S1), relative cell density (Table S2) biovolume (Table S3) and relative biovolume (Table S4). These tables further characterize within-lake conditions and may serve as a set of coefficients suited to further indicator development and monitoring assessments.

Several taxa have high (e.g., *Aulacoseira distans*) or low (e.g., *Cyclotella michiganiana*) optima for TP. However, it is interesting that all of the significant linear species relationships with TP are positive (Table 2, Table S3). A similar negative effect occurred for silica. This appears to be due to the dominating effect of summer in Lake Erie, a time when TP is high, silica is low, and several algal taxa are highly abundant, thus driving positive slopes in the biovolume-TP relationship. A comparative examination of relative densities (Table S2) indicates a greater diversity in species responses to TP because relative abundances were more evenly distributed across lakes and environmental gradients.

Based on weighted abundance across all samples, the centric diatom *Aulacoseira islandica* is the most abundant alga in the Great Lakes phytoplankton ([41]). However, it has a very specific spring bloom period in Lake Erie, with very low occurrence across the other lakes. Its distribution reflects high chlorophyll *a* and fluorescence, which is not surprising as it comprises most of Lake Erie's spring algal bloom. *A. islandica* has a significantly negative relationship with silica, which seemingly counters its very high silica requirement in the formation of its heavily silicified cell walls. This may imply that by the time spring sampling occurs in April of each year diatoms have exhausted the dissolved silica in the water that is available to incorporate into their cell walls.

While structurally similar to *A. islandica*, *Aulacoseira granulata* is a distinctly summer diatom in Lake Erie. *A. granulata* has an apparent tolerance for warmer, late-summer conditions when it coexists with green and blue-green algae [42] that account for high chlorophyll concentrations. *Fragilaria crotonensis* is a pennate diatom that also occurs in the summer in Lake Erie, with lower abundances in Ontario and Michigan. It has high optima for pH and temperature and appears to be associated with a low N:P ratio.

The cyanophyte *Microcystis aeruginosa* is a well-known coastal bloom species in Lake Erie [43], and it has been observed abundantly in recent offshore phytoplankton samples [42]. Our assessment confirms its fairly strict occurrence in the summer in Lake Erie, as well as its occurrence at times of high chlorophyll *a*, temperature and pH, as well as low nitrates and N:P.

Three cryptomonad algae were abundant: *Cryptomonas rostratiformis*, *C. reflexa* and *C. erosa*. These species occurred year-round in all of the lakes. While seemingly cosmopolitan, there were clear distinctions within this genus, such as *C. reflexa* being tolerant of high chloride and temperature while *C. rostratiformis* occurred under conditions of low chloride and temperature.

The diatom-based model

Under the full diatom-inferred TP model, optimal model parameters included use of WA with tolerance down-weighting, which offered minimal improvement. The model was developed using log-transformation of the TP data and it was cross-validated using the jackknifing procedure [35]. There were no transformations of the taxonomic data or removal of taxa or samples as these procedures did not improve apparent model performance. We tested removal of benthic and rare taxa (i.e., occurring in fewer than 5 samples) but such considerations also degraded model performance. Comparisons between observed TP and DI-TP resulted in a jackknifed r^2 of 0.79 and RMSEP of 0.38 (log-TP units) using relative density data and an r^2 of 0.77 and RMSEP of 0.36 using relative biovolumes (Table 3).

Based on GAM analyses of relative density data for each common diatom taxon, 90 out of 118 taxa (76%) taxa had a

Table 2. Environmental indicator values for select algae taxa in the Great Lakes.

	Season		Lake					Chlorophyll a			Chloride			Fluorescence			Nitrates + Nitrites			Temperature			Silica			Total Phosphorus			pH			N:P		
Taxon Full Name	**Spring**	**Summer**	**Superior**	**Michigan**	**Huron**	**Erie**	**Ontario**	**Lin**	**Gau**	**WA**	**Lin**	**Gau**	**WA**	**Lin**	**Gau**	**WA**	**Lin**	**Gau**	**WA**	**Lin**	**Gau**	**WA**	**Lin**	**Gau**	**WA**	**Lin**	**Gau**	**WA**	**Lin**	**Gau**	**WA**	**Lin**	**Gau**	**WA**
Aulacoseira distans (Ehr.) Simonsen	9	1	1	7	1	1	0	+	3*	2	+	5*	5		7	2		10	3	+	3*	3	−	10*	6	+	10*	2		3	3	−	3	3
Aulacoseira granulata (Ehr.) Simonsen	0	10	0	0	0	10	0	+	10*	8		6	6	+	8	6		0*	1	+	10	10		3	4		7	5	+	10	8		0	0
Aulacoseira islandica (O. Mull.) Simonsen	10	0	0	0	0	10	0	+	10*	7	+	7*	7	+	10*	5	−	0*	1		3	2	−	0*	0	+	5*	5	+	6*	5	−	0*	0
Cyclotella michiganiana Skv.	0	10	0	7	2	0	0		0	2	+	5	5		0	1	−	2*	2	+	9*	8	−	4*	4		0	1	+	6*	6	−	3	3
Fragilaria crotonensis Kitton	0	10	0	1	0	8	2	+	6	5	+	7*	7	+	6	4	−	0*	2	+	10*	9	−	1*	2	+	4	3	+	10*	7	−	1*	1
Cryptomonas erosa Ehr.	3	7	1	1	0	6	2	+	6	4	+	7*	6	+	4	3	−	0*	2	+	10*	7	−	0	3	+	5	3	+	10*	6	−	0*	1
Cryptomonas reflexa Skuja	2	8	1	1	0	6	2	+	6	4	+	6*	6	+	4*	3	−	0*	2	+	10*	8	−	4	3	+	5	3	+	10*	6	−	0*	1
Cryptomonas rostratiformis Skuja	3	7	4	0	0	2	4		6	3	+	10*	4		1*	2	−	1	2	+	8*	6	−	0*	4		3	2	+	0*	5		10	3
Microcystis aeruginosa (Kütz.) emend. Elenkin	0	10	0	0	0	10	0		10	9		6*	6	+	8	6		0*	1	+	10*	10		3*	4	+	7*	6		10	9		0*	0

Select algae taxa in the Great Lakes and their environmental indicator values based on biovolume distributions across the environmental gradients. Negative (−) and positive (+) signs indicate significant linear relationships ($P < 0.05$) based on the Pearson product moment correlation. Gaussian optima (Gau) display an asterisk (*) if fitted residuals were significantly smaller (t-test; $P < 0.05$) than residuals from the linear fit. WA indicates optima based on weighted-averaging. "Season" and "Lake" coefficients reflect seasonal and spatial occurrence of the taxa. Indicator values from 0 through 10 respectively represent whether a given taxon had its greatest modeled abundance at a low or high region of a variable's gradient. Indicator numbers for seasons or lakes represent the relative proportion that each taxon achieved in 2007–2011, again scaled from 0 to 10.

Table 3. Results of DI-TP model testing using diatom data based on relative cell densities and biovolumes.

	Density	Biovolume
Model $r^2_{jackknife}$	0.79	0.77
Model RMSEP (log-TP units)	0.38	0.36
Number of taxa with significant GAM relationship to TP (out of 118)	90	89
Number of taxa with significant relationship to TP after factoring out Alk (out of 118)	65	67
Percentile of actual model in simulated remapping of training set	99.8%	99.8%
RDA 1 eigenvalue	16.7%	20.0%
RDA 2 eigenvalue	6.4%	7.4%
Total variance explained by TP	6.8%	9.1%
Unique variance explained by TP	4.8%	6.0%
Downcore DI-TP versus PCA 1 r	0.98	0.68

significant response across the TP gradient. After removing the effects of alkalinity, 65 out of 118 (55%) had a significant response to TP, indicating that 25 of the taxa with significant relationships with TP had a signal that was confounded with alkalinity. Comparable results using relative biovolumes revealed 89 taxa with significant responses to phosphorus, 67 after accounting for alkalinity. Basing the model on diatom densities or biovolumes appeared to have little effect on model performance. It is more typical in paleolimnological studies to use count data without the additional benefit of cell measurements to calculate biovolumes, so subsequent figures focus on the use of relative diatom valve densities.

In a check of the effects of sample removal in the training set, random removal of samples had a small effect on apparent model performance (Figure 2). For instance, the observed-inferred r^2 decreased from 0.77 for the full model to 0.66 after removal of 80% of samples. However, removal of spatially and environmentally similar sites had a more substantial impact on performance, indicating autocorrelation among sites. Removal of sites that were geographically close to the test site in cross-validation resulted in a sudden drop in the r^2, as low as 0.13 after sites within 300 km were removed. This result is not surprising given the known uniqueness of the phytoplankton assemblages in each lake [44], so removal of training set samples in the same lake as the test site undoubtedly had a substantial effect on assemblage analogs. A similarly precipitous drop in performance occurred due to removal of sites that were environmentally similar to the test site, but model degradation was not noticeably different from that due to spatial removal. The relative importance of geographical and environmental neighbors shows how important adjacent sites are for the performance of the transfer function. As explained by Telford and Birks [21], the occurrence of autocorrelation indicates potential problems in a transfer function, and so the modern analog technique (MAT [37]) of inference is not recommended for this model. However, the model may still have predictive power under weighted-averaging scenarios. Based on models derived by random re-mapping of the environmental data over the species map, the actual model performed better 99.8% of the time (Table 3; Figure S1), indicating that the diatom-TP WA transfer function is statistically significant, suggesting that TP can be reconstructed from this training set.

An RDA of the diatom assemblages constrained by the 11 significant environmental variables indicated that these variables account for a major part (16.7% on the first axis; Table 3, Figure 3) of the underlying gradients in the diatom relative density data. Axis 1 of an unconstrained PCA captured 20.6% of the variation in the diatom data, indicating that the significant environmental variables accounted for a large portion of the underlying species gradients. Environmental data accounted for a larger portion of the variance in diatom relative biovolume data (20.0%; Table 3). TP (and the related TDP) were strongly correlated to axis 1, the main gradient of species turnover. There is also a clear alkalinity/chloride gradient that diagonally represents axes 1 and 2. Axis 2 captures a smaller proportion of variance (6.4%), although it is closely correlated to the NOx gradient. These proportions of explained variance are typical and informative in multivariate analyses of biological abundance data that tend to be noisy (Gauch [45]).

Figure 4 illustrates variance partitioning of the diatom relative density data by selected environmental variables. The total and unique effects of TP were 6.8% and 4.8% respectively (Table 3), a 29% reduction in explained variance. Based on relative biovolumes the total and unique effects were 9.1% and 6.0%, indicating a closer linkage between TP and diatom biovolumes than densities. That the unique TP effect is lower than total is due to redundancies, such as correlation to variables like chlorophyll *a* and turbidity (Figure 1). While it is difficult to confirm the relative importance of these variance partitioning values, the unique variance explained by TP surpasses that observed for TP training sets developed in Europe [46] and Minnesota [47], [48]; 3.9% and 2.5% respectively, as presented by Juggins et al. [22]. Of the selected variables, TP independently explained the most variation in the assemblage data, in contrast to alkalinity and chloride which had more than 70% reduction from total to uniquely explained variance. NOx explained the least variation but was minimally confounded by other variables. For TP to be confounded with other correlated variables is not surprising, but these results suggest that changes in the diatom assemblages across the Great Lakes are meaningfully related to TP.

Downcore testing of the diatom-based model

Figure 5 presents the common diatom taxa (>5% in any sample) in a sediment core from eastern Lake Superior (unpublished data). The stratigraphy is based on relative diatom density, including the DI-TP results which indicate a temporary enrichment event. This inference was largely driven by a short-term abundance of *A. islandica* which temporarily replaces lower-nutrient *Cyclotella* and *Discostella* species through much of the 20th century. Passively plotting the Lake Superior fossil samples on the RDA traces the path of historical changes in the diatom

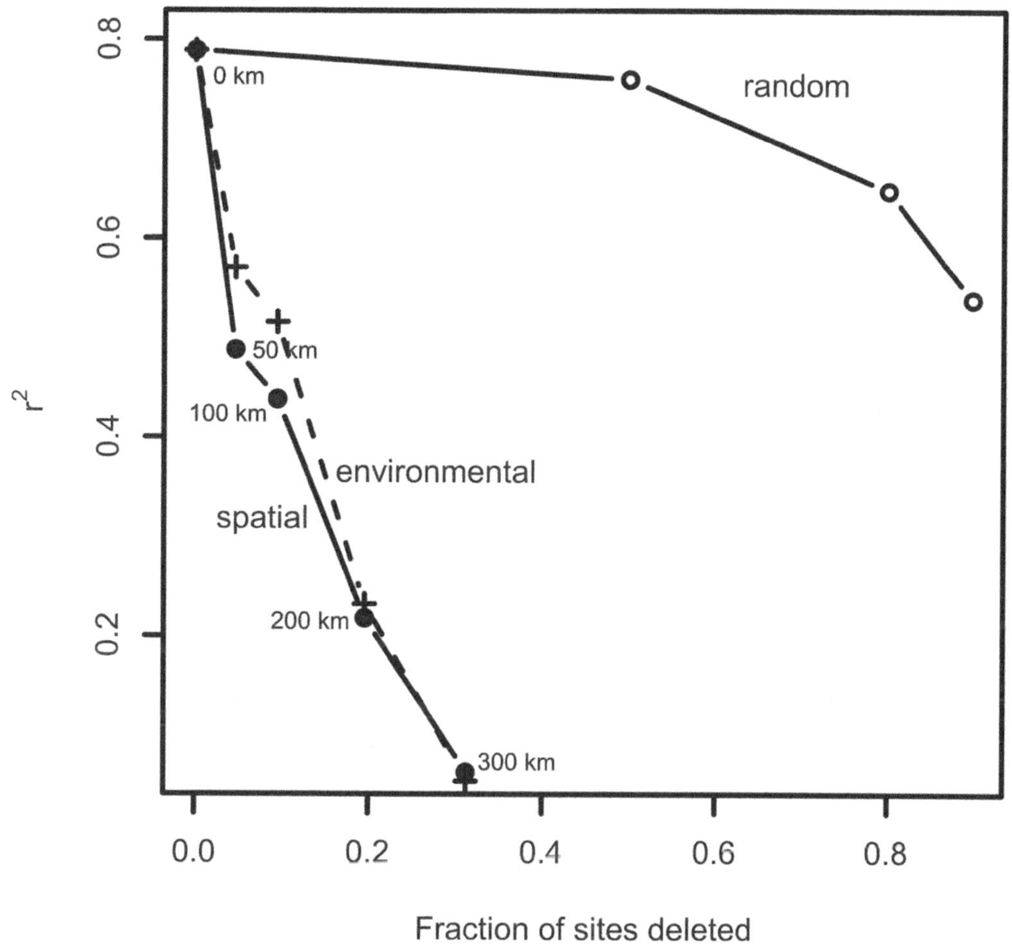

Figure 2. Effect of deleting sites on diatom-TP model performance. The effect on transfer function r^2 of deleting sites at random (open circles), from the geographical neighborhood of the test site (filled circles), and those that are environmentally similar (crosses) during cross-validation of the Great Lakes diatom training set. The radius of each distance is labelled.

assemblages (Figure 3). The long-term trend indicates a temporary movement of the scores in the direction of the TP vector from 1922 through 1970, a period that reflects nutrient enrichment in Lake Superior in association with the increased abundance of *Aulacoseira islandica* (unpublished data; [49]). Subsequent to 1970 the sample scores revert to locations close to pre-Euro-American samples, indicating TP reductions as the diatom assemblage shifted to a greater dominance by *Cyclotella* (unpublished data).

There was a high correlation ($r = 0.98$; $N = 26$) between DI-TP and the main gradient in the Lake Superior sedimentary diatom data (Figure S2), indicating TP was a strong driver of Superior's assemblages. The fraction of variance explained by DI-TP, represented as a fraction of the maximum explainable variance, was also high ($\lambda_R/\lambda_P = 0.97$). These results suggest that phosphorus has been an important determinant of diatom assemblages in Lake Superior, and that the TP transfer function was well-suited to reconstructing the nutrient history of the lake.

General Discussion

We provide a suite of environmental coefficients for the dominant phytoplankton taxa in the Laurentian Great Lakes. We hope that these data increase understanding of taxon-specific ecology, and expect these data will be used to develop inference models such as the diatom-based model tested herein. We endorse using these species coefficient tables (Table 2, Tables S1, S2, S3, S4) as checklists for future indicator studies. For instance, Kelly and Whitton [50] provided a checklist of trophic coefficients for diatom specimens, calibrated to nutrient and organic pollution variables, from rivers in the United Kingdom. Their coefficients have been used repeatedly in ecological assessments intended to manage river ecosystems [51]. Model-specific coefficients (linear, Gaussian, WA) and their statistical significance may also be used selectively in development of an indicator.

We recommend some cautions in using these data to infer environmental conditions. The coefficients are taxon-environmental relationships based on simultaneous sampling of algae and water quality. A phytoplankton assemblage may be a product of water quality conditions prior to sampling. For instance, despite the low silica optimum for *Aulacoseira islandica*, we know this taxon requires high silica concentrations for cell wall development [52], and so the Lake Erie spring diatom assemblages are undoubtedly a product of high silica concentrations prior to spring sampling. So, while high densities of *A. islandica* may infer low dissolved silica, the taxon may not represent a prevailing condition of low silica. We expect coefficients based on water quality variables that were not limiting at the time of sampling (e.g., spring nutrients [53]) are more reliable.

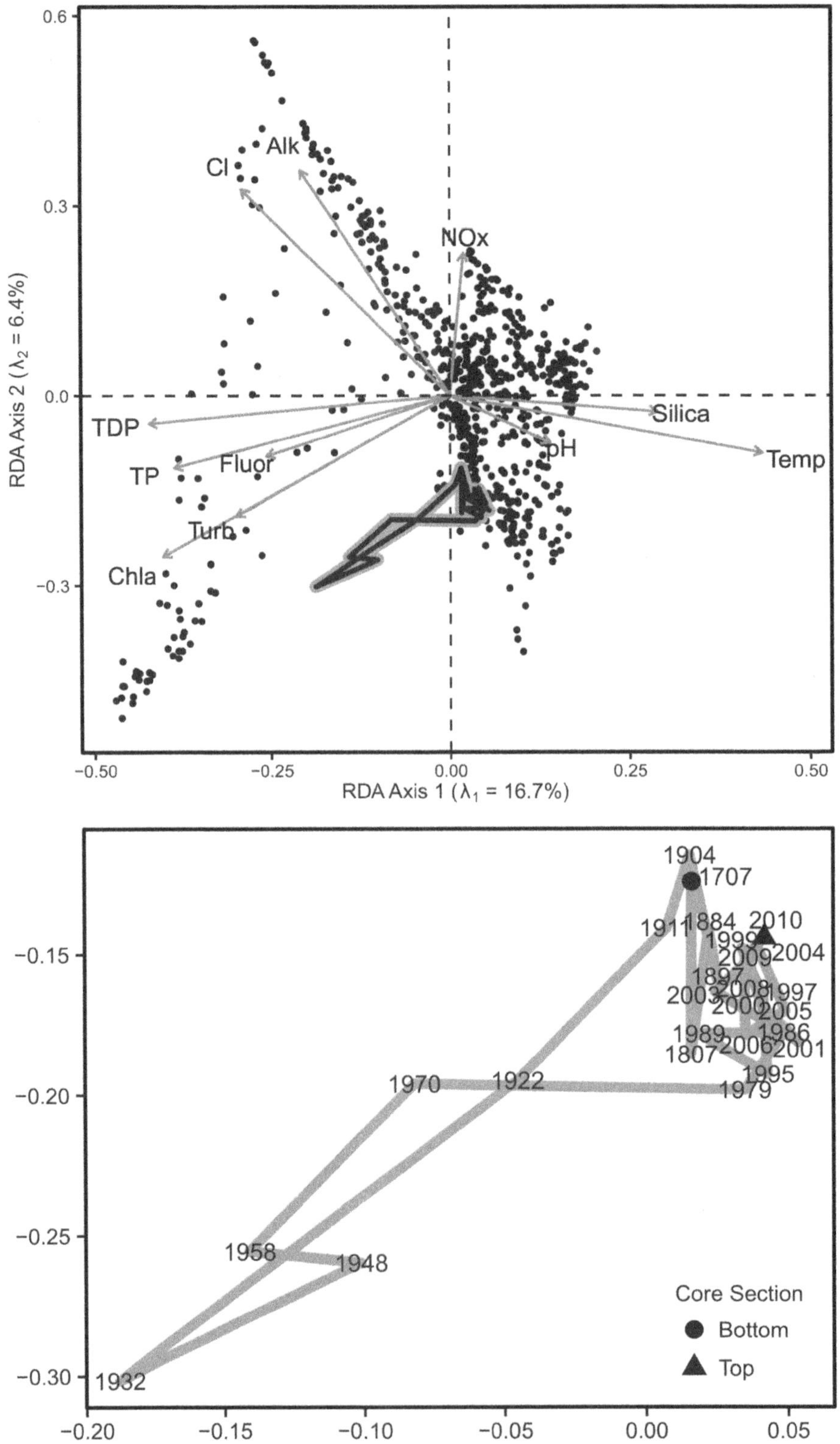

Figure 3. Redundancy analysis of the diatom training set constrained to significant water quality variables. Eigenvalues indicate variance explained by each axis. Passive plotting of the Lake Superior sedimentary assemblages is indicated by the black line set in a thicker gray line. The lower panel is a zoomed-in plot of the passive ordination of Lake Superior sedimentary diatom assemblages. The core top (2010) and bottom (~1707) are indicated and the date of each interval (based on ^{210}Pb dating) is provided for each interval.

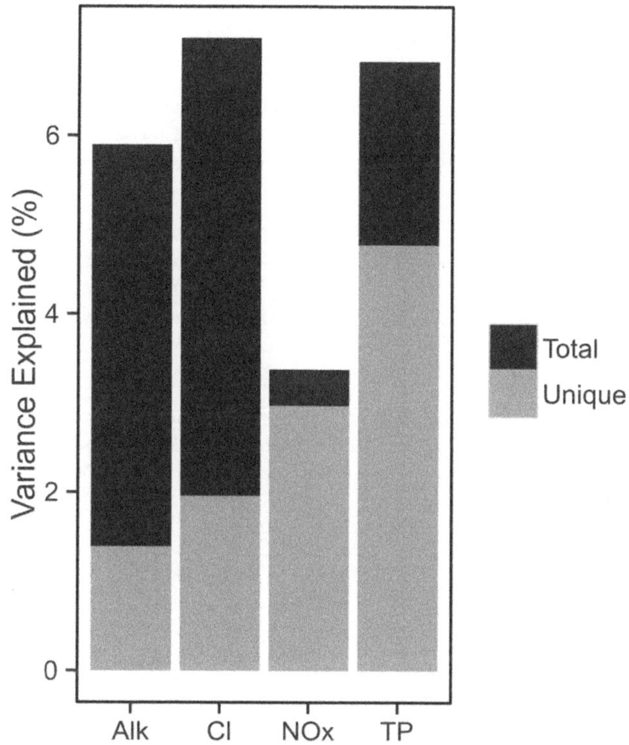

Figure 4. Histograms show percentage of variation in data explained by four water quality gradients. Total height indicates the total variation explained while gray height indicates the unique fraction of variance captured by that variable.

The provided transfer function, a diatom-based total phosphorus model, shows promise for future applications in the Great Lakes. Although internal testing of the model showed substantial spatial and environmental autocorrelation among sample locations, which is to be expected in large lakes that are likely homogeneous across vast areas, further testing indicates that phosphorus is an important driver of diatom assemblages in the Great Lakes, and that paleolimnological studies may benefit from this indicator model.

Unlike the snapshot approach of phytoplankton sampling, diatom assemblages retrieved from sediment cores reflect year-round integration of valves. For our Lake Superior example we confirm good analogues between fossil and training set assemblages, as well as a credible long-term TP reconstruction. Past assemblages, which included relatively high abundances of *A. islandica* during the mid-20th century, differ greatly from the phytoplankton in Lake Superior today, so the TP reconstruction relied on taxon coefficients that were largely derived from other lakes that presently contain abundant *A. islandica* (e.g., Lake Erie). It is unknown whether we may assume similar performance from cores collected elsewhere in the Great Lakes system, so we recommend thorough downcore validation of the model as we have done here, and as others recommend [22], before making conclusions about paleoecological nutrient trends. Also, the reconstructed range of TP for Lake Superior was on the low end of the range for the Great Lakes as a whole, so future testing should consider whether reconstructions in more nutrient-rich lakes are possible.

Lake Superior has encountered substantial diatom assemblage reorganization in the last 200 years, and as speculated by others [49] these changes were largely driven by anthropogenic nutrients (i.e. the rise and fall of pollutant flux to the lake). A sedimentary record containing similarly strong historical revisions in assem-

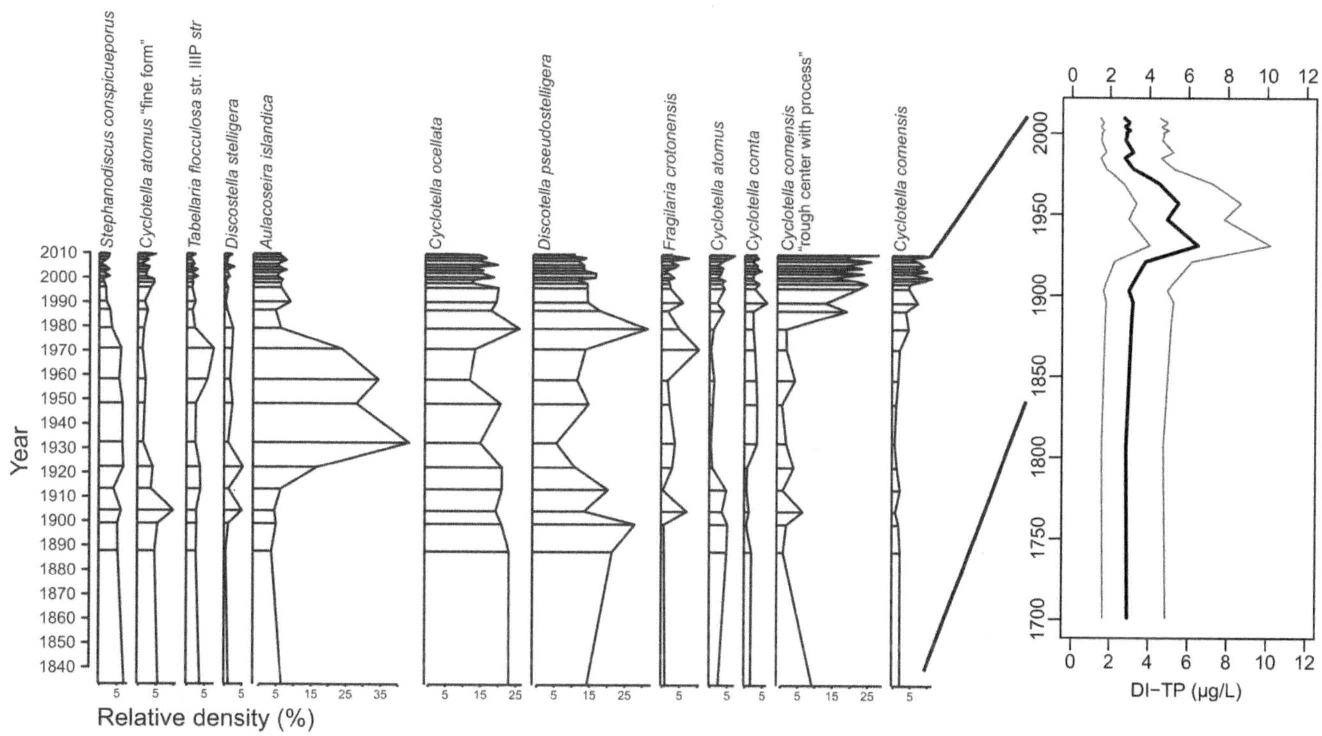

Figure 5. Dominant diatom species (>5% relative abundance) for the eastern core of Lake Superior. A plot of corresponding DI-TP for the fossil assemblages is shown on the right. Black line indicates inferred TP and grey lines indicate the range of model error (RMSEP).

blages should have a strong assemblage-DI relationship as we observed for TP in Lake Superior, otherwise an investigator must explore other variables as the cause of historical shifts. However, monotonous stratigraphic records of diatoms will have narrow ecological breadth, and a probable narrow range of DI-TP. Such cases may result in a weak correlation between assemblage patterns and DI data, so other means would be necessary to confirm the importance of TP to the sedimentary record.

In conclusion, the taxon-specific autecological information provided herein may be used to inform future observations of these species within the Great Lakes and elsewhere. More importantly, Great Lakes phytoplankton assemblages can make a contribution to indicator studies as suggested by Stoermer [10]. We present one of many indicator possibilities; the transfer function presented here appears to be suitable for downcore reconstructions. Using similar rule-based models with clear validation protocols should lead to refined, defensible environmental predictions and reconstructions.

Supporting Information

Figure S1 Performance of 1000 transfer functions with simulated environmental variables based on re-mapping of the TP dataset. 998 simulated transfer functions had poorer performance than the actual model r^2 based on the observed-inferred TP relationship, as indicated by the dotted line.

Figure S2 Relationship between Lake Superior downcore DI-TP and PCA axis 1 for diatom data.

Table S1 Environmental indicator values for the common algae taxa (>5%) in the Great Lakes based on distribution of cell densities across the environmental gradients.

Table S2 Environmental indicator values for the common algae taxa (>5%) in the Great Lakes based on distribution of relative cell densities across the environmental gradients.

Table S3 Environmental indicator values for the common algae taxa (>5%) in the Great Lakes based on biovolume distribution across the environmental gradients.

Table S4 Environmental indicator values for the common algae taxa (>5%) in the Great Lakes based on distribution of relative biovolume across the environmental gradients.

Acknowledgments

M. Agbeti supported assessments of the recent GLNPO phytoplankton samples. The U.S. Government is authorized to reproduce and distribute reprints for government purposes, notwithstanding any copyright notation that may appear hereon.

Author Contributions

Conceived and designed the experiments: EDR AJH VLSC. Performed the experiments: EDR AJH VLSC. Analyzed the data: EDR AJH VLSC. Contributed reagents/materials/analysis tools: EDR. Wrote the paper: EDR AJH.

References

1. Chapra SC, Dove A, Warren GJ (2012) Long-term trends of Great Lakes major ion chemistry. J Great Lakes Res 38:550–560.
2. Vanderploeg HA, Pothoven SA, Fahnenstiel GL, Cavaletto JF, Liebig JR, et al. (2012) Seasonal zooplankton dynamics in Lake Michigan: Disentangling impacts of resource limitation, ecosystem engineering, and predation during a critical ecosystem transition. J Great Lakes Res 38: 336–352.
3. Barbiero RP, Schmude K, Lesht BM, Riseng CM, Warren GJ, et al. (2011) Trends in Diporeia populations across the Laurentian Great Lakes, 1997–2009. J Great Lakes Res 37: 9–17.
4. Michalak AM, Anderson EJ, Beletsky D, Boland S, Bosch NS, et al. (2013) Record-setting algal bloom in Lake Erie caused by agricultural and meteorological trends consistent with expected future conditions. Proc Nat Acad Sci 110: 6448–6452.
5. McCormick PV, Cairns J Jr (1994) Algae as indicators of environmental change. J Appl Phycol 6: 509–526.
6. Holt EA, Miller SW (2011) Bioindicators: Using organisms to measure environmental impacts. Nat Educ Knowl 3: 8.
7. Brazner JC, Danz NP, Niemi GJ, Regal RR, Trebitz AS, et al. (2007) Evaluation of geographic, geomorphic and human influences on Great Lakes wetland indicators: a multi-assemblage approach. Ecol Indic 7: 610–635.
8. Reavie ED, Axler RP, Sgro GV, Danz NP, Kingston JC, et al. (2006) Diatom-based weighted-averaging transfer functions for Great Lakes coastal water quality: relationships to watershed characteristics. J Great Lakes Res 32: 321–347.
9. Dixit SS, Smol JP, Kingston JC, Charles DF (1992) Diatoms: Powerful indicators of environmental change. Environ Sci Technol 26: 22–33.
10. Stoermer EF (1978) Phytoplankton assemblages as indicators of water quality in the Laurentian Great Lakes. Trans Am Microsc Soc 97: 2–16.
11. U.S. Environmental Protection Agency (2010) Sampling and Analytical Procedures for GLNPO's Open Lake Water Quality Survey of the Great Lakes. Great Lakes National Program Office, U.S. Environmental Protection Agency, document EPA 905-R-05-001, Chicago, IL.
12. Munawar M, Munawar IF (1982) Phycological studies in Lakes Ontario, Erie, Huron, and Superior. Can J Bot 60: 1837–1858.
13. Imbrie J, Kipp NG (1971) A new micropaleontological method for quantitative paleoceanography based on microfossils. Dev Mar Geol 1:523–563.
14. Tibby J (2004) Development of a diatom–based model for inferring total phosphorus in southeastern Australian water storages. J Paleolim 31: 23–36.
15. Meriläinen JJ, Hynynen J, Palomäki A, Mäntykoski K, Witick A (2003) Environmental history of an urban lake: A palaeolimnological study of Lake Jyväsjärvi, Finland. J Paleolim 30:387–406.
16. Dixit SS, Smol JP (1994) Diatoms as indicators in the Environmental Monitoring and Assessment Program–Surface Waters. Environ Monit Assess 31: 275–306.
17. Siver PA, Ricard R, Goodwin R, Giblin AE (2003) Estimating historical in–lake alkalinity generation from sulfate reduction and its relationship to lake chemistry as inferred from algal microfossils. J Paleolim 29: 179–197.
18. Smol JP (2002) Pollution of Lakes and Rivers: A Paleoenvironmental Perspective. London, UK: Arnold Publishers.
19. Charles DF (1990) A checklist for describing and documenting diatom and chrysophyte calibration data sets and equations for inferring water chemistry. J Paleolim 3: 175–178.
20. Last WM, Smol JP, Birks HJ, editors (2001) Tracking environmental change using lake sediments, vol 3: Terrestrial, algal, and siliceous indicators. Springer.
21. Telford RJ, Birks HJB (2009) Evaluation of transfer functions in spatially structured environments. Quat Sci Rev 28: 1309–1316.
22. Juggins S, Anderson NJ, Hobbs JMR, Heathcote AJ (2013) Reconstructing epilimnetic total phosphorus using diatoms: statistical and ecological constraints. J Paleolim 49: 373–390.
23. Utermöhl H (1958) Zur Vervollkommnung der quantitativen Phytoplankton-Methodik. Mitt Int Verein Limnol 9:1–38.
24. Shaw Chraïbi VL, Kireta AR, Reavie ED, Brown TN, Cai M (2014) A Paleolimnological Assessment of Human Impacts on Lake Superior. J Great Lakes Res (in press).
25. R Core Team (2013) R: A language and environment for statistical computing. R Foundation for Statistical Computing, Vienna, Austria. URL http://www.R-project.org/
26. Shaw Chraïbi VL (2013) A 250-year assessment of human impacts on Lake Superior: an updated paleolimnological perspective. M.Sc. Thesis, University of Minnesota Duluth. Available: http://conservancy.umn.edu/bitstream/11299/157356/1/Chraibi_umn_0130M_13654.pdf. Accessed 2014 June18.
27. Oksanen J, Blanchet FG, Kindt R, Legendre P, Minchin PR, et al. (2013) Vegan: Community ecology package. R package version 2.0-9. http://CRAN.R-project.org/package = vegan
28. Juggins S, Birks HJB (2012) Quantitative environmental reconstructions from biological data. In: Birks HJB, Lotter AF, Juggins S, Smol JP, editors. Tracking environmental change using lake sediments. Springer Netherlands. pp 431–494.

29. Battarbee RW, Jones VJ, Flower RJ, Cameron NG, Bennion H, et al. (2001) Diatoms. In Smol JP, Birks HJB, Last WM, editors. Tracking environmental change using lake sediments. vol 3: Terrestrial, algal, and siliceous indicators. Dordrecht Kluwer Academic Publishers. pp. 155–202.
30. Hastie TJ (2013) gam: Generalized additive models. R package version 1.08. http://CRAN.R-project.org/package = gam
31. Yee TW, Mitchell ND (1991) Generalized additive models in plant ecology. J Veg Sci 2: 587–602.
32. Wood SN (2006) Generalized additive models: an introduction with R. Chapman Hall/CRC, Boca Raton.
33. Juggins S (2012) rioja: Analysis of quaternary science data, R package version (0.8–5). (http://cran.r-project.org/package = rioja).
34. Birks HJB, Line JM, Juggins S, ter Braak CJF (1994) WACALIB version 3.3 – a computer program to reconstruct environmental variables from fossil assemblages by weighted averaging and to derive sample–specific errors of prediction. J Paleolim 10: 147–152.
35. Juggins S, Birks HJB (2012) Quantitative environmental reconstructions from biological data. In: Birks HJB, Lotter AF, Juggins S, Smol JP, editors. Tracking environmental change using lake sediments: Data handling and statistical techniques. Dordrecht: Springer. pp.431–494.
36. Efron B, Gong G (1983) A leisurely look at the bootstrap, the jackknife, and cross-validation. Am Stat 37: 36–48.
37. Overpeck JT, Webb T, Prentice IC (1985) Quantitative interpretation of fossil pollen spectra: Dissimilarity coefficients and the method of modern analogs. Quat Res 23: 87–108.
38. Telford RJ (2013) palaeoSig: Significance tests of quantitative palaeoenvironmental reconstructions. R package version 1.1-2. http://cran.r-project.org/web/packages/palaeoSig
39. Reavie ED, Juggins S (2011) Exploration of sample size and diatom-based indicator performance in three North American phosphorus training sets. Aquat Ecol 45:529–538.
40. Pebesma EJ (2004) Multivariable geostatics in S: the gstat package. Comput Geosci 30:683–691.
41. Reavie ED, Barbiero RP, Allinger LE, Warren GJ (2014) Phytoplankton trends in the Laurentian Great Lakes: 2001–2011. J Great Lakes Res (in press).
42. Allinger LE, Reavie ED (2013) The ecological history of Lake Erie as recorded by the phytoplankton community. J Great Lakes Res 39: 365–382.
43. Millie DF, Fahnenstiel GL, Dyble Bressie J, Pigg RJ, Rediske RR, et al. (2009) Late-summer phytoplankton in western Lake Erie (Laurentian Great Lakes): bloom distributions, toxicity, and environmental influences. Aquat Ecol 43: 915–934.
44. Barbiero RP, Tuchman ML (2001) Results from the US EPA's biological open water surveillance program of the Laurentian Great Lakes: I. Introduction and phytoplankton results. J Great Lakes Res 27: 134–154.
45. Gauch HG (1982) Multivariate analysis in community ecology. Cambridge: Cambridge University Press. 298 p.
46. Bennion H, Juggins S, Anderson NJ (1996) Predicting epilimnetic phosphorus concentrations using an improved diatom-based transfer function and its application to lake eutrophication management. Environ Sci Technol 30: 2004–2007.
47. Ramstack JM, Fritz SC, Engstrom DR, Heiskary SA (2003) The application of a diatom-based transfer function to evaluate regional water-quality trends in Minnesota since 1970. J Paleolim 29: 79–94.
48. Reavie ED, Robbins JA, Stoermer EF, Douglas MS V, Morehead NR, et al. (2005) Paleolimnology of a fluvial lake downstream of Lake Superior and the industrialized region of Sault Saint Marie. Can J Fish Aquat Sci 62: 2586–2608.
49. Stoermer EF, Kociolek CL, Schelske CL, Conley DJ (1985) Siliceous microfossil succession in the recent history of Lake Superior. Proc Acad Nat Sci Phila 137: 106–118.
50. Kelly MG, Whitton BA (1995) The trophic diatom index: a new index for monitoring eutrophication in rivers. J Appl Phyc 7: 433–444.
51. Szczepocka E, Szulc B (2009) The use of benthic diatoms in estimating water quality of variously polluted rivers. Oceanol Hydrobiol Stud 38: 17–26.
52. Barbiero RP, Carrick HJ, Volerman JB, Tuchman ML (2000) Factors affecting temporal and spatial distribution of diatoms in Lake Michigan. Verh Int Verein Limnol 27: 1788–1794.
53. Lean DRS, Abbott AP, Charlton MN, Rao SS (1983) Seasonal phosphate demand for Lake Erie plankton. J Great Lakes Res 9: 83–91.

9

Biomarkers of Dissolved Oxygen Stress in Oysters: A Tool for Restoration and Management Efforts

Heather K. Patterson[1,2]*, Anne Boettcher[3], Ruth H. Carmichael[1,2]

1 Department of Marine Sciences, University of South Alabama, Mobile, Alabama, United States of America, **2** Dauphin Island Sea Lab, Dauphin Island, Alabama, United States of America, **3** Department of Biology, University of South Alabama, Mobile, Alabama, United States of America

Abstract

The frequency and intensity of anoxic and hypoxic events are increasing worldwide, creating stress on the organisms that inhabit affected waters. To understand the effects of low dissolved oxygen stress on oysters, hatchery-reared oysters were placed in cages and deployed along with continuously recording environmental data sondes at a reef site in Mobile Bay, AL that typically experiences low oxygen conditions. To detect and measure sublethal stress, we measured growth and survival of oysters as well as expression of three biomarkers, heat shock protein 70 (HSP70), hypoxia inducible factor (HIF) and phospho-p38 MAP kinase, in tissues from juvenile and adult oysters. Survival rates were high for both juvenile and adult oysters. Expression levels of each of the 3 isoforms of HSP 70 were negatively correlated to dissolved oxygen (DO) concentrations, suggesting that HSP 70 is useful to quantify sublethal effects of DO stress. Results for HIF and phospho-p38 MAP kinase were inconclusive. Test deployments of oysters to assess expression of HSP 70 relative to environmental conditions will be useful, in addition to measuring abiotic factors, to identify appropriate sites for restoration, particularly to capture negative effects of habitat quality on biota before lethal impacts are incurred.

Editor: Kay C. Vopel, Auckland University of Technology, New Zealand

Funding: The funders had no role in study design, data collection and analysis, decision to publish, or preparation of the manuscript. Funding provided by University of South Alabama Oyster Restoration Program, and the Northern Gulf Institute - BP Gulf Research Intiative (Phase 1) to MESC - to R. Carmichael.

* Email: hkpatt@uab.edu

Introduction

Restoration and management efforts are ongoing worldwide to enhance the area and function of oyster reefs to compensate for those that have gone functionally extinct [1]. Declines are linked to overharvesting, presence of non-native species, and disease as well as to degradation of environmental conditions, including increasing incidence of hypoxia [1,2]. Pollack et al. [3] created a suitability index for oyster restoration based on salinity, temperature, turbidity, dissolved oxygen (DO) and depth, which demonstrated that identifying suitable sites for restoration is critical to success. This suitability index and associated reef quality index, however, could be taken a step further to include measurements of physiological stress, particularly to capture sublethal stress, further avoid species loss, and improve restoration success.

Anoxic and hypoxic conditions are pervasive and likely to be enduring because they are engendered by natural, as well as anthropogenic causes [4]. Low DO concentration conditions occur naturally due to water column stratification, normal biological processes and lack of wind mixing [5]. Anthropogenic low DO concentration conditions are typically related to increased inputs of land-derived nitrogen (N) to coastal waters that lead to phytoplankton blooms and their subsequent decomposition that consumes oxygen [6,7]. Excessive or sustained low DO conditions contribute to loss of commercially important fishery stocks [8–10] and hinder associated restoration efforts [11,12]. There are ongoing efforts world-wide to understand and manage the causes and effects of hypoxia on coastal ecosystems [7,13]. Currently, the threshold for hypoxia is typically set at 2 mg L^{-1} DO concentration due to the significant impacts to ecosystems in terms of species diversity and abundance at and below this level, with 4 mg L^{-1} DO concentration set as the minimum level for maintenance of a healthy system (for review see [14]).

Low DO stress can have lethal and sublethal effects on many species, including oysters [15,16]. Sublethal affects can lead to reduced growth, reduced feeding, and increased susceptibility to disease [16,17]. Sublethal stress from low DO in relatively tolerant species such as oysters is not well understood, especially cumulative stress; given low DO concentration events are often episodic in nature [5,8]. Several molecular biomarkers have been identified that may prove useful in examining low DO induced sublethal stress [18]. The term "biomarker" was first used in relation to environmental contamination, a biological response to chemical(s) given some amount of exposure [19]. Heat shock protein 70 (Hsp70) has been used broadly as a biomarker for environmental stress, and while first discovered in response to heat shock, it has been shown to respond to a variety of stressors [20]. Additional biomarkers for DO stress include hypoxia inducible factor (HIF) and phospho-p38 MAPK. Hsp70 has been increasingly used to study effects of a variety of abiotic stressors, but is not well studied in invertebrates [21].

Hsp70 is considered a molecular chaperone, it is responsible for maintaining folding, minimizing proteins aggregating in a non-native state and targeting denatured proteins for removal during normal cell functioning and during stress (reviewed in [22–25]). Hsp70 is highly conserved through evolutionary history [26]. In

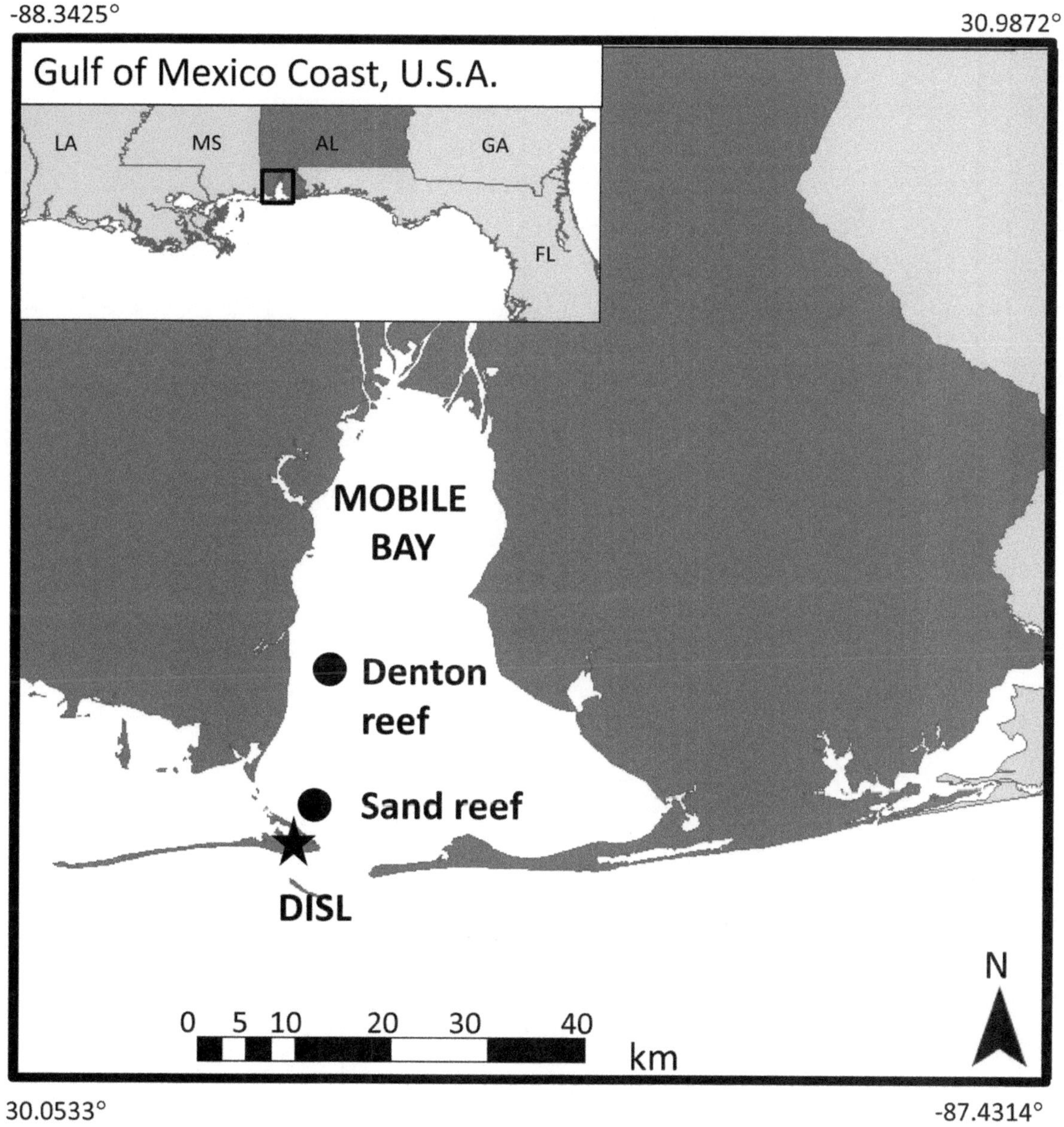

Figure 1. Sand Reef and Denton reef study site in Mobile Bay, AL near the Dauphin Island Sea Lab (DISL).

oysters, there are 3 isoforms of Hsp70 (Hsc77, Hsc72, Hsp69). The first two are constitutively expressed but there can be increased expression in response to stress, while the third is induced as a result of stress [27]. HIF is a transcription factor, which under low oxygen conditions leads to an increase in the transcription of genes involved in responses to low oxygen conditions [28,29]. The signaling pathway involving HIF has been well studied in vertebrates, but initial studies in invertebrate systems suggested it might not be important or even present. The genetic sequence for HIF and associated transcriptional changes in response to hypoxia have been documented in the oyster *Crassostrea virginica* (Gmelin, 1791), suggesting that it may be a valuable marker for low DO concentration stress in oysters as in vertebrates [30]. p38 MAPKs are part of a group of protein kinases that are intermediaries between extracellular stress and the resulting increase in expression of other pathways in the cell nucleus [31,32]. In marine mussels and snails, low DO concentration exposure was shown to increase p38 MAPK expression [33,34]. In vertebrate systems the four isoforms of phospho-p38 MAPK responded to different stressors and not every isoform is present in every tissue type [31], with alpha and gamma isoforms implicated in hypoxia response [35]. The role of different isoforms in invertebrates has not been examined. To date there have been mixed results reported for the usefulness of biomarkers for hypoxia in invertebrates [36,37], indicating that more detailed study is needed and suggesting that multiple biomarkers might be important.

To examine the response of oysters to low DO concentration, a field study was performed in Mobile Bay, AL (Fig. 1), an area that experiences periodic extended low DO concentrations. To capture responses to low DO relative to ontogeny and relative to the range of bottom relief typical to reef sites, adult and juvenile oysters were

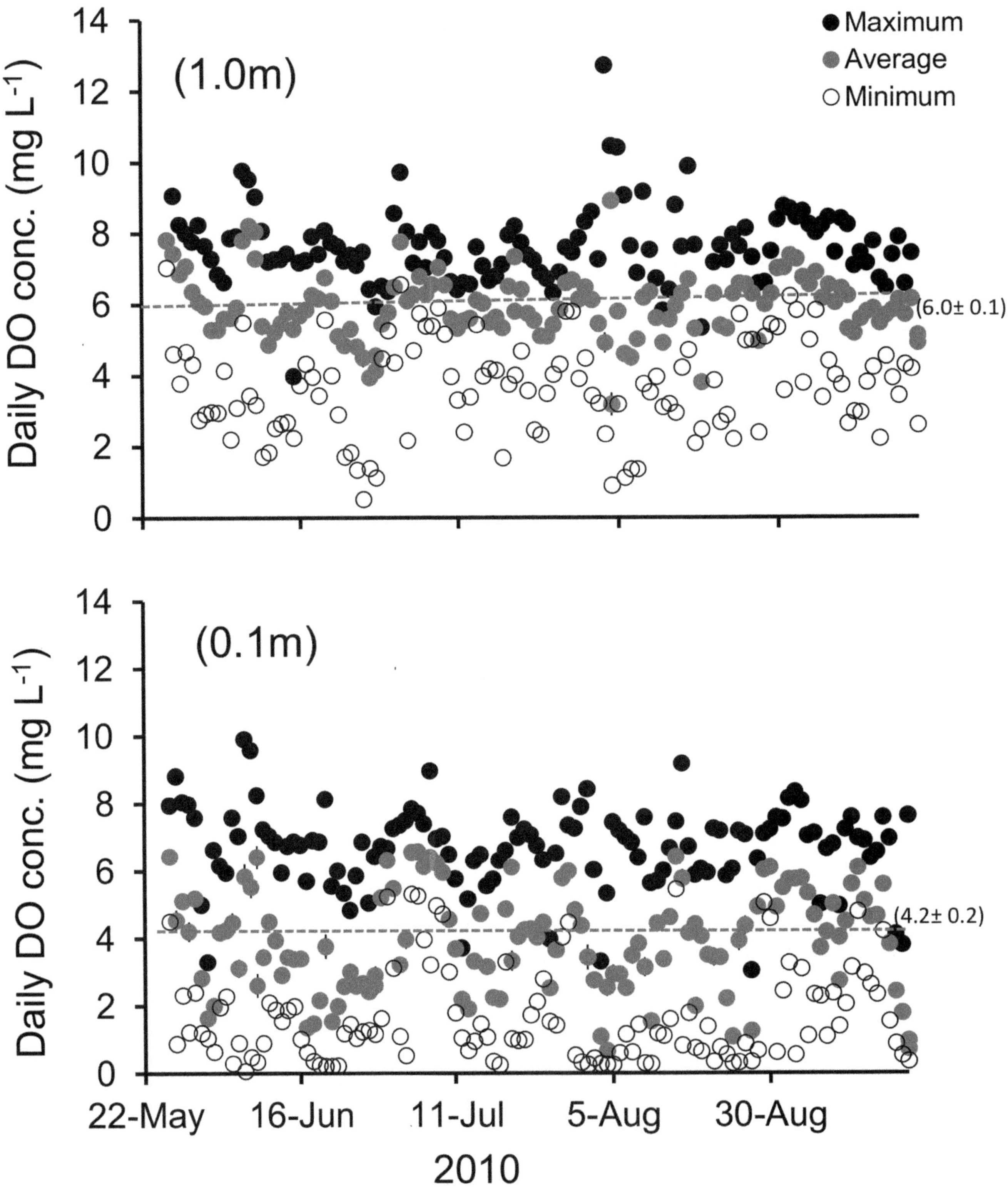

Figure 2. Daily maximum, average, and minimum dissolved oxygen (DO) concentrations for the top (1.0 m off bottom) and bottom (0.1 m off bottom) depth at Sand Reef, Mobile, AL.

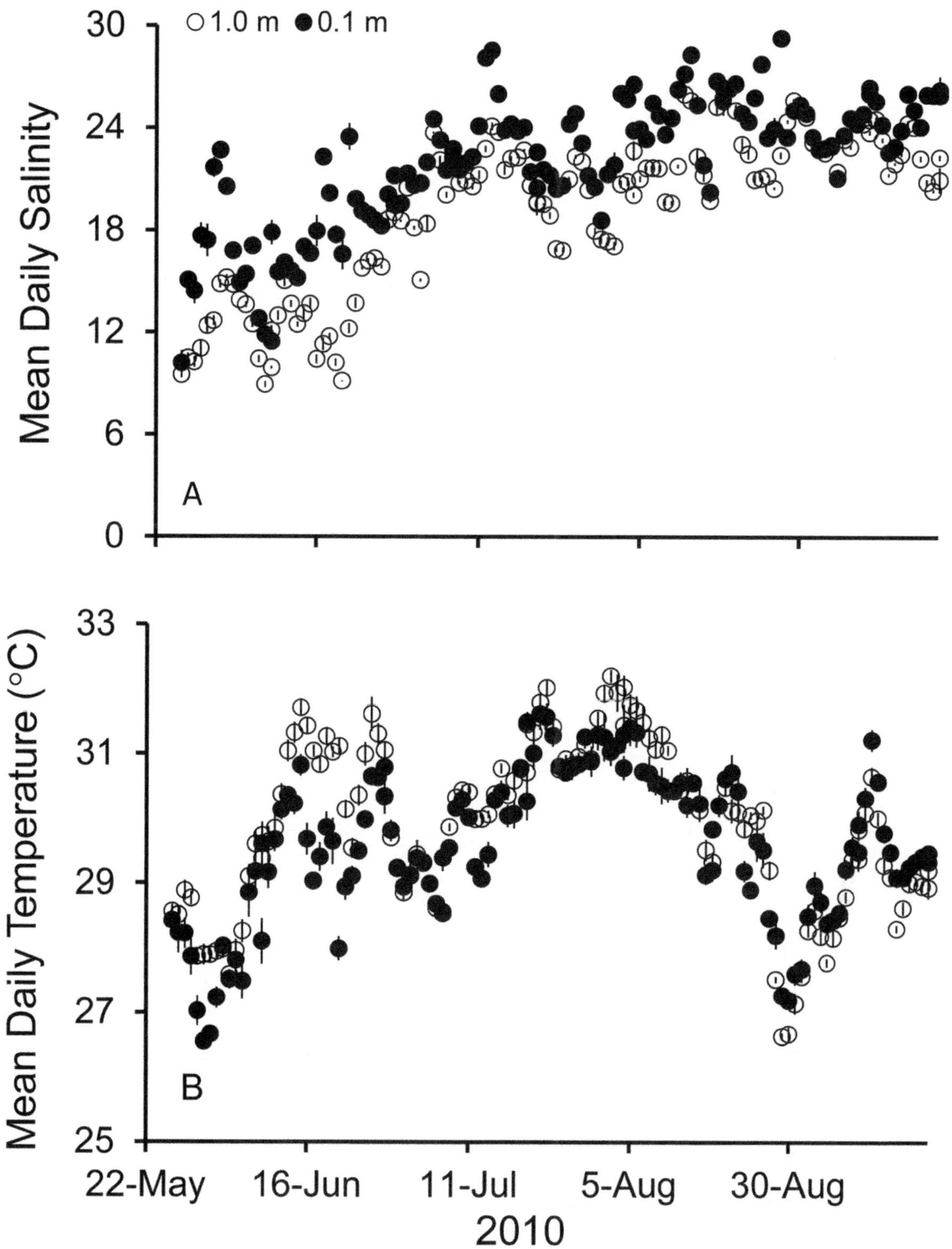

Figure 3. Daily mean salinity (A) and temperature (B) for the top depth (1.0 m off bottom) and the bottom depth (0.1 m off bottom) at Sand Reef, Mobile, AL.

deployed at two depths (0.1 m and 1.0 m from the sediment surface). Oysters were sampled at various time points, and traditional measurements of growth and survival were coupled to examination of the expression levels of 3 biomarkers, HSP70 (3 isoforms), HIF and p38 MAPK.

Mobile Bay presents an ideal location to study the effects of sublethal DO concentration stress on oysters. Although it experiences periodic low DO concentrations, there is still a standing stock of oysters and restoration efforts are currently underway [1,38]. Historically oyster restoration in Mobile Bay has

Table 1. Daily mean (± standard error) growth rates for juvenile and adult oysters and top and bottom depths, along with regression statistics.

	Growth rate (mm day^{-1})			
	Top (1.0 m)		Bottom (0.1 m)	
Juvenile	0.14±0.02	$F_{reg\ 1,\ 0.003}=40.82$	0.12±0.02	$F_{reg\ 1,\ 0.003}=41.45$
Adult	0.12±0.02	$F_{reg\ 1,\ 0.002}=35.80$	0.07±0.03	$F_{reg\ 1,\ 0.034}=8.44$

been dominated by shell planting with varied success [39]. More recently projects have used various types of cement blocks and focused on the economic benefits of oyster reef restoration to the Gulf of Mexico [40]. Native stocks and restoration projects will benefit from better understanding of the environmental conditions that lead to sublethal and lethal stress on oysters.

Methods

Oyster transplants

To assure that previous exposure was similar and that oysters were of the same genetic stock, oysters were obtained from the Auburn University Shellfish Hatchery and transplanted at Sand reef (Fig. 1), a historical oyster reef site in Mobile Bay, AL. Because the relief at this reef (largely remnant) varied gradually up to a height of 1.0 m above the sediment bottom across several meters, we simulated this natural relief by deploying caged oysters at fixed heights (0.1 m (bottom) and 1.0 m (top)) above the bottom on two existing pilings at each reef. This range of relief is also typical to oyster restoration projects [9]. To capture ontogenic variation in response to low DO conditions, 50 juvenile (40.1±0.3 mm) and 45 adult (62.8±0.4 mm) hatchery-reared oysters were placed in three replicate aquaculture cages (30 cm wide×40 cm long×10 cm deep). Cages were cleaned and oyster survival was noted every two weeks, at which time, up to 15 individuals of each age group were collected for growth measurement, mortality assessment, dry weight and protein analysis. To test cumulative effects, oysters were deployed for 87 days, thereby allowing the oysters to be exposed to several periods of low DO concentrations. To test for the co-occurring stressors of low DO concentration and low salinity, oysters were also deployed at nearby Denton reef, in Mobile Bay, AL (Fig. 1). Collection permits were attained from the Alabama Department of Conservation and Natural Resources.

Environmental parameters

YSI 6600 data sondes were deployed to take measurements every 15 minutes, creating a 5 minute average of temperature, salinity, and DO. Sondes were deployed at the two cage depths, 1.0 m (top) and 0.1 m (bottom) off the bottom. Sondes were cleaned and calibrated ~biweekly, using the manufacturer's standard methods for YSI 6-Series Sondes, one point saturated air method.

Protein analyses

Gill, mantle and adductor muscle tissues were dissected from sampled oysters, frozen in liquid nitrogen, and stored at −80°C until analysis. Mantle and gill tissues have been shown to exhibit rapid response to stress [26]. We also chose to include adductor muscle because it is commonly sampled among all types of biological studies and it is important in the catch response, allowing the valves to open or close [41]. Tissues were extracted in lithium dodecyl sulfate (LiDS) sample buffer (1:4 w:v) (125 mM Tris pH 6.8, 10% glycerol, 2% LiDS); samples were centrifuged at 14,000×g for 10 minutes after which the supernatant was collected and re-centrifuged for another 10 minutes. Total protein concentration was determined using the BioRad DC assay (BioRad, Hercules, CA, USA) with bovine serum albumin standards. Proteins were analyzed by SDS-polyacrylamide gel electrophoresis. Samples of 100 µg total protein concentration were run on 8% gels at 100 V for 2.5 hours. Proteins were transferred to a nitrocellulose membrane (Amersham Hybond-ECL, GE Healthcare) at 100 V for 1 hour. Membranes were blocked for 1 hour with 50/50 Odyssey Buffer (LiCOR) and PBS (pH 7.4). Membranes were analyzed by immunoblotting with anti-HSP 70 antibodies (1:10,000, Sigma St Louis, MO), HIF (created by Thermo Fisher from Genbank sequence HM441076, 1:500 Waltham, MA) and p38 MAP kinase (1:1,000, Cell Signaling Beverly, MA). Actin (1:10,000, Neo Markers Fremont, CA) was used as a loading control, and an internal standard (anoxic treated mantle – mantle from oysters exposed to 4 days of anoxia) was run on all gels to allow between gel comparisons. Membranes were probed with GAR (1:4,000) and GAM (1:20,000) fluorescent secondary antibodies (LiCOR). Blots were viewed under infrared fluorescence (LiCOR, Lincoln, NE) and images analyzed with a LICOR Odyssey Imager and associated software (LiCOR, Lincoln, NE). Bands were quantified by drawing a box around and adjusted to the loading control (actin) and an internal standard (anoxic treated mantle). Band intensity is expressed as relative intensity.

Statistical analysis

Minitab (version 16) was used for all statistical analyses. A GLM with date as a covariate was run on the environmental data and the growth and survival data. For the protein expression data all intensity values were square root and arc sin transformed before analysis. Liner regression analysis was used for protein expression versus 2-week average DO concentrations [27]. An α value of 0.05 was used for all tests.

Results

Site parameters

Average daily DO concentrations differed through time at study sites on Sand reef in Mobile Bay, AL, with DO concentration at the top (1.0 m) significantly higher than at the bottom (0.1 m) depth (Fig. 2, $p<0.001$). Absolute DO concentrations ranged from 0.52 to 12.72 mg L^{-1} at the top and 0.07–9.91 mg L^{-1} at the bottom depth (Fig. 2). Average daily salinity also differed between the top (1.0 m) and bottom (0.1 m) depths (Fig. 3A, $p<0.001$), but temperature did not (Fig. 3B, $p=0.36$). Salinity at the bottom depth ranged from 10.2 to 29.3 and at the top depth ranged from 8.9 to 26.7. Temperature ranged from 26.5–31.9°C overall. Sand reef had significantly higher salinity than comparison site Denton

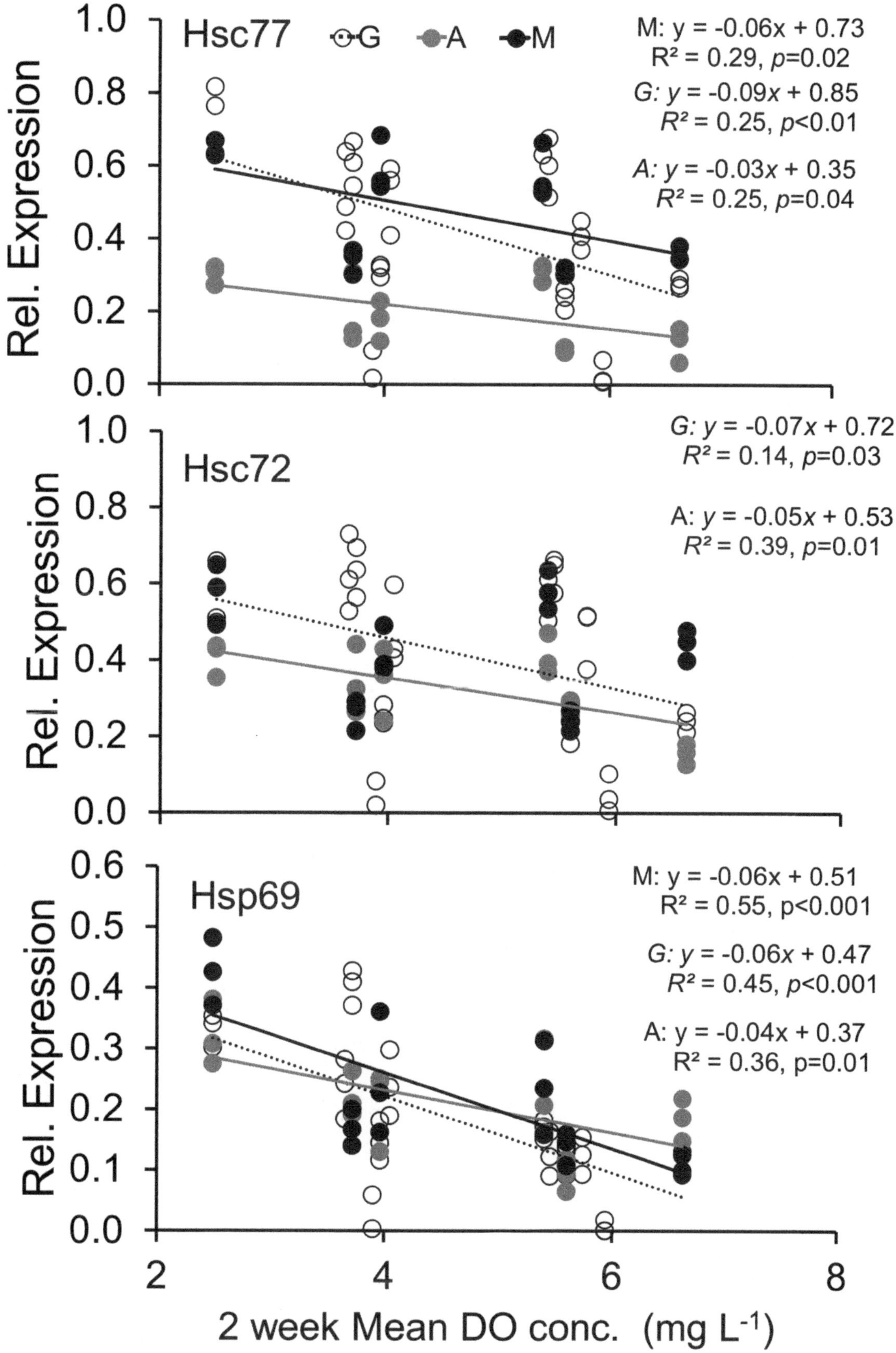

Figure 4. Regression analysis of juvenile gill (G, open circle, dashed line), adductor muscle (A, grey circle, grey line), and mantle (M, black circle, black line) tissue relative protein expression for the 3 isoforms of Hsp70 versus the average of previous 2 weeks daily dissolved oxygen (DO) concentrations. Regression lines are only present for significant regressions.

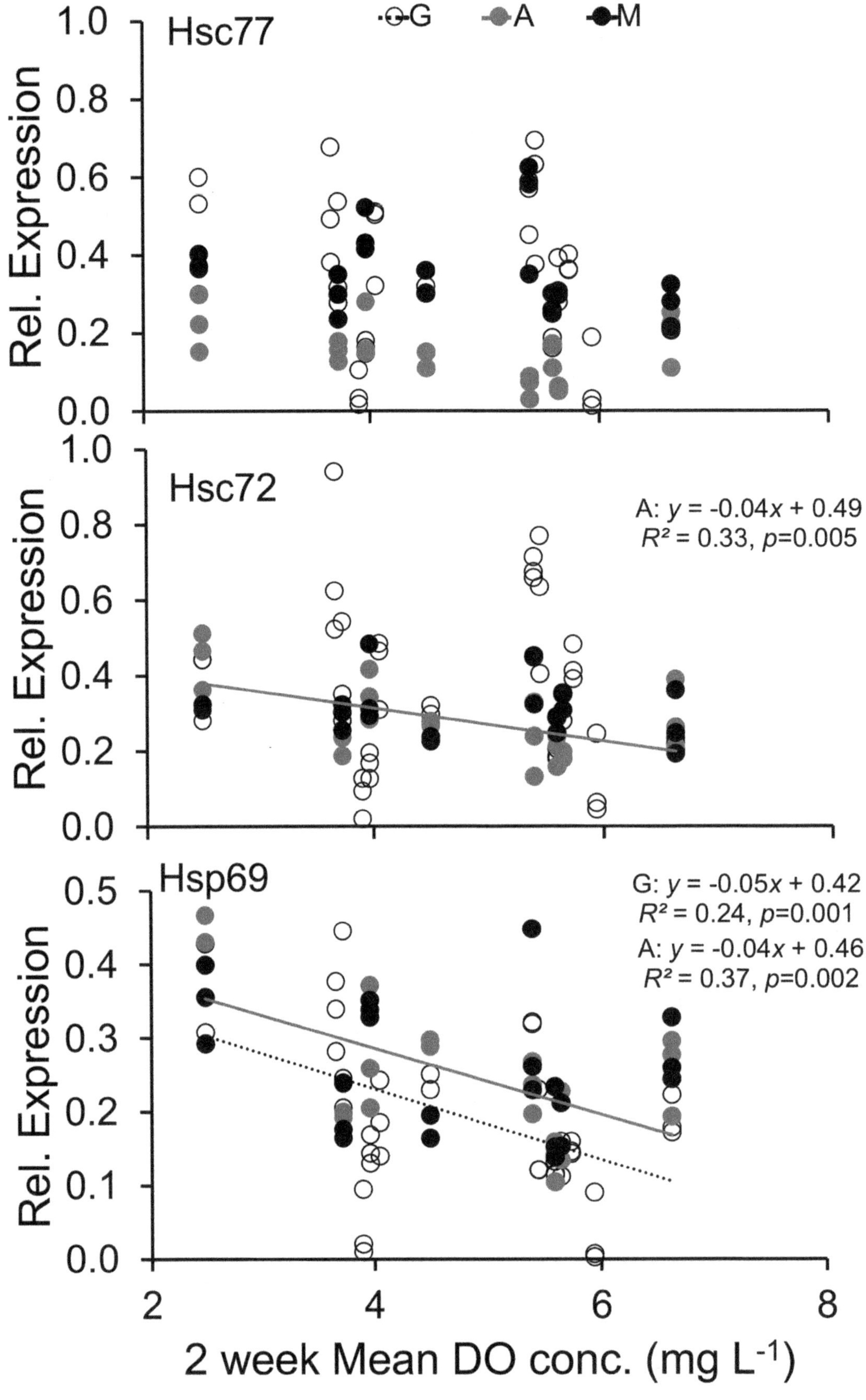

Figure 5. Regression analysis of adult gill (G, open circle, dashed line), adductor muscle (A, grey circle, grey line), and mantle (M, black circle, black line) tissue relative protein expression for the 3 isoforms of Hsp70 versus the average of previous 2 weeks daily dissolved oxygen (DO) concentrations. Regression lines are only present for significant regressions.

reef (GLM, $F = 505.38$, df = 1. $p<0.001$; Sand 20.68±0.02, Denton 14.76±0.02) and higher DO concentration (GLM, $F = 24.67$, df = 1, $p<0.001$ Sand 4.92±0.01 mg L^{-1}, Denton 4.06±0.01 mg L^{-1}).

Growth and Survival

Juveniles had higher growth rates than adult oysters at their respective depths (Table 1). Oysters at the top depth had higher growth rates than those at the bottom depth (GLM, $F = 27.63$, df = 3. $p<0.001$). Survival did not differ with depth or between age classes and was greater than 90% at all depths and ages.

Biomarkers

Hsp70. The three isoforms of Hsp70 were present in all tissues sampled from both age groups and transplant depths (example blot, Figure S1). For juveniles, 45% of the variation in relative protein expression of the inducible form of Hsp70 (Hsp69) in gill tissue was accounted for by DO concentration (Fig. 4, bottom), with higher expression corresponding to lower DO concentration. In the constitutive forms, there was also significantly higher expression with lower DO, although less of the variation was explained (25% and 14% respectively for Hsc 77 and 72, Fig. 4). Similar patterns were seen for expression of all three isoforms in juvenile adductor muscle (25, 39 and 36% for Hsc77, 72 and Hsp69 respectively, Fig. 4). However, with mantle tissue only Hsc77 and Hsp69 showed a relationship with DO concentration, with 29% (Hsc77) and 55% (Hsp 69) of variation in expression explained by DO concentration (Fig. 4).

For adult oysters, 24% of Hsp69 expression in gill tissue was explained by DO concentration (Fig. 5, bottom). In adult gill tissue, there was no significant relationship between expression of the constitutive isoforms (Hsc77, Hsc72) and DO concentration (Fig. 5, top panels). In adult adductor muscle Hsc72 and Hsp69 expression showed a positive relationship with DO concentration (Fig. 5, middle and bottom). For the adult mantle tissue, there was no relationship between the expression of any of the Hsp70 isoforms and DO concentration (Fig. 5).

Low DO also accounted for 60% of the variation Hsp69 in juvenile gill tissue at nearby Denton reef. Interestingly, when oysters were subjected to dual stressors (low salinity and low DO at Denton reef), Hsp69 expression remains high even with higher DO concentration (Fig. 6).

HIF. We detected HIF expression, but did not observe any change in HIF expression in relation to different DO concentrations, regardless of age or depth (data not shown, Figure S2).

Phospho-p38 MAPK. We were not able to separate the alpha and beta isoforms, the combined isoform alpha/beta was expressed, along with the gamma isoform, in oysters from both

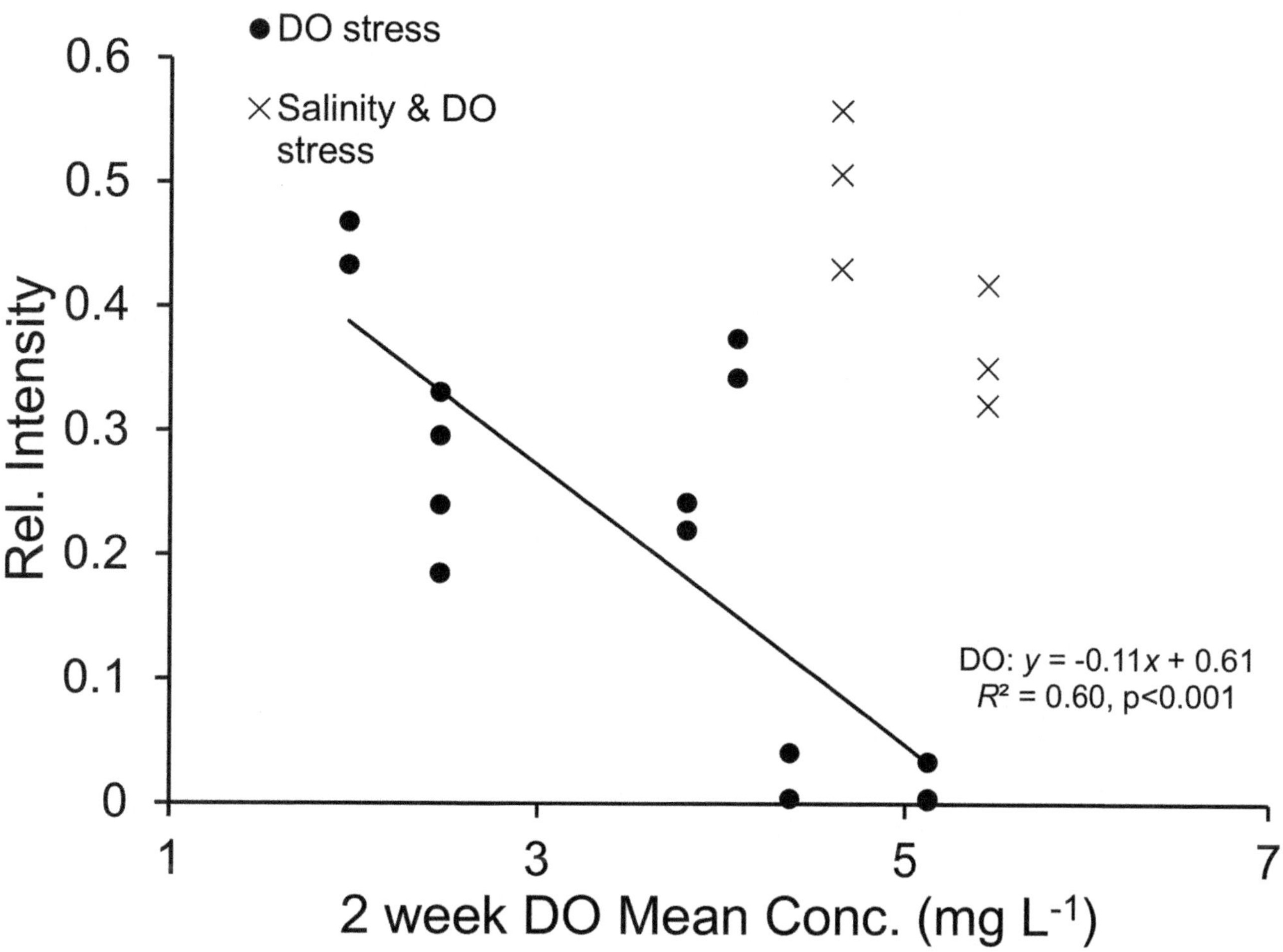

Figure 6. Regression analysis of juvenile gill tissue for the inducible Hsp70 isoform, Hsp69 versus the average of the previous 2 weeks daily dissolved oxygen (DO) concentrations. DO data from Denton reef (circles), and data when there was both salinity and DO stress as defined by [41] (X).

depths, ages and in all three tissue types. We detected differences in expression between ages, depths and isoforms, but we found no pattern related to DO concentration or any other stressor (data not shown, Figure S3).

Discussion

With the exception of periodic low DO concentrations (down to 3.2% saturation) most environmental parameters at study sites in Mobile Bay, AL were within the ranges suitable for oyster survival and growth [41,42]. Accordingly oysters at our study sites survived well and had growth rates typical to the location and season [43], but showed physiological responses (protein expression) consistent with periodic stress associated with low DO concentrations. We found the best relationship between Hsp70 expression and the 2-week average DO concentration, indicating that the recent history of exposure to low DO through time is important to detect stress responses in oysters. Helmuth & Hofmann [44] found a similar pattern for temperature response. These data demonstrate the benefit of monitoring oysters in the field for several months while continuously measuring environmental attributes to test conditions oysters would experience in restoration programs.

There were differences in protein expression among the Hsp70 isoforms between the two age classes of oysters. There was greater expression of all three isoforms of Hsp70 in the juveniles in response to low DO concentration. However, for adults the primary effects were only seen for the inducible form Hsp69. Interestingly, there also were tissue-specific differences in protein expression between age classes. Such differences have been observed by other authors (e.g., [26]); however, the exact reason for these differences is not yet clear. Differences between age classes may indicate an ontogenetic effect due to either differences in relative growth rate or differences in physiological or cell functioning between age classes. Other authors have demonstrated that age can affect organism response to stress [45,46]. These findings highlight the need to measure cumulative stress effects and consider endogenous factors (tissue type and age class) to identify a dominant environmental stressor.

Finding a biomarker that specifically responds to low DO conditions in oysters is a work in progress. We did not detect differences in expression of hypoxia inducible factor (HIF), a known hypoxia biomarker in other species, in relation to changes in DO concentration or with age in oysters, despite antibody detection of the protein in oysters and embryonic mouse lung (Figure S2). Our findings may result from detection of the constitutive expression and not expression of the different subunits as in vertebrate systems [47–49]. Kawabe & Yokoyoma [49] found changes in protein and m-RNA expression, using a different peptide that was not available when we performed this study. This peptide may be more appropriate for hypoxia investigations. We were able to detect phosphorylation of at least 3 isoforms of phospho-p38 MAPK, indicating the protein expression increased and that oysters were experiencing stress. We however, did not detect significant patterns in any of the isoforms, in either age class. Since DO concentration only accounted for up to 50% of the variation in Hsp70 isoform expression, the expression of the phosphorylated isoforms of p38-MAPK may be due to other stressors such as low salinity, infection, or environmental pollution [50,51]. Many studies have detected a response to thermostress, contaminant exposure, and multiple stressors by induction of Hsp70 in oysters (e.g., [26,27,52–56]). Hsp70 responded to the dominant stressor in our system, DO concentration; based on our findings, Hsp70 isoforms, particularly Hsp69, show promise to indicate sublethal stress before traditional growth and survival measurements are affected.

Experiments to test site suitability prior to selection have been recognized as key to successful restoration of oyster reefs [3,43,57,58]. The suitability index created by Pollack et al. [3] to aid in oyster restoration site management, stressed the importance of adaptive management strategies for site selection. In addition to the information suggested by Pollack et al. [3] (abiotic and oyster abundance), using stress proteins as broad response biomarkers has been recommended for initial environmental monitoring [59,60]. We have shown that Hsp70 has the ability to respond to an abiotic stressor (low DO concentration) within two weeks, even in relatively stress-tolerant bivalves [61], prior to detectable effects on growth or survival. Understanding sublethal stress to predict and avoid species loss has broad applications, and has been identified as important to management of communities from recreational fisheries to coral reefs as well as to contaminant monitoring [62–64]. Our data support the idea that protein biomarkers, measured in sentinel species such as oysters, provide a promising approach to identify suitable sites for restoration, particularly to capture negative effects of habitat quality on biota before lethal impacts are incurred.

Supporting Information

Figure S1 Example blot of Hsp70 showing the 3 isoforms (Hsc77, Hsc72, Hsp69) and Actin in Sand Bottom (SB) Adult mantle and AM (adductor muscle) tissues and in the internal standard (Std).

Figure S2 Example blot of HIF with mouse embryonic lung (lung), juvenile gill tissue from hypoxic site 6/9/10 (DBJ), juvenile gill tissue from normoxic site 8/4/10 (STJ) and internal standard (STD).

Figure S3 Example blot of phopho-p38 MAPK showing the gamma (γ) and alpha/beta (α/β) isoforms in adult and juvenile gill tissue 9/21/10 for the top and bottom depths.

Acknowledgments

We thank the Carmichael Lab and Dauphin Island Sea Lab Technical support staff for field assistance, and members of the Boettcher and Gerhoffer lab's for laboratory assistance.

Author Contributions

Conceived and designed the experiments: HKP AB RHC. Performed the experiments: HKP. Analyzed the data: HKP AB RHC. Contributed reagents/materials/analysis tools: AB RHC. Wrote the paper: HKP AB RHC.

References

1. Beck MW, Brumbaugh RD, Airoldi L, Carranza A, Coen LD, et al. (2011) Oyster reefs at risk and recommendation for conservation, restoration, and management. BioScience 61(2): 107–116.
2. Grabowski JH, Peterson CH (2007) Restoring Oyster Reefs to Recover Ecosystem Services. In: Cuddington K, editor. Ecosystem Engineers: Plants to Protists: Elsevier.

3. Pollack JB, Cleveland A, Palmer TA, Reisinger AS, Montagna PA (2012) A restoration suitability index model for the Eastern oyster (*Crassostrea virginica*) in the Mission-Aransas Estuary, TX, USA. PLoS ONE 7(7): e40839.
4. Bishop M, Powers S, Porter H, Peterson C (2006) Benthic biological effects of seasonal hypoxia in a eutrophic estuary predate rapid coastal development. Estuarine, Coastal and Shelf Science 70: 415–422.
5. Park K, Kim C-K, Schroeder WW (2007) Temporal variability in summertime bottom hypoxia in shallow areas of Mobile Bay, Alabama. Estuaries and Coasts 30(1): 54–65.
6. Cloern J (2001) Our evolving conceptual model of the coastal eutrophication problem. Marine Ecology Progress Series 210: 223–253.
7. Kemp W, Testa J, Conley D, Gilbert D, Hagy J (2009) Temporal responses of coastal hypoxia to nutrient loading and physical controls. Biogeosciences 6 2985–3008.
8. Gray J, Wu R, Or Y (2002) Effects of hypoxia and organic enrichment on the coastal marine environment. Marine Ecology Progress Series 238: 249–279.
9. Gregalis K, Powers S, Heck K (2008) Restoration of oyster reefs along a bio-physical gradient in Mobile Bay, Alabama. Journal of Shellfish Research 27(5): 1163–1169.
10. Wallace R, Rikard F, Howe J, Rouse D (2002) Optimum size for planting hatchery produced oyster seed. Final Technical Report Mississppi-Alabama Sea Grant Consortium. Grant No. NA86RG0073. Project No. R/LR-46-NSI-2. 48 p.
11. Saoud IG, Rouse DB, Wallace RK, Supan JE, Rikard S (2000) An *in situ* study on the survival and growth of *Crassostrea virginica* juveniles in Bon Secour Bay, Alabama. Journal of Shellfish Research 19(2): 809–814.
12. Saoud I, Rouse D, Wallace R, Howe J, Page B (2000) Oyster *Crassostrea virginica* spat settlement as it relates to the restoration of Fish River Reef in Mobile Bay, Alabama. Journal of World Aquaculture Society 31(4): 640–650.
13. Cerco C, Noel M (2007) Can oyster restoration reverse cultural eutrophication in Chesapeake Bay? Estuaries and Coasts 30(2): 331–343.
14. NRC (2000) Clean Coastal Waters: Understanding and Reducing the Effects of Nutrient Pollution. Washington, DC: The National Academies Press.
15. Stickle WB, Kapper MA, Liu L-L, Gnaiger E, Wang SY (1989) Metabolic adaptations of several species of crustaceans and molluscs to hypoxia: Tolerance and microcalorimetric studies. Biological Bulletin 177: 303–312.
16. Baker S, Mann R (1994) Description of metamorphic phases in the oyster *Crassostrea virginica* and effects of hypoxia on metamorphosis. Marine Ecology Progress Series 104: 91–99.
17. Burnett LE, Stickle WB (2001) Physiological Response to Hypoxia. In: Rabalais N, Turner RE, editors. Coastal Hypoxia: Consequences for Living Resources and Ecosystems: American Geophysical Union.
18. Karouna-Reneir NK, Zehr JP (1999) Ecological implications of molecular markers: assaying sub-lethal stress in the midge *Chironomus tentans* using heat stock protein 70 (HSP-70) expression. Hydrobiologia 401: 255–264.
19. Peakall DB (1994) The role of biomarkers in environmental assessment (1). Introduction. Ecotoxicology 3(3): 157–160.
20. McCarthy JF, Shugart LR (1990) Biomarkers of environmental contamination. Boca Raton, FL: Lewis Publishers. 457 p.
21. Roberts R, Agius C, Saliba C, Bossier P, Sung Y (2010) Heat shock proteins (chaperones) in fish and shellfish and their potential role in relation to fish health: a review. Journal of Fish Diseases 33: 789–901.
22. Feder ME, Hofmann GE (1999) Heat-shock proteins, molecular chaperones, and the stress response: Evolutionary and ecological physiology. Annual Review of Physiology 61: 243–282.
23. Fink AL (1999) Chaperon-mediated protein folding. Physiological Reviews 79(2): 425–449.
24. Hendrick JP, Hartl FU (1993) Molecular chaperone functions of heat-shock proteins. Annual Review of Physiology 62: 349–384.
25. Mayer MP, Bukau B (2005) Hsp70 chaperones: Cellular functions and molecular mechanism. CMLS Cellular and Molecular Life Sciences 62: 670–684.
26. Fabbri E, Valbonesi P, Franzellitti S (2008) HSP expression in bivalves. Invertebrate Survival Journal 5: 135–161.
27. Clegg JS, Uhlinger KR, Jackson SA, Cherr GN, Rifkin E, et al. (1998) Induced thermotolerance and the heat sock protein-70 family in the Pacific oyster *Crassostrea gigas*. Molecular Marine Biology and Biotechnology 7(1): 21–30.
28. Gorr TA, Wichmann D, Hu J, Hermes-Lima M, Welker AF, et al. (2010) Hypoxia tolerance in animals: Biology and application. Physiological and Biochemical Zoology 83(5): 733–752.
29. Wu RSS (2002) Hypoxia: from molecular responses to ecosystem responses. Marine Pollution Bulletin 45: 35–45.
30. Piontkivska H, Chung JS, Ivanina AV, Sokolov EP, Techa S, et al. (2011) Molecular characterization and mRNA expression of two key enzymes of hypoxia-sensing pathways in eastern oysters *Crassostrea virginica* (Gmelin): Hypoxia-inducible factor α (HIF-α) and HIF-proyl hydroxylase (PHD). Comparative Biochemistry and Physiology, Part D 6(2): 103–114.
31. Korb A, Tohidast-Akrad M, Cetin E, Axmann R, Smolen J, et al. (2006) Differential tissue expression and activation of p38 MAPK α, β, γ, and δ isoforms in rheumatoid arthritis. Arthritis & Rheumatism 54(9): 2745–2756.
32. Kyriakis JM, Avruch J (2001) Mammalian Mitogen-Activated Protein Kinase Signal Transduction Pathways Activated by Stress and Inflammation. Physiological Reviews 81: 807–869.
33. Gaitanaki C, Kefaloyianni E, Marmari A, Beis I (2004) Various stressors rapidly activate the p38-MAPK signaling pathways in *Mytilus galloprovincialis* (Lam.). Molecular and Cellular Biochemistry 260: 119–127.
34. Larade K, Storey KB (2006) Analysis of signal transduction pathways during anoxia exposure in a marine snail: A role of p38 MAP kinase and downstream signaling cascades. Comparative Biochemistry and Physiology, Part B 143: 85–91.
35. Conrad PW, Rust RT, Han J, Millhorn DE, Beinter-Johnson D (1999) Selective activation of p38α and p38γ by hypoxia. The Journal of Biological Chemistry 274(33): 23570–23576.
36. Ueda N, Ford C, Rikard S, Wallace RK, Boettcher A (2009) Heat shock protein 70 expression in juvenile Eastern oysters, *Crassostrea virginica* (Gmelin, 1791), exposed to anoxic conditions. Journal of Shellfish Research 28(4): 849–854.
37. Ma E, Haddad GG (1997) Anoxia regulates gene expression in the central nervous system of *Drosophila melanogaster*. Molecular Brain Research 46: 325–328.
38. zu Ermgassen PSE, Spalding MD, Grizzle RE, Brumbaugh RD (2013) Quantifying the loss of a marine ecosystem service: Filtration by the Eastern oyster in US estuaries. Estuaries and Coasts 36: 36–43.
39. Wallace RK, Heck KLJ, van Hoose M (1999) Oyster restoration in Alabama. In: Luckenbach MW, Mann R, Wesson JA, editors. Oyster Reef Habitat Restoration: A Synopsis and Synthesis of Approaches. Gloucester Point, VA: Virginia Institute of Science Press.
40. Kroeger T (2012) Dollars and Sense: Economic Benefits and Impacts from two Oyster Reef Restoration Projects in the Northern Gulf of Mexico. The Nature Conservancy. 101 p.
41. Kennedy VS, Newell RIE, Eble AF (1996) The Eastern Oyster *Crassostrea virginica*. College Park: Maryland Sea Grant College.
42. Eastern Oyster Biological Review Team (2007) Status review of the eastern oyster (*Crassostrea virginica*). Report to the National Marine Fisheries Service, Northeast Regional Office February 16, 2007. NOAA Tech. Memo. NMFS F/SPO-88, 105 p.
43. Johnson MW, Powers SP, Senne J, Park K (2009) Assessing *in situ* tolerances of Eastern Oysters (*Crassostrea virginica*) under moderate hypoxic regimes: Implications for Restoration. Journal of Shellfish Research 28(2): 185–192.
44. Helmuth BST, Hofmann GE (2001) Microhabitats, thermal heterogeneity, and patterns of physiological stress in the rocky intertidal zone. Biological Bulletin 201: 374–384.
45. Clark MS, Husmann G, Thorne MAS, Burns G, Truebano M, et al. (2013) Hypoxia impacts large adults first: consequences in a warming world. Global Change Biology 19: 2251–2263.
46. Husmann G, Abele D, Rosenstiel P, Clark MS, Kraemer L, et al. (2014) Age dependent expression of stress and antimicrobial genes in the hemocytes and siphon tissue of the Antarctic bivalve, *Laternula elliptica*, exposed to injury and starvation. Cell Stress and Chaperones 19: 15–32.
47. Soñanez-Organis JG, Peregrino-Uriate AB, Gómez-Jiménez S, López-Zavala A, Forman HJ, et al. (2009) Molecular characterization of hypoxia inducible factor-1 (HIF-1) from the white shrimp *Litopenaeus vannaemei* and tissue specific expression under hypoxia. Comparative Biochemistry and Physiology, Part C 150: 395–405.
48. Li T, Brouwer M (2007) Hypoxia-inducible factor, gsHIF, of the grass shrimp *Palaemonetes pugio*: Molecular characterization and response to hypoxia. Comparative Biochemistry and Physiology, Part B 147: 11–19.
49. Kawabe S, Yokoyama Y (2012) Role of hypoxia-inducible factor α in response to hypoxia and heat shock in the Pacific Oyster *Crassostrea gigas*. Marine Biotechnology 14: 106–119.
50. Canesi L, Betti M, Ciacci C, Scarpato A, Citterio B, et al. (2002) Signaling pathways involved in the physiological response of mussel hemocytes to bacterial challenge: the role of stress-activated p38 MAP kinases. Developmental and Comparative Immunology 26: 325–334.
51. Châtel A, Hamer B, Talarmin H, Dorange G, Schröder HC, et al. (2010) Activation of MAP kinase signaling pathway in the mussel *Mytilus galloprovincialis* as biomarker of environment pollution. Aquatic Toxicology 96: 247–255.
52. Jackson SA, Uhlinger KR, Clegg JS (2011) Duration of induced thermal tolerance and tissue-specific expression of hsp/hsc70 in the eastern oyster, *Crassostrea virginica* and the pacific oyster, *Crassostrea gigas*. Aquaculture 317: 168–174.
53. Ivanina AV, Taylor C, Sokolova IM (2009) Effects of elevated temperature and cadium exposure on stress protein response in eastern oysters *Crassostrea virginica* (Gmelin). Aquatic Toxicology 91: 245–254.
54. Li Y, Qin JG, Abbott CA, Li X, Benkendorff K (2007) Synergistic impacts of heat shock and spawning on the physiology and immune health of *Crassostrea gigas*: an explanation for summer mortality in Pacific oysters. American Journal of Physiology - Regulatory, Integrative and Comparative Physiology 293: R2353–R2362.
55. Piano A, Valbonesi P, Fabbri E (2004) Expression of cytoprotective proteins, heat shock protein 70 and metallothioneins, in tissues of *Ostrea edulis* exposed to heat and heavy metals. Cell Stress & Chaperones 9(2): 134–142.
56. Ringwood AH, Conners DE, Keppler CJ, Dinovo AA (1999) Biomarker studies with juvenile oysters (*Crassostrea virginica*) deployed in-situ. Biomarkers 4(6): 400–414.

57. Coen LD, Luckenbach MW (2000) Developing success criteria and goals for evaluating oyster reef restoration: Ecological function or resource exploitation? Ecological Engineering 15: 323–343.
58. Peyre MKL, Nix A, Laborde L, Piazza BP (2012) Gauging state-level and user group views of oyster reef restoration activities in the northern Gulf of Mexico. Ocean & Coastal Management 67: 1–8.
59. Sanders BM (1993) Stress proteins in aquatic organisms: An environmental perspective. Critical Reviews in Toxicology 23(1): 49–75.
60. Bierkens JGEA (2000) Applications and pitfalls in stress-proteins in biomonitoring. Toxicology 153: 61–72.
61. Joyner-Matos J, Andrzejewski J, Briggs L, Baker SM, Downs CA, et al. (2009) Assessment of cellular and functional biomarkers in bivalves exposed to ecologically relevant abiotic stressors. Journal of Aquatic Animal Health 21: 104–116.
62. Downs CA, Ostrander GK, Rougee L, Rongo T, Knutson S, et al. (2012) The use of cellular diagnostics for identifying sub-lethal stress in reef corals. Ecotoxicology 21: 768–782.
63. Cooke SJ, Schramm HL (2007) Catch-and-release science and its application to conservation and management of recreational fisheries. Fisheries Management and Ecology 14: 73–79.
64. Nascimento IA, Leite MBN, Sansone G, Pereira SA, Smith DH (1998) Stress protein accumulation as an indicator of impact by the petroleum industry in Todos os Santos Bay, Brazil. Aquatic Ecosystem Health and Management 1: 101–108.

10

Long-Term Citizen-Collected Data Reveal Geographical Patterns and Temporal Trends in Lake Water Clarity

Noah R. Lottig[1]*, Tyler Wagner[2], Emily Norton Henry[3,4], Kendra Spence Cheruvelil[3,5], Katherine E. Webster[6], John A. Downing[7], Craig A. Stow[8]

1 Center for Limnology Trout Lake Station, University of Wisconsin Madison, Boulder Junction, Wisconsin, United States of America, **2** U.S. Geological Survey, Pennsylvania Cooperative Fish and Wildlife Research Unit, Pennsylvania State University, University Park, Pennsylvania, United States of America, **3** Department of Fisheries and Wildlife, Michigan State University, East Lansing, Michigan, United States of America, **4** Department of Outreach and Engagement, Oregon State University, Tillamook, Oregon, United States of America, **5** Lyman Briggs College, Michigan State University, East Lansing, Michigan, United States of America, **6** School of Natural Sciences, Trinity College, Dublin, Ireland, **7** Department of Ecology, Evolution, and Organismal Biology, Iowa State University, Ames, Iowa, United States of America, **8** Great Lakes Environmental Research Laboratory, National Oceanic and Atmospheric Administration, Ann Arbor, Michigan, United States of America

Abstract

We compiled a lake-water clarity database using publically available, citizen volunteer observations made between 1938 and 2012 across eight states in the Upper Midwest, USA. Our objectives were to determine (1) whether temporal trends in lake-water clarity existed across this large geographic area and (2) whether trends were related to the lake-specific characteristics of latitude, lake size, or time period the lake was monitored. Our database consisted of >140,000 individual Secchi observations from 3,251 lakes that we summarized per lake-year, resulting in 21,020 summer averages. Using Bayesian hierarchical modeling, we found approximately a 1% per year increase in water clarity (quantified as Secchi depth) for the **entire population** of lakes. On an **individual lake** basis, 7% of lakes showed increased water clarity and 4% showed decreased clarity. Trend direction and strength were related to latitude and median sample date. Lakes in the southern part of our study-region had lower average annual summer water clarity, more negative long-term trends, and greater inter-annual variability in water clarity compared to northern lakes. Increasing trends were strongest for lakes with median sample dates earlier in the period of record (1938–2012). Our ability to identify specific mechanisms for these trends is currently hampered by the lack of a large, multi-thematic database of variables that drive water clarity (e.g., climate, land use/cover). Our results demonstrate, however, that citizen science can provide the critical monitoring data needed to address environmental questions at large spatial and long temporal scales. Collaborations among citizens, research scientists, and government agencies may be important for developing the data sources and analytical tools necessary to move toward an understanding of the factors influencing macro-scale patterns such as those shown here for lake water clarity.

Editor: Peter S. Petraitis, University of Pennsylvania, United States of America

Funding: Funding for this research was provided by the National Science Foundation (MSB-1065786, EF-1065818, EF-1065649) and NTL-LTER (DEB-0822700). K.E.W. also thanks STRIVE grant 2011-W-FS-7 from the Environmental Protection Agency, Ireland for support. The funders had no role in study design, data collection and analysis, decision to publish, or preparation of the manuscript.

Competing Interests: The authors have declared that no competing interests exist.

* E-mail: nrlottig@wisc.edu

Introduction

Macrosystems ecology has emerged as a new ecology sub-discipline aimed at understanding ecosystem patterns and processes resulting from broad-scale environmental changes such as land use and climate change. This sub-discipline studies multi-scaled patterns and processes in biological, geophysical, and social components, and their interactions with each other and processes operating at finer and coarser scales [1]. One of the approaches that macrosystems ecologists are taking is integrating large datasets from a variety of sources. There may be many inherent challenges in assembling appropriate datasets for analysis in a macrosystems ecology context [2], however, we feel that well-established citizen monitoring programs can play an important role in macrosystems ecology, especially those that collect data across large spatial and temporal extents with standardized measurement, data entry, and quality control protocols [3]. The Christmas Bird Count is a well-known example of a citizen-based program (http://birds.audubon.org/about-christmas-bird-count), long recognized as a successful model of how to integrate citizens into meaningful science and conservation [4,5].

In freshwater ecosystems, the indicator of water quality with the longest history of standardized measurement by scientists and citizens alike is Secchi depth [6–8]. The first transparency measurements were made in the ocean during the early 1800s, followed by more careful investigations by Pietro Angelo Secchi in the mid 1800s, which led to his recognition as the method's founder (Fig. 1) [8]. Water clarity is easy and inexpensive to measure using a Secchi disk, and comparisons have shown that citizen-collected Secchi depth readings are nearly identical to those taken by professionals [9,10]. Consequently, the frequent, citizen-collected Secchi depth readings that are publically available online (e.g., Secchi Dip-In; http://www.secchidipin.org) often

Figure 1. Monument to Pietro Angelo Secchi in the Villa Borghese Park in Rome. Pietro Secchi, who lived in Italy during the 1800s, is recognized as the founder of the method of using a Secchi transparency disk to measure water clarity (Photo by K.E. Webster).

span multiple decades for individual lakes and can be used to complement data collected by state and federal agencies.

The U.S. Clean Water Act of 1972 regulates anthropogenic inputs of pollutants to receiving waters [11]. Therefore, Secchi depth records from citizen monitoring programs provide a potentially valuable resource for examining broad-scale responses of lakes to nutrient control efforts instituted in the subsequent 40 years. However, there have been few attempts at compiling and using these data to concurrently analyze spatial and temporal trends at broad scales. Previous studies generally have focused at the scale of a single U.S. state [7,12,13].

We compiled citizen-collected, publically available Secchi depth measurements to answer two questions: (1) what are the long-term trends in lake-water clarity across a broad geographic region?; and (2) how do trends differ as a function of spatial location, lake size, and when Secchi depth records were collected? To our knowledge, this is the first study to look at long-term water clarity trends using (1) a broad spatial extent (eight states with a total land area greater than 1.1 million km^2 in the Upper Midwest region of the United States), (2) a long temporal window (1938–2012), (3) a large number of lakes (>3,000), and (4) publically-available citizen-collected Secchi depth measurements (>140,000). We answer these questions using Bayesian hierarchical modeling, an approach that is useful because it can incorporate disparate data from multiple sources to examine relationships across multiple spatial and temporal extents [14,15]. Our results can be used as a basis for forming hypotheses to test the mechanisms driving large-scale spatial patterns and temporal trends in water clarity as well as informing local, regional, or federal policies and management actions.

Methods

Data Acquisition

We obtained 239,741 citizen-collected Secchi depth measurements from monitoring networks in eight states in the Upper Midwest region of the United States (Fig. 2; Table 1) [16]. Observations spanned the time period from 1938–2012 with the oldest records from the state of Minnesota; the other citizen monitoring networks began collecting data during the early 1970s, and the majority (99%) of measurements were taken post-1972. The number of waterbodies monitored varied widely among states, ranging from 5 in Iowa to 1,507 in Wisconsin (Table 1).

Prior to analysis, Secchi data were put through a series of QA/QC checks, limited to a consistent seasonal window, and summarized as annual averages. First, we converted readings to meters and removed 31 Secchi depth observations that exceeded 30 m. Given the distribution of all Secchi observations (data not shown), these observations appeared to be unrealistic for the region and were likely the result of transcription errors. Second, we limited our analysis to Secchi observations collected during June through August, which removed a source of inter-seasonal variation [17] and generally was the season when most of the monitoring networks collected data (e.g., Secchi Dip-In), providing consistency across sampling networks. Finally, the response variable used for all analyses was average summer Secchi depth calculated for each lake-year, estimated using raw-untransformed Secchi depth values. We also calculated the median sampling date of each lake for use as a predictor variable. For example, for a lake that was continuously monitored by citizens from 1980 through 2000, the median date of the lake record was 1990. Therefore, median sample date represents a surrogate for the time period a given lake was sampled over the entire 1938–2012 span of this study.

Using the latitude, longitude, and unique lake identifiers provided by the individual sampling networks, we merged the study lakes with the high-resolution National Hydrography Dataset (NHD; http://nhd.usgs.gov). We then verified the identity and location (latitude and longitude) of lakes in the dataset and obtained lake surface areas (ha) from the NHD. The final dataset included 148,991 Secchi observations, from which we calculated 21,020 summer lake specific average Secchi depth values for 3,251 lakes that spanned monitoring record lengths from 1 to 34 years (Fig. 2).

Statistical Analysis

To estimate the overall temporal trend (i.e., trend across all lakes) and assess the variability among lakes in temporal trends, we first fit a hierarchical model with log$_e$-transformed Secchi depth as a linear function of year allowing the intercepts, slopes, and model error variances to differ among lakes. The form of the hierarchical model was as follows:

$$y_i \sim N\left(\alpha_{j(i)} + \beta_{j(i)}x_i, \sigma_j^2\right), \text{for } i = 1, \ldots n$$

$$\begin{pmatrix} \alpha_j \\ \beta_j \end{pmatrix} \sim N\left(\begin{pmatrix} \mu_\alpha \\ \mu_\beta \end{pmatrix}, \begin{pmatrix} \sigma_\alpha^2 & \rho\sigma_\alpha\sigma_\beta \\ \sigma_\alpha\sigma_\beta & \sigma_\beta^2 \end{pmatrix}\right), \text{for } j = 1, \ldots J$$

$$log(\sigma_j) \sim N\left(\mu_\sigma, \omega_\sigma^2\right), \text{for } j = 1, \ldots J \quad (1)$$

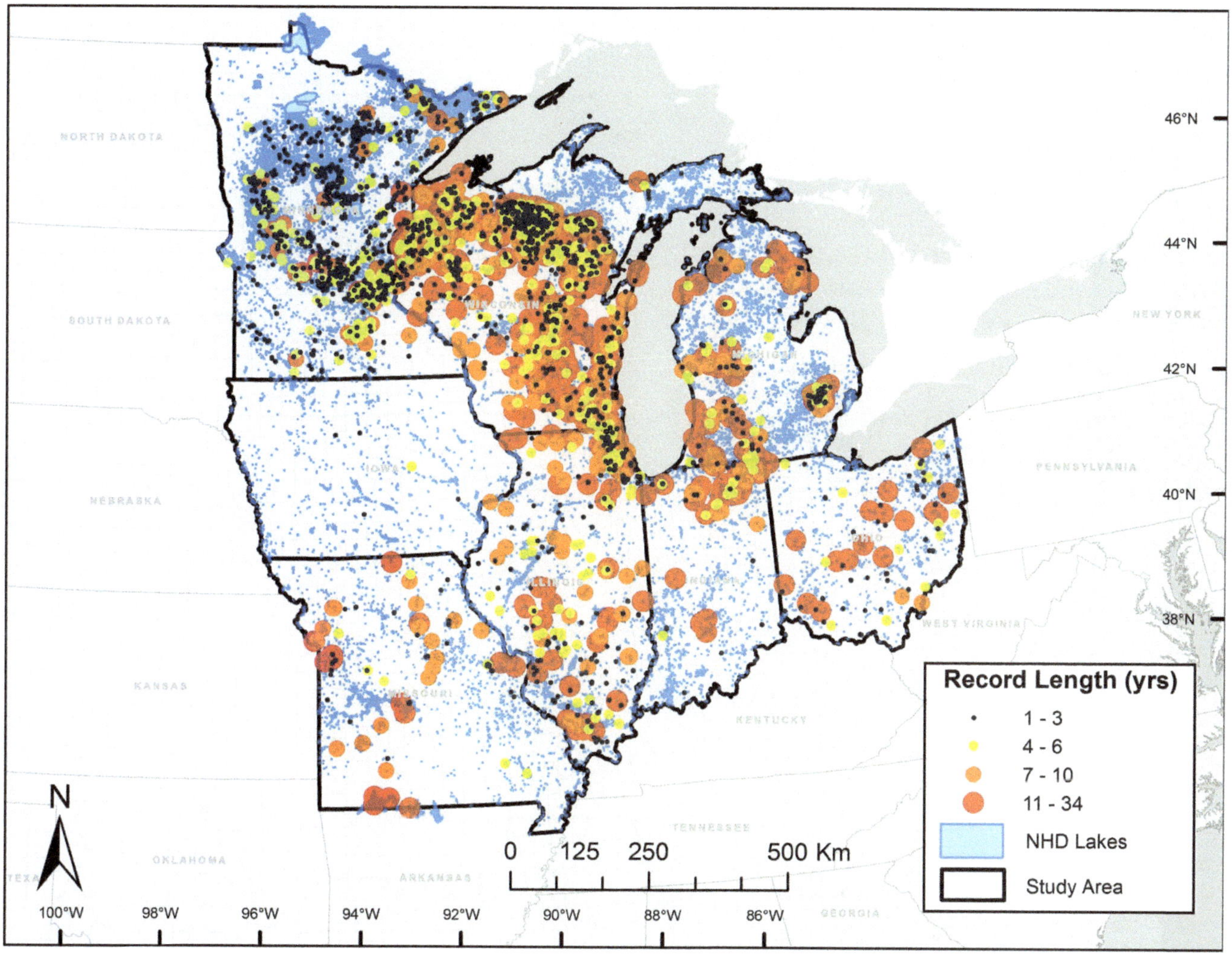

Figure 2. Upper Midwest states in the US and Secchi record lengths (years) for all citizen monitored lakes included in the study. Size and color of symbol designate record lengths. See Table 1 for information on citizen monitoring networks by state.

Where y_i is observation i of log$_e$-transformed Secchi depth from lake j; α_j and β_j are the intercept and slope (temporal trend) for lake j, respectively; x is the year covariate (standardized; $(x_i-\bar{x})/SD(x_i)$) corresponding to the year Secchi observation i was measured; μ_α is the population-average intercept and μ_β is the population-average slope; σ_α^2 and σ_β^2 are the variances of the intercepts and slopes, respectively; and $\rho\sigma_\alpha\sigma_\beta$ describes the covariance between α_j and β_j, with ρ describing the correlation between α_j and β_j. We assumed a normal probability distribution for log$_e$-transformed σ_j (lake-specific residual variances), with mean μ_σ and variance ω_σ^2. Non-informative normal priors were used for μ_α, μ_β, and μ_σ (i.e., $\mu_x \sim N(0{,}1000)$). Non-informative uniform priors were used for σ_α, σ_β, ω_σ (i.e., uniform distribution on [0,100]), and ρ (i.e., uniform distribution on [−1,1]).

Secondly, we fit models that were similar to Eq. 1, but that included predictor variables (i.e., latitude, lake surface area, median sample date) to model variation among lake-specific intercepts, slopes, and residual standard deviations. For example, using latitude (*lat*) as a lake-specific predictor, the model was as follows:

$$y_i \sim N\left(\alpha_{j(i)} + \beta_{j(i)}x_i, \sigma_j^2\right), \text{for } i = 1, \ldots n$$

$$\begin{pmatrix} \alpha_j \\ \beta_j \end{pmatrix} \sim N\left(\begin{pmatrix} \gamma_0^\alpha + \gamma_1^\alpha \times lat_j \\ \gamma_0^\beta + \gamma_1^\beta \times lat_j \end{pmatrix}, \begin{pmatrix} \sigma_\alpha^2 & \rho\sigma_\alpha\sigma_\beta \\ \sigma_\alpha\sigma_\beta & \sigma_\beta^2 \end{pmatrix}\right), \text{for } j = 1, \ldots \text{J}$$

$$log(\sigma_j) \sim N\left(\gamma_0^\sigma + \gamma_1^\sigma \times lat_j, \omega_\sigma^2\right), \text{for } j = 1, \ldots \text{J} \quad (2)$$

Where γ_0^x and γ_1^x are intercepts and slopes, respectively. Non-informative normal priors were used for γ_0^x and γ_1^x and priors on other parameters were as described above. All covariates were standardized prior to analysis. Using an assessment method similar to classical significance testing, we examined 95% credible intervals (CRIs) for γ_1^x to determine the importance of predictor variables (i.e., examined if 95% CRIs overlapped zero). The models were estimated using Bayesian estimation and the program JAGS was used for all analyses [18]. Three parallel chains were run with different initial values. After discarding the first 10,000

Table 1. Overview of the data sources and records available from the citizen monitoring networks included in this study.

State	Data Source	# Lakes[1]	# Lakes Monitored	% Lakes Monitored	# Secchi Observations[2]
Illinois	Volunteer Lake Monitoring Program[3]	2,857	303	10.6	40,681
Indiana	Indiana Clean Lakes Program[4]	1,879	95	5.06	8,610
Iowa	Secchi Dip-In Project[5]	1,081	5	0.46	47
Michigan	Michigan Clean Water Corps[6]	6,641	225	3.39	32,222
Missouri	Lakes of Missouri Volunteer Program[7]	1,886	79	4.19	7,581
Minnesota	Citizen Lake Monitoring Program[8]	14,043	961	6.84	13,392
Ohio	Citizen Lake Awareness and Monitoring[9]	1,257	75	5.97	7,819
Wisconsin	Citizen Lake Monitoring Network[10]	5,914	1507	25.5	129,389

Numbers represent totals through 2012.
[1]Number of lakes determined from 2007 National Hydrography Dataset (NHD) with surface area equal to or greater than 4 ha.
[2]Includes all citizen Secchi observations available from designated monitoring network.
[3]http://www.epa.state.il.us/water/vlmp.
[4]http://www.indiana.edu/~clp.
[5]http://www.Secchidipin.org.
[6]http://www.micorps.net/.
[7]http://www.lmvp.org.
[8]http://www.pca.state.mn.us/index.php/water/water-types-and-programs/surface-water/lakes/citizen-lake-monitoring-program/index.html.
[9]http://www.olms.org/citizen-lake-awareness-and-monitoring.
[10]http://dnr.wi.gov/lakes/CLMN.

samples, we retained every 3rd sample for a total of 20,000 samples. We examined the scale reduction factor $(\hat{R})$, a convergence statistic, for each parameter, trace plots, and plots of posterior distributions to assess convergence.

Results

Average lake specific summer Secchi depth for the 3,251 study lake population covered a wide range from 0.2 to 16 m, with a sample mean of 2.4 m (median 2.1 m; Fig 3A). For a majority of lakes (75%) the year-to-year variability of Secchi depth values within a specific lake was small (coefficient of variation <30%). However, at the population-level, the range was large (coefficient of variation between 6.5 and 184%).

For the **entire lake population**, temporal trends in Secchi depth from 1938 to 2012 were positive (average of all lake specific long-term trends), suggesting an overall increase in water clarity. The posterior estimate for the mean slope, μ_β, (Eq. 1) was 0.0092, indicating that Secchi depth, on average, has been increasing by 0.92% per year (95% CRI = 0.61–1.79%; Fig. 3B). On an **individual lake** basis, however, there was little evidence for long-term increases or decreases in water clarity for a majority of lakes (Fig. 3C). Of the 3,251 lakes examined, 128 (3.8%) and 224 (6.9%) had a 90% or greater probability of long-term declines or increases in Secchi depth, respectively (i.e., the number of the $\beta_j's$ estimated in Eq. 1 had a greater than 90% probability of being either <0 or >0; Fig. 3C).

Lake surface area was not related to average annual summer Secchi depth, long-term Secchi depth trend, or inter-annual variability in Secchi depth (posterior means and 95% CRIs for $\gamma_1^\alpha = -0.018$ [−0.034, −0.002]; $\gamma_1^\beta = 0.001$ [−0.005, 0.006]; $\gamma_1^\sigma = 0.004$ [−0.012, 0.020]). We found relationships between these response variables and the latitude of the monitored lakes (posterior mean and 95% CRI for $\gamma_1^\alpha = 0.249$ [0.224, 0.274]; $\gamma_1^\beta = 0.019$ [0.009, 0.029]; $\gamma_1^\sigma = -0.060$ [−0.086, −0.034]). Average annual summer Secchi depth (Fig. 4A) and long-term trend slope (Fig. 4B) increased and inter-annual variability in Secchi decreased (Fig. 4C) across the southern to northern latitudinal gradient (Fig. 2).

In addition to detecting spatial patterns in Secchi trends, we found that trends were related to when the data were collected within the multi-decadal time interval of 1938–2012 (i.e., the median year of sample collection). Median sample date was negatively related to both long-term Secchi depth trend and inter-annual variability (posterior mean and 95% CRI for $\gamma_1^\beta = -0.033$ [−0.043, −0.023]; $\gamma_1^\sigma = -0.137$ [−0.164, −0.108]; Fig. 5B, C). These results suggest that historical records were characterized by more positive annual Secchi depth trends relative to records collected more recently (Fig. 5B) and that inter-annual variability of Secchi depths was greater historically than what is observed from more current records (Fig. 5C). There was no evidence that relationships were due to differences in the water clarity of lakes being monitored by citizen volunteers over time, which has not changed (posterior mean and 95% CRI for $\gamma_1^\alpha = -0.012$ [−0.043, 0.017]; Fig. 5A).

Discussion

This study relied solely on citizen monitoring data to assess long-term water clarity (quantified as Secchi depth) trends and their relationships with spatial location, size of lake monitored, and time period monitoring took place. Overall, water clarity has been increasing slightly for the **entire population** of monitored lakes (average trend slope $[\mu_\beta] >$ zero). Eleven percent of lakes have strong trends in water clarity (90% probability slope is different than zero), with two-thirds corresponding to increased water clarity. We observed compelling spatial and temporal patterns with respect to the water clarity, long-term trends, and inter-annual variability of water clarity. Other studies have observed similar patterns in citizen monitored and remotely sensed Secchi depth [12,13] but over no more than two decades. We are unaware of any studies conducted with citizen-collected Secchi data at the spatial and temporal scales considered in this study.

Previous analyses of satellite-derived long-term Secchi depth patterns in the study region found that on average water clarity

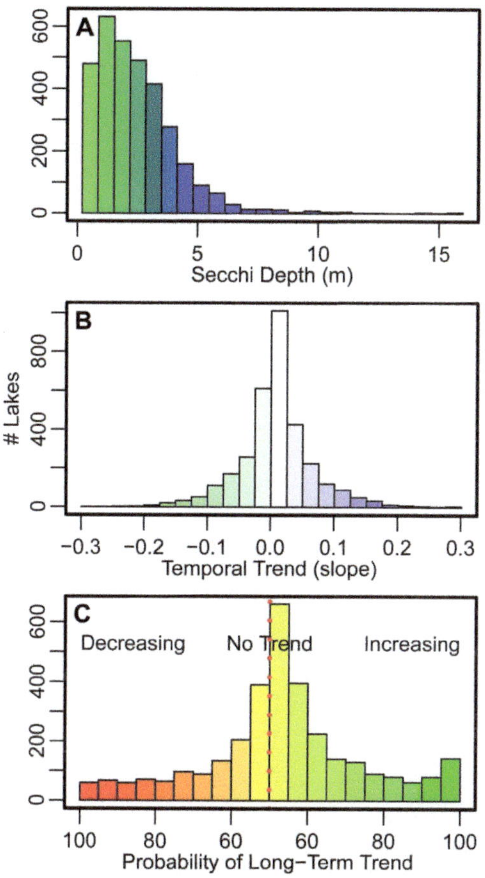

Figure 3. Distribution of lake-specific overall average Secchi depths (α_j from Eq. 1; A), long-term trends (β_j from Eq. 1; B), and probability of lake specific long-term trend (probability of β_j's estimated in Eq. 1 of being either <0 or >0; C) based on a hierarchical Bayesian analysis. Long-term trend (B) positive values indicate long-term increases in water clarity (i.e., increased water clarity) and negative values indicate long-term declines in water clarity (i.e., decreased water clarity) for individual lakes. Probability of lake specific long-term trends (C) left of the dotted red line indicate the number of lakes with long-term annual declines in water clarity (Secchi depth becoming shallower) and associated probability of those long-term declines; values to the right indicate the number of lakes with long-term increases in annual water clarity (Secchi depth becoming deeper) and associated probability of those long-term increases.

trends have been relatively stable (Minnesota) [13,19] or increasing (Wisconsin) [12]. We observed that clarity is increasing on average by about 1% per year for the **entire population of lakes** included in this study. In large part, the long-term average increases in water clarity reported here were driven by data from lakes in Minnesota and Wisconsin, which comprise 76% of total number of lakes in the dataset (30% and 46% respectively). Excluding these two states, the average annual increase in clarity was negligible (0.17%). Interestingly, we found larger annual increases in Minnesota (1.59%) relative to Wisconsin (0.88%), which is contrary to the satellite-derived results that suggested relatively stable long-term patterns in Minnesota and more substantial increases in Wisconsin [12,13]. It is important to note, however, that the different approaches taken for these studies (e.g., citizen-collected Secchi data or satellite derived data, frequentist or Bayesian statistical approaches) to identify long-term trends are not suggesting drastically different patterns at regional scales (e.g., increasing versus decreasing), but rather subtle differences in the magnitudes of overall changes.

On an **individual lake** basis, results from our large multi-state study (e.g., Fig. 3) are generally consistent with small-scale lake-specific studies [7,20,21]. The analytical approaches employed in our study, however, provided an opportunity to combine space and time to analyze temporal trend patterns across a broad spatial extent. A majority of our lakes exhibited relatively stable (i.e., no significant long-term directional trend) water clarity with minimal inter-annual variability, whereas a small percentage of lakes (10.7%) exhibited strong evidence of long-term directional trends. These results are similar in magnitude to a recent analysis of satellite derived Secchi depths for >10,000 lakes, which found significant directional trends over 20 years in 10.8% of all Minnesota lakes [19]. Prior state-wide analyses of citizen volunteer data have also found similar patterns in water clarity over time. For example, over 50, 80, and 60% of the 50–122 lakes monitored by citizens in Minnesota, Wisconsin, and Michigan, respectively, showed no trend in water clarity over time [7,20,21]. Increasing trends in water clarity were found for 36, 31, and 10% of lakes in these three states, respectively, with a much smaller percentage of lakes in these studies showing decreasing water clarity over time [7,20,21]. Our study of a much larger number of lakes located across a very large spatial extent corroborates these previous studies and demonstrates that (1) most citizen monitored lakes in the Upper Midwest of the United States have had a relatively stable long-term water clarity record and (2) long-term increases in water clarity for individual lakes occur more frequently than do declines.

Some of the most compelling results emerging from our analyses are the relationships between long-term water clarity trends and the spatial location of lakes. Our results suggest that lakes situated at more southern latitudes tended to have trends of long-term decline in water clarity, whereas lakes situated at more northern latitudes had more of an overall shift towards long-term increases in water clarity (Fig. 5B). Although the water clarity patterns presented here were correlated with latitude, we suspect that these patterns were likely more directly influenced by a secondary, confounded, variable such as land use/cover [7,22] or climate [23]. For example, an analysis of satellite water clarity data from Minnesota over the last 20 years found similar patterns emerged at the state-level that were correlated with urban/agriculture (decreasing water clarity) and forested (increasing water clarity) land use types [19]. There is substantial spatial heterogeneity across our study extent, with extremes represented by large differences between the southern-most and northern-most regions in local and regional land use/cover patterns, as well as the underlying geology and soil patterns that are confounded with those human land uses. Consequently, understanding these regional water clarity patterns requires information about physical, chemical, and biological lake properties as well as data on watershed and regional landscape features at broad spatial and temporal scales [24]. However, a broad-scale, multi-thematic database containing such variables does not exist.

We also examined water clarity trends across the time period that encompassed passage of the U.S. Clean Water Act, leading to the expectation that an increase in water clarity would have occurred during the past 30–40 years. Our study did not find this pattern, at least not at the aggregate scale and for this population of lakes; in fact, the analyses provided evidence of stronger increases in the early part of the record. There are three possible reasons for this result. First, the data in our study are not continuous for all lakes pre- and post-Clean Water Act. Second, it is critical to acknowledge that lakes monitored by citizens are likely

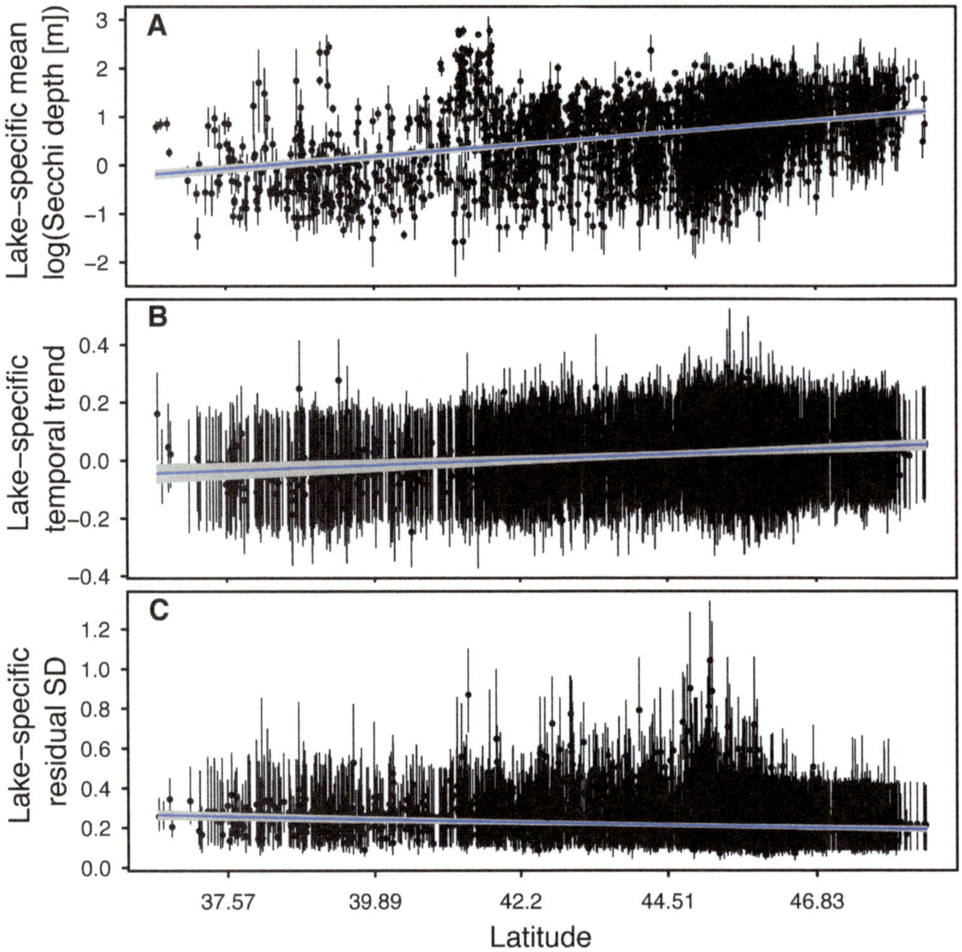

Figure 4. Relationships between average annual summer Secchi depth (A), long-term trend (B), and inter-annual variability (C) and spatial location (latitude) for 3,251 lakes in the Upper Midwest region of the United States. Solid points represent the parameter estimate for each individual lake and vertical lines represent the 95% credible interval of the estimates. Solid blue line represents the trend between each parameter and latitude. Grey shading around the solid blue lines represents the 95% credible interval of the latitudinal trend.

not representative of all lakes within the study extent. Although a substantial number of water clarity measurements have been made, only a fraction of all lakes in each state had water clarity actively monitored by citizen scientists (Table 1). Additionally, the distribution of surface areas of citizen-monitored lakes diverged from that of the entire distribution of lakes in our study extent (Fig. 6). The lakes that citizens monitor are disproportionately larger than would be expected based on the size distribution of lakes in general [25]. Similar biases towards overrepresentation of large lakes have been observed when examining the types of lakes monitored by state agencies [26]. We believe that this is the largest dataset of citizen-monitored Secchi data to be analyzed to date and suspect this bias may also exist for datasets from beyond our study region. The patterns presented here are characteristic of the suite of lakes included in our dataset and, thus, they may not be representative of all lakes in our study region or those beyond it. Consequently, careful consideration needs to be made should these results be extrapolated to lake populations other than the those considered in this study [26]. Third, water clarity is affected by a number of natural features that alter light attenuation including dissolved organic carbon, chlorophyll, and suspended sediments [17] and is altered by human-induced nutrient inputs [27], invasion by non-native species (e.g., zebra mussels) [28], acidification [29] and climate change (browning of waters) [23]. The complex set of factors affecting lake water clarity, many of which are correlated, makes it difficult to interpret the causes of trends in water clarity and has, perhaps, led to Secchi depth being considered a relatively coarse indicator of overall water quality. However, its use as an approximation of water quality status [30] and as a measure of the reliability of satellite-derived water clarity data [13], combined with ease of collection by citizen scientists, suggests this 150-year measure of water clarity will continue to provide valuable insights into the condition of lake ecosystems.

Conclusions

Although there has been a long history of examining water clarity trends and patterns augmented with citizen collected data, the results presented here incorporate over 140,000 citizen monitoring water clarity values, extending back to 1938, collected at greater spatial and temporal scales than any previous study we are aware of. It is impossible for aquatic management organizations to monitor all lakes all the time; citizen water clarity data have been critical in augmenting traditional agency monitoring efforts and provide vital data for identifying and understanding water quality patterns at broad spatial and temporal scales (e.g.,

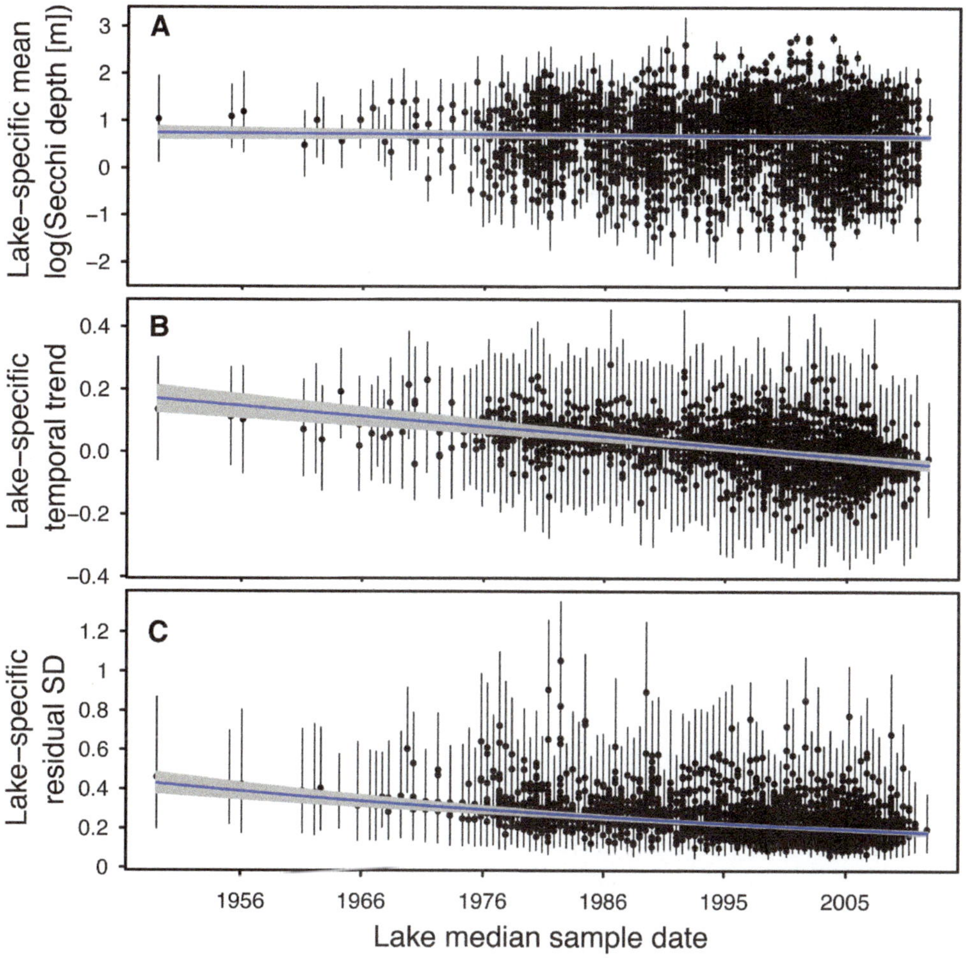

Figure 5. Relationships between average annual summer Secchi depth (A), long-term trend (B), and inter-annual variability (C) and median sample date (see text for details) for 3,251 lakes in the Upper Midwest region of the United States. Solid points represent the parameter estimate for each individual lake and vertical lines represent the 95% credible interval of the estimates. Solid blue line represents the trend between each parameter and latitude. Grey shading is the 95% credible interval of the latitudinal trend.

calibration of satellite-derived water clarity estimates). Here, we demonstrate the effectiveness of citizen data for quantifying long-term water clarity trends in individual lakes and identifying compelling population-level patterns in lake water clarity at the macro-scale. In an era of limited resources and during a push toward applying data-intensive science to environmental management and policy, the importance of citizen data such as these is likely to increase. Therefore, there is a clear need to continue to learn how to effectively leverage these types of data in the future.

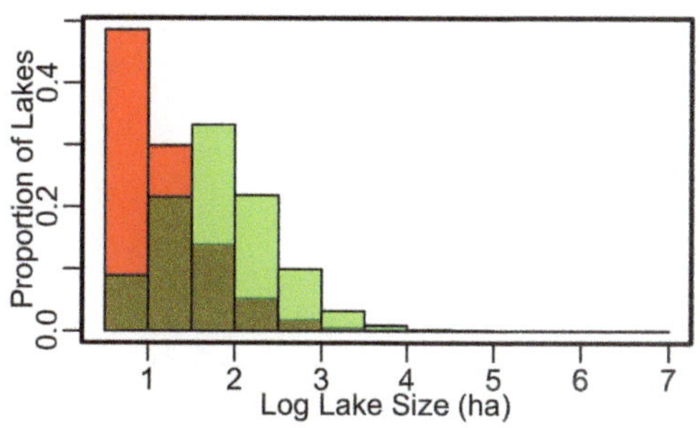

Figure 6. Distribution of lake sizes (surface areas) >4 hectare in the study region based on 2007 National Hydrography Dataset (red) relative to the distribution of lake sizes included in citizen science monitoring networks (green). Data are Log_{10} transformed.

These data needs coincide with the increasing recognition that stressors occurring at local, regional, and global scales interact to influence ecological patterns and processes from local to macro-scales [24]. At the scale of an **individual lake**, our data indicate that a majority of lakes in the Upper Midwest region have had relatively stable long-term water clarity patterns. However at the **regional scale**, we observe emerging macro-scale patterns in long-term water clarity trends and inter-annual variability. Understanding the linkages between the multi-scale water clarity patterns presented here and the factors like climate, land use/cover, and biotic interactions that have been shown to influence water clarity [7,23,31] have been elusive. Although citizen science affords a unique opportunity to obtain data such as water clarity at unprecedented spatial and temporal scales, we generally lack the large, multi-thematic data necessary to explain these patterns and fully apply a macrosystems ecology approach. Consequently, there is a need to compile multi-thematic data across multiple spatial and temporal scales in order to link patterns observed at local and regional scales to processes occurring across the landscape. We believe that collaborations among citizens, research scientists, and local, state and national agencies are important for developing the data sources and analytical tools necessary to move beyond identifying patterns toward an understanding of the critical factors influencing macro-scale patterns such as those shown here for lake water clarity.

Acknowledgments

We are grateful to the countless numbers of citizens who contributed Secchi data and the organizations that managed and provided us with data. It is only with the dedication of each citizen volunteer who diligently collected observations, state coordinators who trained volunteers and maintained databases, and high profile programs that have stimulated participation in lake monitoring at broad spatial extents that these invaluable and unique long-term data records have become available. We would also like to acknowledge the helpful contributions and suggestions from the individuals that reviewed this manuscript. We also thank the Macrosystems CSI Limnology research group for their contributions to the effort, especially Ed Bissell for all his assistance managing the data and GIS analyses. Any use of trade, firm, or product names is for descriptive purposes only and does not imply endorsement by the U.S. Government. GLERL contribution number 1703.

Author Contributions

Conceived and designed the experiments: NRL TW ENH KSC KEW JAD CAS. Analyzed the data: TW NRL. Wrote the paper: NRL TW ENH KSC KEW. Collect, compiled, and QAQCd data: NRL JAD ENH. Conducted Bayesian hierarchical modeling: TW. Created tables and figures: NRL TW. Contributed critical reviews and edited manuscript prior to submission: NRL TW ENH KSC KEW JAD CAS.

References

1. Heffernan J, Soranno PA, Angilletta MJ, Buckley LB, Gruner DS, et al. (2014) Macrosystems ecology: understanding ecological patterns and process at continental scales. Front Ecol Environ 12: 5–14 doi:10.1890/130017.
2. Ruegg J, Griese C, Bond-Lamberty B, Bowen GJ, Felzer BS, et al. (2014) Completing the data life cycle: using information management in macrosystems ecology research. Front Ecol Environ 12: 24–30 doi:10.1890/120375.
3. Hochachka WM, Fink D, Hutchinson RA, Sheldon D, Wong W-K, et al. (2011) Data-intensive science applied to broad-scale citizen science. Trends Ecol Evol 27: 130–137 doi: 10.1016/j.tree/2011.11.006.
4. Link WA, Sauer JR, Niven DK (2006) A hierarchical model for regional analysis of population change using Christmas Bird Count data, with application to the American Black Duck. Condor 108: 13–24.
5. Wilson S, Anderson EM, Wilson ASG, Bertram DF, Arcese P (2013) Citizen science reveals an extensive ehift in the winter distribution of migratory Western Grebes. PLoS ONE 8: e65408 doi:10.1371/journal.pone.0065408.s001.
6. Lathrop RC, Carpenter SR, Rudstam LG (1996) Water clarity in Lake Mendota since 1900: responses to differing levels of nutrients and herbivory. Can J Fish Aquat Sci 53: 2250–2261.
7. Bruhn LC, Soranno PA (2005) Long Term (1974–2001) Volunteer monitoring of water clarity trends in Michigan lakes and their relation to ecoregion and land use/cover. Lake Reserv Manag 21: 10–23.
8. Wernand MR (2010) On the history of the Secchi disc. JEOS:RP 5: 10013s doi:10.2971/jeos.2010.10013s.
9. Obrecht DV, Milanick M, Perkins BD, Ready D, Jones JR (1998) Evaluation of data generated from lake samples collected by volunteers. Lake Reserv Manag 14: 21–27. doi:10.1080/07438149809354106.
10. Canfield DE Jr., Brown CD, Bachmann RW, Hoyer MV (2002) Volunteer lake monitoring: testing the reliability of data collected by the Florida LAKEWATCH Program. Lake Reserv Manag 18: 1–9 doi:10.1080/07438140209353924.
11. Andreen WL (2013) Success and Backlash: The remarkable (continuing) story of the Clean Water Act. Geo Wash J Energy & Envtl L 4: 25–26.
12. Peckham SD, Lillesand TM (2006) Detection of spatial and temporal trends in Wisconsin lake water clarity using Landsat-derived estimates of Secchi depth. Lake Reserv Manag 22: 331–341 doi:10.1080/07438140609354367.
13. Olmanson LG, Bauer ME, Brezonik PL (2008) A 20-year Landsat water clarity census of Minnesota's 10,000 lakes. Remote Sense Environ 112: 4086–4097 doi:10.1016/j.rse.2007.12.013.
14. Stow CA, Lamon EC, Qian SS, Soranno PA, Reckhow KH (2009) Bayesian hierarchical/multilevel models for inference and prediction using cross-system lake data. In: Miao S, Carstenn S, Nungesser MK, editors. Real world ecology: large-scale and long-term case studies and methods. New York: Springer. pp. 111–136 doi:10.1007/978-0-387-77942-3_5.
15. Qian SS, Cuffney TF, Alameddine I, McMahon G, Reckhow KH (2010) On the application of multilevel modeling in environmental and ecological studies. Ecology 91: 355–361.
16. Lottig N, Wagner T, Henry E, Webster K, Downing J, et al. (2014) Upper Midwest Great Lake region citizen science Secchi data 1938–2012. Long Term Ecological Research Network doi:10.6073/pasta/5f94d3db0f3b7f1648579bf319aefacc.
17. Wetzel RG (2001) Limnology: lake and river ecosystems. 3rd ed. New York: Academic Press. 1006 p.
18. Plummer M (2011) JAGS Version 3.3.0 user manual. http://iweb.dl.sourceforge.net/project/mcmc-jags/Manuals/3.x/jags_user_manual.pdf.
19. Olmanson LG, Brezonik PL, Bauer ME (2013) Geospatial and temporal snalysis of a 20-year record of Landsat-based water clarity in Minnesota's 10,000 lakes. J Am Water Resour Assoc doi:10.1111/jawr.12138.
20. Webster KE (1998) Responses of lakes to drought: geomorphic and landscape controls. Ph.D. University of Wisconsin-Madison.
21. Heiskary S, Lindbloom J, Wilson CB (1994) Detecting water quality trends with citizen volunteer data. Lake Reserv Manag 9: 4–9.
22. Cross TK, Jacobson PC (2013) Landscape factors influencing lake phosphorus concentrations across Minnesota. Lake Reserv Manag 29: 1–12 doi:10.1080/10402381.2012.754808.
23. Gunn JM, Snucins E, Yan ND, Arts MT (2001) Use of water clarity to monitor the effects of climate change and other stressors on oligotrophic lakes. Environ Monit Assess 67: 69–88.
24. Soranno PA, Cheruvelil KS, Bissell EG, Bremigan MT, Downing JA, et al. (2014) Cross-scale interactions: quantifying multi-scaled cause-effect relationships in macrosystems. Front Ecol Environ 12: 65-73 doi:10.1890/120366.
25. Downing JA, Prairie YT, Cole JJ, Duarte CM, Tranvik LJ, et al. (2006) The global abundance and size distribution of lakes, ponds, and impoundments. Limnol Oceangr 51: 2388–2397.
26. Wagner T, Soranno PA, Cheruvelil KS, Renwick WH, Webster KE, et al. (2007) Quantifying sample biases of inland lake sampling programs in relation to lake surface area and land use/cover. Environ Monit Assess 141: 131–147 doi:10.1007/s10661-007-9883-z.
27. Jeppesen E, Sondergaard M, Jensen JP, Havens KE, Anneville O, et al. (2005) Lake responses to reduced nutrient loading - an analysis of contemporary long-term data from 35 case studies. Freshw Biol 50: 1747–1771 doi:10.1111/j.1365-2427.2005.01415.x.
28. Kissman CE, Knoll LB, Sarnelle O (2010) Dreissenid mussels (*Dreissena polymorpha* and *Dreissena bugensis*) reduce microzooplankton and macrozooplankton biomass in thermally stratified lakes. Limnol Oceangr 55: 1851–1859 doi:10.4319/lo.2010.55.5.1851.
29. Schindler DW, Curtis PJ, Parker BR, Stainton MP (1996) Consequences of climate warming and lake acidification for UV-B penetration in North American boreal lakes. Nature 379: 705–708.
30. Peeters ET, Franken RJ, Jeppesen E, MOSS B, Bécares E, et al. (2009) Assessing ecological quality of shallow lakes: Does knowledge of transparency suffice? Basic Appl Ecol 10: 89–96.
31. Carpenter SR, Kitchell JF, Hodgson JR (1985) Cascading trophic interactions and lake productivity. BioScience 35: 634–639.

11

Preliminary Studies on Membrane Filtration for the Production of Potable Water

Gomotsegang F. Molelekwa[1,2]*, Murembiwa S. Mukhola[2], Bart Van der Bruggen[1], Patricia Luis[3]

1 ProcESS, Department of Chemical Engineering, KU Leuven, Leuven, Belgium, **2** Department of Environmental Health, Tshwane University of Technology, Pretoria, South Africa, **3** Materials and Process Engineering (iMMC-IMAP), Université Catholique de Louvain, Louvain-la-Neuve, Belgium

Abstract

Ultrafiltration (UF) systems have been used globally for treating water from resources including rivers, reservoirs, and lakes for the production of potable water in the past decade. UF membranes with a pore size of between 0.1 and 0.01 micrometres provide an effective barrier for bacteria, viruses, suspended particles, and colloids. The use of UF membrane technology in treating groundwater for the supply of potable water in the impoverished and rural village, Tshaanda (i.e., the study area) is demonstrated. The technical and administrative processes that are critical for the successful installation of the pilot plant were developed. Given the rural nature of Tshaanda, the cultural and traditional protocols were observed. Preliminary results of the water quality of untreated water and the permeate are presented. *Escherichia coli* in the untreated water during the dry season (i.e., June and July) was 2 cfu/100 ml and was <1 cfu/100 ml (undetected) following UF, which complied with the WHO and South African National Standards and Guidelines of <1 cfu/100 ml. During the wet/rainy season (February) total coliform was unacceptably high (>2419.2 cfu/100 ml) before UF. Following UF, it dramatically reduced to acceptable level (7 cfu/100 ml) which is within the WHO recommended level of <10 cfu/100 ml. Additionally, during the wet/rainy season *E. coli* and enterococci were unacceptably high (40.4 cfu/100 ml and 73.3 cfu/100 ml, respectively) before UF but were completely removed following UF, which are within the WHO and SANS recommended limit. The values for electrical conductivity (EC) and turbidity were constantly within the WHO recommended limits of 300 μS/cm corrected at 25°C and <5 NTU, respectively, before and after UF, during dry season and wet season. This suggests that there is no need for pre-treatment of the water for suspended particles and colloids. Considering these data, it can be concluded that the water is suitable for human consumption, following UF.

Editor: Vishal Shah, Dowling College, United States of America

Funding: The funding was received from VLIR-UOS, Belgium (Project No.: ZEIN2011PR386, URL:http://www.vliruos.be/). The funders had no role in study design, data collection and analysis, decision to publish, or preparation of the manuscript.

Competing Interests: The authors have declared that no competing interests exist.

* Email: gomotsegangfred.molelekwa@cit.kuleuven.be

Introduction

Water is a precious and valuable resource that plays a critical role in the economic and social growth of any population. Most countries around the world are still struggling to meet the demand for potable water. This is evident by the uneven distribution of and accessibility to potable water between developed and developing countries as well as between urban and rural communities [1].

About 780 million people do not have access to improved drinking water sources and 2.5 billion people lack improved sanitation [2]. Most of these people reside in rural settings where piped and other improved water supplies have failed to reach the dwellers [3], thus forcing rural communities to collect drinking water from untreated sources such as rivers and springs [4]. Unsafe drinking water, with poor sanitation and hygiene, accounts for nearly 10% of the total burden of disease worldwide. This includes an estimated 4 billion cases of diarrhea annually, causing 1.8 million deaths, mostly among children under 5 years of age [5].

South Africa is home to 51.8 million people [6] and approximately 40% of this population lives in small rural villages and scattered settlements [7]. Universal access to potable water and decent sanitation is a governmental objective in South Africa [8]; thus, the government has set a target of 100% coverage of potable water supply by 2014 [9]. Since 1994, access to potable water supply has improved significantly, from 59% of the population [10] to 95.2% in 2011 [11]. Still, around five million South Africans, mainly residing in rural areas, do not have a reliable source of safe drinking water [6,11]. This situation exposes the rural communities to risks of acquiring water-borne diseases with resultant negative financial implications on the government when providing medical treatment to sick people, as these people predominantly use government clinics or hospitals for treatment [12]. For example, the South African government loses about $340 millionannually, due to approximately three million diarrhoeal cases and 50 000 related mortalities [13].

A systematic review by Wright et al. [14] showed that water contamination occurs between source and point-of-use. This pattern has been confirmed by subsequent studies of water contamination in rural Sierra Leone [15] and rural Honduras [16]. The extent of water contamination varies across the water supply chain (i.e., from source to cup used in a household), whereby more contamination is at the source than in a household setting [17].

Studies by Mpenyana-Monyani et al. [18] and Zamxaka et al. [19] reported concentrations of total coliform and *E. coli* which were higher than the limits set by WHO [20], SANS [21] and DWAF [22] for drinking water quality. As a result of the potential disease-causing characteristics of certain *E. coli*, the removal of *E. coli* from raw water is a major step in all drinking water purification systems [23].

Several studies have been done on the bacteriological water quality in some of the villages next to Tshaanda [17,24–30]. The results of these studies showed that the levels of bacteriological concentration, which included *E. coli*, were above the recommended WHO and South African National Standards for drinking water quality and thus suggesting that the water was not fit for human consumption unless treated. Subsequently, ceramic filters [24,31] were applied in an attempt to address water pollution at household level in the affected villages. Although this intervention significantly improved the quality of water and also reduced the incidence of diarrhoea, there was no follow up study to determine sustained use of filters [24]. The back drop of household water treatment options is that the responsibility of treating the water rests on the individual household. This could present variation in the quality of drinking water in the same village. For ceramic filters, tasks such as daily backwashing and cleaning of the filters could be considered tedious by some households and they may decide not to follow the instructions resulting in compromised quality of drinking water. Furthermore, some households may not have time to filter the water and would therefore drink untreated water. Monitoring of household water quality may be difficult to conduct due to restricted access to some households, for instance when there is no one in the house, thus increasing the duration of monitoring. On the other hand, population-based or source treatment is easy to monitor, can be accessed at any time and it saves time because the focus is on a single source. Furthermore, it provides uniform water quality at a given time thus becoming easy to determine which intervention should be implemented to improve water quality. Membrane filtration technology, particularly ultrafiltration (UF) system, can be used for source water treatment to provide potable water. The application of membrane filtration technology for potable water supply is well documented [32–40]. However, fouling is the most important factor that has limited the widespread use of membrane technology for the removal of microorganisms from drinking water. Feed pretreatment and membrane cleaning [41] are common strategies to minimize the fouling effect. Ultrafiltration systems have been used globally for treating various water resources including rivers, reservoirs, and lakes [42–44] for the production of potable water in the past decade. Ultrafiltration membranes with a pore size of between 0.1 and 0.01 micrometres provide an effective barrier for bacteria, viruses, suspended particles and colloids [45–47]. Ultrafiltration can be used on its own for treating drinking water where the feed water concentration is not too high in terms of organic contents [48–50]; for instance, if water turbidity is <100 NTU [51]. Accordingly, it has been used as pre-treatment for nanofiltration to reduce its fouling [50,52] and for reverse osmosis (RO) to provide an excellent treated water quality [46,47]. This has led to UF membranes replacing conventional pre-treatment procedures that consist of acid addition, coagulant/flocculant addition, chlorination, media filtration, and cartridge filtration [47].

Conventional large-scale ultrafiltration systems are operated at a transmembrane pressure of between 0.5–1.0 bar and require pumps for operation and backflushing. However, if operated by gravity, pump costs are avoided, and this can be an attractive option for decentralized, small-scale applications [45]. Presently, only few gravity-driven ultrafiltration systems for decentralized application exist (SkyJuice for community water supply and LifeStraw Family for household water treatment [51,53]. These systems can be operated at ultra-low pressure (100–150 mbar) and require little maintenance compared to the conventionally operated UF [48]. Boulestreau et al. [51] successfully conducted trials at Ogunjini village in Kwa-Zulu Natal, South Africa with a decentralized, gravity-driven membrane system to treat river water and to produce 5 m^3/d of drinking water. However, unlike our study, they used flat-sheet UF membranes. Furthermore, their system was not incorporated into any water supply infrastructure and treated water was not used for human consumption, instead it was conveyed back into the river.

Population-level potable water supply interventions such as UF, that are based on infrastructure network received less attention compared to point-of-use water treatment interventions which are being applied to improve water quality within homes and personal hygiene practices within households in rural areas of developing countries [3,4,14–17,24,54]. Furthermore, treatment of groundwater using decentralized gravity-fed ultrafiltration membrane technology for village-wide supply of potable water in South Africa's impoverished rural areas has not been tested. The aim of this study is to ascertain the use of ultrafiltration membrane for the provision of potable water in Tshaanda. The purpose of this article is to present the preliminary results of the ultrafiltration of groundwater for potable water supply in Tshaanda. The objectives of this study are to:

1. Demonstrate the feasibility of using membrane technology to provide potable water in a rural area;
2. Determine the bacteriological quality of raw water during low and high rainfall periods;
3. Determine the quality of the permeate against the WHO and South African national standards for drinking water quality.

Methods

Contextualization of Study Area

Tshaanda village is situated in Vhembe District Municipality, in the Limpopo province, South Africa. The location of Tshaanda is shown in Figure 1. Tshaanda has approximately 800 people and 150 households. The profile of Tshaanda is presented in Table 1. The community gets water from an unprotected spring, which is the only source of water supply in the area. Tshaanda was chosen because it is one of the small villages that do not have safe drinking water infrastructure in Vhembe District Municipality jurisdiction. Furthermore, the village is impoverished and excluded from the mainstream water network in Vhembe and it is unlikely that the village will be connected to the mainstream water infrastructure in the future because the mountains that are surrounding it make Tshaanda difficult to access. The spring is situated deep in the mountain valleys and it can only be accessed by foot, thus posing a challenge, especially to older persons, people with disability, women, children and AIDS sufferers. The community shares this source with domestic animals. Furthermore, the community uses this source for bathing and washing clothes. Water supply infrastructure in Tshaanda is constituted of a diesel-driven engine, 60 000 L concrete water reservoir and five street taps.

No research study was done on water quality before in Tshaanda, therefore this study presented an opportunity to maximize the use of the water source and to also show case,

Figure 1. Map of South Africa showing Tshaanda.

based on the success of the project, the possibility of using membrane technology to provide safe drinking water in a rural area, especially since drinking water sources in rural areas are not monitored.

Preliminary water sampling and analysis

On 18th June 2012, three water samples were taken in Tshaanda in order to create baseline data of the quality of drinking water before treatment in order to compare it with the quality of water after treatment with UF membrane.

The quality of drinking water in Tshaanda was analysed in relation to the WHO [20] and South African [22] guidelines for drinking water quality. The water samples were analyzed for turbidity, pH, electrical conductivity, total coliform, enterococci and *E. coli* [55]. One water sample (500 ml) was taken at each of the three source points (*i.e.*, *Spring*, *Tap 1* and *Tap 2*). Distilled water (500 ml) which was obtained from the laboratory of the Department of Agriculture, Veterinary section in Makhado, Limpopo province, was used as a control.

Defined Substrate Technology was used for the bacteriological analysis of the water samples. The IDEXX Colilert was used to process the Total coliform and *E. coli* water samples, whereas IDEXX Enterolert was used to process enterococci samples in the laboratory. Water samples of IDEXX Colilert and IDEXX Enterolert were incubated for 22 hours at 37°C and 41°C, respectively. After incubation, total coliform was determined visually by identifying all test wells that turned yellow and

Table 1. Tshaanda community profile [39].

Number of households	153
Population	835
Number of households with government subsidized houses	44
Number of households with Ventilated Improved Pit Toilets	63
Number of households without toilets	21
Source of water supply (spring)	1
Number of street water taps	8
Number of People with Matric (Grade 12)	50
Number of People with Tertiary qualification	12
Primary School	1
Clinic	0

subsequently, the Most Probable Number (MPN) was determined by counting the number of test wells that showed yellow colour. After determining the MPN, all the IDEXX Colilert Quanti-Tray plates were put individually under long wave (366 nm) ultraviolet light in order to detect any wells that had fluorescence [56]. The yellow wells that fluoresced were counted to determine the MPN for *E. coli*. A multi-meter (HACH) was used to measure the pH and electrical conductivity of the water samples. A portable turbidity meter (HACH) was used to measure the turbidity of the samples.

Social and administrative processes (non-technical)

Figure 2 highlights critical and fundamental issues that need to be adhered to in order to efficiently and successfully plan, implement and wind-up a water supply and management project in rural villages which, in most cases, are administered by traditional authorities in conjunction with local and/or district municipalities. Maintaining constant communication with the municipality, traditional authority and the community is very critical towards ensuring seamless implementation of the project and to promptly address concerns of all the key stakeholders. Furthermore, the water group will play a critical role of ensuring that the membrane pilot plant operates properly and that water related issues are resolved promptly and amicably. The water group will also ensure that the community takes ownership of the pilot plant, use it properly and protect it from vandalism and abuse. The group will educate the community about prevention of water pollution, responsible use of water and promotion of public health and personal hygiene.

Ethics statement

The process started with seeking and obtaining approval to conduct the research in Tshaanda from Vhembe District Municipality, Mutale Local Municipality and Tshitandani Traditional Council, all of which have authority over the Tshaanda jurisdiction. The first meeting was held with the local chief of Tshitandani Traditional Authority in order to explain the purpose of the project to him and to obtain approval from the traditional authority to conduct the study in Tshaanda. Subsequent meetings were held with other key stakeholders (e.g., Vhembe District Municipality, Mutale Local Municipality and Tshaanda community) in order to present the research background, obtain approval and support from them and to discuss their roles during the entire duration of the research.

Written ethics approval for this research was obtained from Vhembe District Municipality, which has overall jurisdiction over Mutale Local Municipality, where Tshaanda is situated. Tshitandani Traditional Authority also gave a written ethics approval for the study to be conducted in Tshaanda, as it has sole traditional authority over the Tshaanda jurisdiction. The participants (whom were adults) from the community of Tshaanda gave their verbal consent to participate in this study during a community meeting, which was organized by the the local chief of Tshitandani Traditional Authority. They did so in the presence of the members of Tshitandani Traditional Authority. The community gave consent for instance, to participate in the discussions during community meetings and when establishing a local water group. Their consent was then recorded as part of the minutes of that meeting. It was sensible to ask for consent from the entire community (excluding any minor) in a meeting, because the study focuses on solving a community problem rather than an individual problem. Furthermore, the engagements with the participants were mainly through community meetings.

This study also received ethical approval from the institution review board committees at Tshwane University of Technology (TUT) and Katholieke Universiteit Leuven (KU Leuven).

Public participation

A public participation process for the installation of a membrane filtration unit in Tshaanda was conducted by with the community of Tshaanda, Vhembe District Municipality, Mutale Local Municipality, Tshitandani Traditional Authority and Tshitandani Primary School Governing Body. Some of the key aspects that needed community involvement were the identification of a suitable location where the membrane filtration system should be installed, matters related to the security, operation and maintenance of the filtration system and the need for the establishment of local 'Water Group'.

Technical implementation

Raw water supply. Raw water was drawn from the local spring and pumped into a 2500 L raw water storage tank, which was placed at the top of the membrane filtration system. An existing diesel driven engine, which belongs to Vhembe District Municipality, was used to pump water from the spring to the raw water storage tank.

Membrane filtration system design, operation and maintenance. Gravity-driven ultrafiltration pilot plant was purchased from Ikusasa Water (A Division of Ikusasa Chemicals (Pty) Ltd and Chris Swartz Eng. (Water Utilization Engineers)) and it was installed in April 2013 at Tshaanda. The membrane filtration system was housed inside a 6 m maritime steel container. A hollow fibre UF membrane (Multibore membrane housed in a dizzer XL 1.5 MB 40 module), supplied by Inge watertechnologies AG, Germany), was used to drive the experiment in Tshaanda. The design of the pilot plant is shown in Figure 3 and Figure 4.

The plant was set to operate automatically even though it can be operated manually. The water flow rate was maintained between 300 and 400 L/h at a pressure of 2 kPa. The plant can produce 5 m^3/day of potable water.

Due to the operational simplicity of the pilot plant, the operation and maintenance of the pilot plant was carried out by a member of the community, who was recommended by the community. He was then trained (1-day training) on how to operate and maintain the pilot plant by a technician from Ikusasa Water. Subsequent training of the plant operator was coupled with the main maintenance schedules. The plant maintenance included backflushing of the membrane module, cleaning of bagfilter,

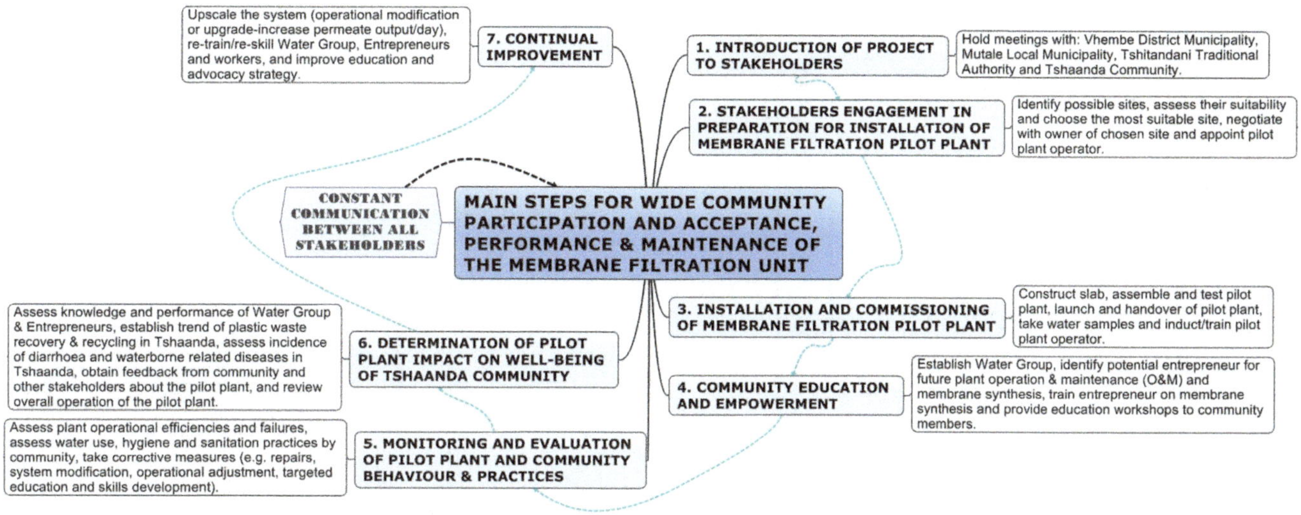

Figure 2. Main steps for wide community participation and acceptance, performance and maintenance of membrane filtration pilot plant.

calibration of the control panel and correcting the flow rate. The main maintenance of the pilot plant was done once, every two months between April 2013 and February 2014 by Ikusasa Water, however; cleaning of the bagfilter was done once every week by the pilot plant operator. The first main maintenance was done in June 2013.

Water sampling and analysis

Three months after the pilot plant was installed, water samples of raw (untreated) and permeate (treated) were taken for analysis on the 25th July 2013 in order to determine whether the pilot plant was able to produce water that met the requirements of the WHO and South African water quality guidelines [20,22] and standards [21]. Two sets of raw water samples (one from the spring, labelled as ***Spring*** and another sample at the inlet of the membrane filtration plant just before the bagfilter, labelled as ***Plant inlet before filtration***) and four samples of treated water were taken (one from the storage tank inside the filtration plant, labelled as ***Permeate*** and three from the three taps outside the plant, labelled as ***Tap 1***, ***Tap 2*** and ***Tap 3***, respectively). Furthermore, other samples were taken on the 25th February 2014 in order to determine the quality of water during wet season. The water samples were taken aseptically and sterile zip-lock water bags were used. The water samples were put inside a cooler bag with ice to maintain the temperature at ~4°C, until they were processed for analysis in the laboratory. The distance between the area where the samples were taken and the laboratory where they were analysed was 160 km. Accordingly, the samples were processed within 6 hours after collection. The samples were processed according to the same procedure that was used during the preliminary investigation of the water quality.

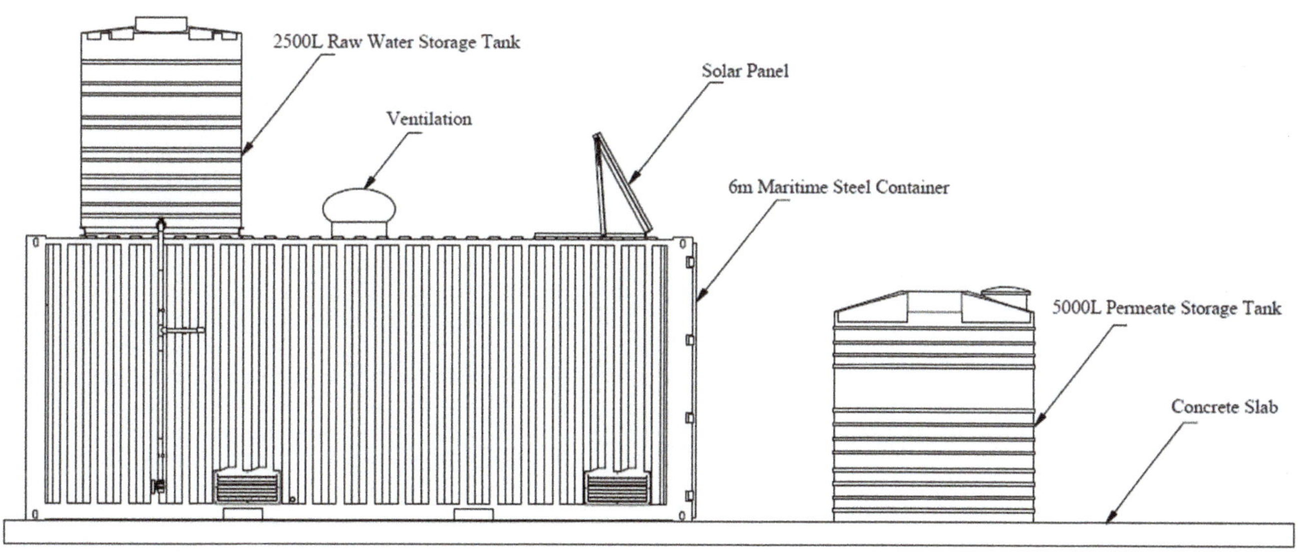

Figure 3. External design of the pilot plant in Tshaanda.

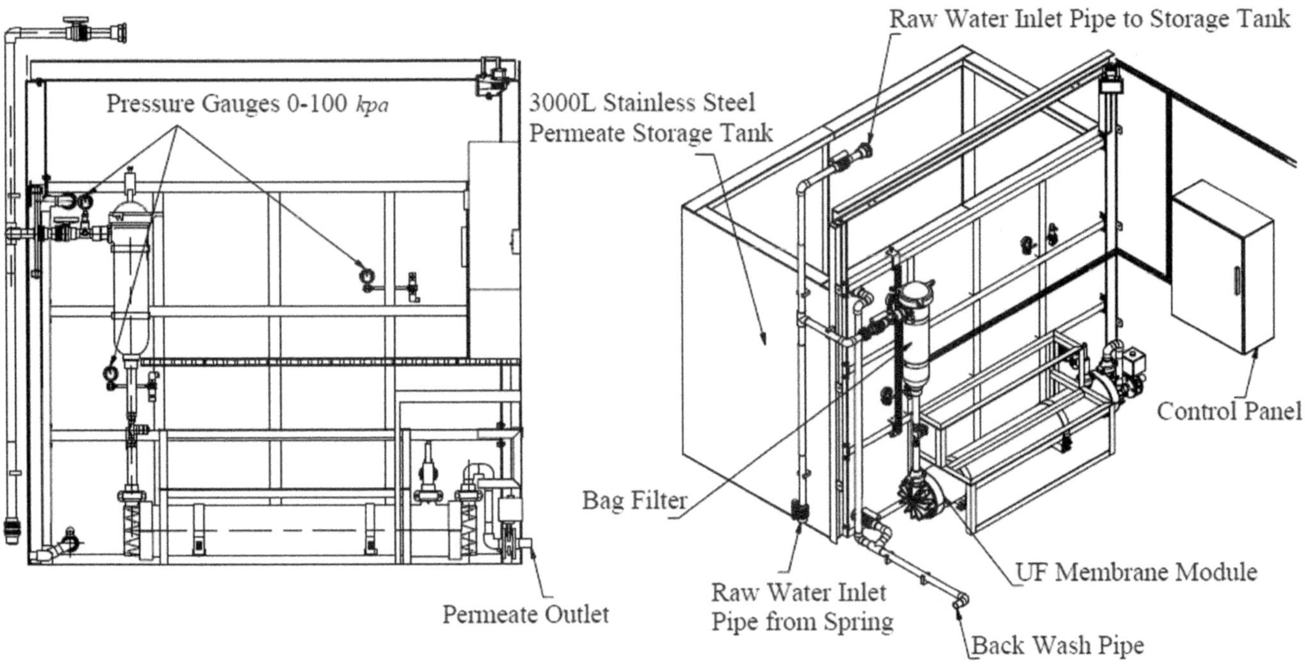

Figure 4. Internal design of the pilot plant in Tshaanda.

Results and Discussion

Installation and operation of pilot plant

The pilot membrane filtration plant was installed on the premises of the local primary school following an agreement between the community and the school governing body. The pilot plant was installed approximately 600 m away from the water source (*i.e.*, spring), construction and commissioning phases were completed within one month. The system operated through gravity for 24 h/day and produced 5000 L of potable water per day for the community of Tshaanda.

Water quality analysis

The water samples were processed and analyzed for microbiological and physico-chemical parameters. Microbiological analysis was carried out in order to determine the presence of *E. coli*, enterococci and total coliform in the water samples. Physico-chemical analysis was carried out to determine the levels of pH, turbidity and electrical conductivity, respectively.

The results of the preliminary water quality investigation are presented in Table 2.

Critical analysis of Table 2 shows that turbidity values of all the samples were within the WHO and South African standards limits for drinking water quality. The results also show the presence of *E. coli*, enterococci and total coliform in concentrations above the recommended limits.

The results of the raw water and treated water samples are shown in Table 3 and Table 4, respectively.

Total Coliform

Detection of total coliform in water indicates potential faecal pollution and thus provide information on treatment efficiency and after growth [21]. After a period of more than 20 months (June 2012–February 2014), total coliform counts of raw water was above the acceptable SANS limit. Furthermore, there was a significant increase in the total coliform count in February 2014 (>2419.2 cfu/100 ml) due to heavy rains that were experienced in the area two weeks before sampling. In July 2013, total coliform of the permeate was about 4 times (200.5 cfu/100 ml) more than that of the untreated water source (48 cfu/100 ml). This was because the permeate tank was not covered on top and therefore it presented an easy passage of dust into the permeate. The tank was subsequently cleaned and covered with a canvas. In February 2014, the coliform count of the permeate was 7.4 cfu/100 ml, following the cleaning of the permeate tank, thus being within the acceptable limit.

Escherichia coli and enterococci

Escherichia coli is the best indicator microorganism of faecal pollution [57]. The overall quality of the water at the spring was found to be consistently unsafe for drinking purposes due to the presence of *E.coli* in the water samples taken. After a period of more than 20 months (June 2012–February 2014), *E. coli* in raw water increased by more than 2 times (17.4 cfu/100 ml in June 2012 as against 40.4 cfu/100 ml in February 2014). The number of enterococci at the water source increased ten folds between June 2012 and February 2014 (7 cfu/100 ml in June 2012 and 73.3 cfu/100 ml) due to rain. This was expected as increased levels of microbial pollution are usually experienced during rainy seasons, since large numbers of microorganisms are washed from various point- and non-point pollution sites [58]. However, the concentration of enterococci was significantly lower at the filtration unit (Plant inlet before filtration). This could probably be attributed to die-off as bacteria compete for limited oxygen and nutrients in the water [59].

This water poses a public health risk to the community of Tshaanda if consumed untreated and could lead to waterborne diseases, such as: hepatitis A and E, cholera, typhoid and poliomyelitis [3]. Consuming water contaminated with *E. coli* produce a 15% chance of of having diarrhea.

Following ultrafiltration, *E. coli* and enterococci were undetected (<1 cfu/100 ml) in the permeate samples taken in July 2013 and February 2014 respectively. Maintaining this water

Table 2. Results of the preliminary water samples taken in Tshaanda on 18 June 2012.

SOURCE	CHEMICAL	GUIDELINES *(DWAF 1996 & WHO 2011)*	BACTERIOLOGICAL *(Indicator of fecal pollution)*	MPN	GUIDELINES *(DWAF 1996 & WHO 2011)*
SPRING	pH = 7.78	5 to 8	Total Coliform = (29/4)	35	<10 cfu/100 ml
	Conductivity = 46.7 µs/cm		*E. Coli* = (15/2)	17.4	0 cfu/100 ml
	Turbidity = 0.52	1 to 5 NTU	Enterococci = (7)	7	0 cfu/100 ml
TAP 1 *(street tap)*	pH = 7.38	5 to 8	Total Coliform = (26/2)	29	<10 cfu/100 ml
	Conductivity = 43.1 µs/cm		*E. Coli* = (16)	16.1	0 cfu/100 ml
	Turbidity = 0.51	1 to 5 NTU	Enterococci = (0/0)	<1	0 cfu/100 ml
TAP 2 *(street tap)*	pH = 7.39	5 to 8	Total Coliform = (15)	15.1	<10 cfu/100 ml
	Conductivity = 40.8 µs/cm		*E. Coli* = (3)	3	0 cfu/100 ml
	Turbidity = 4.31	1 to 5 NTU	Enterococci = (0)	<1	0 cfu/100 ml
CONTROL			Total Coliform = (0)	<1	0 cfu/100 ml
			E. Coli = (0)	<1	0 cfu/100 ml
			Enterococci = (0)	<1	0 cfu/100 ml

Note: **DWAF**-Formerly, Department of Water Affairs & Forestry, now Department of Water Affairs (DWA). **MPN**-Most Probable Number.

quality and ensuring that potable water is available in sufficient quantity will help to reduce the diarrhea episodes [12,60–62].

Turbidity

The turbidity values recorded for all the water samples were within the WHO [20] and SANS [21] acceptable limits for drinking water (<5 NTU) during dry and rainy seasons (June 2012, July 2013 and February 2014). Interestingly, the turbidity of the permeate was almost two times lower than that of the spring between July 2013 and February 2014. These observed lower turbidity values suggest that there is no need for pre-treatment of the water for suspended particles and colloids [51].

Table 3. Results of the water samples taken in Tshaanda on 25 July 2013.

SOURCE	CHEMICAL	GUIDELINES *(DWAF 1996 & WHO 2011)*	BACTERIOLOGICAL *(Indicator of fecal pollution)*	MPN	GUIDELINES *(DWAF 1996 & WHO 2011)*
SPRING	pH = 6.45	5 to 8	Total Coliform = (46)	118.4	<10 cfu/100 ml
	Conductivity = 49.3 µs/cm		*E. Coli* = (6)	6.4	0 cfu/100 ml
	Turbidity = 1.01	1 to 5 NTU	Enterococci = (3)	3.1	0 cfu/100 ml
PLANT INLET (Before filtration)	pH = 6.31	5 to 8	Total Coliform = (21)	27.1	<10 cfu/100 ml
	Conductivity = 47.1 µs/cm		*E. Coli* = (0)	<1.0	0 cfu/100 ml
	Turbidity = 1.69	1 to 5 NTU	Enterococci = (1)	1.0	0 cfu/100 ml
PERMEATE	pH = 6.14	5 to 8	Total Coliform = (50)	200.5	<10 cfu/100 ml
	Conductivity = 53.6 µs/cm		*E. Coli* = (0)	<1	0 cfu/100 ml
	Turbidity = 0.61	1 to 5 NTU	Enterococci = (0)	<1	0 cfu/100 ml
TAP 1	pH = 6.10	5 to 8	Total Coliform = (50)	200.5	<10 cfu/100 ml
	Conductivity = 48.5 µs/cm		*E. Coli* = (0)	<1	0 cfu/100 ml
	Turbidity = 2.09	1 to 5 NTU	Enterococci = (0)	<1	0 cfu/100 ml
TAP 2	pH = 6.11	5 to 8	Total Coliform = (50)	200.5	<10 cfu/100 ml
	Conductivity = 48.1 µs/cm		E. Coli = (0)	<1	0 cfu/100 ml
	Turbidity = 1.34	1 to 5 NTU	Enterococci = (0)	<1	0 cfu/100 ml
TAP 3	pH = 6.08	5 to 8	Total Coliform = (50)	200.5	<10 cfu/100 ml
	Conductivity = 48.2 µs/cm		E. Coli = (0)	<1	0 cfu/100 ml
	Turbidity = 0.88	1 to 5 NTU	Enterococci = (0)	<1	0 cfu/100 ml
CONTROL			Total Coliform = (0)	<1	0 cfu/100 ml
			E. Coli = (0)	<1	0 cfu/100 ml
			Enterococci = (0)	<1	0 cfu/100 ml

Note: **DWAF**-Formerly, Department of Water Affairs & Forestry, now Department of Water Affairs (DWA). **MPN**-Most Probable Number.

Table 4. Results of the water samples taken in Tshaanda on 25 February 2014.

SOURCE	CHEMICAL	GUIDELINES *(DWAF 1996 & WHO 2011)*	BACTERIOLOGICAL *(Indicator of fecal pollution)*	MPN	GUIDELINES *(DWAF 1996 & WHO 2011)*
SPRING	pH = 7.04	5 to 8	Total Coliform = (49/48)	>2419.2	<10 cfu/100 ml
	Conductivity = 45.4 µs/cm		*E. Coli* = (27/2)	40.4	0 cfu/100 ml
	Turbidity = 0.82	1 to 5 NTU	Enterococci = (37/5)	73.3	0 cfu/100 ml
PLANT INLET (Before filtration)	pH = 7.17	5 to 8	Total Coliform = (49/10)	204.6	<10 cfu/100 ml
	Conductivity = 46.8 µs/cm		*E. Coli* = (24/0)	31.7	0 cfu/100 ml
	Turbidity = 0.72	1 to 5 NTU	Enterococci = (7/1)	8.5	0 cfu/100 ml
PERMEATE	pH = 6.93	5 to 8	Total Coliform = (15)	15.1	<10 cfu/100 ml
	Conductivity = 51.9 µs/cm		*E. Coli* = (0)	<1	0 cfu/100 ml
	Turbidity = 0.46	1 to 5 NTU	Enterococci = (0)	<1	0 cfu/100 ml
CONTROL			Total Coliform = (0)	<1	0 cfu/100 ml
			E. Coli = (0)	<1	0 cfu/100 ml
			Enterococci = (0)	<1	0 cfu/100 ml

Note: **DWAF**-Formerly, Department of Water Affairs & Forestry, now Department of Water Affairs (DWA). **MPN**-Most Probable Number.

pH and electrical conductivity

The pH and electrical conductivity levels of all the water samples were within the recommended limits of WHO [20] and SANS [21] for drinking water quality.

Challenges of the project

The main challenge encountered was that there was no formal procedure for ordering and supplying diesel which at times delayed the supply of the diesel to Tshaanda. This affected pumping of raw water to the pilot plant, consequently leading to operational stoppage of the pilot plant. In order to address these challenges, a plan was developed, whereby Vhembe District Municipality supplied 220 L of diesel on a monthly basis. Furthermore, the plant operator would notify the local councillor when it was time to order diesel, the local councillor would request the responsible manager at Vhembe District Municipality to deliver the diesel to Tshaanda. The plan also included an institutional arrangement whereby other parties (i.e., KU Leuven or Tshwane University of Technology) would provide assistance to repair the engine or the water pump in case Vhembe District Municipality needed support.

Conclusions

A gravity driven ultrafiltration membrane pilot plant was used for ten (10) months (from April 2013 to February 2014) to produce potable water to the rural community of Tshaanda. The preliminary results showed that the microbiological quality of the permeate was within recommended and acceptable limits of the WHO and South African standards for drinking water quality.

Following the successful removal of *E. coli*, it can be inferred that the membrane system in Tshaanda was able to produce safe drinking water. This further confirms that the integrity of the filtration membrane was still maintained and that it was still highly effective in removing indicator microorganisms, suspended solids and colloidal matter [63] after ten (10) months of operation.

Though the UF pilot plant in Tshaanda could produce safe drinking water, there is still much work to be done before this technology could be rolled-out country wide. Subsequent research will be expanded to include aspects of evidence of sustained use, positive health impact, water quality over extended periods of use, operation and maintenance, cost-effectiveness, system performance and optimization (considering seasonal variations), and water quality monitoring, whereby the frequency of water testing will be increased. Furthermore, future research will concentrate on the water distribution network to determine bacterial regrowth and water quality deterioration, public perception, household health behaviors, acceptability of the technology, community outreach and awareness raising on water, sanitation and hygiene. Acquiring this information would help in ensuring that the system continues to deliver safe water in the long run.

Acknowledgments

We thank Professor Sadiku R, Professor Maree J and Professor Mathipa E.R for providing invaluable comments on the drafts of this paper. We would also like to thank Ikusasa Water (A Division of Ikusasa Chemicals (Pty) Ltd and Chris Swartz Eng. - Water Utilization Engineers) for their contribution in conceptualizing the design of the pilot plant and technical assistance during the operation and maintenance of the pilot plant. Thanks also go to Mr. David Nemakonde and Mrs. Maria Mufamadi from Vhembe District Municipality and Councillor Esther Mudzielwana from Mutale Local Municipality for providing logistical support in arranging meetings with the Tshitandani Traditional Authority and the community of Tshaanda. We also thank Tshitandani Traditional Authority for arranging the initial community meeting with the people of Tshaanda and for providing logistical support during the preparation for the official launch of the pilot plant. Our gratitude also goes to the community of Tshaanda for allowing us to conduct this study in their village and also for agreeing to participate in the study. We thank Mr. Mike Mokwena for his assistance during the collection and interpretation of the results of the water samples. The findings and conclusions of this paper are those of the authors and do not necessarily represent the views of the organizations they individually represent.

Author Contributions

Conceived and designed the experiments: GFM MSM BVB PL. Performed the experiments: GFM. Analyzed the data: GFM MSM BVB PL. Contributed reagents/materials/analysis tools: MSM. Wrote the paper: GFM. Reviewed and contributed to manuscript: MSM BVB PL.

References

1. World Health Organization and UNICEF (2013) Progress on Drinking Water and Sanitation: 2013 Update. Available: http://www.wssinfo.org/fileadmin/user_upload/resources/JMPreport2013.pdf. Accessed: 2013 September 07.
2. World Health Organization, UNICEF (2012) Progress on Drinking Water and Sanitation: 2012 Update. Available: http://www.unicef.org/media/files/JMPreport2012.pdf. Accessed: 2013 May 24.
3. Rosa G, Clasen T (2010) Estimating the scope of household water treatment in low-and medium-income countries. The American Journal of Tropical Medicine and Hygiene 82: 289–300.
4. Sobsey MD, Water S (2002) Managing water in the home: accelerated health gains from improved water supply. Geneva: World Health Organization.
5. World Health Organization (2005) The World Health Report 2005: Make every mother and child count. Geneva: World Health Organization.
6. Statistics South Africa (2012) Census 2011 Statistical release – P0301.4/Statistics South Africa. Pretoria. Available: http://www.statssa.gov.za/Publications/P03014/P030142011.pdf. Accessed: 2013 May 24.
7. de la Harpe J (2011) South Africa: Lessons for Rural Water Supply; Assessing progress towards sustainable service delivery. The Hague: IRC International Water and Sanitation Centre. Available: http://www.washinschools.info/docsearch/title/179989. Accessed: 2013 April 30.
8. Republic of South Africa (1996) Constitution of the Republic of South Africa Act, 1996 (Act No. 108 of 1996). Pretoria: Government Printing Works.
9. Department of Water Affairs (DWA) (2012) Briefing of the Portfolio Committee: Water & Environmental Affairs: Progress and Perspectives on Water Supply Millennium Development Goals. Available: http://www.pmg.org.za/report/20121018-department-water-affairs-attainment-millennium-development-goals. Accessed: 2013 May 24.
10. Wall K (2011) SAICE infrastructure report card for South Africa 2011. SAICE. Available: http://www.csir.co.za/enews/2011_jun/download/infrastructure_report_card_sa_2011.pdf. Accessed: 2013 May 24.
11. Department of Water Affairs (2013) Speech by the Minister of Water and Environmental Affairs, Mrs BEE Molewa, on the occasion of the presentation of Budget Vote Number 37 on behalf of the Department of Water Affairs at the Old Assembly Chamber, Parliament, Cape Town, 21 May 2013. Available: http://www.dwa.gov.za/Communications/PressReleases/2013/STATEMENT%20BY%20MINISTER%20EDNA%20MOLEWA%20-BUDGET%20VOTE%20%20MEDIA%20BRIEFING%202013%20L%20PA.pdf. Accessed: 2013 May 24.
12. Cameron J, Jagals P, Hunter PR, Pedley S, Pond K (2011) Economic assessments of small-scale drinking-water interventions in pursuit of MDG target 7C. Science of the Total Environment 410: 8–15.
13. Momba M, Obi C, Thompson P (2009) Survey of disinfection efficiency of small drinking water treatment plants: Challenges facing small water treatment plants in South Africa. Water SA 35: 485–493.
14. Wright J, Gundry S, Conroy R (2004) Household drinking water in developing countries: a systematic review of microbiological contamination between source and point-of-use. Tropical Medicine & International Health 9: 106–117.
15. Clasen T, Bastable A (2003) Faecal contamination of drinking water during collection and household storage: the need to extend protection to the point of use. J Water Health 1: 109–115.
16. Trevett A, Carter R, Tyrrel S (2005) The importance of domestic water quality management in the context of faecal-oral disease transmission. J Water Health 3: 259–270.
17. Gundry S, Wright J, Conroy R, Du Preez M, Genthe B (2013) Contamination of drinking water between source and point-of-use in rural households of South Africa and Zimbabwe: implications for monitoring the Millennium Development Goal for water. Water Practice and Technology 1: 1–9.
18. Mpenyana-Monyatsi L, Onyango MS, Benteke Momba MN (2012) Groundwater Quality in a South African Rural Community: A Possible Threat to Public Health. Polish Journal of Environmental Studies 21: 1349–1358.
19. Zamxaka M, Pironcheva G, Muyima N (2004) Microbiological and physicochemical assessment of the quality of domestic water sources in selected rural communities of the Eastern Cape Province, South Africa. Water SA 30: 333–340.
20. World Health Organization (2011) Guidelines for Drinking Water Quality. Available: http://whqlibdoc.who.int/publications/2011/9789241548151_eng.pdf. Accessed 2013 October 15.
21. South African Bureau of Standards (2011) Drinking Water - SANS 241-1: 2011. Part 1: Microbiological, physical, aesthetic and chemical determinands. Pretoria: SABS Standards Division.
22. Department of Water Affairs and Forestry (DWAF) (1996) South African Water Quality Guidelines Volume 1: Domestic Water Use Second Edition, Pretoria. Available: http://www.dwaf.gov.za/IWQS/wq_guide/domestic.pdf. Accessed: 2013 April 22.
23. Varkey A, Dlamini M (2012) Point-of-use water purification using clay pot water filters and copper mesh. Water SA 38: 721–726.
24. du Preez M, Conroy RM, Wright JA, Moyo S, Potgieter N, et al. (2008) Short report: use of ceramic water filtration in the prevention of diarrheal disease: a randomized controlled trial in rural South Africa and Zimbabwe. American Journal of Tropical Medicine and Hygiene 79: 696–701.
25. Potgieter N, Obi CL, Bessong PO, Igumbor EO, Samie A, et al. (2005) Bacteriological contamination of vhuswa, a local weaning food, and stored drinking water in impoverished households in the Venda region of South Africa. J Health, Pop Nutr 23: 150–155.
26. Potgieter N, Mudau LS, Maluleke FRS (2006) The microbiological quality of ground water sources used by rural communities in the Limpopo Province, South Africa. Water, Science and Technology 54 (11): 371–377.
27. Wright J, Gundry SW, Conroy R, Du Preez M, Genthe B, et al. (2009) Child dysentery in the Limpopo Valley: a cohort study of water, sanitation, and hygiene risk factors. Journal of Water and Health 7: 259–266.
28. Obi CL, Potgieter N, Bessong PO, Matsaung G (2002) Assessment of the microbiological quality of river water sources in rural Venda communities in South Africa. Water SA 28: 287–291.
29. Obi CL, Potgieter N, Bessong PO, Matsaung G (2003) Scope of potential bacterial agents of diarrhoea and microbial assessment of quality of river water sources in rural Venda communities in South Africa. Wat Sci Tech 47: 59–64.
30. Du Preez M, le Roux W, Potgieter N, Venter SN (2008) The genetic relatedness of E. coli associated with post-collection drinking water contamination in rural households. Water SA 34: 107–111.
31. Potgieter N, Becker PJ, Ehlers MM (2009) Evaluation of the CDC safe water-storage intervention to improve the microbiological quality of point-of-use drinking water in rural communities in South Africa. Water SA 35: 505–516.
32. Watson BM, Hornburg CD (1989) Low-Energy Membrane Nanofiltration for Removal of Color, Organics and Hardness from Drinking-Water Supplies. Desalination 72: 11–22.
33. Cartwright P (1991) Zero discharge/water reuse—The opportunities for membrane technologies in pollution control. Desalination 83: 225–241.
34. Rachwal A, Khow J, Colbourne J, O'Donnell J (1994) Water treatment for public supply in the 1990's—A role for membrane technology? Desalination 97: 427–436.
35. Kunikane S, Magara Y, Itoh M, Tanaka O (1995) A comparative study on the application of membrane technology to the public water supply. Journal of membrane science 102: 149–154.
36. Ventresque C, Bablon G (1997) The integrated nanofiltration system of the Mery-sur-Oise surface water treatment plant (37 mgd). Desalination 113: 263–266.
37. Arnal JMA, Fernandez MS, Martin GV, Garcia JL, Zafrilla JMG, et al. (2002) Design and construction of a water potabilization membrane facility and its application to the third world countries. Preliminary tests. Desalination 145: 305–308.
38. Van der Bruggen B, Vandecasteele C (2003) Removal of pollutants from surface water and groundwater by nanofiltration: overview of possible applications in the drinking water industry. Environmental Pollution 122: 435–445.
39. Muhammad N, Sinha R, Krishnan ER, Patterson CL (2009) Ceramic filter for small system drinking water treatment: Evaluation of membrane pore Size and importance of integrity monitoring. Journal of Environmental Engineering 135: 1181–1191.
40. Fane A, Tang C, Wang R (2011) Membrane technology for water: microfiltration, ultrafiltration, nanofiltration, and reverse osmosis. Treatise on Water Science, Academic Press, Oxford 4: 301–335.
41. Pangarkar BL, Sane MG, Guddad M (2011) Reverse osmosis and membrane distillation for desalination of groundwater: A review. ISRN Materials Science. http://dx.doi.org/10.5402/2011/523124.
42. Laîné JM, Vial D, Moulart P (2000) Status after 10 years of operation—overview of UF technology today. Desalination 131: 17–25.
43. Murthy Z, Chaudhari L (2009) Treatment of distillery spent wash by combined UF and RO processes. Global NEST Journal 11: 235–240.
44. Manjunath D, Madhumala M, Prasad R, Kalyani S, Sridhar S (2013) Processing of surface and ground water by hydrostatic pressure-driven membrane techniques: design and economic aspects. Desalination and Water Treatment 51: 5873–5885.
45. Peter-Varbanets M, Johnston R, Meierhofer R, Kage F, Müller S, et al. (2011) Gravity Driven Membrane Disinfection for household drinking water treatment. 35th WEDC International Conference, Loughborough, UK.
46. Pearce G (2007) The case for UF/MF pretreatment to RO in seawater applications. Desalination 203: 286–295.
47. Fernández G, Plaza F, Garralón G, Garralón A, Pérez JI, et al. (2012) A comparative study of ultrafiltration and physicochemical process as pretreatment of seawater reverse osmosis. Desalination and Water Treatment 42: 73–79.
48. Peter-Varbanets M, Zurbrügg C, Swartz C, Pronk W (2009) Decentralized systems for potable water and the potential of membrane technology. Water Research 43: 245–265.
49. Yu Z, Chu H, Cao D, Ma Y, Dong B, et al. (2012) Pilot-scale hybrid bio-diatomite/dynamic membrane reactor for slightly polluted raw water purification. Desalination 285: 73–82.
50. Zheng X, Yu M, Liang H, Qi L, Zheng H, et al. (2012) Membrane technology for municipal drinking water plants in China: progress and prospect. Desalination and Water Treatment 49: 281–295.
51. Boulestreau M, Hoa E, Peter-Verbanets M, Pronk W, Rajagopaul R, et al. (2012) Operation of gravity-driven ultrafiltration prototype for decentralised water supply. Desalination and water treatment 42: 125–130.

52. Ferella F, De Michelis I, Zerbini C, Veglio F (2013) Advanced treatment of industrial wastewater by membrane filtration and ozonization. Desalination 313: 1–11.
53. Clasen T, Naranjo J, Frauchiger D, Gerba C (2009) Laboratory assessment of a gravity-fed ultrafiltration water treatment device designed for household use in low-income settings. The American Journal of Tropical Medicine and Hygiene 80: 819–823.
54. Brown J, Proum S, Sobsey M (2008) Escherichia coli in household drinking water and diarrheal disease risk: evidence from Cambodia. Water Science & Technology 58: 757–763.
55. World Health Organization (2008) Guidelines for Drinking-water Quality. Available: http://www.who.int/water_sanitation_health/dwq/fulltext.pdf. Accessed: 2013 September 23.
56. Landre J, Gavriel A, Lamb A (1998) False-positive coliform reaction mediated by Aeromonas in the Colilert defined substrate technology system. Letters in applied microbiology 26: 352–354.
57. Edberg S, Rice E, Karlin R, Allen M (2000) Escherichia coli: the best biological drinking water indicator for public health protection. Journal of Applied Microbiology 88: 106S–116S.
58. Dobrowksy P, van Deventer A, De Kwaadsteniet M, Ndlovu T, Khan S, et al. (2013) Prevalence of virulence genes associated with pathogenic Escherichia coli strains isolated from domestically harvested rainwater during low and high rainfall periods. Appl Environ Microbiol. doi:10.1128/AEM.03061-13.
59. Momba MN, Notshe T (2003) The microbiological quality of groundwater-derived drinking water after long storage in household containers in a rural community of South Africa. Journal of Water Supply: Research & Technology-AQUA 52: 67–77.
60. Cameron J, Hunter P, Jagals P, Pond K (2011) Valuing water, valuing livelihoods. London: IWA Publishing.
61. Esrey SA, Habicht J (1985) The impact of improved water supplies and excreta disposal facilities on diarrheal morbidity, growth, and mortality among children. Cornell International Nutrition Monograph Series 15: 1–85.
62. Esrey SA, Feachem RG, Hughes JM (1985) Interventions for the control of diarrhoeal diseases among young children: improving water supplies and excreta disposal facilities. Bulletin of the World Health organization 63: 757–772.
63. Hofman J, Beumer M, Baars E, Van der Hoek J, Koppers H (1998) Enhanced surface water treatment by ultrafiltration. Desalination 119: 113–125.

12

Microbial Community Functional Structures in Wastewater Treatment Plants as Characterized by GeoChip

Xiaohui Wang[1,2], Yu Xia[1], Xianghua Wen[1]*, Yunfeng Yang[1], Jizhong Zhou[1,3,4]

1 Environmental Simulation and Pollution Control State Key Joint Laboratory, School of Environment, Tsinghua University, Beijing, China, 2 Department of Environmental Science and Engineering, Beijing University of Chemical Technology, Beijing, China, 3 Institute for Environmental Genomics and Department of Microbiology and Plant Biology, University of Oklahoma, Norman, Oklahoma, United States of America, 4 Earth Sciences Division, Lawrence Berkeley National Laboratory, Berkeley, California, United States of America

Abstract

Background: Biological WWTPs must be functionally stable to continuously and steadily remove contaminants which rely upon the activity of complex microbial communities. However, knowledge is still lacking in regard to microbial community functional structures and their linkages to environmental variables.

Aims: To investigate microbial community functional structures of activated sludge in wastewater treatment plants (WWTPs) and to understand the effects of environmental factors on their structure.

Methods: 12 activated sludge samples were collected from four WWTPs in Beijing. A comprehensive functional gene array named GeoChip 4.2 was used to determine the microbial functional genes involved in a variety of biogeochemical processes such as carbon, nitrogen, phosphorous and sulfur cycles, metal resistance, antibiotic resistance and organic contaminant degradation.

Results: High similarities of the microbial community functional structures were found among activated sludge samples from the four WWTPs, as shown by both diversity indices and the overlapped genes. For individual gene category, such as *egl, amyA, lip, nirS, nirK, nosZ, ureC, ppx, ppk, aprA, dsrA, sox* and *benAB*, there were a number of microorganisms shared by all 12 samples. Canonical correspondence analysis (CCA) showed that the microbial functional patterns were highly correlated with water temperature, dissolved oxygen (DO), ammonia concentrations and loading rate of chemical oxygen demand (COD). Based on the variance partitioning analyses (VPA), a total of 53% of microbial community variation from GeoChip data can be explained by wastewater characteristics (25%) and operational parameters (23%), respectively.

Conclusions: This study provided an overall picture of microbial community functional structures of activated sludge in WWTPs and discerned the linkages between microbial communities and environmental variables in WWTPs.

Editor: Zhi Zhou, National University of Singapore, Singapore

Funding: This study was supported by National Natural Science Foundation of China funds (51078207 and 51178239) and the special fund of State Key Joint Laboratory of Environment Simulation and Pollution Control (12L03ESPC). This study made use of the GeoChip and associated computational pipelines funded by ENIGMA-Ecosystems and Networks Integrated with Genes and Molecular Assemblies through the Office of Science, Office of Biological and Environmental Research, the U. S. Department of Energy under Contract No. DE-AC02-05CH11231. The funders had no role in study design, data collection and analysis, decision to publish, or preparation of the manuscript.

Competing Interests: The authors have declared that no competing interests exist.

* E-mail: xhwen@tsinghua.edu.cn

Introduction

Biological activated sludge process is the most widely used biological process for treating municipal and industrial wastewater. By enriching selected functional microorganisms in the activated sludge of wastewater treatment plant (WWTP), microbial activities in the sludge community are accelerated, enabling removal of oxygen-depleting organics, toxics, and nutrients. Biological WWTPs must be functionally stable to continuously and steadily remove contaminants which rely upon the activity of complex microbial communities [1,2]. Therefore, a better understanding of the microbial community structure and functional genes of activated sludge in WWTPs can help elucidate the mechanisms of biological pollutant removal and improve the treatment performance and operational stability [3,4].

Although microbial communities of activated sludge in WWTPs have been intensively studied [5–11], most of such efforts have been focused only on microbial phylogenetic composition. Knowledge is still lacking in regard to microbial community functional structures and their linkages to environmental variables. To date, there have been only a limited number of studies that characterized the overall functional profiles and metabolic pathways in the activated sludge of WWTPs using high-throughput sequencing [12,13].

Functional gene arrays, which contain genes encoding key enzymes involved in a variety of biogeochemical cycling processes [14], can be used to study the overall microbial functional potentials of activated sludge in WWTPs. GeoChip 4.2, a powerful functional gene array, contains 83,992 oligonucleotide (50-mer) probes targeting 152,414 genes in 410 gene categories involved in nitrogen, carbon, sulfur and phosphorus cycling, metal resistance, antibiotic resistance as well as organic contaminant degradation [15,16]. GeoChip has been widely used to examine microbial community functional structures in various environmental samples, such as soil and sediments [17–19], groundwater [20], acid mine drainage[21], deep-sea water [15], ocean crust [15], and hydrothermal vent [22].

In this study, 12 activated sludge samples from four WWTPs in Beijing were analyzed by GeoChip 4.2 to address the following two questions: (i) What were the microbial community functional structures of activated sludge in WWTPs? (ii) How do environmental factors and operational parameters affect microbial community functional structures? Our results revealed high similarities of microbial functional communities among activated sludge samples from the four WWTPs, and microbial functional potentials were highly correlated with water temperature, dissolved oxygen (DO), ammonia concentrations and loading rate of chemical oxygen demand (COD).

Materials and Methods

WWTPs and Sampling

Anaerobic/anoxic/aerobic (A^2O) is the most commonly used process for wastewater treatment in China, thus we examined four A^2O WWTPs in Beijing. Activated sludge samples were taken from the aeration tank of each WWTP once a day for three consecutive days in the summer of 2011. Therefore, three replicate samples were available from each WWTP, resulting in a total of 12 activated sludge samples. No specific permits were required for the described field studies. We confirm that: i) the locations were not privately-owned or protected in any way; and ii) the field studies did not involve endangered or protected species.

For microbial community analysis, each sample was dispensed into a sterile Eppendorf tube and centrifuged at 14,000 *g* for 10 min. After decanting the supernatant, the pellet was stored at −80°C prior to further analysis. Meanwhile, various chemical parameters such as COD, total nitrogen (TN), ammonia, total phosphorus (TP), pH and conductivity were analyzed according to standard methods [23]. Concentrations of chromium, cobalt, nickel, copper, zinc, and cadmium were measured by inductively coupled plasma mass spectrometry (ICP-MS)(Thermo Fisher Scientific, Waltham, MA, USA).

DNA Purification and GeoChip Hybridization

Microbial genomic DNA was extracted from the activated sludge samples by combining freeze-thawing and sodium dodecyl sulfate (SDS) for cell lysis as previously described [24]. Crude DNA was purified using the Wizard SV Genomic DNA Purification Kit (Promega, Madison WI, USA). The quality of purified DNA was assessed based on the ratios of 260/280 nm and 260/230 nm absorption measured by an ND-1000 spectrophotometer (Nanodrop Inc., Wilmington, DE, USA) and agarose gel electrophoresis. The quantity of community DNA was determined with Quant-It PicoGreen kit (Invitrogen, Carlsbad, CA, USA).

Purified DNA (1 μg) was labeled with Cy5, purified and resuspended in 10 μl hybridization solution as previously described (Lu *et al.* 2012). Then it was hybridized with GeoChip 4.2 on a MAUI hybridization station (BioMicro, Salt Lake City, UT, USA) at 42°C with 40% formamide for 16 hours. After hybridization, microarrays were scanned (NimbleGen MS200, Madison, WI, USA) at a laser power of 100%. The ImaGene version 6.0 (Biodiscovery, El Segundo, CA) was then used to determine the intensity of each spot, and identify poor-quality spots. The GeoChip raw data has been uploaded to the NCBI GEO with the accession number GSE54055.

Data Analysis

Raw data from ImaGene were submitted to Microarray Data Manager (http://ieg.ou.edu/microarray/) and analyzed using the data analysis pipeline with the following major steps: (i) The spots flagged as 1 or 3 by Imagene and with a signal to noise ratio (SNR) less than 2.0 were removed as low quality spots; (ii) After removing the bad spots, normalized intensity of each spot was calculated by dividing the signal intensity of each spot by the mean intensity of the microarray; (iii) If any of replicates had more than two times the standard deviation, this replicate was moved as an outlier [18].

Hierarchical cluster analysis was performed using the unweighted pairwise average-linkage clustering algorithm in the CLUSTER software and visualized in TREEVIEW software [25]. Canonical correspondence analysis (CCA) was used to examine the relationships of microbial communities and environmental variables. The significant environmental variables were identified by a forward selection procedure using Monte Carlo permutations ((999 permutations with a p-value$<$0.05). In addition, Variance inflation factors (VIFs; a measure for cross-correlation of explanatory variables) were checked and eliminated if VIFs were more than 20 [26].Based on partial CCA, variance partitioning analysis (VPA) was performed to attribute the variation observed in the microbial communities to the environmental variables. CCA and VPA were performed by the vegan package in R 2.14.0 (R Development Core Team, 2011).

Results

Performance of WWTPs

Details of the influent and effluent characteristics and operational parameters of the four WWTPs are listed in Table S1 and S2 in File S1. Chemical oxygen demand (COD) in the influents ranged from 257 to 452 mg/L, while total nitrogen (TN) concentrations were between 46 and 64 mg/L. The COD removal efficiencies were more than 90% in all systems, and TN removal efficiencies varied from 68% to 75%. The removal efficiencies of COD and TN indicated that the four WWTPs were functionally stable.

Overall Microbial Community Functional Structure

Microbial community functional structures of 12 activated sludge samples from four WWTPs were analyzed with GeoChip 4.2 A total of 29,124 functional genes were detected. For individual samples, the numbers of detected genes were between 18,847 and 24,268 (Table 1). Hierarchical cluster analysis showed that the three replicate samples from each WWTP were clustered together, and the distances among the three replicate samples was always <2 (Figure 1). The samples from different WWTPs were well separated, and the distances among the four WWTPs were generally>2.

To assess α-diversity of microbial communities, Shannon-Weaver index (H), Simpson's (1/D) and evenness were calculated (Table 1). The values of H were very similar across the 12 samples, ranging from 9.9 to 10.1. Consistently, the Simpson's (1/D) and evenness did not show significant differences among these four plants.

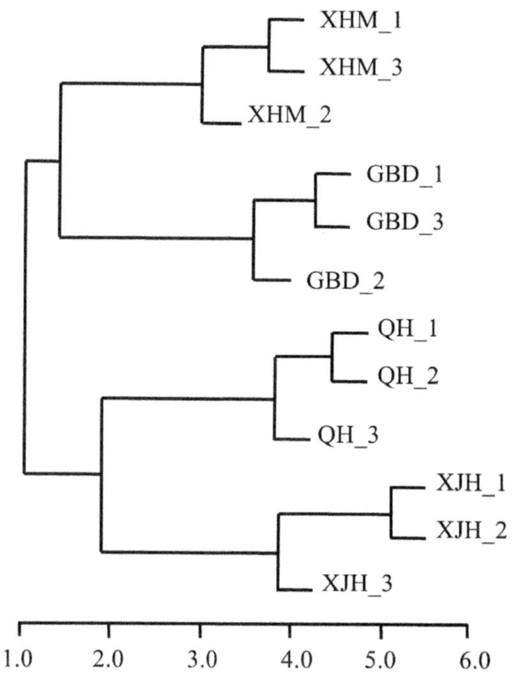

Figure 1. Hierarchical cluster analysis of functional genes in 12 activated sludge samples from four WWTPs named GBD, QH, XHM and XJH. Each sample is named after WWTP plant with "_1", "_2", or "_3" that indicates one of three replicate samples from each plant.

The overlapping genes were also determined. A large percentage (65–89%) of detected genes were shared among the samples (Table 1), indicative of high similarities of microbial community functional structures among activated sludge of four WWTPs. In consistency to the results of hierarchical clustering analysis, the percentages of the overlapping genes in samples from the same WWTP were higher than those from different WWTPs. For example, sample GBD_1 had 85% and 89% of the genes in common with sample GBD_2 and GBD_3 respectively, but 65% in common with QH_3.

At the phylogenetic level, 87.6–88.7% genes were derived from bacteria, 1.9–2.0% from archaea, and 8.7–9.7% from fungi. *Proteobacteria* was the predominant phylum of bacteria, constituting 50.8–52.2% of all detected microorganisms. *Actinobacteria*, *Firmicutes*, *Cyanobacteria* and *Bacteroidetes* were the subdominant bacterial groups, each containing 14.8–15.5%, 4.6–5.1%, 1.8–2.1% and 1.3–1.6% of detected microorganisms, respectively. *Ascomycota* was the most abundant phylum of fungi, constituting 6.3–7.4% of all detected microorganisms, followed by *Basidiomycota* (1.5–1.7%) as the second most abundant fungal group. In addition, *Euryarchaeota*, containing about 1.5% of detected microorganisms, was the most predominant phylum within archaea. (Table S3 in File S1). Taken together, these results indicated overall functional structures as well as phylogenetic diversities of microbial communities within WWTPs appeared to be quite high.

Carbon Cycling Genes

Carbon cycling, especially carbon degradation, is a crucial biochemical process in WWTPs. The rate of carbon degradation depends on a number of factors, including the availability and types of carbon substrates as well as the microbial community [27]. Among detected carbon cycling genes, 3,375 genes were involved in the degradation of complex carbon substrates such as cellulose, starch, hemicelluloses, chitin and lignin. Generally, relative abundances and numbers of all detected carbon degradation genes were similar among activated sludge of four WWTPs (Figure 2 and Table S4 in File S1).

Table 1. Percentage of overlapping genes and diversity indices of all functional genes in activated sludge samples.

sample	GBD_1	GBD_2	GBD_3	QH_1	QH_2	QH_3	XHM_1	XHM_2	XHM_3	XJH_1	XJH_2	XJH_3
GBD_1												
GBD_2	84.67%											
GBD_3	89.16%	86.65%										
QH_1	69.41%	71.45%	71.96%									
QH_2	67.39%	69.49%	69.22%	85.78%								
QH_3	65.12%	68.03%	67.38%	81.68%	80.53%							
XHM_1	79.10%	80.08%	81.10%	72.82%	69.85%	67.72%						
XHM_2	75.56%	78.38%	77.87%	71.42%	68.90%	68.33%	85.82%					
XHM_3	77.18%	78.64%	79.68%	74.95%	72.89%	71.38%	88.14%	82.43%				
XJH_1	73.92%	77.03%	74.81%	71.96%	71.38%	71.08%	75.72%	76.30%	76.37%			
XJH_2	74.11%	76.22%	74.91%	72.62%	72.29%	72.01%	74.94%	75.19%	76.71%	89.15%		
XJH_3	67.12%	69.41%	67.89%	70.21%	71.17%	70.94%	67.74%	69.71%	69.96%	79.99%	80.27%	
Richness[a]	23064	22434	23554	20862	20610	19688	24268	22831	23413	20774	20838	18847
H[b]	10.1	10.0	10.1	10.0	9.9	9.9	10.1	10.0	10.1	10.0	10.0	9.9
1/D[c]	23301.2	22682.7	23805.7	21049.3	20777.9	19855.9	24452.9	23022.1	23624.7	20959.6	21033.5	18984.4
Evenness[d]	0.996	0.996	0.996	0.996	0.995	0.995	0.994	0.995	0.995	0.995	0.995	0.995

[a]Detected gene number.
[b]Shannon-Weiner index, higher number represents higher diversity.
[c]Reciprocal of Simpson's index, higher number represents higher diversity.
[d]Evenness index.

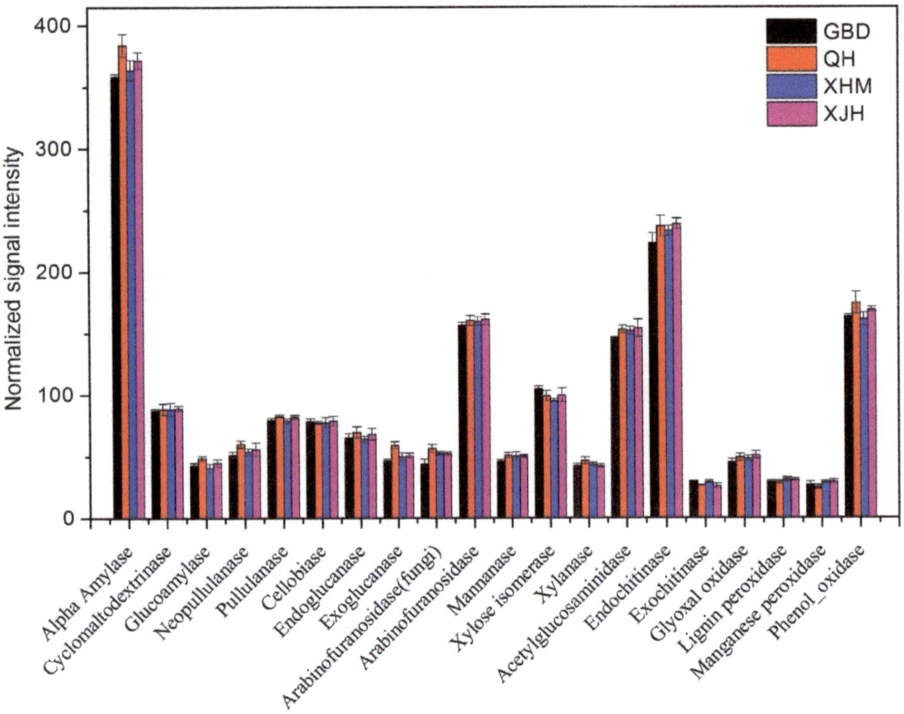

Figure 2. The normalized signal intensity of detected key genes involved in carbon cycling. The signal intensity for each functional gene was the average of the signal intensity from all the replicates. All data are presented as mean ± SE.

Endoglucanase are responsible for the initial steps of cellulose degradation. A total of 41 *egl* genes encoding endoglucanase were detected, and about half of them (20 genes) were commonly shared by all samples (Figure S1 in File S1). Nearly all of the *egl* genes were derived from isolated microorganisms, including the probes from the sequenced 149121623 (protein id number) of *Methylobacterium* sp. 4–46, 197934254 of *Streptomyces sviceus* ATCC 29083, and 171059025 of *Leptothrix cholodnii* SP-6.

α-Amylase functions to hydrolyze starch to yield glucose and maltose, which plays an important role in starch degradation. A total of 366 *amyA* genes encoding α-amylase were detected, among which 156 genes were commonly shared across all samples. Almost all the detected *amyA* genes were derived from isolated organisms such as 113897923 of *Herpetosiphon aurantiacus* ATCC 23779, 50842594 of *Propionibacterium acnes* KPA171202, 240169491 of *Mycobacterium kansasii* ATCC 12478.

Xylanase is an enzyme that breaks down hemicellulose, which is a major component of plant cell walls. 33 xylanase genes were detected, and 13 of them were shared by all samples (Figure S2 in File S1). Most of xylanase genes were from isolates such as 113935746 of *Caulobacter* sp. K31, 116098389 of *Lactobacillus brevis* ATCC 367, 68230042 of *Frankia* sp. EAN1pec, etc. Also a total of 422 genes encoding chitin degradative enzymes, including endochitinase, acetylglucosaminidase and exochitinase, were found in all 12 samples. For genes encoding endochitinase, 99 out of 239 genes were shared by all samples.

A total of 268 genes related to lignin degradation were detected, including genes encoding glyoxal oxidase (*glx*), ligninase (*lip*), manganese peroxidase (*mnp*) and phenol oxidase (*lcc*). Many of these genes were detected across all samples. For example, 13 out of 28 *lip* genes were shared by all samples and all of them were derived from isolates (Figure S3 in File S1). Similarly, almost all genes of *glx*, *mnp* and *lcc* were from isolated microorganisms rather than uncultured microorganisms.

Nitrogen Cycling Genes

As nitrogen removal is one of the most important functions in WWTPs, we examined the nitrogen cycling genes and their relationship to nitrogen concentrations. A total of 2,055 genes involved in ammonification, nitrification, denitrification and assimilatory nitrate/nitrite reduction were detected by GeoChip 4.2. Similar to carbon degradation genes, the relative abundances of detected nitrogen cycling genes were similar among activated sludge of four WWTPs (Figure 3).

Of 197 detected *nirS* genes encoding nitrite reductases, 65 were present in all samples. 7 genes were derived from isolates, while the majority of genes were derived from environmental clones in various environments. The results of Mantel test showed that the abundance of *nirS* genes was positively correlated to the influent concentrations of TN ($r = 0.489$, $P<0.01$) and ammonia ($r = 0.498$, $P<0.01$) (Table S5 in File S1).

For *nirK* genes encoding nitrite reductases, 184 genes were detected. Among them, 62 genes were present in all samples. Most of them were from laboratory clones, while only 10 genes were derived from isolates. The results of Mantel test indicated that *nirK* genes were positively correlated to the concentrations of influent TN ($r = 0.421$, $P<0.01$) and ammonia ($r = 0.358$, $P<0.05$) (Table S5 in File S1).

For *nosZ* genes encoding nitrous oxide reductases, 110 genes were detected. Among them, 43 were shared by all samples. 94 *nosZ* genes were derived from uncultured microorganisms and 16 were from cultured organisms, including 39935130 of *Rhodopseudomonas palustris* CGA009, 83816541 of *Salinibacter ruber* DSM 13855, 82947026 of *Magnetospirillum magneticum* AMB-1, etc (Figure 4). The results of Mantel test analysis showed abundances

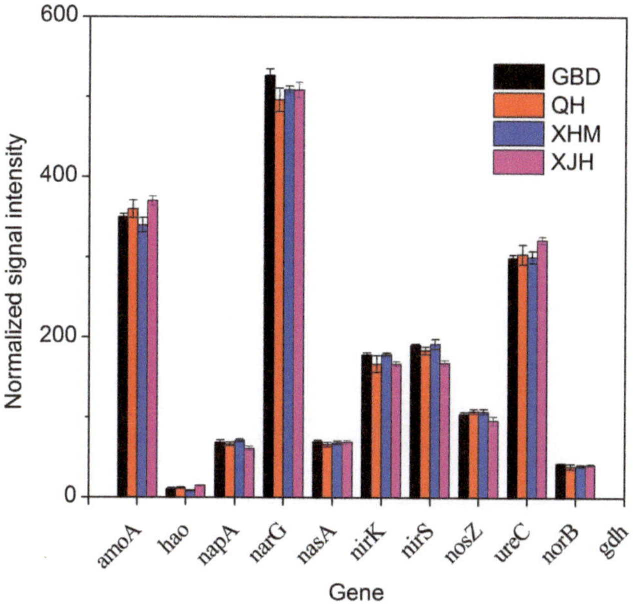

Figure 3. The normalized signal intensity of detected key genes involved in nitrogen cycling. The signal intensity for each functional gene was the average of the signal intensity from all the replicates. All data are presented as mean ± SE.

of *nosZ* genes were positively correlated with the concentrations of influent TN (r = 0.348, $P<0.05$) and ammonia (r = 0.338, $P<0.01$) (Table S5 in File S1).

295 *ureC* genes encoding ureases were detected, and 128 genes were shared by all samples. Unlike other nitrogen genes, most of the *ureC* genes were form cultured microorganisms and only 9 from uncultured bacteria. The results of Mantel test showed a positive correlation between *ureC* genes and the concentrations of influent TN (r = 0.463, $P<0.01$) and ammonia (r = 0.393, $P<0.01$) (Table S5 in File S1).

Only four *hzo* genes encoding hydrazine oxidoreductases were detected, including two genes (208605292 and 224712632) from *Candidatus Brocadia* sp. and two genes (224712622 and 208605290) from uncultured *Planctomycete*. No significant correlation between *hzo* genes and the concentrations of influent TN ($P>0.05$) and ammonia ($P>0.05$) (Table S5 in File S1) was observed.

Phosphorus Cycling Genes

Phosphorus removal from WWTPs is a key factor in preventing eutrophication of surface waters. Genes encoding exopolyphosphatase (*ppx*) for inorganic polyphosphate degradation, polyphosphate kinase (*ppk*) for catalyzing the formation of polyphosphate from ATP and phytase for phytate degradation were detected among all samples (Figure 5).

A total of 90 of 246 detected *ppx* genes were shared by all samples, including genes derived from *Leptothrix cholodnii* SP-6 (171060286), *Streptomyces coelicolor* A3(2) (21221778), *Erythrobacter* sp. NAP1 (85709358). Nevertheless, there was no significant correlation between *ppx* genes and the concentrations of influent TP (Table S5 in File S1). For *ppk*, 57 of 165 detected genes were shared by all samples. Significantly positive correlation was observed between *ppk* gene abundances and the concentrations of influent TP (r = 0.242, $P<0.05$) (Table S5 in File S1). In addition, a total of 27 phytase genes were detected. Among them, 7 were shared by all samples (Figure S4 in File S1). The results of Mantel tests suggested that there was no significant correlation between phytase genes and influent TP (Table S5 in File S1).

Sulfur Cycling Genes

A total of 1,241 genes involved in sulfite reductase, adenylylsulfate reductase, sulphur oxidation, sulfide oxidation were detected. The relative abundances of sulfur cycling genes were presented in Figure 6. Among them, there were 46 *aprA* genes encoding dissimilatory adenosine-5′-phosposulfate reductase, a key enzyme involved in microbial sulfate reduction and sulfur oxidation. 15 of 22 *aprA* genes present in all 12 samples were derived from isolates (Figure S5 in File S1). In addition, 327 *dsrA* genes encoding dissimilatory sulfite reductase were detected. Of the 100 genes shared by all samples, most genes were derived from uncultured microorganisms except for 18 genes from isolates such as 107784967 of *Desulfovibrio desulfuricans*, 13992710 of *Bilophila wadsworthia* and 83573380 of *Moorella thermoacetica* ATCC 39073. For *sox* genes encoding sulfite oxidase, 77 of 205 detected genes were shared by all samples, of which most were derived from isolated microorganisms such as 56677633 of *Silicibacter pomeroyi* DSS-3, 255258527 of *Sideroxydans lithotrophicus* ES-1, 214028622 of *Ruegeria* sp. R11.

Degradation of Organic Contaminants

Organic contamination is a concern and a lot of research has been undertaken to examine the role of microorganisms in the degradation and remediation of organic contaminants [14]. In total, 7,012 genes from 165 gene categories involved in the degradation of organic contaminants related to aromatics, chlorinated solvents, herbicides and pesticides were detected. The relative abundances of keg genes involved in organic contaminant degradation were present in Figure 7. Among them, 24 *benAB* genes encoding benzoate 1, 2-dioxygenases were detected across all samples and 19 were present in all samples. The most abundant genes were derived from *Burkholderia xenovorans* LB400 (91779181), *Mycobacterium* sp. MCS (108768762), *Rhodococcus* sp. RHA1 (110818905).

Metal Resistance Genes

Various heavy metals were often detected in WWTPs as a result of societal industrialization and modernization. Resistance genes of Ag, As, Cd, Co, Cr, Cu, Hg, Pb, Ni, Te and Zn were detected in this study. A substantial number (3,381) of genes involved in metal resistance were detected in all the samples, of which the vast majority was derived from isolated microorganisms. Notably, relative abundances of detected metal resistance genes were similar among activated sludges of four WWTPs (Figure S6 in File S1). In addition, the results of Mantel test showed that most of metal resistance genes, such as *chrA*, *nreB*, *copA*, *cueO*, *zntA*, *cadA* and *cadBD*, were positively correlated to the concentrations of Cr, Ni, Cu, Cu, Zn, Cd and Cd in WWTPs, respectively. In contrast, other metal resistance genes (*corC*, *cusA*, *zitB*) did not show significant correlations to the respective metal concentrations (Table S5 in File S1).

Antibiotic Resistance

Antibiotic resistance is a growing concern worldwide as more and more pathogens develop resistance to commonly used antibiotics. 9 gene categories families related to antibiotic resistance were detected in this study, including five transporters (ATP-binding cassette, multidrug toxic compound extrusion, major facilitator superfamily, Mex, and small multidrug resistance efflux pumps), three β-lactamase genes and several genes involved

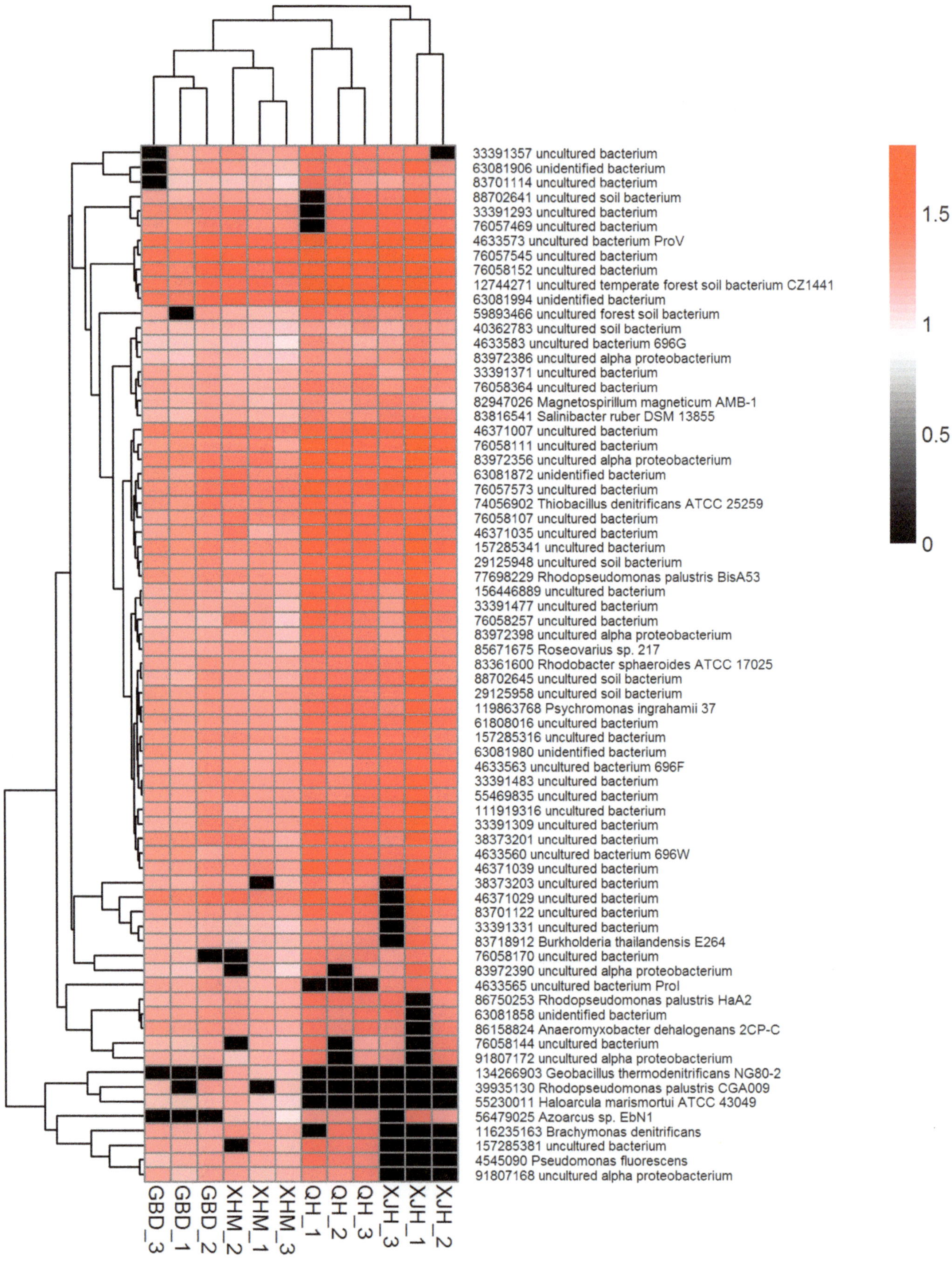
1.5
1
0.5
0
33391357 uncultured bacterium
63081906 unidentified bacterium
83701114 uncultured bacterium
88702641 uncultured soil bacterium
33391293 uncultured bacterium
76057469 uncultured bacterium
4633573 uncultured bacterium ProV
76057545 uncultured bacterium
76058152 uncultured bacterium
12744271 uncultured temperate forest soil bacterium CZ1441
63081994 unidentified bacterium
59893466 uncultured forest soil bacterium
40362783 uncultured soil bacterium
4633583 uncultured bacterium 696G
83972386 uncultured alpha proteobacterium
33391371 uncultured bacterium
76058364 uncultured bacterium
82947026 Magnetospirillum magneticum AMB-1
83816541 Salinibacter ruber DSM 13855
46371007 uncultured bacterium
76058111 uncultured bacterium
83972356 uncultured alpha proteobacterium
63081872 unidentified bacterium
76057573 uncultured bacterium
74056902 Thiobacillus denitrificans ATCC 25259
76058107 uncultured bacterium
46371035 uncultured bacterium
157285341 uncultured bacterium
29125948 uncultured soil bacterium
77698229 Rhodopseudomonas palustris BisA53
156446889 uncultured bacterium
33391477 uncultured bacterium
76058257 uncultured bacterium
83972398 uncultured alpha proteobacterium
85671675 Roseovarius sp. 217
83361600 Rhodobacter sphaeroides ATCC 17025
88702645 uncultured soil bacterium
29125958 uncultured soil bacterium
119863768 Psychromonas ingrahamii 37
61808016 uncultured bacterium
157285316 uncultured bacterium
63081980 unidentified bacterium
4633563 uncultured bacterium 696F
33391483 uncultured bacterium
55469835 uncultured bacterium
111919316 uncultured bacterium
33391309 uncultured bacterium
38373201 uncultured bacterium
4633560 uncultured bacterium 696W
46371039 uncultured bacterium
38373203 uncultured bacterium
46371029 uncultured bacterium
83701122 uncultured bacterium
33391331 uncultured bacterium
83718912 Burkholderia thailandensis E264
76058170 uncultured bacterium
83972390 uncultured alpha proteobacterium
4633565 uncultured bacterium ProI
86750253 Rhodopseudomonas palustris HaA2
63081858 unidentified bacterium
86158824 Anaeromyxobacter dehalogenans 2CP-C
76058144 uncultured bacterium
91807172 uncultured alpha proteobacterium
134266903 Geobacillus thermodenitrificans NG80-2
39935130 Rhodopseudomonas palustris CGA009
55230011 Haloarcula marismortui ATCC 43049
56479025 Azoarcus sp. EbN1
116235163 Brachymonas denitrificans
157285381 uncultured bacterium
4545090 Pseudomonas fluorescens
91807168 uncultured alpha proteobacterium
GBD_3
GBD_1
GBD_2
XHM_2
XHM_1
XHM_3
QH_1
QH_2
QH_3
XJH_3
XJH_1
XJH_2

Figure 4. The hierarchical cluster analysis of *nosZ* genes encoding nitrous oxide reductases. The protein id number and its derived organism for each gene are indicated. The color intensity of each panel shows the normalized signal intensity of individual genes, referring to color key at the top right.

in tetracycline resistance (Figure S7 in File S1). A total of 1,072 antibiotic resistance genes were detected. Of 92 tetracycline resistance genes, 39 were shared by all samples and most of them were derived from isolated microorganisms, such as 214028433 of *Ruegeria* sp. R11, 259487609 of *Aspergillus nidulans* FGSC A4, 254407454 of *Streptomyces sviceus* ATCC 29083, etc. Since antibiotic resistance is often associated with metal resistance, Mantel test was performed to examine the correlation between heavy metal resistance genes and antibiotic resistance genes. The results indicated that the whole metal resistance genes were positively correlated to the whole antibiotic resistance genes ($r = 0.325$, $P<0.01$).

Relationship of Environmental Factors to the Microbial Communities.

CCA was performed to correlate environmental variables with microbial community functional structure and determine major environmental variables in shaping the microbial community structure. On the basis of variance inflation factors, four variables were selected: DO, temperature, ammonia and COD (Figure 8). The results of CCA showed that the model was significant ($P<0.05$), suggesting that four variables were major environmental factor influencing microbial community functional structure. DO showed a strong positive correlation with the first axis and a negative correlation with the second axis, while ammonia and COD loading rate showed a strong positive correlation with the second axis and a negative correlation with the first axis. Temperature showed a strong negative correlation with both the first and second axes. Microbial community functional structures in WWTPs GBD and XHM were primarily linked to DO, while QH was linked to temperature. Ammonia and COD loading rate seemed to be major factors linking to microbial community functional structures in XJH.

Variance partitioning analyses (VPA) was further performed to assess contributions of wastewater characteristics and operational parameters to microbial community variances. Wastewater characteristics included influent concentrations of COD, TN, ammonia, TP, pH and conductivity. Operational parameters consisted of DO, temperature, hydraulic retention time (HRT) and mixed liquor suspended solids (MLSS). The results showed that 53% of total variances could be explained by these variables (Figure S8 in File S1). Wastewater characteristics and operational parameters can independently explain 25% and 23% of the variation of microbial communities, respectively. Interactions among the two major components (5%) appeared to be small.

Discussion

In this study, microbial community functional structures in activated sludges of four WWTPs were analyzed by GeoChip 4.2. A total of 29,124 functional genes involved in carbon, nitrogen, phosphorous and sulfur cycles, metal resistance, antibiotic resistance and organic contaminant degradation were detected. It is reasonable to expect that the high diversity of functional genes can increase the functional redundancy, which is ensured by the presence of a pool of species able to perform the same ecological function. That is to say, many species share similar traits in activated sludge, therefore the loss of a few species has little impact on the system performance. Functional redundancy of microbial community in activated sludge is important to ensure the system stability when individual species are lost due to environmental changes, such as significant changes in the influent wastewater composition [4].

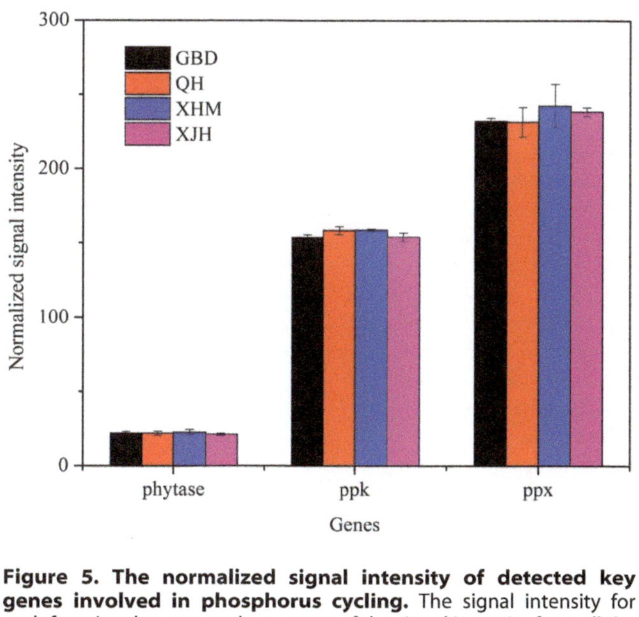

Figure 5. The normalized signal intensity of detected key genes involved in phosphorus cycling. The signal intensity for each functional gene was the average of the signal intensity from all the replicates. All data are presented as mean ± SE.

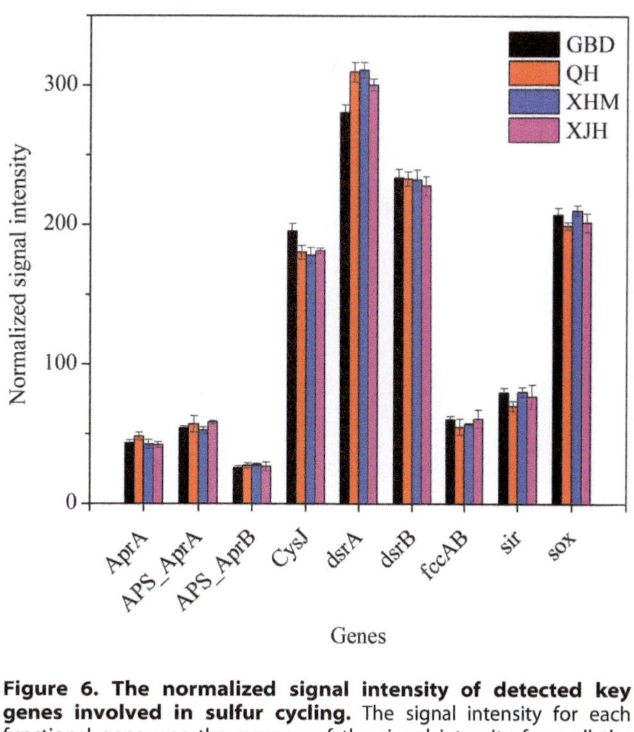

Figure 6. The normalized signal intensity of detected key genes involved in sulfur cycling. The signal intensity for each functional gene was the average of the signal intensity from all the replicates. All data are presented as mean ± SE.

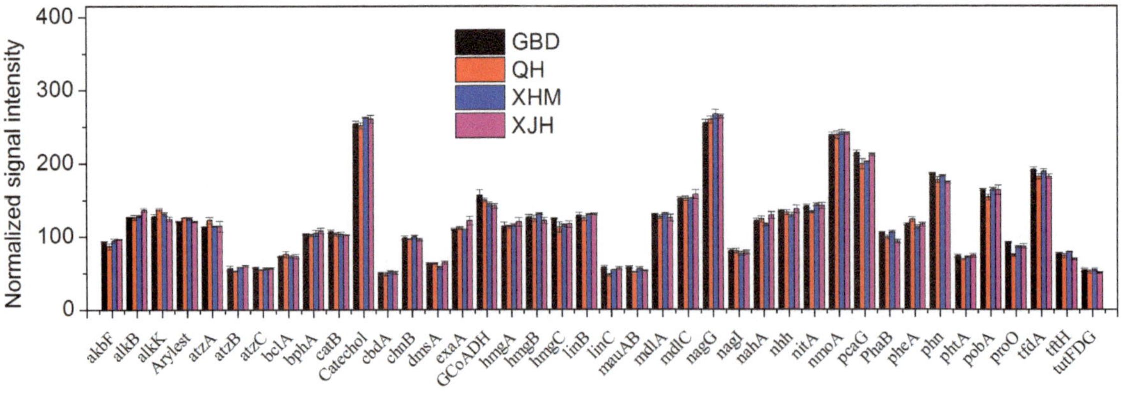

Figure 7. The normalized signal intensity of detected key genes involved in organic contaminant degradation. The signal intensity for each functional gene was the average of the signal intensity from all the replicates. All data are presented as mean ± SE.

For individual gene category, such as *egl, amyA, lip, nirS, nirK, nosZ, ureC, ppx, ppk, aprA, dsrA, sox* and *benAB*, high diversity were detected. For example, 110 *nosZ* genes were detected in this study. *nosZ* gene encodes nitrous oxide reductase that is the only enzyme to mediate the conversion of N_2O to N_2 during denitrification process, which makes the *nosZ* gene an important molecular marker to trace complete denitrification [28]. This high diversity of *nosZ* genes ensures the presence of different species with similar function mediating the conversion of N_2O to N_2, which could lead to a more stable denitrification in WWTPs. Similarly, the high diversity of *egl, amyA*, etc. can increase the functional redundancy, and ensure the stable operation of plants.

GeoChip data revealed high similarities among microbial community functional structures within activated sludge of four WWTPs as measured by diversity indices and overlapped genes. This can be partly explained by the fact that all the WWTPs studied were located in the same city (Beijing), used to treat similar domestic wastewater, and operated with the same anaerobic/anoxic/aerobic (A^2O) process of similar operating parameters (COD and TN loading rate and MLSS concentration). For each gene category such as *egl, amyA, lip, nirS, nirK, nosZ, ureC, ppx, ppk, aprA, dsrA, sox, benAB*, there were a number of microorganisms commonly shared by all 12 samples, suggesting that there could be a core microbial community in activated sludge of four WWTPs. This agrees with the findings of zhang et al.[29], who suggested, based on pyrosequencing surveys of 14 WWTPs, that some core populations were shared by multiple samples. Similar results were also reported by other studies [30,31].

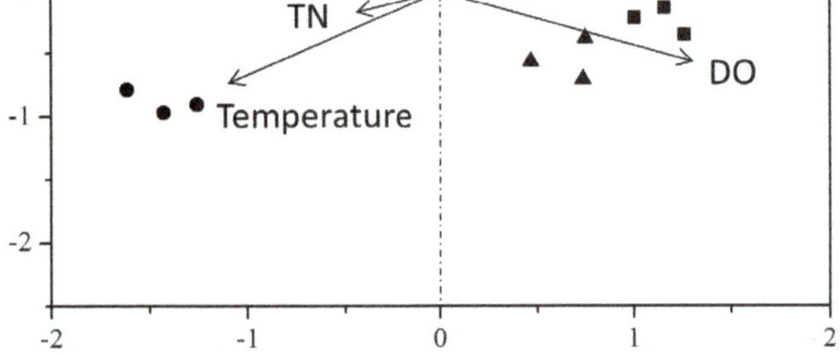

Figure 8. Canonical correspondence analysis (CCA) of GeoChip data and environmental variables. Environmental variables were chosen based on significance calculated from individual CCA data and variance inflation factors (VIFs).

Wastewater contains a broad spectrum of organic contaminants resulting from the mixing of wastewaters from different sources. Starch, cellulose, chitin and lignin could be the major carbon and energy sources for various microbial populations. Thus, we hypothesized that microbial communities would be abundant in functional genes involved in starch, cellulose, chitin and lignin. Our GeoChip data validated the hypothesis by evidence of detecting a wide variety of these genes across all samples such as *egl, amyA, lip*, xylanase genes, etc. Also, in WWTPs, biological nitrification coupled with denitrification is widely used to remove ammonia from wastewater. Nitrification is the main supply pathway of NO_3^- for denitrification, and thus it is reasonable to expect the genes involved in nitrogen cycle would be tightly linked with NH_4^+ and NO_3^- concentration. The results of Mantel test supported this expectation (Table S5 in File S1).

Understanding the factors that shape microbial community structure in WWTPs could potentially enhance treatment performance and control [11]. Of the 9 operational and environmental variables tested in this study, CCA ordination analysis indicated that DO was an important variable influencing microbial community structures. This agreed with the findings of Wells *et al.* [32], which suggested that DO was one of the most influential variables on microbial community structures of ammonia-oxidizing bacteria (AOB). Similar results have also been obtained in several lab-scale bioreactors by Park *et al.* [33].

Our results indicated that water temperature was also significantly linked to the microbial community structures in WWTPs, which was consistent with previous observations that water temperature was an important factor in shaping the microbial community structures. Ebrahimi *et al.* [34] found that, in two sequencing batch reactors (SBR) operated at 20°C, 30°C and 35°C, bacterial richness and diversity were clearly altered. In a membrane bioreactor treating domestic sewage, Ebie *et al.* [35] revealed shifts in AOB lineages in response to temperature changes by using denaturing gradient gel electrophoresis. Several other studies also demonstrated that temperature imposed a strong selective pressure on microbial communities of activated sludge [36–38].

In addition to DO and water temperature, ammonia concentrations within WWTPs were strongly and significantly linked to microbial community functional structures. Some previous studies have demonstrated the importance of ammonia concentration within bioreactor. In three gravel biofilters treating saline wastewater, Gregory *et al.* [39] observed that increasing ammonia concentration significantly influenced community structures of total bacteria and AOB. Miura *et al.* [40] found that ammonia concentration was an important structuring factor based on the DGGE analysis of AOB communities in four pilot-scale wastewater treatment systems with different ammonia concentrations.

The importance of COD loading rate for microbial community structures has been shown by previous studies. Pholchan *et al.* [41] reported that in laboratory-scale activated sludge reactors, high COD loading rate resulted in an increase and decrease of community diversity of total bacteria and AOB, respectively. In addition, Zhou *et al.* [42] showed that organic carbon (equivalent to COD loading rates) significantly affected microbial diversity in soil.

VPA was performed to attribute variations observed in microbial community functional structures to wastewater characteristics and operational parameters. VPA results showed that the deterministic wastewater characteristics and operational factors explained 53% of variances of microbial community functional structures, thus 47% of the variance was attributed to unknown factors. Some of them may result from unknown environmental variables or additional factors such as stochastic dispersal and immigration processes [43,44], protozoan grazing [45], phage predation [46] and chaotic behavior [47,48], which may play an influential role in shaping microbial communities in WWTPs.

Like other microarray technologies, GeoChip only detects genes that are represented on the array. Therefore, GeoChip 4.2 used in this study may not cover all of functional genes of microbial communities, though it contains as many as 83,992 probes targeting 152,414 genes involved in major microbial biogeochemical processes. Also, in our study, the DNA rather than cDNA from RNA was used for GeoChip hybridization. It is argued that DNA-based community analysis detects microbes irrespective of their viability or metabolic activity. Furthermore, DNA could persist extracellularly in environments after cell death, and it could result in biased population profiles [49].

In conclusion, GeoChip 4.2 was used to evaluate the microbial functional genes of activated sludge in four WWTPs. Our results revealed high similarities of microbial community functional structures among activated sludges of four WWTPs as measured by diversity indices and overlapped genes. CCA analysis indicated that microbial community functional structures were highly correlated to the water temperature, DO, ammonia concentrations and COD loading rate. Finally, a total of 53% of microbial community variations from GeoChip data were explainable by wastewater characteristics and operational parameters.

Supporting Information

File S1 This includes Tables S1–S5 and Figures S1–S8. Table S1 Summary of overall operating parameters of four WWTPs. Table S2 Metal concentrations within the four WWTPs. Table S3 Phylogenetic classification based on GeoChip data. Table S4 Numbers of detected genes involved in carbon cycling and total gene numbers present on GeoChip. Table S5 The relationship of different gene categories to related environmental variables revealed by Mantel test. Figure S1 Hierarchical cluster analysis of egl genes. Figure S2 The hierarchical cluster analysis of Xylanase genes. Figure S3 The hierarchical cluster analysis of lip genes. Figure S4 Hierarchical cluster analysis of phytase genes. Figure S5 The hierarchical cluster analysis of aprA genes. Figure S6 The normalized signal intensity of detected key genes involved in metal resistance. Figure S7 The normalized signal intensity of detected key genes involved in Antibiotic resistance. Figure S8 Variation partitioning analysis of microbial diversity explained by wastewater characteristics (W) and operational parameters (O).

Author Contributions

Conceived and designed the experiments: X. Wen YY JZ. Performed the experiments: X. Wang YX. Analyzed the data: X. Wang. Contributed reagents/materials/analysis tools: X. Wang YX. Wrote the paper: X. Wang YX.

References

1. Gentile ME, Jessup CM, Nyman JL, Criddle CS (2007) Correlation of functional instability and community dynamics in denitrifying dispersed-growth reactors. Appl Environ Microb 73: 680–690.
2. Wang XH, Wen XH, Yan HJ, Ding K, Hu M (2011) Community dynamics of ammonia oxidizing bacteria in a full-scale wastewater treatment system with nitrification stability. Front Environ Sci Engin China 5: 92–98.
3. Rittmann BE, Hausner M, Loffler F, Love NG, Muyzer G, et al. (2006) A vista for microbial ecology and environmental biotechnology. Environ Sci Technol 40: 1096–1103.
4. Briones A, Raskin L (2003) Diversity and dynamics of microbial communities in engineered environments and their implications for process stability. Curr Opin Biotech 14: 270–276.
5. Akarsubasi AT, Eyice O, Miskin I, Head IM, Curtis TP (2009) Effect of sludge age on the bacterial diversity of bench scale sequencing batch reactors. Environ Sci Technol 43: 2950–2956.
6. Dytczak MA, Londry KL, Oleszkiewicz JA (2008) Activated sludge operational regime has significant impact on the type of nitrifying community and its nitrification rates. Water Res 42: 2320–2328.
7. Huang ZH, Gedalanga PB, Asvapathanagul P, Olson BH (2010) Influence of physicochemical and operational parameters on Nitrobacter and Nitrospira communities in an aerobic activated sludge bioreactor. Water Res 44: 4351–4358.
8. López-Vázquez CM, Hooijmans CM, Brdjanovic D, Gijzen HJ, van Loosdrecht MCM (2008) Factors affecting the microbial populations at full-scale enhanced biological phosphorus removal (EBPR) wastewater treatment plants in The Netherlands. Water Res 42: 2349–2360.
9. Miura Y, Hiraiwa MN, Ito T, Itonaga T, Watanabe Y, et al. (2007) Bacterial community structures in MBRs treating municipal wastewater: relationship between community stability and reactor performance. Water Res 41: 627–637.
10. Wang XH, Wen XH, Yan HJ, Ding K, Zhao F, et al. (2011) Bacterial community dynamics in a functionally stable pilot-scale wastewater treatment plant. Bioresource Technol 102: 2352–2357.
11. Wang XH, Wen XH, Criddle C, Wells G, Zhang J, et al. (2010) Community analysis of ammonia-oxidizing bacteria in activated sludge of eight wastewater treatment systems. J Environ Sci-China 22: 627–634.
12. Sanapareddy N, Hamp TJ, Gonzalez LC, Hilger HA, Fodor AA, et al. (2009) Molecular Diversity of a North Carolina Wastewater Treatment Plant as Revealed by Pyrosequencing. Appl Environ Microb 75: 1688–1696.
13. Ye L, Zhang T, Wang T, Fang Z (2012) Microbial structures, functions, and metabolic pathways in wastewater treatment bioreactors revealed using high-throughput sequencing. Environ Sci Technol 46: 13244–13252.
14. He Z, Deng Y, Van Nostrand JD, Tu Q, Xu M, et al. (2010) GeoChip 3.0 as a high-throughput tool for analyzing microbial community composition, structure and functional activity. ISME J 4: 1167–1179.
15. Hazen TC, Dubinsky EA, DeSantis TZ, Andersen GL, Piceno YM, et al. (2010) Deep-Sea Oil Plume Enriches Indigenous Oil-Degrading Bacteria. Science 330: 204–208.
16. Lu Z, Deng Y, Van Nostrand JD, He Z, Voordeckers J, et al. (2012) Microbial gene functions enriched in the Deepwater Horizon deep-sea oil plume. ISME J 6: 451–460.
17. Zhou JZ, Kang S, Schadt CW, Garten CT (2008) Spatial scaling of functional gene diversity across various microbial taxa. Proc Natl Acad Sci USA 105: 7768–7773.

18. He ZL, Xu MY, Deng Y, Kang SH, Kellogg L, et al. (2010) Metagenomic analysis reveals a marked divergence in the structure of belowground microbial communities at elevated CO(2). Ecol Lett 13: 564–575.
19. Van Nostrand JD, Wu LY, Wu WM, Huang ZJ, Gentry TJ, et al. (2011) Dynamics of Microbial Community Composition and Function during In Situ Bioremediation of a Uranium-Contaminated Aquifer. Appl Environ Microb 77: 3860–3869.
20. Hemme CL, Deng Y, Gentry TJ, Fields MW, Wu LY, et al. (2010) Metagenomic insights into evolution of a heavy metal-contaminated groundwater microbial community. ISME J 4: 660–672.
21. Xie JP, He ZL, Liu XX, Liu XD, Van Nostrand JD, et al. (2011) GeoChip-Based Analysis of the Functional Gene Diversity and Metabolic Potential of Microbial Communities in Acid Mine Drainage. Appl Environ Microb 77: 991–999.
22. Wang F, Zhou H, Meng J, Peng X, Jiang L, et al. (2009) GeoChip-based analysis of metabolic diversity of microbial communities at the Juan de Fuca Ridge hydrothermal vent. Proc Natl Acad Sci USA 106: 4840–4845.
23. Association) AAPH (2012) Standard Methods for the Examination of Water and Waste Water. 22th ed. APHA, Washington DC, USA.
24. Zhou JZ, Bruns MA, Tiedje JM (1996) DNA recovery from soils of diverse composition. Appl Environ Microb 62: 316–322.
25. Eisen MB, Spellman PT, Brown PO, Botstein D (1998) Cluster analysis and display of genome-wide expression patterns. Proc Natl Acad Sci USA 95: 14863–14868.
26. ter B, F CJ, Smilauer P (2002) CANOCO Reference Manual and CanoDraw for Windows User's Guide: Software for canonical community ordination (version 4.5). Microcomputer Power, Ithaca, New York, , USA.
27. Wu L, Kellogg L, Devol AH, Tiedje JM, Zhou J (2008) Microarray-based characterization of microbial community functional structure and heterogeneity in marine sediments from the gulf of Mexico. Appl Environ Microb 74: 4516–4529.
28. Bergaust L, Bakken LR, Frostegard A (2011) Denitrification regulatory phenotype, a new term for the characterization of denitrifying bacteria. Biochem Soc Trans 39: 207–212.
29. Zhang T, Shao MF, Ye L (2012) 454 Pyrosequencing reveals bacterial diversity of activated sludge from 14 sewage treatment plants. ISME J 6: 1137–1147.
30. Wang X, Hu M, Xia Y, Wen X, Ding K (2012) Pyrosequencing analysis of bacterial diversity in 14 wastewater treatment systems in china. Appl Environ Microb 78: 7042–7047.
31. Xia SQ, Duan LA, Song YH, Li JX, Piceno YM, et al. (2010) Bacterial Community Structure in Geographically Distributed Biological Wastewater Treatment Reactors. Environ Sci Technol 44: 7391–7396.
32. Wells GF, Park HD, Yeung CH, Eggleston B, Francis CA, et al. (2009) Ammonia-oxidizing communities in a highly aerated full-scale activated sludge bioreactor: betaproteobacterial dynamics and low relative abundance of Crenarchaea. Environ Microbiol 11: 2310–2328.
33. Park HD, Noguera DR (2004) Evaluating the effect of dissolved oxygen on ammonia-oxidizing bacterial communities in activated sludge. Water Res 38: 3275–3286.
34. Ebrahimi S, Gabus S, Rohrbach-Brandt E, Hosseini M, Rossi P, et al. (2010) Performance and microbial community composition dynamics of aerobic granular sludge from sequencing batch bubble column reactors operated at 20 A degrees C, 30 A degrees C, and 35 A degrees C. Appl Microbiol Biot 87: 1555–1568.
35. Ebie Y, Matsumura M, Noda N, Tsuneda S, Hirata A, et al. (2002) Community analysis of nitrifying bacteria in. an advanced and compact Gappel-Johkasou by FISH and PCR-DGGE. Water Sci Technol 46: 105–111.
36. Li T, Bo L, Yang F, Zhang S, Wu Y, et al. (2012) Comparison of the removal of COD by a hybrid bioreactor at low and room temperature and the associated microbial characteristics. Bioresource Technol 108: 28–34.
37. Szukics U, Abell GCJ, Hodl V, Mitter B, Sessitsch A, et al. (2010) Nitrifiers and denitrifiers respond rapidly to changed moisture and increasing temperature in a pristine forest soil. Fems Microbiol Ecol 72: 395–406.
38. Urakawa H, Tajima Y, Numata Y, Tsuneda S (2008) Low temperature decreases the phylogenetic diversity of ammonia-oxidizing archaea and bacteria in aquarium biofiltration systems. Appl Environ Microb 74: 894–900.
39. Gregory SP, Shields RJ, Fletcher DJ, Gatland P, Dyson PJ (2010) Bacterial community responses to increasing ammonia concentrations in model recirculating vertical flow saline biofilters. Ecol Eng 36: 1485–1491.
40. Lydmark P, Almstrand R, Samuelsson K, Mattsson A, Sorensson F, et al. (2007) Effects of environmental conditions on the nitrifying population dynamics in a pilot wastewater treatment plant. Environ Microbiol 9: 2220–2233.
41. Pholchan MK, Baptista JD, Davenport RJ, Curtis TP (2010) Systematic study of the effect of operating variables on reactor performance and microbial diversity in laboratory-scale activated sludge reactors. Water Res 44: 1341–1352.
42. Zhou JZ, Xia BC, Treves DS, Wu LY, Marsh TL, et al. (2002) Spatial and resource factors influencing high microbial diversity in soil. Appl Environ Microb 68: 326–334.
43. Curtis TP, Sloan WT (2006) Towards the design of diversity: stochastic models for community assembly in wastewater treatment plants. Water Sci Technol 54: 227–236.
44. Sloan WT, Lunn M, Woodcock S, Head IM, Nee S, et al. (2006) Quantifying the roles of immigration and chance in shaping prokaryote community structure. Environ Microbiol 8: 732–740.
45. Petropoulos P, Gilbride KA (2005) Nitrification in activated sludge batch reactors is linked to protozoan grazing of the bacterial population. Can J Microbiol 51: 791–799.
46. Kunin V, He S, Warnecke F, Peterson SB, Martin HG, et al. (2008) A bacterial metapopulation adapts locally to phage predation despite global dispersal. Genome Res 18: 293–297.
47. Graham DW, Knapp CW, Van Vleck ES, Bloor K, Lane TB, et al. (2007) Experimental demonstration of chaotic instability in biological nitrification. ISME J 1: 385–393.
48. Ofiteru ID, Lunn M, Curtis TP, Wells GF, Criddle CS, et al. (2010) Combined niche and neutral effects in a microbial wastewater treatment community. Proc Natl Acad Sci USA 107: 15345–15350.
49. Nogales B, Moore ER, Llobet-Brossa E, Rossello-Mora R, Amann R, et al. (2001) Combined use of 16 S ribosomal DNA and 16 S rRNA to study the bacterial community of polychlorinated biphenyl-polluted soil. Appl Environ Microbiol 67: 1874–1884.

13

Predicted Deep-Sea Coral Habitat Suitability for the U.S. West Coast

John M. Guinotte[1]*, Andrew J. Davies[2]

1 Marine Conservation Institute, Seattle, Washington, United States of America, **2** School of Ocean Sciences, Bangor University, Menai Bridge, Anglesey, United Kingdom

Abstract

Regional scale habitat suitability models provide finer scale resolution and more focused predictions of where organisms may occur. Previous modelling approaches have focused primarily on local and/or global scales, while regional scale models have been relatively few. In this study, regional scale predictive habitat models are presented for deep-sea corals for the U.S. West Coast (California, Oregon and Washington). Model results are intended to aid in future research or mapping efforts and to assess potential coral habitat suitability both within and outside existing bottom trawl closures (i.e. Essential Fish Habitat (EFH)) and identify suitable habitat within U.S. National Marine Sanctuaries (NMS). Deep-sea coral habitat suitability was modelled at 500 m×500 m spatial resolution using a range of physical, chemical and environmental variables known or thought to influence the distribution of deep-sea corals. Using a spatial partitioning cross-validation approach, maximum entropy models identified slope, temperature, salinity and depth as important predictors for most deep-sea coral taxa. Large areas of highly suitable deep-sea coral habitat were predicted both within and outside of existing bottom trawl closures and NMS boundaries. Predicted habitat suitability over regional scales are not currently able to identify coral areas with pin point accuracy and probably overpredict actual coral distribution due to model limitations and unincorporated variables (i.e. data on distribution of hard substrate) that are known to limit their distribution. Predicted habitat results should be used in conjunction with multibeam bathymetry, geological mapping and other tools to guide future research efforts to areas with the highest probability of harboring deep-sea corals. Field validation of predicted habitat is needed to quantify model accuracy, particularly in areas that have not been sampled.

Editor: Erik Sotka, College of Charleston, United States of America

Funding: This project was supported by funding from NOAA's Deep Sea Coral Research and Technology Program. The funders had no role in study design, data collection and analysis, decision to publish, or preparation of the manuscript.

Competing Interests: The authors have declared that no competing interests exist.

* E-mail: john.guinotte@marine-conservation.org

Introduction

Predictive habitat suitability modelling is a cost effective approach to assist scientific research, conservation and management of vulnerable marine ecosystems (VMEs) in the deep sea. To date, the majority of studies using predictive models in the deep sea have focused on deep-sea corals, mostly due to the conservation status of this group and the relative abundance of data compared to other VMEs (e.g. [1,2–7]). These models identify areas with the highest probability of containing deep-sea corals and enhance our knowledge of the factors that control the distribution of these organisms. Whitmire and Clarke [8] reviewed the state of deep coral ecosystems in the waters of California, Oregon, and Washington and reported 101 species of corals from six cnidarian orders have been identified in the region. These included 18 species of stony corals (Class Anthozoa, Order Scleractinia) from seven families, seven species of black corals (Order Antipatharia) from three families, 36 species of gorgonians (Order Gorgonacea) from 10 families, eight species of true soft corals (Order Alcyonacea) from three families, 27 species of pennatulaceans (Order Pennatulacea) from eleven families, and five species of stylasterid corals (Class Hydrozoa, Order Anthoathecatae, Family Stylasteridae). The U.S. West Coast has been relatively well sampled for deep-sea corals in comparison to many other regions of the world's oceans, but the spatial distribution of deep-sea corals in unsurveyed areas within the EEZ remains largely unknown.

Predictive habitat models work by extrapolating potential species' distributions from presence data and a range of environmental variables. These two components are critical, as incomplete or erroneous data can reduce confidence in the approach and can potentially lead to predictions of limited conservation or management value [9]. When considering the utility of a model, one further consideration is the selection of an appropriate spatial scale. For example, poor spatial resolution of environmental data continues to hinder the spatial accuracy of deep-sea habitat modelling at global scales [3,5,7]. To address this, several studies have focused on improving smaller-scale, local models (i.e. 10 to 100 km2) by integrating terrain variables derived from multibeam bathymetry (e.g. [10,11,12]). Whilst smaller-scale modelling produces valuable data on species distributions in localised areas, it often requires intensive sampling effort and is of limited use in the identification of unknown habitat for cruise planning, management and conservation initiatives. Regional-scale models are needed to predict habitat suitability for corals in areas that have not been surveyed and have to be accurate enough to guide a research vessel towards a clearly defined area where sampling can be targeted [7]. Recent approaches have investigated the overlap between areas of protection and models of the distribution of vulnerable marine species [13,14].

This manuscript presents a predictive habitat suitability modelling effort for deep-sea corals within U.S. Exclusive Economic Zone (EEZ) waters off the coasts of California, Oregon and Washington. The objectives of this manuscript are to; 1) develop predictive habitat suitability models at the highest possible spatial and taxonomic resolution for deep-sea corals, 2) use model results, in addition to other tools and data, to help guide field research efforts to areas with the highest probability of harboring deep-sea corals, and 3) integrate model results with existing bottom trawl closures (i.e. essential fish habitat (EFH) area closures) and National Marine Sanctuary boundaries to determine high probability habitat areas that remain at risk from human activity.

Methods

Coral presence data

Coral distribution data were gathered from several sources including: Monterey Bay Aquarium Research Institute (MBARI), NOAA Fisheries, NOAA National Marine Sanctuaries, Smithsonian Institute's National Museum of Natural History, and Washington State University. These records were obtained from a variety of gear types: remotely operated vehicles (ROVs), manned submersibles, cameras, grabs and bottom trawls. Over 90,000 coral records were collected for the U.S. West Coast region. However, only of a fraction of these could be retained for use in the habitat suitability models. Coral observations were eliminated if they matched the following criteria: 1) records were collected as bycatch in bottom trawls as they have inherent spatial and taxonomic accuracy issues, creating uncertainties that stem from both the method in which they were collected and the taxonomic knowledge of observers on fishing vessels. Bottom trawls can be several kilometers in length and it can be difficult, if not impossible, to determine the position of the actual coral occurrence [15]. 2) Records were located in waters of less than 50 m depth. This depth cutoff was based on the fact that most zooxanthellate corals are found in shallow waters and this study is focused on deep-sea azooxanthellate corals, which tend to occur in waters deeper than 50 m. 3) The taxonomy of coral records was uncertain at the family level. 4) If more than one coral record of the same taxon (order or suborder) was located within the same 500 m grid cell. The spatial resolution of the bathymetry, environmental data, and model results was 500 m×500 m. If more than one coral record from the same taxon occurred in the same 500 m grid cell, it was treated as a 'spatial duplicate' and removed. Spatial duplicates skew models towards the environmental conditions found in those cells resulting in distorted model predictions. Some sampling approaches, such as ROVs, drop cameras or manned submersibles, document numerous coral records along relatively short distances and can introduce significant spatial bias into the analysis if all records are retained.

There were several issues which prevented models from being performed at the species level: 1) taxonomic disagreement, 2) varying degrees of taxonomic knowledge among observers and collectors, and 3) many coral presences are documented without a sample being collected to conclusively determine coral taxonomy to species. These are concerns that have been similarly noted in global models for octocoral habitat suitability [16]. For these reasons coral records were binned and modelled at the Suborder and Order levels. Suborders for which coral presence data were obtained included Alcyoniina, Calcaxonia, Filifera, Holaxonia, Scleraxonia, Stolonifera. Order level data included Antipatharia and Scleractinia. A total of 2,120 coral records were retained for analysis (Table 1 and Figure 1). Predictive models were not performed for Suborders Filifera (n = 12) and Stolonifera (n = 30) due to a paucity of coral records. It should be noted that nine of the 203 scleractinian presence records used in the predictive models were habitat-forming scleractinians (e.g. *Lophelia pertusa* and *Oculina profunda*). All other scleractinian records were solitary, non-branching corals. Most scleractinian records used in this analysis were not structure forming, but habitat suitability was modelled due to the high level of research interest for this taxon.

Bathymetry

The bathymetry of the U.S. West Coast shelf consists of a complex series of canyons, ridges and seamounts [8]. There has been significant effort in the acquisition of reliable bathymetry in this region and several data products are available. The most prominent is the Coastal Relief Model (CRM) generated by NOAA's National Geophysical Data Centre (NGDC). This is a publically available dataset with a stated cell resolution of 3-arc seconds (~90 m). The bathymetric component of the model is constructed from soundings obtained from National Oceanographic Service (NOS) hydrographic soundings, the NGDC multibeam database, and recently digitized soundings from NOS [17]. Soundings are gridded into a continuous grid for much of the shelf using the Generic Mapping Tools program Surface. However, the final CRM output is highly smoothed, omits smaller-scale features, and is of limited extent due to the high density of soundings in the shallower waters of the shelf (Figure 2a).

A custom bathymetry with a resolution of 500 m×500 m was produced from NOS hydrographic soundings, the NGDC

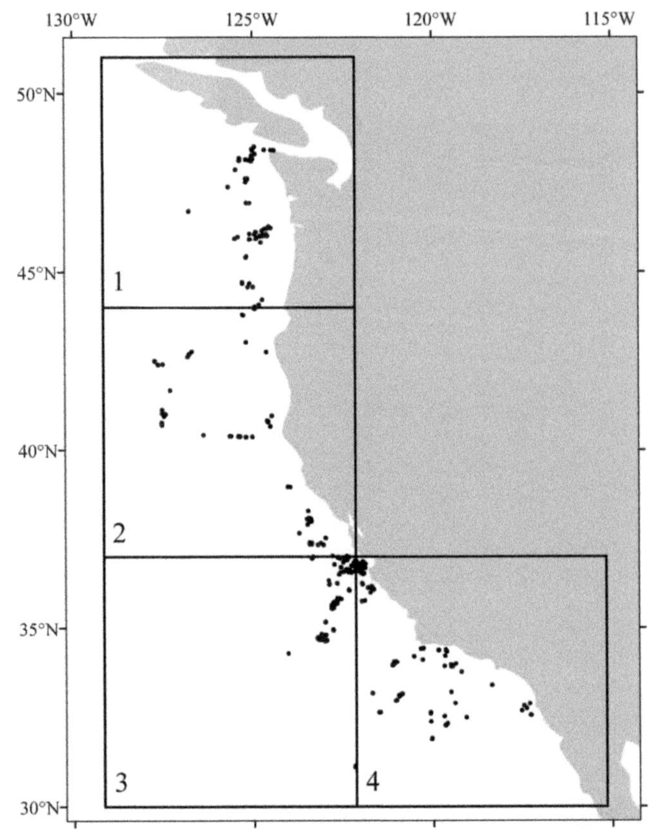

Figure 1. Location of all coral records (black markers) and regions 1–4 used for Maxent model cross-validation for each coral taxon.

Table 1. Coral records retained for habitat suitability modelling by taxon.

Taxa	Records retained
Order	
Antipatharia	128
Scleractinia	203
Suborder	
Alcyoniina	791
Calcaxonia	413
Holaxonia	308
Scleraxonia	277
Total	2120

Families of coral records listed included Order Antipatharia: Antipathidae, Cladopathidae, and Schizopathidae; Order Scleractinia: Fungiacyathidae, Micrabaciidae, Oculinidae, Caryophylliidae, Flabellidae, and Dendrophyllida; Suborder Alcyoniina: Alcyoniidae; Suborder Calcaxonia: Chrysogorgiidae, Isididae, and Primnoidae; Suborder Holaxonia: Acanthaogorgiidae, Gorgoniidae, and Plexauridae; Suborder Scleraxonia: Anthothelidae, Coralliidae, Paragorgiidae, and Plexauridae).

multibeam database, and Trackline data [17–19]. Raw soundings were extracted into XYZ for the target area of interest (Figure 2b) using MB System [20]. As with the CRM dataset, the raw sounding data was not corrected to the same vertical and horizontal datum, but this has little effect on the accuracy of the final grid output [21]. Null values and erroneous soundings were removed using either a PERL script or were removed manually. The final grid was created using the spline interpolation program MB Grid. In total, 35% of the area of interest was covered by sounding data with additional data used to infill areas with sparse soundings from Smith and Sandwell's global seafloor topography version 14.1 [22]. The final resolution of the custom grid was 500 m×500 m; smaller cell sizes showed little improvement in the quality of the bathymetry as it is limited by the spatial coverage and density of soundings. The custom grid was highly correlated with CRM data (Pearson's correlation $= 0.999$, $p<0.001$ based on 500 random points within the extent of CRM), spanned the entire study region, and retained more topographic complexity than

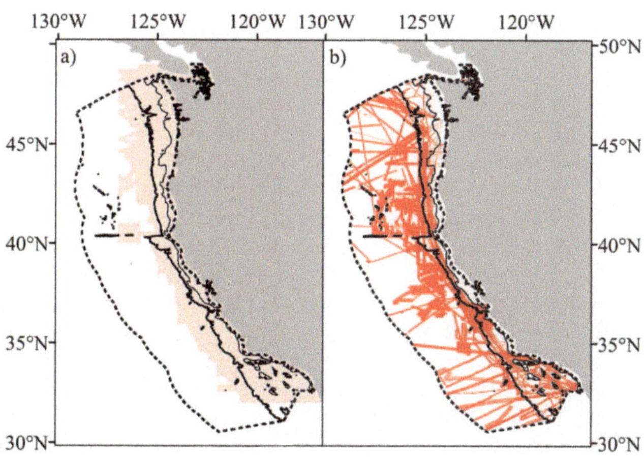

Figure 2. Custom bathymetry. a) Extent of the Coastal Relief Model data from NOAA (faint red) and the target model domain (dashed line). Faint black line indicates the 200 m contour, bold black line the 2000 m contour. b) Extent of soundings used to construct the custom 500 m×500 m bathymetry used in this study (red). Faint black line indicates the 200 m contour, bold black line the 2000 m contour and the dashed line indicates the analysis extent.

CRM. However, considering this is the first development of a custom bathymetry, care should be taken when interpreting data in areas that contain sparsely distributed or no soundings as these areas rely on satellite derived altimetry for the bathymetry (Figure 2b).

Environmental, physical and chemical data

Environmental layers were collated and constructed from sources that included ship-based CTD casts, satellites and global climatologies such as World Ocean Atlas (Table 2). The majority of source data was available as gridded datasets partitioned into standardized depth-bins ranging from 0 to 5500 m. Other data were available only as single layers from the surface (e.g. surface primary productivity) (Table 2). For depth-binned datasets, it was assumed that the conditions found at a specific gridded depth were representative of conditions on the seafloor. This allowed for the creation of continuous representations of seafloor conditions by extrapolating each depth-bin to the corresponding area of seafloor at that depth. This approach was initially developed for the creation of global environmental, physical, and chemical datasets [7].

Converting depth-binned datasets into representations of seafloor conditions involved several computer intensive processes that were conducted within a series of Python scripts. Firstly, each depth-bin of the gridded data is extracted into a single layer and interpolated at a higher spatial resolution (usually 0.1°) using inverse distance weighting. The interpolation was required to reduce gaps that appear between adjacent depth bins due to a lack of overlap when extrapolated to the bathymetry. Each of these layers was then resampled to match the extent and resolution of the bathymetry with no further interpolation. Secondly, these layers were resampled to match the extent and cell resolution of the bathymetry. Thirdly, each resampled depth-bin was clipped by the area of seafloor that was available at that particular depth. Each bin did not overlap and all were merged to produce a continuous representation of the variable on the seafloor.

This approach essentially develops a model of potential conditions for each variable. It uses the best available data, but makes several assumptions: 1) environmental conditions from the gridded CTD data are representative of the conditions at the seafloor. The majority of CTD casts do not normally reach the seafloor as they are usually stopped between 5 and 10 m from the bottom to reduce the chance of damage to the CTD system

Table 2. Environmental, physical, and chemical layers developed for this study.

Variable	Native resolution	Source
Terrain variables		
Depth[b]	0.0083°	Custom bathymetry
Slope[1, b]	0.0083°	Custom bathymetry
Chemical variables		
Apparent oxygen utilisation[b], dissolved oxygen[b], percent oxygen saturation[b].	1°	Garcia et al. [51]
Aragonite[b] and calcite[b] saturation states	1°, 3.6°x0.8–1.8°	Orr et al. [52][2], Steinacher et al. [53][3]
Nitrate[b], phosphate[b], silicate[b]	1°	Garcia et al. [54]
Salinity[b], temperature[b]	0.25°	Boyer et al. [55]
Biological variables		
Particulate organic carbon[a]	0.08°	Lutz et al. [24]

[1]Derived using ArcGIS spatial analyst and in several layers created using moving windows of 500 m, 1 km, 2.5 km, 5 km, 20 km. [2]Extracted from OCMIP2 model data for 1995. [3]Extracted from SRES B1 scenario model; mean 2000–2009. [a]Indicates a surface variable. [b]indicates a seafloor variable.

through impact, and 2) seafloor conditions are relatively stable. Annual mean values were used to maximize the amount of environmental data incorporated. While much of the deep sea is relatively stable below 200 m, there is still significant temporal variability in shelf and coastal areas and caution should be taken when interpreting predictions in shallow-water areas. However, the longevity of many deep-sea coral species far exceeds the measuring period of most oceanographic variables. Surface datasets were not up-scaled by the above process, as they were only available as a single depth-bin. Surface variables were interpolated to a higher spatial resolution using the data-interpolating variational analysis approach (DIVA; [23]) that is written into Ocean Data View version 4. For this study, we selected particulate organic carbon flux to the seafloor from Lutz et al. [24] as the productivity variable. Slope was calculated within ArcGIS Spatial Analyst using a moving window to extrapolate both fine scale slopes (1 km, 5 km) and broad-scale slopes (10 km, 20 km) using Horn's algorithm [25].

The accuracy of the up-scaled environmental variables was tested using quality controlled water bottle data obtained from the 2009 version of the World Ocean Database (WOD) [26]. Only points collected post-2001 were used in the statistical comparison and null values were removed as these were not used in the development of the temperature or salinity grids. WOD data were filtered to ensure 1) values contained a bottom depth meta-data flag, and 2) data values were within 5% of the total depth from the custom bathymetry for a cast location. Four variables contained adequate data for statistical comparison with environmental layers: temperature, salinity, phosphate and dissolved oxygen (Figure 3). The four up-scaled environmental variables that were assessed with WOD water bottle data were highly correlated at each sampling location (Pearson's correlation, R2, temperature $=0.98$ ($n=108$), salinity $=0.93$ ($n=105$), dissolved oxygen $=0.91$ ($n=100$) and phosphate $=0.88$ ($n=101$), all values significant at $p<0.001$). The phosphate comparison showed an artifact at bottle concentrations above 3.5 mg l-1 (Figure 3). This occurred in regions that have low topographic relief, resulting in environmental variables not being upscaled by the bathymetry and the original resolution of the environmental variable being visible (in this case 1°). Several other CTD datasets from the U.S. West Coast were assessed for suitability, but many did not penetrate into the deep sea and did not include bottom depth as meta-data making it impossible to determine whether a cast went near the seafloor.

Variable selection

Variables were selected based on a literature search of environmental, physical, and chemical factors known or thought to influence deep-sea coral growth and survival. Temperature, salinity, aragonite saturation state, and topographic complexity have been shown to be strong predictors of bioherm forming scleractinian coral distribution in recent deep-sea modelling efforts [3,5,7,27]. Calcite saturation state was chosen over aragonite saturation state for use in this study as the majority of coral taxa that were modelled use calcite to build their calcium carbonate spicules and structures [16]. Scleractinians have aragonitic skeletons, but the vast majority of scleractinian corals used in this analysis were solitary, non-reef forming species. Living specimens of these solitary species have been collected in highly undersaturated waters with respect to aragonite, which led to the deduction that aragonite saturation state would not be a strong predictor for determining their potential distribution at a regional scale [28]. Slope was calculated at a variety of spatial resolutions ranging from 1 km–20 km and is a useful proxy for current acceleration and mixing, which are known to influence coral distribution and abundance [29–31].

Covariation between environmental datasets is a complication that must be addressed in many predictive modelling efforts. Environmental datasets used in this analysis were assessed for covariation in correlation matrices (See Figures S1–S10). Although Maxent is reasonably robust with respect to covariation, an *a priori* variable selection process was used to reduce covariation by removing variables that were highly correlated and likely to adversely affect final predictions. Covariation was assessed using correlation matrices in R. Strong correlations between variables (>0.7) were addressed by omitting one of the environmental variables (except for calcite saturation state, temperature, and depth; see Results and Discussion). The importance of each variable in the model was assessed using a jack-knifing procedure that compared the contribution of each variable (when absent from the model) with a second model that included the variable. The final habitat suitability maps were produced by applying the calculated models to all cells in the study region, using a logistic link function to yield a habitat suitability index (HSI) between zero and one [32].

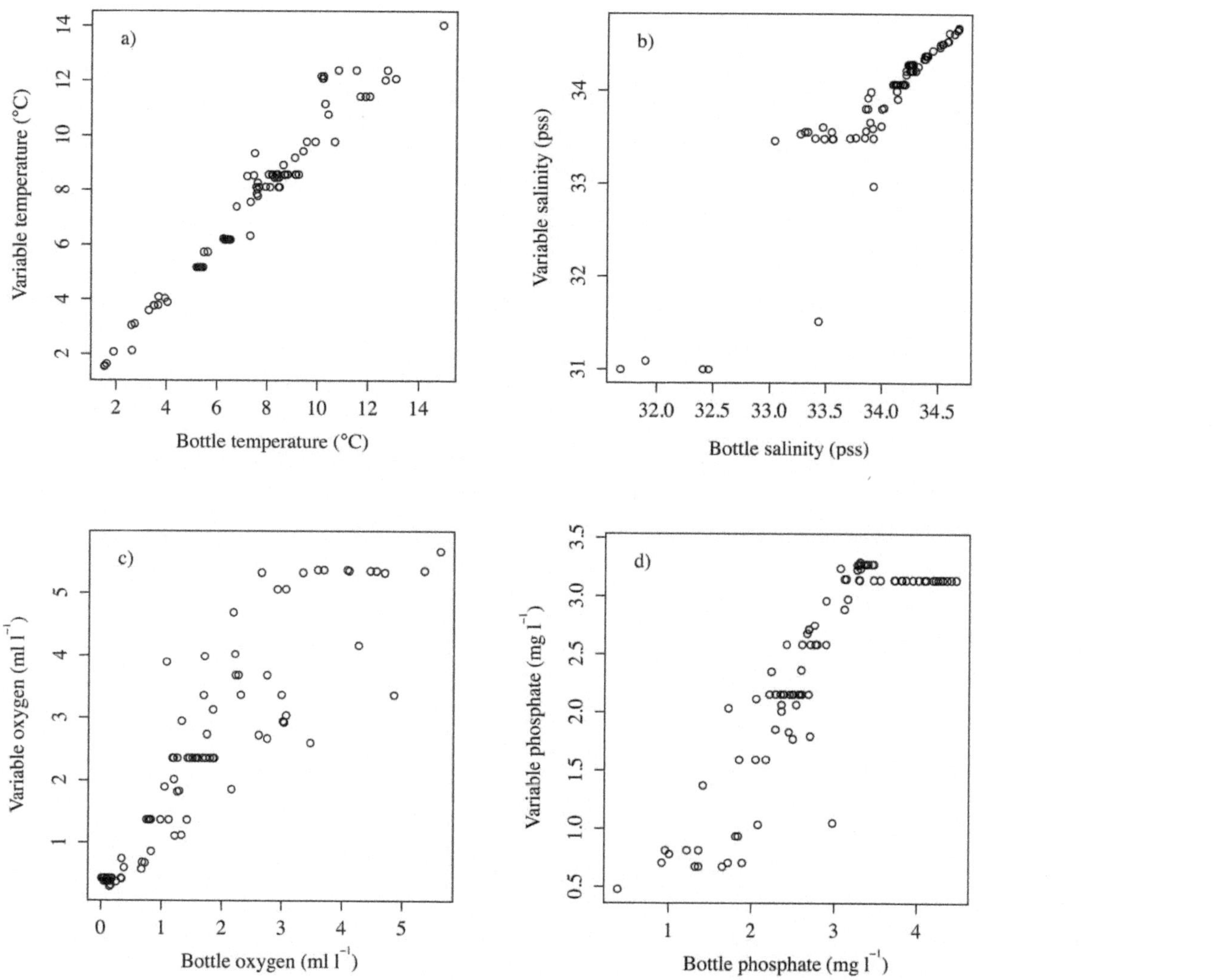

Figure 3. Validation of the environmental layer creation process by comparing the variable value on y axis (Variable), against WOD 2009 CTD data (Bottle, on x axis). a) Temperature, b) salinity, c) dissolved oxygen and d) phosphate.

Modelling Methods

Maxent version 3.2.1 (http://www.cs.princeton.edu/~schapire/maxent) was used to model predicted deep-sea coral distributions for the U.S. West Coast. Maxent (maximum entropy modelling) consistently outperforms other presence-only modelling packages including Ecological Niche Factor Analysis (ENFA) [5,33]. Presence-only modelling is one of the only methods available for modelling species distributions in the deep sea because documented absence data is sparse and when available can be unreliable. Maxent's underlying assumption is the best way to determine an unknown probability distribution is to maximise entropy based on constraints derived from environmental variables [32]. Default model parameters were used as they have performed well in other studies (a convergent threshold of 10^{-5}, maximum iteration value of 500 and a regularisation multiplier of 1 [34]).

Model accuracy between the test data and the predicted suitability models was assessed using a threshold-independent procedure that used a receiver operating characteristic curve with area under curve (AUC) for the test localities and a threshold-dependent procedure that assessed misclassification rate. With presence-only data, Phillips et al. [32] define the AUC statistic as the probability that a presence site is ranked above a random background site. In this situation, AUC scores of 0.5 indicate that the discrimination of the model is no better than random, with the maximum achievable AUC value being 1, which implies perfect discrimination of validation data. To develop the models in this study, coral presence data was spatially partitioned into four regions to calculate validation metrics and assess whether or not spatial sampling bias of coral records was influencing model performance (regions depicted in Figure 1). Four Maxent models were performed for each coral taxon (order/suborder) so models could be spatially cross-validated. For example, model 1 for any given taxon used coral records from regions 2, 3, and 4 as training records and region 1 coral records as test data to assess model performance using AUC. Model 2 used coral records from regions 1, 3, and 4 as training records and region 2 coral records as test records. The same procedure was performed for models 3 and 4. The cross-validation of models across the four regions was necessary due to the high number of coral presence records provided by MBARI for Monterrey Canyon and Davidson seamount and has the benefit of testing models with spatially independent data (see regions 3 and 4 in Figure 1). In this study, spatially cross-validated models with AUC scores >0.7 were retained for further analysis and used in the production of thresholded predictions, models scoring lower than this were omitted.

There is ongoing debate regarding the interpretation of Maxent's logistic prediction values (0–1) for habitat suitability [35,36], but it represents the best metric at present. Several studies have defined a binary threshold, which states that a species is likely to be found in an area with a habitat suitability value above a given threshold, but not likely to be found below it [37–39]. To create usable predictions from the cross-validations we used a 0.5 logistic presence value threshold for each taxa and all taxa to provide a cut-off point for suitability in this study (below which was considered unsuitable and above suitable), we also used a 0.75 logistic presence value threshold to further constrain the model output for EFH management applications. These values are higher than used in previous studies (i.e. the 10th percentile training presence used in [7]) because the main application of this modelling effort is to use predictions to help target areas for future field research and provide an assessment of EFH area closures. Summary maps were generated for each order and suborder by creating thresholded predicted outputs for each of the four regions (using 0.5 or 0.75 as the cutoff presence/absence value). If predicted logistic suitability was greater than 0.5 or 0.75 for any given cell, that cell was assigned a value of 1. If predicted habitat suitability was less than the cutoff value, the cell was assigned a value of 0. The binary models were summed for each coral taxon resulting in final consensus grids that had cell values ranging 0–4 depending on the number of retained spatially-partitioned models (maximum = 4).

Fishing intensity data

To estimate the amount of fishing activity within the U.S. West Coast, a map of fishing intensity was acquired that showed the relative intensity of commercial bottom trawling from 12 June 2006 to 31 December 2010 [40]. The map was developed from a commercial logbook program administered by coastal states and records for bottom-contact fishing gear (e.g., "small" footrope, "large" footrope, flatfish, selective flatfish, and roller trawl) collated by the Pacific Fisheries Information Network (PacFIN). This is not a fully comprehensive dataset, as some states do not submit full data, state-managed fisheries such as pink shrimp, ridgeback prawn and sea urchin are not included in PacFIN and cells with data from less than three fishing vessels were omitted from available maps to protect privacy. The final layer represents the total bottom trawl lines that fall within a 3 km radius neighbourhood centered on cells within a 500×500 m grid similar to the custom bathymetry (represented as km of trawl per km^2). These data were contrasted against a thresholded prediction for all taxa, with the threshold raised to 0.75 to locate areas that are highly suitable habitat for cold-water corals.

Substrate data

Substrate data was obtained for a limited subset of the model domain that mostly covered the shallower shelf of the US West Coast [41]. This layer was built from a variety of archived data including limited multibeam echosounder bathymetry and back-scatter and was provided as a 25 m×25 m grid [42]. The substrate types described included, probable soft sediment, probable rock (including predicted rock based on expert knowledge) and a mixture of soft sediment and rock. This layer was also accompanied with a confidence layer that shows that for the majority of the data, the probable substrate type was of low confidence with only shallower water being granted medium and high confidence levels. This data layer and any output from it must be considered with substantial caution especially given that a wide variety of mapping approaches were used in its creation and these sources were interpreted into the three coarse categories. Due to the low confidence in the substrate data, an exploratory analysis was undertaken to determine how the habitat suitability models for each species fared in light of the available substrate in the region.

Results

Species' niche

From the available environmental data, an *a priori* variable selection process that took into account closely related and highly correlated variables, identified seven variables that were likely to influence the probability of species presence (temperature, salinity, particulate organic carbon, depth, dissolved oxygen, calcite saturation state, slope 1 km) (Tables 3, 4 and S1–S10 in File S1). The jack-knife of variable contribution showed slope 1 km, temperature, and salinity were the strongest predictors for Suborder Alcyoniina, Order Antipatharia, Suborder Calcaxonia, and Suborder Scleraxonia. Temperature and salinity were consistently strong predictors in models for Suborder Holaxonia, whereas, salinity and depth were the strongest predictors for Order Scleractinia. For all taxa combined the strongest predictors were salinity, temperature and depth. Three highly correlated variables (depth, calcite saturation state, and temperature) were retained due to ecophysiological importance and the strength of their contributions. This must be interpreted with caution as these layers covary and may contain similar information, which can artificially inflate variable contribution scores. However, the test AUC scores for models generated with a single variable reinforced that these variables were top predictor variables regardless of covariation (Tables 3 and 4). Suborder Holaxonia was the only group to have calcite saturation state as one of the top three predictor variables (two of the four models) indicating some species within this Suborder could be sensitive to changes in carbonate chemistry.

It was possible to gain insight into the species niches and the factors that are most important in driving their distribution by intersecting the distribution of coral records with the environmental, physical, and chemical layers (Figure 4). For Ω_{CALC}, all coral records were found in waters supersaturated with respect to calcite ($\Omega_{CALC} > 1$) with the majority (82%) being found in waters with Ω_{CALC} between 1 and 2. Most coral records were found in waters with a temperature range of 1.5–8°C and salinity in the range of 33.5–34.7. Coral records were found in depths ranging from 50–4,129 m, but the majority (88%) were found between 50 and 2500 m. Slope 1 km values were widely distributed across taxa, but was an important predictor variable (see jack-knife of variable importance in Tables 3 and 4) for all taxa combined, Alcyoniina, Antipatharia, Calcaxonia, and Scleraxonia. Slope 1 km was not in the top three predictor variables for Holaxonia and Scleractinia. Dissolved oxygen values ranged from 0.3–5.9 ml l-1 with 89% of records having values in the 0.3–3.1 ml l-1 range. Particulate organic carbon values were widely distributed across taxa with Antipatharia having notably lower POC values (80% of Antipatharian records were found in waters with POC<7 g Corg m-2 yr-1).

Model evaluation

The coral habitat models performed well across all the metrics used to validate the modeled outputs. All, bar two AUC scores, were >0.7 and were significantly different from that of a random prediction of AUC = 0.5 (Wilcoxon rank-sum test, $p<0.01$). High AUC scores were supported by high test gains and low omission rates across many of the modeled taxa indicating most presences

Table 3. Validation statistics and jack-knife analysis of variable contributions to the models for all taxa (50th percentile), Alcyoniina, Antipatharia and Calcaxonia.

	All Taxa				Alcyoniina				Antipatharia				Calcaxonia			
Cross-validation cell	**1**	**2**	**3**	**4**	**1**	**2**	**3**	**4**	**1***	**2**	**3**	**4**	**1**	**2**	**3**	**4**
Validation statistics																
Test AUC	0.82	0.846	0.815	0.943	0.871	0.861	0.881	0.95	0.577	0.835	0.929	0.912	0.787	0.923	0.922	0.896
Test AUC standard deviation	0.016	0.012	0.016	0.005	0.026	0.015	0.013	0.004	0.204	0.033	0.017	0.037	0.036	0.016	0.01	0.011
10th percentile training presence	0.4866	0.5028	0.4545	0.4222	0.515	0.577	0.4962	0.4586	0.434	0.4829	0.2917	0.4321	0.6061	0.5946	0.4228	0.4631
Maximum test sensitivity plus specificity	0.181	0.262	0.412	0.379	0.153	0.162	0.098	0.35	0.57	0.054	0.375	0.26	0.046	0.195	0.117	0.323
Jack-knife of variable importance (jack of regularized training gain)																
Depth	**0.7343**	0.8258	**1.0306**	**0.6101**	0.9197	1.0217	**1.2143**	0.7726	1.0188	1.1635	0.528	1.0441	1.1326	1.1511	1.3354	1.1089
Dissolved Oxygen	0.5784	0.5817	0.6399	0.5601	0.8359	0.8946	0.9726	**0.9285**	1.2431	1.3786	0.4208	**1.4051**	1.0852	1.0692	1.1143	**1.3166**
Calcite Saturation State	0.6823	0.7536	0.9572	0.5682	0.9203	1.0206	1.1697	0.8571	1.2913	1.4405	**0.5438**	1.3341	1.1359	1.1417	1.2385	1.2377
Particulate Organic Carbon	0.6244	0.7207	0.9181	0.5533	0.7423	0.8689	1.0206	0.6541	0.798	0.7914	0.2422	1.0263	0.5591	0.55	0.9316	0.7141
Salinity	0.815	**0.8898**	**1.0605**	**0.6733**	**0.9825**	**1.1053**	**1.2755**	0.8786	1.2911	**1.4415**	0.5035	**1.3727**	**1.2273**	**1.2426**	**1.3415**	1.296
Slope (1 km)	**0.9349**	**0.9133**	0.8904	0.5992	**1.2309**	**1.3442**	**1.3521**	**1.0203**	1.4655	**1.5807**	**1.3852**	1.3643	**1.5481**	**1.4591**	**1.4562**	**1.3894**
Temperature	**0.749**	**0.8451**	**1.0054**	**0.6698**	**0.9414**	**1.0662**	1.1859	**0.8904**	1.4282	**1.627**	**0.7199**	**1.4986**	**1.2905**	**1.2951**	**1.3924**	**1.3747**
Test AUC for a single variable																
Depth	**0.9045**	0.7761	0.7048	**0.8734**	**0.9403**	**0.8161**	0.784	0.8639	0.6869	0.6496	0.8357	0.7562	**0.8125**	0.8368	0.8067	**0.8675**
Dissolved Oxygen	0.7588	**0.8111**	**0.7253**	0.7746	0.8504	0.8013	**0.7904**	0.8379	0.5127	**0.7592**	0.7807	0.6544	0.7517	**0.853**	0.7973	0.7244
Calcite Saturation State	**0.8791**	0.7738	**0.7312**	0.8449	**0.9165**	**0.8303**	**0.8193**	0.845	0.5853	0.6969	**0.8488**	**0.8689**	**0.8115**	0.8512	**0.8235**	0.8435
Particulate Organic Carbon	0.8558	0.7159	0.6523	0.7759	0.7956	0.7	0.673	0.6805	0.611	0.771	0.6574	0.5644	0.693	0.6625	0.6173	0.6079
Salinity	0.8185	**0.8036**	0.7082	0.8618	0.9113	0.7971	0.7875	0.8921	0.6308	**0.8093**	0.8468	0.5811	0.7874	**0.8524**	**0.8133**	0.8549
Slope (1 km)	0.6041	**0.7819**	**0.8298**	**0.9205**	0.7739	**0.8421**	**0.8948**	**0.9515**	0.7553	**0.7945**	**0.9116**	**0.9571**	0.727	**0.9184**	**0.9324**	**0.9436**
Temperature	**0.8882**	0.7622	0.6842	**0.8656**	**0.9331**	0.7594	0.7889	0.8476	0.6478	0.6091	**0.8535**	**0.7822**	**0.8095**	0.8448	0.8045	**0.8603**

Higher values for the regularized training gain of the jack-knife test indicates greater contribution to the model for a variable (these values are not directly comparable between the different taxa). Test AUC numbers in parentheses are the standard deviation of the Test AUC scores. The top three variables are highlighted in bold for each taxon, both for the jack-knife variable contribution and test AUC values for Maxent models generated using a single variable. *indicates cross-validation cells that were eliminated due to low Test AUC scores.

Table 4. Validation statistics and jack-knife analysis of variable contributions to the models for Holaxonia, Scleractinia and Scleraxonia.

	Holaxonia				Scleractinia				Scleraxonia			
Cross-validation cell	**1**	**2**	**3**	**4**	**1**	**2**	**3***	**4**	**1**	**2**	**3**	**4**
Validation statistics												
Test AUC	0.854	0.898	0.851	0.934	0.926	0.95	0.396	0.959	0.824	0.957	0.942	0.895
Test AUC standard deviation	0.011	0.018	0.011	0.006	0.014	0.006	0.059	0.009	0.021	0.01	0.009	0.01
10th percentile training presence	0.4679	0.3864	0.3992	0.4575	0.544	0.4343	0.4062	0.483	0.6404	0.556	0.4058	0.5849
Maximum test sensitivity plus specificity	0.05	0.17	0.031	0.174	0.122	0.338	0.515	0.087	0.019	0.244	0.004	0.092
Jack-knife of variable importance (jack of regularized training gain)												
Depth	1.2116	**1.244**	**1.3494**	1.1917	**1.1782**	**1.1851**	1.5965	**1.1525**	1.4298	1.3367	**1.6165**	1.3282
Dissolved Oxygen	0.8588	0.7853	0.7656	0.7875	0.3822	0.3561	0.4062	0.6344	1.4176	1.2455	1.3594	1.4416
Calcite Saturation State	**1.192**	1.2233	1.2653	**1.2093**	0.9057	0.889	1.4655	0.7286	1.4166	1.3256	1.4452	**1.5113**
Particulate Organic Carbon	0.9251	1.0044	**1.3606**	0.8145	1.1713	**1.1957**	1.5983	0.9781	0.7525	0.7048	1.1665	0.8166
Salinity	**1.26**	**1.3094**	**1.3832**	**1.297**	**1.3867**	**1.387**	1.7	**1.3008**	**1.5009**	**1.395**	**1.5007**	**1.5334**
Slope (1 km)	1.1797	0.9897	0.7825	0.5465	0.6968	0.8238	0.932	0.1163	**1.6012**	**1.416**	1.3632	1.1158
Temperature	**1.2693**	**1.3011**	1.328	**1.3376**	**1.2203**	1.1763	1.5474	**1.2191**	**1.5832**	**1.5469**	**1.7183**	**1.5909**
Test AUC for a single variable												
Depth	**0.8897**	**0.8617**	0.7838	**0.8747**	**0.9606**	0.9258	0.4169	**0.898**	**0.8057**	**0.9097**	0.8256	0.8735
Dissolved Oxygen	0.758	0.8098	**0.8116**	0.8161	0.6083	0.8045	0.3762	0.5607	0.6921	**0.9164**	**0.8485**	0.7649
Calcite Saturation State	**0.8652**	0.8545	0.7811	0.8398	**0.9398**	**0.9279**	0.3778	0.8443	**0.7799**	0.91	**0.8437**	0.8363
Particulate Organic Carbon	0.8273	0.7954	0.597	0.8424	0.9024	0.8826	0.5526	0.8698	0.6085	0.6775	0.4183	0.6408
Salinity	0.8436	**0.8631**	**0.7988**	0.8686	**0.9441**	**0.9398**	0.4193	**0.9038**	0.7735	0.8932	0.8038	0.8598
Slope (1 km)	0.5424	0.7232	**0.927**	**0.9077**	0.7569	0.4317	0.4468	0.7678	0.6157	0.8681	**0.9331**	0.9469
Temperature	**0.8742**	**0.8641**	0.7633	**0.8897**	0.905	**0.9402**	0.3001	**0.9024**	**0.869**	**0.9115**	0.8395	0.8912

Higher values for the regularized training gain of the jack-knife test indicates greater contribution to the model for a variable (these values are not directly comparable between the different taxa). Test AUC numbers in parentheses are the standard deviation of the Test AUC scores. The top three variables are highlighted in bold for each taxon, both for the jack-knife variable contribution and test AUC values for Maxent models generated using a single variable. *indicates cross-validation cells that were eliminated due to low Test AUC scores.

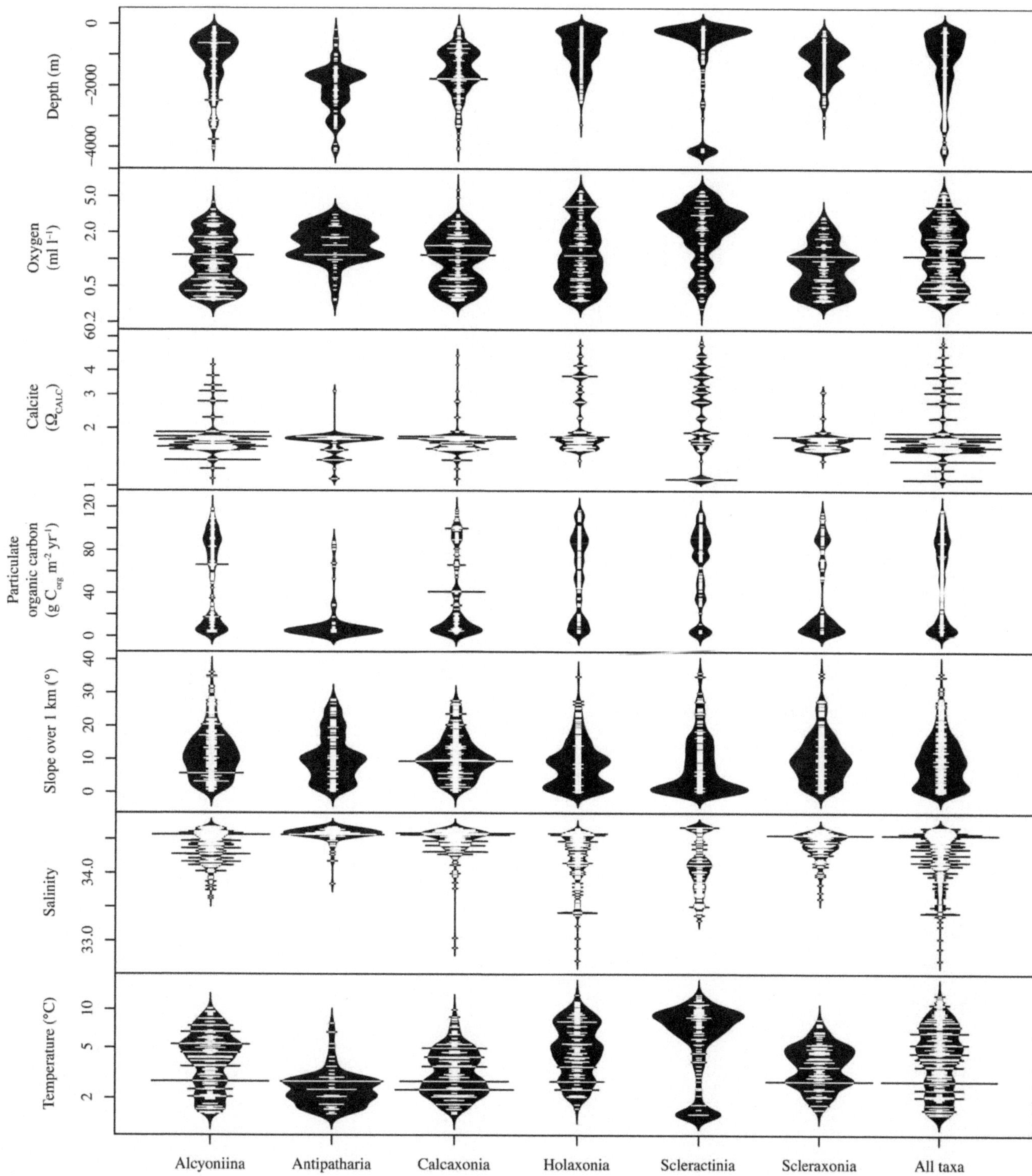

Figure 4. Bean plots of coral presences intersected with the environmental, physical and chemical variables used in the models (the small lines in the center of each bean shows individual coral presence points). The bean itself is a density trace that is mirrored to show as a full bean [56].

were accounted for in the predictions (Tables 3 and 4). Models generated for the spatial partitions of Antipatharia (Model 1) and Scleractinia (Model 4) were excluded from the summed grids for each taxon as the AUC scores for these partitions were <0.7, so suitability was ranked between 0–3, rather than 0–4 as for all other taxa (Figures 5 and 6).

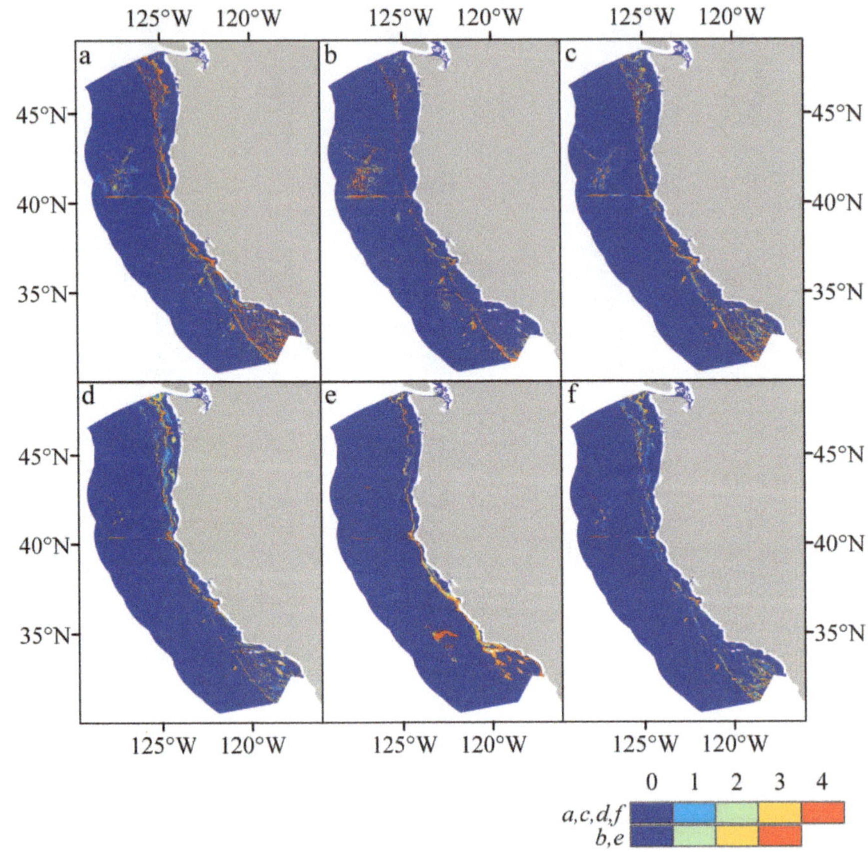

Figure 5. Predicted habitat suitability. Suborder Alcyoniina (a), Order Antipatharia (b), Suborder Calcaxonia (c), Suborder Holaxonia (d), Order Scleractinia (e), Suborder Scleraxonia (f). Legend shows the summed values for 0.5 cut off for each model validated by the spatial cross validation approach, the legend is on a scale of 0–4 for most variables, however, some only had 3 valid cells incorporated and the summed value is a maximum of 3.

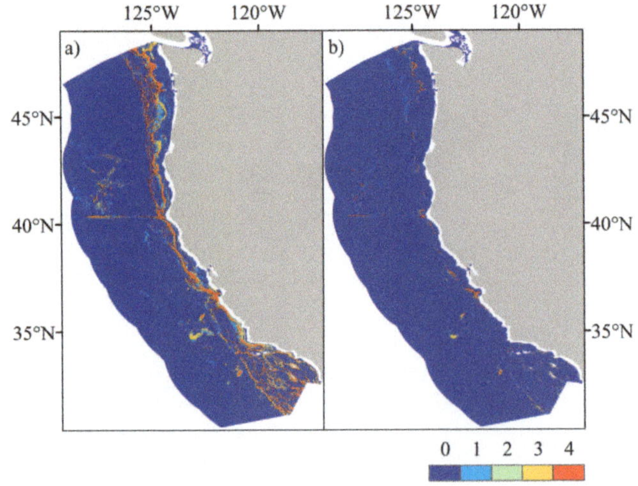

Figure 6. Predicted habitat suitability for all taxa. a) 0.5 threshold as per the previous figures, b) 0.75 threshold. The higher threshold greatly constrains the output, producing predictions that are more focused on areas of the highest suitability.

Habitat suitability maps for National Marine Sanctuaries

Habitat suitability maps for each taxon and all taxa combined are presented in Figures 5 and 6 and are available to download as GeoTIFF files (File S2). The locations of the Olympic Coast, Cordell Bank, Gulf of the Farallones, Monterey Bay and Channel Islands National Marine Sanctuaries overlain with predicted suitability for selected taxa are shown in Figure 7, additional and higher resolution digital figures are available as electronic supplementary materials (Figures S1–S7).

Olympic Coast National Marine Sanctuary (OCNMS): Areas within the OCNMS that have the highest probability of containing coral habitat include the eastern regions of Juan de Fuca Canyon, Quillayute Canyon, and Quinalt Canyon. Habitat suitability probabilities were highest in these areas for Alcyoniina (Figure 7b and S1a), Calcaxonia (Figure S1c), Holaxonia (Figure S1d), Scleractinia (Figure S1e), and Scleraxonia (Figure S1f). Predicted habitat suitability and the areal extent of suitable habitat were low in these areas for Antipatharia (Figure S1b). Large areas of suitable deep-sea coral habitat were predicted in areas adjacent to the western boundary of the OCNMS. This area of high habitat suitability extends approximately 30 km westward of the sanctuary's western boundary (Figure S1).

Cordell Bank (CBNMS), Gulf of the Farallones (GFNMS), and Monterey Bay National Marine Sanctuaries (MBNMS): The boundaries of these sanctuaries encompass the vast majority of suitable coral habitat in the region from the coastline westward to

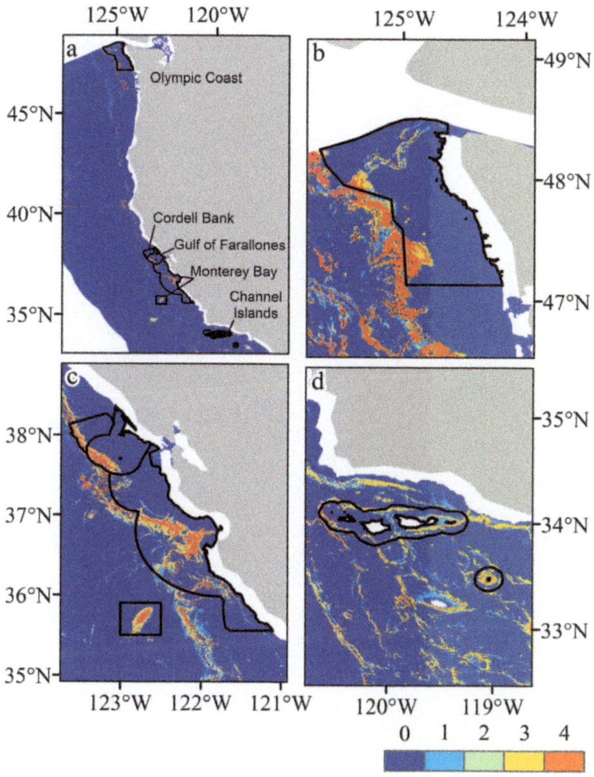

Figure 7. Locations of National Marine Sanctuaries overlain with predicted suitability. a) Overview with all taxa constrained to a 0.75 threshold, b) Olympic Coast and prediction for Alcyoniina, c) Cordell Bank, Gulf of the Farallones and Monterey Bay with prediction for Calcaxonia, d) Channel Islands and prediction for Holaxonia.

approximately 90 km offshore (from Monterey Bay). Habitat suitability probabilities and areal extent of predicted habitat were highest in these sanctuaries for Alcyoniina (Figure S2a), Calcaxonia (Figure 7c and S2c), Holaxonia (Figure S2d), and Scleractinia (Figure S2e). In contrast, suitability probabilities and areal extent were low in these areas for Antipatharia (Figure S2b) and Scleraxonia (Figure S2f). High probability areas were modeled to the west of MBNMS's northwest boundary. This area of highly suitable habitat extends approximately 50 km to the west of MBNMS's NW boundary (Figure S2).

Channel Islands National Marine Sanctuary (CINMS): Habitat suitability probabilities and areal extent of predicted habitat were generally low across taxa within the CINMS boundary when compared to surrounding waters. Predicted habitat probabilities and areal extent of suitable habitat were low for Alcyoniina (Figure S3a), Antipatharia (Figure S3b), Calcaxonia (Figure S3c), Holaxonia (Figure 7d and Figure S3d), and Scleraxonia (Figure S3f). Modeled habitat probabilities and areal extent of predicted habitat were high in waters surrounding all islands within the sanctuary for Scleractinia (Figure S3e). It is worth noting that probabilities and areal extent were high in the waters surrounding the islands of San Nicolas, Santa Catalina and San Clemente for Alcyoniina, Calcaxonia, Holaxonia, Scleractinia and to a lesser extent Scleraxonia. Probabilities and areal extent were very low in all island waters, both within and outside the sanctuary boundaries, for Antipatharia (Figure S3).

To evaluate the effectiveness of each NMS in encompassing predicted coral habitat, the total area of suitable habitat within each was calculated (Table 5). Overall, a large proportion of total area within each sanctuary encompassed habitat that was classed as suitable (a habitat suitability value greater than zero) for multiple taxa. In some NMS', certain taxa dominated. For example, suitable habitat for Holaxonia was predicted to occur within 45% of the area of OCNMS, whereas in the Gulf of Farallones, Antipatharia was predicted to occur in 60% of the area. To avoid over-prediction skewing the effectiveness, the highly constrained model for all taxa (cut-off above 0.75 logistic suitability) showed that the most effective NMS may be the Gulf of Farallones, which contained 16% suitable habitat, followed by Monterey Bay that contained 14%, Olympic Coast at 12% and the Channel Islands at 9%. The least effective may be Cordell Bank that encompassed 6% suitable habitat.

Spatial distribution of predicted habitat suitability and bottom trawl closures: Essential Fish Habitat (EFH) and Cowcod Conservation Area West (CCA-West)

Coral habitat suitability maps for selected taxa and all taxa combined are depicted in Figure 8 (all taxa are shown in Figure S4) with overlays of essential fish habitat area closures and Cowcod Conservation Areas sourced from the Pacific Fishery Management Council. These area closures depict areas with fishing gear restrictions off Washington, Oregon, and California. Gear restrictions were established under NMFS' Final Rule to implement Amendment 19 to the Pacific Coast Groundfish Fishery Management Plan (71 FR 27408; May 11, 2006). Fishing with bottom trawl gear within these areas was prohibited to minimise adverse effects from fishing. All bottom contact gear was prohibited in waters surrounding Thompson Seamount, President Jackson Seamount, and several sites in the Channel Islands and Cordell Bank National Marine Sanctuaries. All bottom contact gear and any gear deployed deeper than 500 fathoms (914 m) was prohibited in the waters surrounding Davidson Seamount. There are two Cowcod Conservation Areas (East and West), but only CCA-East is designated as EFH. CCA-West is large in size and has high levels of habitat suitability for many of the coral taxa modelled. The EFH and the CCA-West areas are not the only bottom trawl closures present on the U.S. West Coast. There are also Rockfish Conservation Areas (RCAs) that extend along the entire length of the U.S. West Coast, but their boundaries can change throughout the year and are based on approximate depth contours between ~100–150 fathoms (183–274 m). Both of these factors make quantitative assessment of RCA closures with predicted habitat suitability highly uncertain. In addition, California and Washington prohibit bottom trawling within their state territorial seas (out to 3 nautical miles). These trawl closures were not included in this analysis as the majority of suitable coral habitat is found in deeper areas outside state territorial waters.

Northern Region (42°N to 48°N, Washington and Oregon)

Significant areas of high probability coral habitat for Alcyoniina were predicted off the coast of Washington state (Figure 5a and Figure S4). The areas with high probabilities that remain open to bottom contact gear include the entire western boundary of the OCNMS. These include regions adjacent to existing EFH area closures Biogenic 1 and Biogenic 2. Highly suitable habitat also was identified between existing EFH area closures Grays Canyon and Nehalem Bank/Shale Pile, along the western boundary of Hecata Bank, along the western boundary of Bandon High Spot, and areas between Bandon High Spot and Rouge Canyon EFH. The majority of Antipatharian predicted habitat was located in

Table 5. Total area (nm²) of modeled habitat suitability within existing National Marine Sanctuary boundaries.

	Habitat suitability					
	0	**1**	**2**	**3**	**4**	**Suitable (>0)**
Olympic Coast (2,854 nm²)						
Alcyoniina	2,478	53	61	129	134	376 (13%)
Antipatharia	2,587	210	43	14	-	266 (9%)
Calcaxonia	2,351	119	103	230	51	503 (18%)
Holaxonia	1,571	487	252	388	155	1,282 (45%)
Scleractinia	2,262	277	188	127	-	592 (21%)
Scleraxonia	2,307	64	82	346	54	546 (19%)
All Taxa -0.5	1,820	184	216	327	306	1,033 (36%)
All Taxa -0.75	2,515	126	63	56	94	338 (12%)
Monterey Bay (5,465 nm²)						
Alcyoniina	3,408	343	234	547	934	2,057 (38%)
Antipatharia	4,488	335	407	235	-	977 (18%)
Calcaxonia	3,959	235	227	628	417	1,507 (28%)
Holaxonia	3,951	377	264	459	415	1,514 (28%)
Scleractinia	3,731	362	805	567	-	1,735 (32%)
Scleraxonia	4,662	203	163	278	159	803 (15%)
All Taxa -0.5	2,637	479	418	704	1,227	2,828 (52%)
All Taxa -0.75	4,716	131	137	204	278	749 (14%)
Gulf of Farallones (1,149 nm²)						
Alcyoniina	898	11	10	30	200	251 (22%)
Antipatharia	460	308	266	115	-	689 (60%)
Calcaxonia	917	12	17	84	118	232 (20%)
Holaxonia	788	109	35	67	149	360 (31%)
Scleractinia	663	150	210	125	-	486 (42%)
Scleraxonia	970	17	37	94	31	179 (16%)
All Taxa -0.5	598	130	97	85	239	551 (48%)
All Taxa -0.75	964	33	26	44	81	184 (16%)
Cordell Bank (469 nm²)						
Alcyoniina	346	13	15	32	64	123 (26%)
Antipatharia	400	32	33	4	-	69 (15%)
Calcaxonia	350	4	9	59	48	119 (25%)
Holaxonia	295	50	41	38	46	174 (37%)
Scleractinia	202	108	117	43	-	267 (57%)
Scleraxonia	431	15	15	8	0	38 (8%)
All Taxa -0.5	241	33	33	45	118	228 (49%)
All Taxa -0.75	441	9	5	6	8	28 (6%)
Channel Islands (1,299 nm²)						
Alcyoniina	959	61	45	123	110	339 (26%)
Antipatharia	1,278	16	4	0	-	21 (2%)
Calcaxonia	1,083	41	49	110	16	215 (17%)
Holaxonia	809	136	99	217	37	489 (38%)
Scleractinia	261	138	412	487	-	1,037 (80%)
Scleraxonia	1,143	12	123	20	0	155 (12%)
All Taxa -0.5	507	177	174	203	238	792 (61%)
All Taxa -0.75	1,184	51	31	21	11	114 (9%)
Entire modelled area (284,863 nm²)						
Alcyoniina	257,486	7,315	4,577	5,994	9,491	23,7378 (9.6%)
Antipatharia	269,471	7,477	4,191	3,725	-	15,393 (5.4%)

Table 5. Cont.

	Habitat suitability					
	0	1	2	3	4	Suitable (>0)
Calcaxonia	261,266	6,839	4,081	7,506	5,172	23,598 (8.3%)
Holaxonia	264,281	7,688	4,330	5,462	3,102	20,583 (7.2%)
Scleractinia	270,015	4,651	4,874	5,324	-	14,849 (5.2%)
Scleraxonia	269,606	5,991	2,933	4,470	1,863	15,257 (5.4%)
All Taxa -0.5	248,226	9,352	6,767	8,586	11,932	36,637 (12.9%)
All Taxa -0.75	278,384	2,705	1,391	1,177	1,206	6,480 (2.3%)

existing EFH area closures in this region; the exception to this are the predicted areas located between Grays Canyon and Nehalem Bank/Shale Pile, and within the OCNMS (Figure S4b). The most significant predicted habitat areas for Calcaxonia, which are not currently contained in EFH area closures, include waters north and south of Biogenic 1 and 2 and areas between Grays Canyon and Nehalem Bank/Shale Pile (Figure S4c). The predicted habitat pattern for Holaxonia was similar to that of Alcyoniina with two additional areas being highly suitable. These areas are located directly east of Nehalem Bank/Shale Pile and Heceta Bank (Figure S4d). Predictions for Scleractinia were limited in this region to a narrow depth band almost all of which is in areas open to bottom trawl gear (Figure S4e). High probability areas for Scleractinia were identified within the OCNMS and in a narrow depth band between Biogenic 2, Grays Canyon, and Nehalem Bank/Shale Pile. Areas west of Heceta Bank and Bandon High Spot were also identified, in addition to the area east of Rogue Canyon. The majority of predicted habitat for Scleraxonia that remains in open trawling areas is located within the OCNMS and between Grays Canyon and Nehalem Bank/Shale Pile (Figure S4f).

Central Region (36°N to 42°N, Northern California)

The majority of high probability areas predicted for Alcyoniina occurred in areas currently open to bottom trawl gear (Figure S5a). These include areas between Rogue Canyon and Eel River Canyon and between Blunts reef and Pt. Arena South Biogenic Area. High probability areas for Antipatharia are almost completely contained within existing EFH area closures in this region (Figure S5b). Calcaxonia predicted habitat that remains in open trawl areas appears to follow the 700 fathoms (1,280 m) contour line between Rogue Canyon and Blunts Reef and also between Pt. Arena North and Cordell Bank Biogenic Area (Figure S5c). The majority of predicted Holaxonia and Scleractinian habitat are located in areas currently open to bottom trawl gear (Figure S5d and S5e). Scleraxonia habitat was predicted mostly in existing EFH area closures, but high probability areas were identified in areas open to bottom trawling in a narrow depth band between Rogue Canyon and Blunts Reef and again from Delgada Canyon to Pt. Arena South Biogenic area (Figure S5f).

Southern Region (30°N to 38°N, Central and Southern California)

Predicted habitat for Alcyoniina was limited predominately to the continental shelf in this region. The majority of predicted Alcyoniina habitat in the northern extents of the mapped area is located within the boundaries for CBNMS, GFNMS and MBNMS (Figure S6a). Predicted habitat for Alcyoniina was identified in areas remaining open to bottom trawl gear in the southern extent of the mapped areas including the waters surrounding existing EFH area closures: Harris Point, Potato Bank, Hidden Reef/Kidney Bank, Catalina Island, and Cowcod Conservation Area East. High probability areas for Alcyoniina were also predicted along the northern boundary of the CINMS (north of San Miguel, Santa Rosa and Santa Cruz Islands). Predicted Alcyoniina habitat in the southern extent of the mapped area along the shelf break is mostly contained in existing EFH area closures. Most of the high probability areas for predicted Antipatharian habitat were located in current EFH area closures; the exception was the area approximately 65–200 km south of Davidson Seamount (Figure S6b). The majority of predicted Calcaxonia habitat in the region is located within current EFH area closures, but high probability areas remain open to bottom trawl gear in the southern extent of the mapped region in waters adjacent to Potato Bank, Catalina Island, and Cowcod Conservation Area East EFH area closures (Figure S6c). The majority of Holaxonia predicted habitat was identified in areas open to bottom trawl gear, most of which occurs within the boundaries of the CBNMS, GFNMS, and MBNMS (Figure S6d). High probability areas were also predicted in trawl areas surrounding Cowcod Conservation Area East. The majority of predicted Scleractinian habitat in the region was identified in areas currently open to bottom trawl gear (Figure S6e). Most of this habitat is limited to a depth range of ~50–400 m (along the coast and islands) and a ~3,500 km^2 deep-water area south of Davidson Seamount. Scleractinian coral presence records documented in this deep-water region were *Fungiacyathus marenzelleri*. Most Scleraxonia habitat was identified in areas currently designated as EFH area closures (Figure S6f). High probability habitat results in areas open to bottom trawl gear include the waters surrounding to Farallon Islands/Fanny Shoal, Monterey Bay/Canyon, and Catalina Island.

Bottom-trawl intensity

The bottom-trawl intensity data obtained from Whitmire [40] only covered the northern region between 33°N and 45°N and became fragmented below 37°N (Figure 9). The layer covers the shelf, and is largely constrained to depths shallower than 1,000 m. Much of this area is open for trawling, with EFH closures generally being deeper, however, as the intensity data does not cover the whole shelf, it is difficult to draw parallels between protection and trawling activity. There are areas where trawling activity overlaps with predicted suitable habitat (Figures 9 and 10), some areas are enclosed within closed areas but the majority falls outside areas of protection. Using the 0.75 logistic threshold-all-taxa model provided a spatially constrained prediction that focuses on areas with the highest predicted suitability. In the Northern region, and

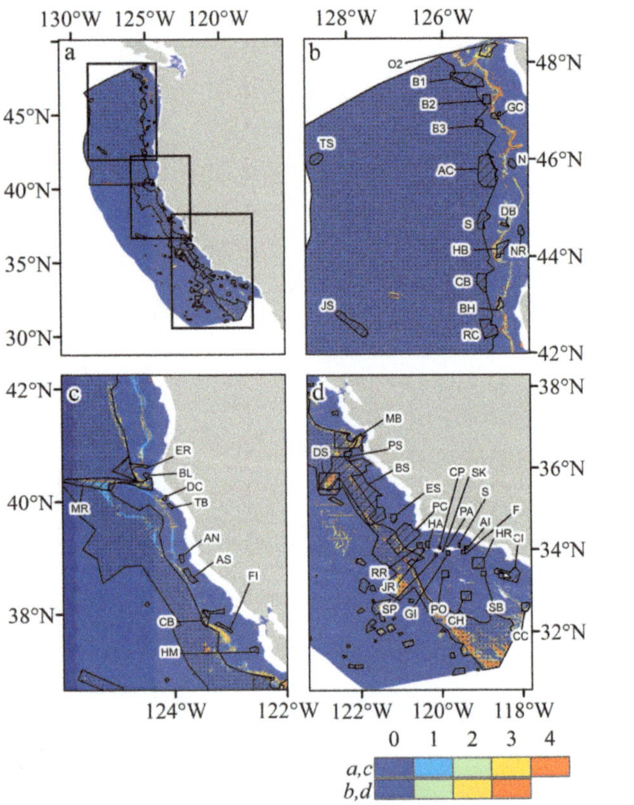

Figure 8. Spatial distribution of predicted habitat suitability and bottom trawl closures for areas designated Essential Fish Habitat (stippled) and CCA-West closures (hatched areas). a) Overview with 0.75 threshold suitability for all taxa, b) northern region (Washington and Oregon) and prediction for Scleractinia, c) central region (northern California) and prediction for Scleraxonia, d) southern region (central and southern California) and prediction for Antipatharia. Location abbreviations: O2: Olympic 2, B1: Biogenic 1, B2: Biogenic 2, B3: Biogenic 3, GC: Grays Canyon, N: Nehalem Bank/Shale Pile, AC: Astoria Canyon, TS: Thompson Seamount, S: Siletz Deepwater, DB: Daisy Bank/Nelson Island, NR: Newport Rockpile/Stonewall Bank, HB: Heceta Bank, CD: Deepwater off Coos Bay, BH: Brandon High Spot, RC: Rogue Canyon, JS: President Jackson Seamount, ER: Eel River Canyon, BL: Blunts Reef, MR: Mendocino Ridge, DC: Delgada Canyon, TB: Tolo Bank, AN: Pt. Arena North, AS: Pt. Arena South, CB: Cordell Bank Biogenic Area, FI: Farallon Islands/Fanny Shaol, HM: Half Moon Bay, MB: Monterey Bay/Canyon, PS: Point Sur Deep, BS: Big Sur Coast/Port San Luis, DS: Davidson Seamount, ES: East San Lucia Bank, PC: Point Conception, RR: Richardson Rock, JR: Judith Rock, HP: Harris Point, CP: Carrington Point, SP: South Point, SK: Skunk Point, S: Scorpion, PA: Painted Cave, AI: Anacopa Island, F: Footprint, HR: Hidden Reef/Kidney Bank, CI: Catalina Island, CC: Cowcod Conservation East Area, SB: Santa Barbara, CH: Cherry Bank, PO: Potato Bank, GI: Gull Island.

the Olympic Coast National Marine Sanctuary, there are several areas with high trawling intensity and high suitability, the majority of which falls outside of any designated area (Figure 9b). Within the Central region there are again areas of high suitability that fall outside of EFH areas such as adjacent to Heceta Bank, Brandon High Spot and Rogue Canyon (Figure 9c and see Figure 8 for locations of EFH areas). In the Southern region, Mendocino Ridge captures an area of high suitability with no trawling intensity data but further south adjacent to Delgada Canyon and Tolo Bank there are large areas of suitability with moderate trawling intensity (Figure 9d). Intersecting the trawling intensity grid with the habitat suitability classes for the whole region shows that there are areas of high suitability that are trawled at moderate levels compared to lower suitability classes, but the level of trawling for habitat suitability classes 1 and 4 are similar in contrast to 2 and 3, indicating that it is possible that trawlers are focusing both on areas that are not likely to contain corals (i.e. suitability values of 1) and are also targeting areas that do contain corals (i.e. suitability values of 4).

Substrate data

The substrate data showed that for the area available, 91% of the shelf was described as probable soft sediment, 1% was probable mixed hard and soft sediment and 8% was classified as hard substrate (Figure 10a). To determine how much suitable habitat fell within areas of each probable substrate type, the modeled suitability layers for each taxa were spatially intersected with each substrate class and the proportion of area enumerated. By contrasting this, it was possible to provide an estimation of the level of over-prediction within the habitat suitability models (Figure 10 and Figure S7). In general, the majority of predicted habitat was found to fall within areas of probable soft sediment for all taxa (Table 6). However, for areas that were predicted as higher suitability (>2 on the habitat suitability scale), the proportion of predicted suitable habitat that fell within hard substrate area

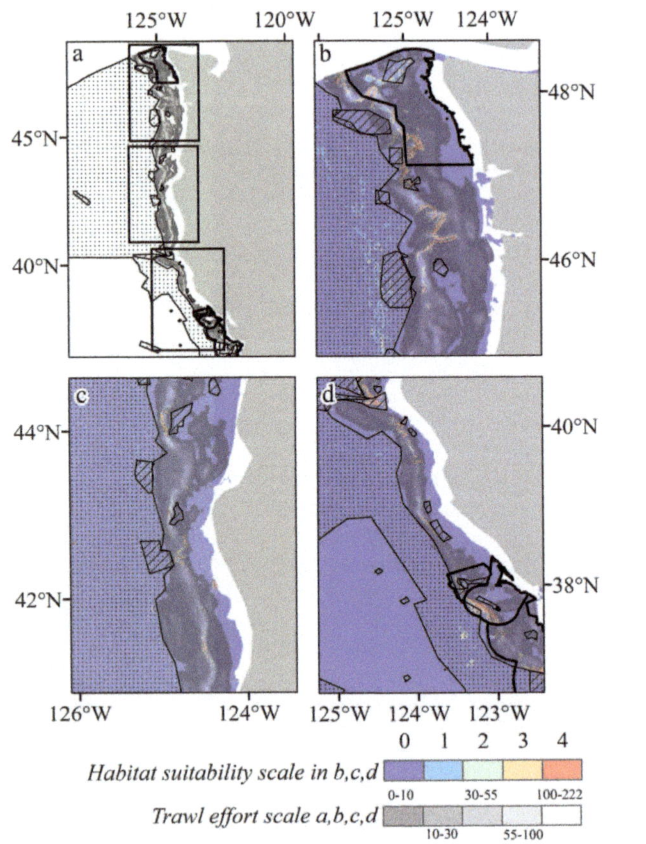

Figure 9. Fishing effort data overlain with predicted habitat from the 0.75 threshold all taxa model. a) Overview with fishing intensity grid, frames indicate location of panels b, c and d. b) Northern region, c) Central region and d) Southern region. Bold black areas in panel b, c, d indicate the locations of National Marine Sanctuaries, the hatched areas indicate locations of CCA-West closures and the stippled areas indicate Essential Fish Habitat closures. The upper legend (0–4) shows habitat suitability, higher is more suitable. The lower scale shows fishing intensity (km per km^2).

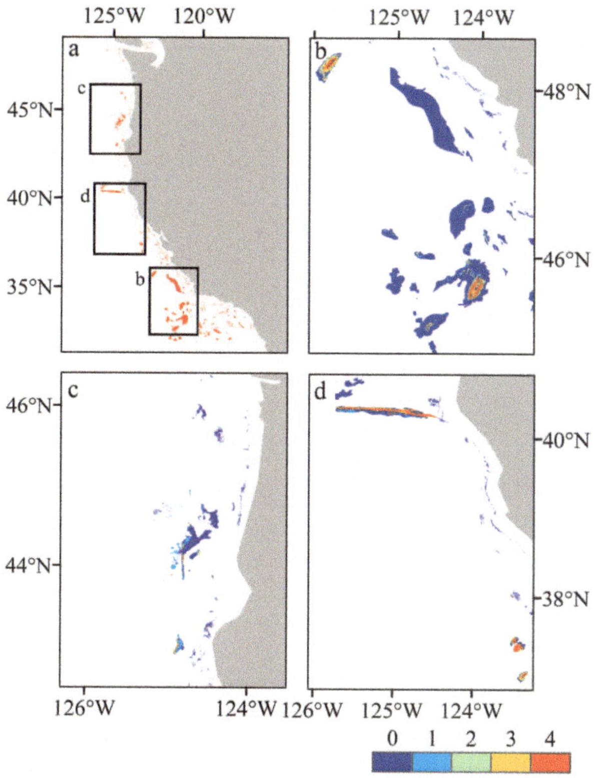

Figure 10. Habitat suitability only in areas with probable hard substrate. a) The distribution of probable hard substrate in red, b) the 0.75 threshold all taxa model, c) the Holaxonia model and d) the Alcyoniina model. The scale bar shows the habitat suitability (0–4, with 4 being high).

generally increased, indicating a link between the variables used in the model, the taxa niche and where hard substrate was found.

Discussion

This study is the first attempt to model the potential distribution of deep-sea coral habitat for the U.S. West Coast EEZ. The approaches presented here are a significant improvement over previous regional efforts such as those in the North East Atlantic [3] and Canadian Shelf [1,2]. In recent studies there have been significant advancements in parallel with the data presented in this study, especially with respects to identifying usable datasets for regional-scale habitat suitability modelling in the deep-ocean [12–14]. However, there still remain several limitations that must be considered when interpreting broad-scale model results.

Unincorporated model variables and model accuracy

There are several variables that are important for coral settlement, growth and survival that were not included in the model because they do not exist at sufficient resolutions, a problem shared with all habitat suitability efforts [43]. These variables include benthic hard substrata, high-resolution current direction/velocity, and the distribution of mobile or benthic sediments. Many corals require hard substrata for colonisation and like depth; substrate tends to be highly variable over small spatial scales. Model results presented here will overpredict the amount of suitable habitat in some areas because fine-scale and moderate scale bathymetric features (10's of metres to 300 metres), substrate, and current data are not available. It is likely that model results indicate suitable coral habitat in areas that are known soft bottom regions with high sediment loadings where corals are likely or known to be absent, as indicated by 91% of the shelf being classed as probable soft sediment [41]. By contrasting the amount of available hard substrate and the predicted suitability for many taxa, the level of over-prediction can be estimated (Table 6). For example, of the area of habitat predicted as unsuitable (habitat suitability of 0) for Alcyoniina, 6.8% fell in areas classed as probable hard substrate. This increased as habitat suitability increased to the highest value (4), where 14.2% of area fell in areas classed as probable hard and mixed substrate. This shows that the model for this species may overpredict in approximately 85.8% of the area, assuming that the hard substrate layer is accurate and that the distribution of this taxa is entirely dependent upon hard substrate that can be mapped at a 25 m×25 m resolution.

Data on the distribution of sediments is unfortunately scarce for much of the world's seafloor, a fact that urgently needs to be addressed by mapping programs around the world. In this study, the IOOS Surficial Geologic Habitat maps for the Washington and Oregon continental shelves (Version 2.2) were initially explored to determine whether or not this information could be used in the habitat suitability models to refine taxon niches but these data were not suitable due to incomplete coverage of the study area. The National Marine Fisheries Service produced a composite substrate dataset as part of their 5-year Essential Fish Habitat Review for the West Coast [42]. This dataset aggregated many sources into a standardised classification (hard, mixed and soft substrate) layer at 25 m×25 m resolution, but covered only the shallower parts of the shelf limiting its utility in this modelling effort. In addition, the layer was also provided with a confidence layer that stated low confidence for most of the area, with only shallower water having medium or high confidence in the probable substrate type. This layer was used to constrain the output of the habitat suitability models for each taxa to produce a focused dataset that highlights areas where the habitat is suitable for coral and areas of probable hard substrate overlap (Figure 10 and Figure S7).

Model results for all taxa combined undoubtedly overpredict as the suborders and orders modeled separately occupy different niches, depth ranges, and caution should be exercised when using all taxa combined model results. However, we introduced a constrained model by increasing the threshold to the 0.75 logistic suitability for all taxa, producing a model that was far more focused on areas of very high suitability. Assessment of model accuracy will be dependent on field operations to validate model predictions. In addition to several unincorporated datasets, the extent, quality, and availability of environmental, chemical and physical data are continually improving and should be incorporated in an iterative process with field surveys to refine predictions and reduce the number of false positives and negatives in habitat suitability models.

Presence records

The limited number of coral presence records used to model habitat distribution for some coral taxa highlights the need for more targeted sampling to document coral locations. For example, very few presence localities for Suborders Filifera and Stolonifera were obtained and preliminary models suffered from significant overprediction and artificially high AUC scores. Low presence numbers could be due to coral rarity among these taxa and/or undersampling. The lack of coral records for these suborders resulted in the omission of these models from the analysis. Several recent studies have investigated the effectiveness and reliability of habitat suitability models constructed with low numbers of

Table 6. Proportion of predicted habitat (higher values indicate more suitable) in relation to different substrate type for each taxa.

	Habitat suitability				
Sediment type	**0**	**1**	**2**	**3**	**4**
Alcyoniina					
Hard	6.8%	7.5%	10.4%	12.2%	13.9%
Mixed	1.2%	1.4%	1.0%	0.6%	0.3%
Soft	91.9%	91.1%	88.6%	87.1%	85.7%
Antipatharia					
Hard	6.6%	16.7%	25.8%	33.3%	-
Mixed	1.2%	1.5%	0.2%	0%	-
Soft	92.2%	81.8%	74.1%	66.7%	-
Calcaxonia					
Hard	6.7%	7.9%	8.6%	12.3%	22.8%
Mixed	1.1%	1.6%	1.5%	0.8%	0.4%
Soft	92.1%	90.5%	89.9%	86.9%	76.8%
Holaxonia					
Hard	7.5%	6.5%	9.4%	13.1%	12.3%
Mixed	0.5%	4.3%	2.4%	4.1%	1.0%
Soft	92.0%	89.2%	88.2%	82.8%	86.7%
Scleractinia					
Hard	8.1%	5.9%	6.2%	9.4%	-
Mixed	0.7%	5.9%	2.7%	1.8%	-
Soft	91.1%	88.2%	91.1%	88.7%	-
Scleraxonia					
Hard	7.2%	7.2%	9.5%	16.4%	26.3%
Mixed	1.1%	1.3%	0.8%	2.3%	0.1%
Soft	91.7%	91.5%	89.7%	81.3%	73.6%
All Taxa -0.5					
Hard	7.0%	6.6%	8.8%	9.4%	12.9%
Mixed	0.7%	2.2%	2.2%	2.7%	1.3%
Soft	92.4%	91.2%	89.0%	87.9%	85.8%
All Taxa -0.75					
Hard	7.3%	17.1%	20.1%	22.7%	15.4%
Mixed	1.1%	2.4%	2.1%	2.1%	0.6%
Soft	91.6%	80.5%	77.9%	75.2%	84.1%

presences, a common problem for difficult to detect species (i.e. deep-sea corals) and those that have had limited systematic survey effort such as records from museum collections [44]. This does not preclude the possibility of modelling species distributions with low sample numbers, as Maxent is capable of producing acceptable models with relatively limited numbers of presences [37]. However, Maxent does appear to overpredict suitable habitat when using small presence datasets compared with other methods [37,45]. In addition, grouping coral records at the order and suborder level undoubtedly combines coral taxa (family, genus, species) with different environmental niches. This is a recognised limitation of the approach, but one that is necessary due to taxonomic uncertainty and total number of coral records available.

Model validation and targeting areas for field operations

Field validation of modeled habitat is needed to 1) Assess the accuracy of model predictions. 2) Refine models by identifying false positives and negatives. 3) Gauge the utility of these modelling methods for identifying deep-sea coral habitat in unsurveyed areas. The predicted habitat suitability results presented here are not meant to identify coral occurrences with pin point accuracy and are unlikely to achieve this based on currently available data. They are more useful for directing research effort to areas that have the highest probability of supporting deep-sea corals and identifying low probability areas that could be avoided to maximise time spent in high probability areas. Broad-scale predictive habitat results should be used in conjunction with multibeam surveys, geologic substrate maps and other tools to determine the most likely areas for harboring deep-sea corals. One additional complication for field validation efforts using these predictions is the current technological limitation of survey vehicles and equipment (i.e.

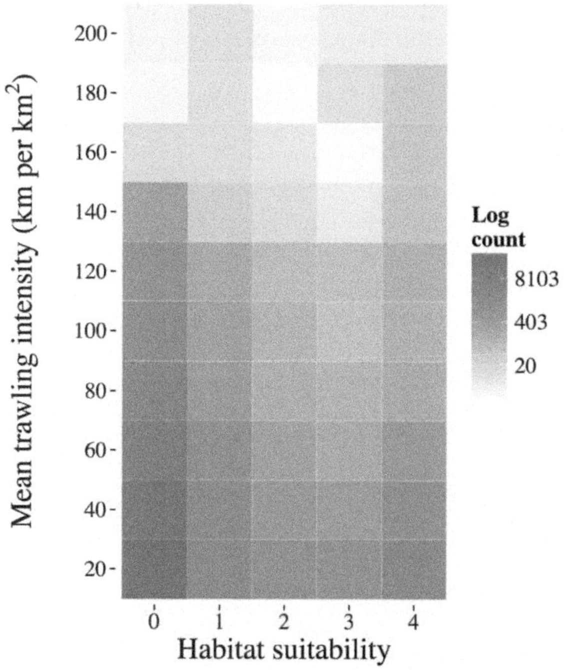

Figure 11. Heat plot of trawling intensity falling within habitat suitability classes for the 0.75 threshold all taxa model. Darker colours indicate higher cell counts compared to lighter, the plot shows that trawling intensity is greatest in cells classified as suitable unsuitable, however suitability classes 1 and 4 tend to have more trawl activity compared to classes 2 and 3.

ROVs, submersibles, drop cameras, etc.). The distribution of deep-sea corals within a single grid cell of these models (500 m×500 m) could be patchy [46] and could be missed on vehicle transects with limited range and narrow fields of view. To address this limitation and to improve the probability of locating undiscovered coral areas, research ships should first use multibeam surveys (in high probability areas) to identify substrate characteristics that can support deep-sea coral growth or identify corals (e.g. emergent hard substrata, coral rubble) and then move towards visual detection methodologies.

Assessment of closures and trawl intensity

Predictive models have been used to assess the suitability of existing protected areas in several areas including the North East Atlantic [14] and South Pacific [47]. Our findings are broadly similar, showing that the boundaries of U.S. National Marine Sanctuaries contained suitable habitat for corals above the average proportions of predicted suitable habitat throughout the entire study area (Table 5). Significant areas of highly suitable deep-sea coral habitat were modeled both within and outside existing NMS, EFH, and CCA-West closure boundaries. However, the majority of suitable habitat for Suborder Holaxonia and Order Scleractinia was predicted in areas outside of existing area closure boundaries. We also, however, identified numerous areas where areas of suitable habitat, from the most constrained model (75th percentile) fell outside of current protection initiatives. This was particularly evident in the EEZ waters off of Washington and Oregon. Overlaying the spatially limited trawling intensity layer from Whitmire [40] revealed that the majority of intense trawling was outside of current closed areas (Figure 11), as expected, but many areas did not have intensity data available. In previous studies, data from vessel monitoring systems have shown that vessels will enter closed areas occasionally and that vessel behaviour may be linked to the establishment of a closed area [48]. There were several high suitability areas that had a higher trawling intensity than other areas (Figure 11), implying that there may be some overlap in areas that are being fished and that contain suitable habitat for corals. Trawling has been shown to be highly damaging to corals, especially reef-forming scleractinans [49]. Emergent epifauna including octocorals will be adversely affected particularly in areas with repeated, high intensity trawling [50].

Conclusion

The U.S. West Coast has been relatively well researched with respect to the distribution of deep-sea coral species when compared to other regions of the world's oceans. However, significant spatial bias in sampling effort exists in the region and future field research efforts should be directed to unsampled areas to improve habitat predictions. Target areas for future field operations should include high probability areas identified in this study in regions of the U.S. West Coast EEZ that have not been visited. Predictive habitat model results are the only data available for areas that have not been sampled and should be used in in conjunction with other tools, data (i.e. geologic maps, multibeam bathymetry, etc.), and field surveys (where available) to help managers identify potential coral areas that remain at risk from human activity.

Supporting Information

Figure S1 Predicted suitability for the area of the Olympic Coast National Marine Sanctuary, a) Alcyoniina, b) Antipatharia, c) Calcaxonia, d) Holaxonia, e) Scleractinia, f) Scleraxonia, g) all taxa (50% threshold), h) all taxa (75% threshold).

Figure S2 Predicted suitability for the area of the Cordell Bank, Gulf of the Farallones, and Monterey Bay National Marine Sanctuaries, a) Alcyoniina, b) Antipatharia, c) Calcaxonia, d) Holaxonia, e) Scleractinia, f) Scleraxonia, g) all taxa (50% threshold), h) all taxa (75% threshold).

Figure S3 Predicted suitability for the area of the Channel Islands National Marine Sanctuary, a) Alcyoniina, b) Antipatharia, c) Calcaxonia, d) Holaxonia, e) Scleractinia, f) Scleraxonia, g) all taxa (50% threshold), h) all taxa (75% threshold).

Figure S4 Predicted habitat suitability in the Northern Region with EFH area closures (stippled areas) and CCA-West closures (hatched areas) for a) Alcyoniina, b) Antipatharia, c) Calcaxonia, d) Holaxonia, e) Scleractinia, f) Scleraxonia, g) all taxa (50% threshold), h) all taxa (75% threshold). For abbreviations, see Figure 8 in the manuscript.

Figure S5 Predicted habitat suitability in the Central Region with EFH area closures (stippled areas) and CCA-West closures (hatched areas) for a) Alcyoniina, b) Antipatharia, c) Calcaxonia, d) Holaxonia, e) Scleractinia, f) Scleraxonia, g) all taxa (50% threshold), h) all taxa

(75% threshold). For abbreviations, see Figure 8 in the manuscript.

Figure S6 Predicted habitat suitability in the Southern Region with EFH area closures (stippled areas) and CCA-West closures (hatched areas) for a) Alcyoniina, b) Antipatharia, c) Calcaxonia, d) Holaxonia, e) Scleractinia, f) Scleraxonia, g) all taxa (50% threshold), h) all taxa (75% threshold). For abbreviations, see Figure 8 in the manuscript.

Figure S7 Predicted habitat suitability for areas identified as probable hard substrate for a) Alcyoniina, b) Antipatharia, c) Calcaxonia, d) Holaxonia, e) Scleractinia, f) Scleraxonia, g) all taxa (50% threshold), h) all taxa (75% threshold).

File S1 Contains Tables S1–S10. Table S1: Correlation matrix for 10000 randomly placed points within the model domain. Table S2: Correlation matrix for points where the taxon Alcyoniina was found (n = 791). Table S3: Correlation matrix for points where the taxon Antipatharia was found (n = 128). Table S4: Correlation matrix for points where the taxon Calcaxonia was found (n = 413). Table S5: Correlation matrix for points where the taxon Filifera was found (n = 11). Table S6: Correlation matrix for points where the taxon Holaxonia was found (n = 308). Table S7: Correlation matrix for points where the taxon Scleractinia was found (n = 203). Table S8: Correlation matrix for points where the taxon Scleraxonia was found (n = 277). Table S9: Correlation matrix for points where the taxon Stolonifera was found (n = 30). Table S10: Correlation matrix for points where the taxon all species were found (n = 1059).

File S2 Model outputs for each taxa as ArcGIS GeoTIFF files with ArcGIS Map Documents and categorical layer files.

Acknowledgments

We would like to thank the scientists and managers from the following institutions for providing coral record locations and other data products: Monterey Bay Aquarium Research Institute, NOAA Fisheries, NOAA National Marine Sanctuaries, Smithsonian Institute's National Museum of Natural History, and Washington State University. Marine Conservation Institute would also like to thank the ESRI Conservation Program for providing ArcGIS software.

Author Contributions

Conceived and designed the experiments: JG AJD. Performed the experiments: JG AJD. Analyzed the data: JG AJD. Contributed reagents/materials/analysis tools: JG AJD. Wrote the paper: JG AJD.

References

1. Leverette TL, Metaxas A (2005) Predicting habitat for two species of deep-water coral on the Canadian Atlantic continental shelf and slope. In: Freiwald A, Roberts JM, editors. Cold-water Corals and Ecosystems. Berlin Heidelberg: Springer-Verlag. p. 467–479.
2. Bryan TL, Metaxas A (2007) Predicting suitable habitat for deep-water coral in the families Paragorgiidae and Primnoidae on the Atlantic and Pacific continental margins of North America. Marine Ecology Progress Series 330: 113–126.
3. Davies AJ, Wisshak M, Orr JC, Roberts JM (2008) Predicting suitable habitat for the cold-water reef framework-forming coral *Lophelia pertusa* (Scleractinia). Deep Sea Research Part I: Oceanographic Research Papers 55: 1048–1062.
4. Guinan J, Brown C, Dolan M, Grehan A (2009) Ecological niche modelling of the distribution of cold-water coral habitat using underwater remote sensing data. Ecological Informatics 4: 83–92.
5. Tittensor DP, Baco AR, Brewin PE, Clark MR, Consalvey M, et al. (2009) Predicting global habitat suitability for stony corals on seamounts. Journal of Biogeography 36: 1111–1128.
6. Woodby D, Carlile D, Hulbert L (2009) Predictive modeling of coral distribution in the Central Aleutian Islands, USA. Marine Ecology Progress Series 397: 227–240.
7. Davies AJ, Guinotte JM (2011) Global habitat suitability for framework-forming cold-water corals. PLoS ONE 6: e18483.
8. Whitmire CE, Clarke E (2007) State of deep coral ecosystems of the U.S. Pacific Coast: California to Washington. In: Lumsden SE, Hourigan TF, Bruckner AW, Dorr G, editors. The State of Deep Coral Ecosystems of the United States NOAA Technical Memorandum CRCP-3. p. 109–154.
9. Etnoyer P, Morgan LE (2007) Predictive habitat model for deep gorgonians needs better resolution: Comment on Bryan & Metaxas (2007). Marine Ecology Progress Series 339: 311–312.
10. Wilson MFJ, O'Connell B, Brown C, Guinan JC, Grehan AJ (2007) Multiscale terrain analysis of multibeam bathymetry data for habitat mapping on the Continental Slope. Marine Geodesy 30: 3–35.
11. Howell KL, Holt RD, Pulido Endrino I, Stewart H (2011) When the species is also a habitat: Comparing the predictively modelled distributions of *Lophelia pertusa* and the reef habitat it forms. Biological Conservation 144: 2656–2665.
12. Rengstorf A, Yesson C, Brown C, Grehan A (2012) Towards high resolution habitat suitability modelling of vulnerable marine ecosystems in the deep-sea: resolving terrain attribute dependencies. Marine Geodesy 35: 343–361.
13. Rengstorf AM, Yesson C, Brown C, Grehan AJ (2013) High-resolution habitat suitability modelling can improve conservation of vulnerable marine ecosystems in the deep sea. Journal of Biogeography 40: 1702–1714.
14. Ross RE, Howell KL (2013) Use of predictive habitat modelling to assess the distribution and extent of the current protection of 'listed' deep-sea habitats. Diversity and Distributions 19: 433–445.
15. Bellman MA, Heppell SA, Goldfinger C (2005) Evaluation of a US west coast groundfish habitat conservation regulation via analysis of spatial and temporal patterns of trawl fishing effort. Canadian Journal of Fisheries and Aquatic Sciences 62: 2886–2900.
16. Yesson C, Taylor ML, Tittensor DP, Davies AJ, Guinotte J, et al. (2012) Global habitat suitability of cold-water octocorals. Journal of Biogeography 39: 1278–1292.
17. NGDC (2011) Marine Geophysical Trackline Data. Available: http://www.ngdc.noaa.gov/mgg/geodas/trackline.html. Accessed: January 2014.
18. NGDC (2011) NOS Hydrographic Survey Data. Available: http://www.ngdc.noaa.gov/mgg/bathymetry/hydro.html. Accessed: January 2014.
19. NGDC (2011) Multibeam Bathymetry. Available: http://www.ngdc.noaa.gov/mgg/bathymetry/multibeam.html. Accessed: January 2014.
20. Caress DW, Chayes DN (2006) MB-System: Mapping the Seafloor, http://www.mbari.org/data/mbsystem and http://www.ldeo.columbia.edu/res/pi/MB-System.Last Accessed: January 2014.
21. NGDC (2011) U.S. Coastal Relief Model Development. Available: http://www.ngdc.noaa.gov/mgg/coastal/model.html. Accessed: January 2014.
22. Smith WHF, Sandwell DT (1997) Global seafloor topography from satellite altimetry and ship depth soundings. Science 277: 1957–1962.
23. Troupin C, Machin F, Ouberdous M, Sirjacobs D, Barth A, et al. (2010) High-resolution climatology of the northeast Atlantic using Data-Interpolating Variational Analysis (Diva). Journal of Geophysical Research 115: C08005.
24. Lutz MJ, Caldeira K, Dunbar RB, Behrenfeld MJ (2007) Seasonal rhythms of net primary production and particulate organic carbon flux to depth describe the efficiency of biological pump in the global ocean. Journal of Geophysical Research 112: C10011.
25. Jones KH (1998) A comparison of algorithms used to compute hill slope as a property of the DEM. Computers & Geosciences 24: 315–323.
26. Boyer TP, Antonov JI, Baranova OK, Garcia HE, Johnson DR, et al. (2009) World Ocean Database 2009. In: Levitus S, editor. NOAA Atlas NESDIS 66, U.S. Government Printing Office, Washington, D.C., 216 pp, DVDs.
27. Guinotte JM, Orr JC, Cairns SD, Freiwald A, Morgan L, et al. (2006) Will human-induced changes in seawater chemistry alter the distribution of deep-sea scleractinian corals? Frontiers In Ecology And The Environment 4: 141–146.
28. Thresher RE, Tilbrook B, Fallon S, Wilson NC, Adkins JE (2011) Effects of chronic low carbonate saturation levels on the distribution, growth and skeletal chemistry of deep-sea corals and other seamount megabenthos. Marine Ecology Progress Series 442: 87–99.
29. Frederiksen R, Jensen A, Westerberg H (1992) The distribution of the scleractinian coral *Lophelia pertusa* around the Faeroe Islands and the relation to internal tidal mixing. Sarsia 77: 157–171.
30. White M, Roberts JM, Van Weering T (2007) Do bottom-intensified diurnal tidal currents shape the alignment of carbonate mounds in the NE Atlantic? Geo-Marine Letters 27: 391–397.

31. White M (2007) Benthic Dynamics at the carbonate mound regions of the Porcupine Sea Bight continental margin. International Journal of Earth Sciences 96: 1–9.
32. Phillips SJ, Anderson RP, Schapire RE (2006) Maximum entropy modeling of species geographic distributions. Ecological Modelling 190: 231–259.
33. Elith J, Graham CH, Anderson RP, Dudík M, Ferrier S, et al. (2006) Novel methods improve prediction of species' distributions from occurrence data. Ecography 29: 129–151.
34. Phillips SJ, Dudík M (2008) Modeling of species distributions with Maxent: new extensions and a comprehensive evaluation. Ecography 31: 161–175.
35. Hernandez PA, Graham CH, Master LL, Albert DL (2006) The effect of sample size and species characteristics on performance of different species distribution modeling methods. Ecography 29: 773–785.
36. Lobo JM, Jiménez-Valverde A, Real R (2008) AUC: a misleading measure of the performance of predictive distribution models. Global Ecology and Biogeography 17: 145–151.
37. Pearson RG, Raxworthy CJ, Nakamura M, Peterson AT (2007) Predicting species distributions from small numbers of occurrence records: a test case using cryptic geckos in Madagascar. Journal of Biogeography 34: 102–117.
38. Raes N, Roos MC, Slik JWF, van Loon EE, ter Steege H (2009) Botanical richness and endemicity patterns of Borneo derived from species distribution models. Ecography 32: 180–192.
39. Rebelo H, Jones G (2010) Ground validation of presence-only modelling with rare species: a case study on barbastelles *Barbastella barbastellus* (Chiroptera: Vespertilionidae). Journal of Applied Ecology 47: 410–420.
40. Whitmire CE (2012) Bottom Trawl Fishing Intensity, data layer. NOAA Fisheries, Northwest Fisheries Science Center, Fishery Resource Analysis and Monitoring Division, http://efh-catalog.coas.oregonstate.edu/effort/. Last Accessed: September 2013.
41. National Marine Fisheries Service (2013) Groundfish Essential Fish Habitat Synthesis: A Report to the Pacific Fishery Management Council. NOAA NMFS Northwest Fisheries Science Center, Seattle, WA, April 2013. 107 pp.
42. Groundfish Essential Fish Habitat Review Committee (2012) Pacific Coast Groundfish 5-Year Review of Essential Fish Habitat Report to the Pacific Fishery Management Council Phase 1: New Information. Agenda Item H6b, EFHRC Report 1, September 2012.
43. Vierod ADT, Guinotte JM, Davies AJ (2014) Predicting the distribution of vulnerable marine ecosystems in the deep sea using presence-background models. Deep Sea Research Part II: Topical Studies in Oceanography 99: 6–18.
44. Graham CH, Ferrier S, Huettman F, Moritz C, Peterson AT (2004) New developments in museum-based informatics and applications in biodiversity analysis. Trends in Ecology & Evolution 19: 497–503.
45. Papeş M, Gaubert P (2007) Modelling ecological niches from low numbers of occurrences: assessment of the conservation status of poorly known viverrids (Mammalia, Carnivora) across two continents. Diversity and Distributions 13: 890–902.
46. Wilson JB (1979) 'Patch' development of the deep-water coral *Lophelia pertusa* (L.) on Rockall Bank. Journal of the Marine Biological Association of the United Kingdom 59: 165–177.
47. Penney AJ, Guinotte JM (2013) Evaluation of New Zealand's High-Seas Bottom Trawl Closures Using Predictive Habitat Models and Quantitative Risk Assessment. PLoS ONE 8(12): e82273.
48. Davies AJ, Roberts JM, Hall-Spencer J (2007) Preserving deep-sea natural heritage: Emerging issues in offshore conservation and management. Biological Conservation 138: 299–312.
49. Hall-Spencer J, Allain V, Fosså JH (2002) Trawling damage to Northeast Atlantic ancient coral reefs. Proceedings of The Royal Society of London Series B-Biological Sciences 269: 507–511.
50. Watling L, Norse EA (1998) Disturbance of the seabed by mobile fishing gear: A comparison to forest clearcutting. Conservation Biology 12: 1180–1197.
51. Garcia HE, Locarnini RA, Boyer TP, Antonov JI (2006) World Ocean Atlas 2005, Volume 3: Dissolved Oxygen, Apparent Oxygen Utilization, and Oxygen Saturation. S. Levitus, Ed. NOAA Atlas NESDIS 63, U.S. Government Printing Office, Washington, D.C.,342 pp.
52. Orr JC, Fabry VJ, Aumont O, Bopp L, Doney SC, et al. (2005) Anthropogenic ocean acidification over the twenty-first century and its impact on calcifying organisms. Nature 437: 681–686.
53. Steinacher M, Joos F, Frölicher TL, Plattner G-K, Doney SC (2009) Imminent ocean acidification projected with the NCAR global coupled carbon cycle-climate model. Biogeosciences 6: 515–533.
54. Garcia HE, Locarnini RA, Boyer TP, Antonov JI (2006) World Ocean Atlas 2005, Volume 4: Nutrients (phosphate, nitrate, silicate). S. Levitus, Ed. NOAA Atlas NESDIS 64, U.S. Government Printing Office, Washington, D.C., 396 pp.
55. Boyer TP, Levitus S, Garcia HE, Locamini RA, Stephens C, et al. (2005) Objective analyses of annual, seasonal, and monthly temperature and salinity for the World Ocean on a 0.25° grid. International Journal of Climatology 25: 931–945.
56. Kampstra P (2008) Beanplot: A boxplot alternative for visual comparison of distributions. Journal of Statistical Software 28: Code Snippet 1.

14

Comparison of the Abiotic Preferences of Macroinvertebrates in Tropical River Basins

Gert Everaert[1,2]*, Jan De Neve[3], Pieter Boets[1], Luis Dominguez-Granda[4], Seid Tiku Mereta[5], Argaw Ambelu[5], Thu Huong Hoang[6], Peter L. M. Goethals[1], Olivier Thas[3,7]

1 Aquatic Ecology Research Unit, Department Applied Ecology and Environmental Biology, Ghent University, Ghent, Belgium, **2** Environmental Toxicology Research Group, Department Applied Ecology and Environmental Biology, Ghent University, Ghent, Belgium, **3** Department of Mathematical Modelling, Statistics and Bioinformatics, Ghent University, Ghent, Belgium, **4** Department of Chemical and Environmental Sciences, Escuela Superior Politécnica del Litoral (ESPOL), Guayaquil, Ecuador, **5** Department of Environmental Health Science and Technology, Jimma University, Jimma, Ethiopia, **6** School of Environmental Science and Technology, Hanoi University of Science and Technology, Hanoi, Vietnam, **7** National Institute for Applied Statistics Research Australia (NIASRA), School of Mathematics and Applied Statistics, University of Wollongong, Wollongong, Australia

Abstract

We assessed and compared abiotic preferences of aquatic macroinvertebrates in three river basins located in Ecuador, Ethiopia and Vietnam. Upon using logistic regression models we analyzed the relationship between the probability of occurrence of five macroinvertebrate families, ranging from pollution tolerant to pollution sensitive, (Chironomidae, Baetidae, Hydroptilidae, Libellulidae and Leptophlebiidae) and physical-chemical water quality conditions. Within the investigated physical-chemical ranges, nine out of twenty-five interaction effects were significant. Our analyses suggested river basin dependent associations between the macroinvertebrate families and the corresponding physical-chemical conditions. It was found that pollution tolerant families showed no clear abiotic preference and occurred at most sampling locations, i.e. Chironomidae were present in 91%, 84% and 93% of the samples taken in Ecuador, Ethiopia and Vietnam. Pollution sensitive families were strongly associated with dissolved oxygen and stream velocity, e.g. Leptophlebiidae were only present in 48%, 2% and 18% of the samples in Ecuador, Ethiopia and Vietnam. Despite some limitations in the study design, we concluded that associations between macroinvertebrates and abiotic conditions can be river basin-specific and hence are not automatically transferable across river basins in the tropics.

Editor: Syuhei Ban, University of Shiga Prefecture, Japan

Funding: Pieter Boets is a postdoctoral research fellow from the Research Foundation Flanders (FWO - http://www.fwo.be/en/). Luis Dominguez-Granda received financial support of the VLIR ESPOL IUC program in Ecuador and SENACYT (http://www.vliruos.be/en/project-funding/programdetail/institutional-university-cooperation_3948/and http://www.educacionsuperior.gob.ec/). Argaw Ambelu was a recipient of an ICP-PhD scholarship from VLIR-UOS(http://www.vliruos.be/en/countries/countrydetail/ethiopia_3854/). Seid Tiku Mereta was a recipient of a IUC-PhD scholarship from VLIR-UOS (http://www.iucju.ugent.be/). Thu Huong Hoang received financial aid from the Belgian Technical Cooperation (BTC - http://www.btcctb.org/en/countries/vietnam). Marie Anne Eurie Forio and Minar Naomi Damanik Ambarita cooperated with financial support of the VLIR Ecuador Biodiversity Network (http://www.vliruos.be/en/ongoing-projects/overview-of-ongoing-projects/network-cooperation/network-biodev-ecuador/). The funders had no role in study design, data collection and analysis, decision to publish, or preparation of the manuscript.

Competing Interests: The authors have declared that no competing interests exist.

* Email: Gert.Everaert@UGent.be

Introduction

Benthic macroinvertebrates have often been used for water quality monitoring and assessment [1,2]. They are direct measures of stream conditions, integrate human and natural stressors over a long period of time and reflect the quality of their surroundings [3]. Benthic macroinvertebrates can be used as bio-indicators since different macroinvertebrate taxa have different tolerances to pollution [4]. Therefore, benthic macroinvertebrates are suitable for assessing the ecological state of aquatic ecosystems. Macroinvertebrate-based water quality assessment methods have increasingly been applied in national monitoring campaigns (e.g., [5,6]). In the United Kingdom, macroinvertebrate-based water quality assessment has been applied since 1970 by means of the Chandler Score [7]. In later years, the biological monitoring working party (BMWP; [8]) was developed and accepted as a standard international biotic index. In various regions biotic indices have been applied using a derivate of the BMWP, including the Iberian BMWP [9] and the South African Scoring System (SASS, [5]). In tropical countries, water quality assessment has been primarily performed based on physical-chemical water quality measurements [10]. However, in recent years, macroinvertebrate-based water quality assessments have also been conducted in tropical countries [11].

Most macroinvertebrate-based water quality assessment methods have been developed in temperate climate regions where relationships between environmental variables and the occurrence of macroinvertebrate taxa are well documented. Due to the infancy of the macroinvertebrate-based monitoring and assessment in the tropics [12], the ecological water quality in tropical countries was often assessed based on indices constructed for temperate climate regions [13,14]. However, it has been shown that applying ecological indices and habitat suitability models from temperate regions to the tropics can lead to uncertain ecological valuations [15,16]. In this perspective, river basin scale habitat suitability models have recently been developed for the tropics, e.g. in Ecuador [17], Ethiopia [18,19] and Vietnam [20,21]. In spite of this progress, limited knowledge is available about the habitat preferences of macroinvertebrates in rivers in the tropics [22]. For instance, it is poorly understood whether associations between macroinvertebrate taxa and environmental conditions vary between river basins and continents. As such, the question remains whether the relations obtained between the macroinvertebrate taxa and the physical-chemical water quality conditions are only valid in the river basin where they have been developed. Therefore, it is questioned whether the abiotic preferences of macroinvertebrates and their responses to environmental pollution are transferable across river basins.

The aim of this study was to assess if abiotic preferences of aquatic macroinvertebrates differed between three river basins in the tropics. We investigated associations between physical-chemical variables and macroinvertebrate occurrences in one Ecuadorian, one Ethiopian and one Vietnamese river basin. Five representative macroinvertebrate taxa, ranging from pollution tolerant to pollution sensitive were selected and their preferences towards environmental conditions were compared using regression-based ecological models.

Materials and Methods

Data collection

Samples were taken in the Chaguana river basin (Ecuador), Gilgel Gibe river basin (Ethiopia) and Cau river basin (Vietnam) (Table 1; Figure S1–S3). In each survey physical-chemical water samples and biological samples (macroinvertebrates) were collected. Per river basin, samples were taken at multiple sampling sites along a pollution gradient (Table 2). Details on sampling locations are provided in Table 1 and Figure S1–S3. The number of sampling sites varied for each river basin, but each sampling site was visited twice in each year; once in the wet season and once in the dry season. In total, 60, 104 and 306 samples were taken at 15, 29 and 47 sampling locations in the Vietnamese, Ecuadorian and Ethiopian river basin, respectively (Table 2). Water samples were analysed according to the ISO standards and only the environmental conditions that were monitored in all three river basins were selected for further analysis, being conductivity ($\mu S.cm^{-1}$), dissolved oxygen concentration ($mg.L^{-1}$), pH (-), stream velocity ($m.s^{-1}$) and water temperature (°C) (Table 2).

Benthic macroinvertebrates were sampled, identified and quantified according to the method described in Gabriels et al. [6] which is an internationally accepted kick-sampling procedure for macroinvertebrate sampling. A conical net with a size of 20×30 cm and a mesh size of 300–500 µm, attached to a stick, was used. With the hand net, all accessible aquatic habitats within a stretch of 10–20 m were sampled using the kick sampling method. The sampling effort was equally divided over the different habitats per sampling site. The organisms were identified to family level and this resulted in binary presence-absence data. Subsequently, five macroinvertebrate families, present in the three river basins and ranging from a pollution tolerant family towards a pollution sensitive family based on the BMWP tolerance list, were selected. The five target macroinvertebrate families that were selected were Chironomidae (tolerance class 2 (TS2)), Baetidae (TS4), Hydroptilidae (TS6), Libellulidae (TS8) and Leptophlebiidae (TS10). For a complete overview of the taxa encountered in each river basin we refer to the related publications [17,18,20]. No specific permission was needed for our sampling activities nor locations since we were only interested in macroinvertebrates. Sampling did not involve endangered or protected species.

Data exploration

Boxplots were used for data exploration. The first series of boxplots visualized the seasonality of the physical-chemical variables per river basin. The second series of boxplots summarized for each macroinvertebrate family and per physical-chemical variable and river basin the conditions under which the taxon was present and absent. Boxplots were constructed with the R software [23].

The univariate associations between explanatory variables were assessed with pairwise Spearman's rank correlations, which is often used in ecology due to its nonparametric nature [24].

Logistic regression model

The occurrence of five macroinvertebrate families was related to the physical-chemical water quality conditions using a regression-based model and these relationships were compared between three river basins situated in the tropics. Logistic regression models (LRMs) were used to infer relationships between occurrences of five aquatic macroinvertebrate families and environmental data. LRMs have been frequently used to model the presence or absence of a species in relation to environmental variables [25–27]. Five logistic regression models (LRMs) were constructed, one for each macroinvertebrate family. In a LRM, a binary response variable (here family presence or absence) is modeled as a function of explanatory variables (here environmental conditions). Since multiple samples were collected on the same sampling site, the responses were not mutually independent. Therefore the LRMs were fitted with Generalized Estimating Equations (GEE, [28]) which account for the dependencies of the clustered sampling scheme. All LRMs were fitted with an independent working correlation matrix.

A hierarchical backward elimination model selection method was carried out to build the LRM. The starting model included five physical-chemical variables (conductivity, dissolved oxygen concentration, pH, stream velocity and water temperature), season and river basin (represented as country). In addition to the main effects, two-way interactions between the physical-chemical variables and the river basin as well as between season and river basin were included (Table S1 and S2). First it was tested whether the interactions were present at a 5% level of significance and insignificant interactions were excluded from the model. A significant interaction between a physical-chemical variable and a river basin suggests that the effect of the physical-chemical variable on the occurrence of the family under study differed between river basins. Furthermore, all main effects were included independent from their significance.

Residual plots and the extended Hosmer-Lemeshow test for LRMs based on GEE [29] were used to assess the goodness-of-fit of the LRM. None of the models showed lack-of-fit. Since the Hosmer-Lemeshow test is insensitive to omitted quadratic terms, quadratic effects of the physical-chemical variables were added to the LRM. However, none of these effects were significant.

The outcome of the LRM per family was visualized as the estimated probability that the family was present as a function of a physical-chemical variable. The explanatory variables different from the one on the x-axis, were set to their river basin specific medians and the season to "dry". The river basin-specific observed range of the corresponding physical-chemical variable were plotted as horizontal boxplots below the response curves and were subdivided between presence and absence points. The gray-

Table 1. Description of the study areas.

	Ecuador	Ethiopia	Vietnam
River basin facts			
Name of river basin	Chaguana	Gilgel Gibe	Cau
Country	Southwest of Ecuador	South to West of Ethiopia	North of Vietnam
Province	El Oro	Oromia	Thai Nguyen
Tributary river	Pagua river [58]	Gibe river [61]	Cau river [60]
Source	Occidental Andes [58]	Ethiopian plateau[62]	Highland of Northern Vietnam[60]
Altitude	2900 m [58]	1096–3259 m [63]	275 m[60]
Surface area	320 km^2 [58]	5371 km^2 [18]	6030 km^2 [60]
Climatic description			
Average annual precipitation	930 mm [59]	2000 mm [63]	2063 mm [60]
Air temperature range	19.9°C–31.4°C [59]	8.3°C–29.1°C [64]	10.0°C–39.0°C [60]
Dry season	May - November [59]	October - April [63]	October - May [60]
Wet season	December-April [59]	May - September [63]	June - September [60]

[58]Dominguez-Granda, 2007, [59] Matamoros, 2004, [60] MONRE, 2006, [61] Demissie et al., 2013, [62] Uhlenbrook et al., 2010, [63] Broothaerts et al., 2012, [18] Ambelu et al., 2010, [64] Colombo and Maran, 2004.

Table 2. Characteristics of biological and physical-chemical sampling.

	Units	Ecuador	Ethiopia	Vietnam
Number of samples		104	306	60
Number of sampling locations		29	47	15
Monitoring years		2005–2006	2006–2011	2009–2010
Seasons		wet	wet	wet
		dry	dry	dry
Physical-chemical variables				
Stream velocity	m/s	0.5±0.4	0.5±0.3	0.5±0.3
Water temperature	°C	23.3±2.8	20.0±2.5	28.6±2.0
Conductivity	µS/cm	149±152	108±59	235±181
pH	-	6.9±0.4	7.4±0.5	7.0±0.8
Dissolved oxygen	mg/L	7.1±1.3	6.5±1.5	6.4±0.8
Macroinvertebrate sampling				
Chironomidae				
Number of samples: absent		9	49	4
Number of samples: present		95	257	56
Baetidae				
Number of samples: absent		18	80	11
Number of samples: present		86	226	49
Hydroptilidae				
Number of samples: absent		88	300	34
Number of samples: present		16	6	26
Libellulidae				
Number of samples: absent		50	157	51
Number of samples: present		54	149	9
Leptophlebiidae				
Number of samples: absent		54	300	49
Number of samples: present		50	6	11

Descriptive statistics of physical-chemical variables are given as mean values ± standard deviations. Presence-absence records per family and river basin (represented as country) are given as the amount of samples in which macroinvertebrate families were present or absent, respectively.

Table 3. P-values for river basin-based interaction effects.

	Stream velocity	Water temperature	Conductivity	pH	Dissolved oxygen
Leptophlebiidae	**0.008**	0.256	**0.033**	**0.038**	0.121
Libellulidae	**0.017**	**<0.001**	0.767	0.055	**0.015**
Hydroptilidae	0.905	0.692	**0.015**	0.420	0.262
Baetidae	0.126	**0.003**	0.113	0.924	0.582
Chironomidae	0.998	**<0.001**	0.999	0.999	0.999

The effect of an explanatory variable was significantly different between the three river basins in case that the p-value for the interaction effect was lower than 0.05. Significant relations are indicated in bold.

colored ends of the response curves indicate extrapolation outside the observed range of the corresponding physical-chemical variable.

For the data exploration, the effects of season and river basin on the continuous physical-chemical variables were assessed using a linear regression model fitted with GEE for accounting for the clustered sampling scheme. Only few p-values were reported in the main text to prevent information overload, all p-values were summarized in Table 3 and 4. All statistical tests were performed at the 5% significance level.

Results

Data exploration and correlation analysis

The observed range of physical-chemical water quality conditions was not always equal between river basins. Seasonal effects were observed for all physical-chemical variables and the seasonal effect of conductivity and stream velocity was similar for the three river basins (Table S1; Figure S4–S8). Dissolved oxygen (DO) concentration and water temperature ranges differed between river basins (Table 2; Figure S9–S33). The average stream velocity was higher during the wet season compared to the dry season ($p < 0.01$; Table S1; Figure S4). The average conductivity in wet season was lower than in dry season in the Ecuadorian river basin ($p < 0.01$) and the Ethiopian river basin ($p < 0.01$). In the Vietnamese river basin, however, this difference was not significant ($p = 0.39$; Figure S6). The seasonal effect differed per river basin for water temperature (Figure S5), pH (Figure S7) and DO concentration (Figure S8). For instance, dry season DO concentrations tended to exceed wet season conditions in the Ecuadorian river basin, but in the Ethiopian river basin the opposite was observed (Table S1; Figure S8). For both seasons there was no significant difference in mean stream velocity between river basins ($p = 0.32$ and $p = 0.78$, Table S2; Figure S4). For DO there was a significant difference

Table 4. Estimates of the main effects for Chironomidae, Baetidae, Hydroptilidae, Libellulidae, Leptophlebiidae.

	Seasonality	Stream velocity	Water temperature	Conductivity	pH	Dissolved oxygen
Chironomidae						
Ecuador	−0.873 (0.265)	**1.358 (0.016)**	0.091 (0.455)	−0.002 (0.316)	0.149 (0.697)	−0.108 (0.290)
Ethiopia	−0.313 (0.402)	**1.358 (0.016)**	−0.078 (0.151)	−0.002 (0.316)	0.149 (0.697)	−0.108 (0.290)
Vietnam	35.838 (>0.999)	**1.358 (0.016)**	**−1.649 (<0.001)**	−0.002 (0.317)	0.149 (0.697)	−0.108 (0.290)
Baetidae						
Ecuador	0.869 (0.184)	**1.289 (0.004)**	0.031 (0.700)	−0.002 (0.121)	**0.554 (0.030)**	−0.065 (0.410)
Ethiopia	**0.704 (0.019)**	**1.289 (0.004)**	−0.049 (0.407)	−0.002 (0.121)	**0.554 (0.030)**	−0.065 (0.410)
Vietnam	0.163 (0.804)	**1.289 (0.004)**	−0.351 (0.133)	−0.002 (0.121)	**0.554 (0.030)**	−0.065 (0.410)
Hydroptilidae						
Ecuador	−0.126 (0.864)	0.829 (0.223)	−0.062 (0.531)	−0.003 (0.297)	**0.963 (0.020)**	0.179 (0.519)
Ethiopia	−1.463 (0.203)	0.829 (0.223)	−0.062 (0.531)	**−0.020 (0.042)**	**0.963 (0.020)**	0.179 (0.519)
Vietnam	−0.408 (0.438)	0.829 (0.223)	−0.062 (0.531)	**−0.009 (0.031)**	**0.963 (0.020)**	0.179 (0.519)
Libellulidae						
Ecuador	**−1.788 (0.005)**	**2.488 (0.005)**	**0.387 (0.002)**	−0.0001 (0.961)	**0.674 (0.006)**	**0.481 (0.023)**
Ethiopia	0.192 (0.445)	−0.279 (0.534)	**−0.094 (0.021)**	−0.0001 (0.961)	**0.674 (0.006)**	−0.046 (0.619)
Vietnam	−1.369 (0.268)	**3.780 (0.021)**	**−0.708 (<0.001)**	−0.0001 (0.961)	**0.674 (0.006)**	0.135 (0.815)
Leptophlebiidae						
Ecuador	**2.147 (0.006)**	**3.679 (0.003)**	**−0.291 (0.048)**	−0.010 (0.072)	0.503 (0.356)	**1.371 (0.002)**
Ethiopia	1.385 (0.232)	0.694 (0.750)	**−0.291 (0.048)**	−0.024 (0.370)	−0.223 (0.840)	**1.371 (0.002)**
Vietnam	1.426 (0.100)	−1.703 (0.410)	**−0.291 (0.048)**	0.002 (0.355)	**1.943 (0.004)**	**1.371 (0.002)**

Corresponding p-values are given between brackets; effects were tested at the 5% significance level. Significant relations are indicated in bold.

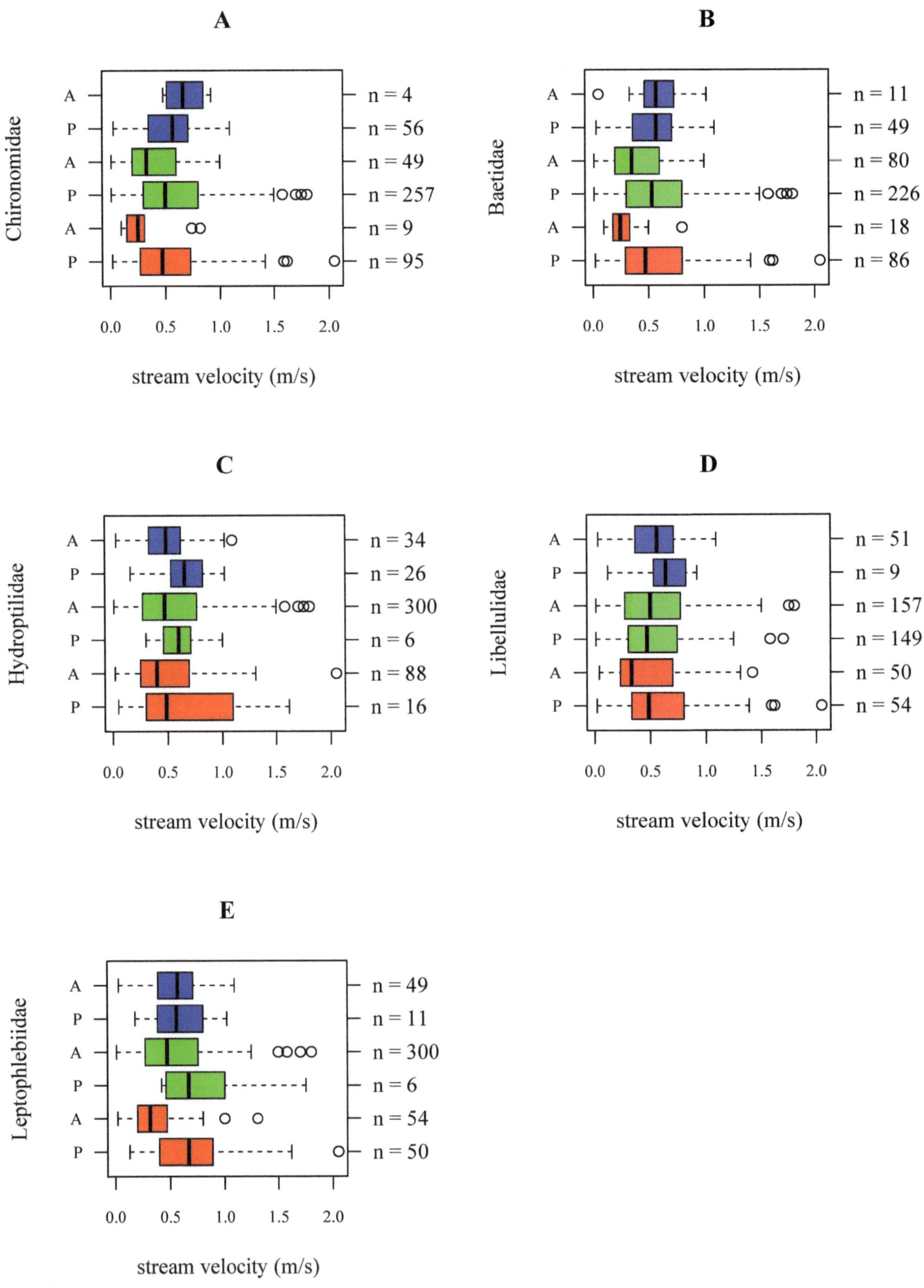
A
Chironomidae
A
P
A
P
A
P
n = 4
n = 56
n = 49
n = 257
n = 9
n = 95
0.0 0.5 1.0 1.5 2.0
stream velocity (m/s)
B
Baetidae
A
P
A
P
A
P
n = 11
n = 49
n = 80
n = 226
n = 18
n = 86
0.0 0.5 1.0 1.5 2.0
stream velocity (m/s)
C
Hydroptilidae
A
P
A
P
A
P
n = 34
n = 26
n = 300
n = 6
n = 88
n = 16
0.0 0.5 1.0 1.5 2.0
stream velocity (m/s)
D
Libellulidae
A
P
A
P
A
P
n = 51
n = 9
n = 157
n = 149
n = 50
n = 54
0.0 0.5 1.0 1.5 2.0
stream velocity (m/s)
E
Leptophlebiidae
A
P
A
P
A
P
n = 49
n = 11
n = 300
n = 6
n = 54
n = 50
0.0 0.5 1.0 1.5 2.0
stream velocity (m/s)

Figure 1. Stream velocity data for which Chironomidae (A), Baetidae (B), Hydroptilidae (C), Libellulidae (D) and Leptophlebiidae (E) are found to be present (denoted by P on the left axis) and absent (denoted by A on the left axis) in Ecuador (red), Ethiopia (green) and Vietnam (blue). The sample sizes per boxplot are shown on the right axis.

between river basins for the dry season ($p<0.01$), but not for the wet season ($p=0.16$). In contrast, the mean water temperature, pH and conductivity were significantly different across river basins for both seasons.

The prevalence values, known as the relative frequencies of occurrence of taxa (here families, [30]), did not always approximate 0.5, i.e. some families occurred rarely or ubiquitously (Table 2). Furthermore, it was found that the range of preferred physical-chemical conditions differed between families (Figure 1A–1E; Figure 2A–2E; Figure S9–S33). For less pollution sensitive families such as Chironomidae (TS2) and Baetidae (TS4) a wide range of suitable physical-chemical conditions has been observed. More sensitive families such as Leptophlebiidae (TS10) were only present within a more narrow range of stream velocity, conductivity and DO concentration (Figure 1A–1E; Figure 2A–2E; Figure S9–S33).

Based on the correlation analysis no variables were discarded from the dataset as most correlations were "weak" since they were smaller than 0.4 in absolute values [21,24]. Only conductivity and water temperature ($r=0.41$, $p<0.01$) and conductivity and DO ($r=-0.42$, $p<0.01$) were moderately correlated (Table 5).

Logistic regression model

Chironomidae (TS2). For Chironomidae there was an interaction effect between river basin and water temperature ($p<0.01$), i.e. the effect of water temperature on the occurrence differed between river basins (Table 3). Although increasing water temperatures were associated with a lower probability that Chironomidae were present in the Vietnamese river basin ($p<0.01$), in the Ecuadorian river basin and in the Ethiopian river basin this effect was not observed ($p=0.46$ and $p=0.15$, Table 4; Figure S35). However, note that Chironomidae were absent in only 4 out of 60 samples from the Vietnamese river basin (Figure S35). We also found a positive association between the stream velocity and the probability of occurrence of Chironomidae ($p=0.02$), i.e. a higher stream velocity was associated with a higher probability that Chironomidae were present (Table 4; Figure 3A). The effect of stream velocity was similar between river basins, i.e. regardless the stream velocity, Chironomidae were always likely to be present in the three river basins (probabilities between 0.8 and 1.0 in Figure 3A and S34). The latter is also reflected in the boxplots, which indicate that absence data for Chironomidae is low (4 out of 60 in Vietnam, 49 out of 306 in Ethiopia and 9 out of 104 in Ecuador; Figure 1A). Within the investigated physical-chemical range, conductivity, pH, and DO concentrations were not associated with the probability of occurrence of Chironomidae (Figure 3B; Figure S36–S38).

Baetidae (TS4). The effect of water temperature on the probability of occurrence of Baetidae differed between river basins ($p<0.01$; Table 3). Indeed, when looking at the response curves (Figure S40) it is clear that within the observed water temperature ranges for each river basin, the slopes of the curves differ. However, probabilities of occurrence always exceeded the value of 0.6 in any river basin regardless the water temperature data. Moreover, within river basins no significant association was found between the presence of Baetidae and water temperature (Table 4). An increased stream velocity and pH was associated with an increased probability of occurrence of Baetidae in all three river basins ($p=0.01$ and $p=0.03$, respectively; Table 4; Figure S39 and S42). Again, similar as for the Chironomidae, we found that Baetidae were likely to be present in all the three river basins (Figure 1B, S39 and S42). In the investigated range, no significant association was found between the probability of occurrence of Baetidae and DO concentration ($p=0.58$, Figure S43) or conductivity ($p=0.11$, Figure S41).

Hydroptilidae (TS6). Associations between the DO concentration, pH, stream velocity and water temperature relative to the probability of occurrence of Hydroptilidae were similar across river basins (Table 3). Only the effect of conductivity differed between river basins ($p=0.02$; Table 4), i.e. higher conductivities were associated with a lower probability that Hydroptilidae were present; this was observed for the Vietnamese river basin ($p=0.03$) and the Ethiopian river basin ($p=0.04$; Table 4; Figure S46). Furthermore, a more alkaline pH was associated with a higher occurrence of Hydroptilidae ($p=0.02$; Table 4; Figure S47). No significant associations were found between water temperature ($p=0.69$; Figure S45), stream velocity ($p=0.90$; Figure S44) and DO concentration ($p=0.26$; Figure S48) and the probability of occurrence of Hydroptilidae (Table 4).

Libellulidae (TS8). The associations between water temperature ($p<0.01$), stream velocity ($p=0.02$) and DO concentration ($p=0.02$) and the probability of occurrence of Libellulidae differed between river basins (Table 3). Whereas in the Ecuadorian river basin, the probability that Libellulidae occurred increased with increasing water temperatures ($p<0.01$), in the Ethiopian and Vietnamese river basin inverse associations were observed ($p=0.02$ and $p<0.01$, respectively; Table 4; Figure S50). Concerning stream velocity, Libellulidae were likely to occur at higher stream velocities in the Ecuadorian and Vietnamese river basin ($p<0.01$ and $p=0.02$, respectively), but in the Ethiopian river basin this was not observed (Table 4; Figure 4A and S49). Increasing DO concentrations were associated with a higher probability of occurrence of Libellulidae in the Ecuadorian river basin ($p=0.02$), but for the Vietnamese ($p=0.82$) and Ethiopian river basin ($p=0.62$) this association was not significant (Figure 4B and S53). The associations between conductivity and pH and the probability of occurrence of Libellulidae were similar between river basins (Table 4; Figure S51 and S52).

Leptophlebiidae (TS10). Associations between stream velocity (Figure 1E, 5A and S54), pH (Figure S57) and conductivity (Figure S56) and the probability of occurrence of Leptophlebiidae were different amongst river basins ($p=0.01$, $p=0.04$ and $p=0.03$; Table 3). However, when interpreting the response curves it was found that the prevalence of Leptophlebiidae was not balanced in the Vietnamese and Ethiopian river basin (Table 2). For streams situated in the Ecuadorian river basin Leptophlebiidae were more likely to occur at higher stream velocities ($p<0.01$). In the Vietnamese and Ethiopian river basin, however, these effects were not observed (Table 4; Figure 5A and S54), probably due the limited amount of presence data in these river basins. The association between conductivity and the probability of occurrence of Leptophlebiidae suggested the highest probability of occurrence at lowest conductivities in the Vietnamese river basin (Figure S56). For all three river basins, an increase in DO concentration was positively associated with the occurrence of Leptophlebiidae ($p<0.01$; Table 4; Figure 5B and S58). Associations between Leptophlebiidae and water temperature were similar between the three river basins (Figure S55).

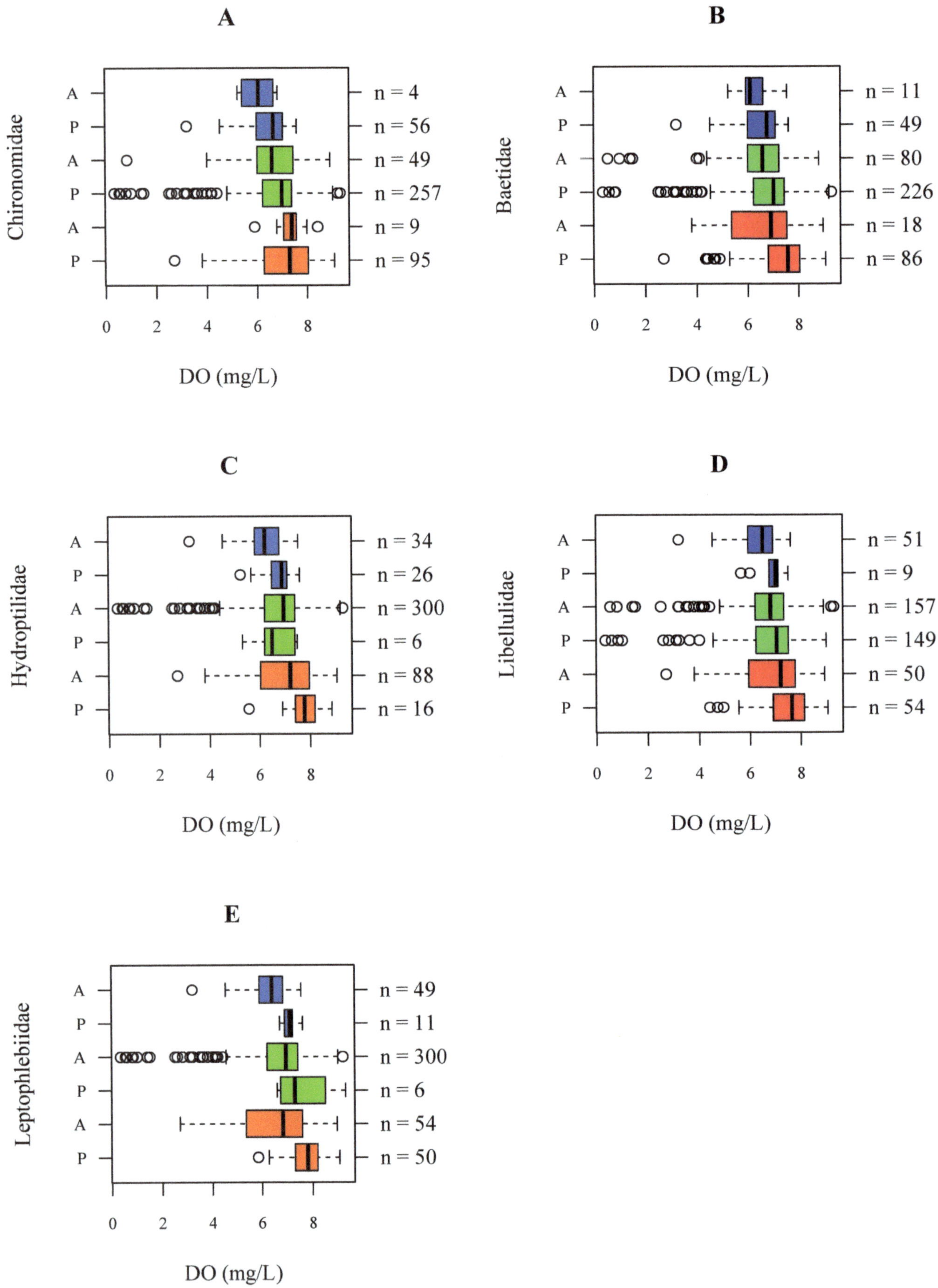
A
B
C
D
E
Chironomidae
Baetidae
Hydroptilidae
Libellulidae
Leptophlebiidae
A
P
A
P
A
P
n = 4
n = 56
n = 49
n = 257
n = 9
n = 95
n = 11
n = 49
n = 80
n = 226
n = 18
n = 86
n = 34
n = 26
n = 300
n = 6
n = 88
n = 16
n = 51
n = 9
n = 157
n = 149
n = 50
n = 54
n = 49
n = 11
n = 300
n = 6
n = 54
n = 50
0
2
4
6
8
DO (mg/L)

Figure 2. Dissolved oxygen concentrations for which Chironomidae (A), Baetidae (B), Hydroptilidae (C), Libellulidae (D) and Leptophlebiidae (E) are found to be present (denoted by P on the left axis) and absent (denoted by A on the left axis) in Ecuador (red), Ethiopia (green) and Vietnam (blue). The sample sizes per boxplot are shown on the right axis.

Discussion

For pollution tolerant macroinvertebrate families (e.g. Chironomidae (TS2)) only few significant associations between physical-chemical conditions and the occurrence of the families were found (Table 4) suggesting their occurrence at a wide range of physical-chemical conditions. This was also reflected in the data exploration as Chironomidae were present in 91%, 84% and 93% of the samples taken in the Ecuadorian, Ethiopian and Vietnamese river basin, respectively (Table 2). Chironomidae are known to be tolerant to disturbance, allowing them to occur in impacted environments [32]. Due to their physiological adaptations they have the ability to survive low oxygen conditions (Figure 2A and 3B, [33,34]). Because of their pollution tolerant nature, it was surprising to conclude that high stream velocities are associated with an increased occurrence of Chironomidae (Figure 1A, 3A and S34). However, in this study we only identified macroinvertebrates to family level. Chironomidae are represented by many species that show a different sensitivity to environmental pollution [33], which might explain the weak association between environmental variables and the occurrence of Chironomidae. Baetidae, present in 83%, 74% and 82% of the samples taken in the Ecuadorian, Ethiopian and Vietnamese river basin, respectively (Table 2), were more sensitive to pollution compared to Chironomidae. The habitat preference of Baetidae was mainly determined by stream velocity and pH. Baetidae have often been reported as one of the most acid-sensitive macroinvertebrate families [35]. As taxa sensitivity to pollution increased, more significant associations were found between environmental conditions and the occurrence of macroinvertebrate families (Table 4). For instance, a significant positive association between DO concentrations and stream velocities and the occurrence of Libellulidae (TS8) and Leptophlebiidae (TS10) was found (Figure 4 and 5). DO concentration is a general indicator of water quality [34,36] and also in other studies it was found that DO concentrations play a crucial role when analyzing the occurrence of macroinvertebrates (e.g., [37–39)]. In streams in Malaysia, Rawi et al. [40] found that DO concentration and stream velocity were crucial variables when explaining macroinvertebrate diversity. For other biological communities DO levels are also important, e.g. most fish require a DO concentration of at least 5 $mg.L^{-1}$ for optimal health [41].

Habitat suitability of macro-invertebrates probably depends on more factors than those included in our statistical analysis. For instance, additional to the variables that were included in our analysis, Al-Shami et al. [42] also integrated geographical variables such as longitude, latitude and altitude. Blanchette and Pearson [43] related macroinvertebrate assemblages to chlorophyll, suspended solids, turbidity and nutrient data. Kasangaki et al. [44] studied benthic macroinvertebrates in Uganda and used DO concentration, conductivity, pH, turbidity and water temperature together with some hydromorphological stream characteristics. As such, physical-chemical variables used by Kasangaki et al. [44] are similar to those included in our statistical analysis. Conductivity can be seen as a general measure for disturbance as it integrates pollution related variables like minerals and inorganic pollutants [37]. For instance, Melo [45] concluded that stream size data and conductivity explained most of the variability in the macroinvertebrate community in a stream in the tropics. In Colombia, Holguin-Gonzalez et al. [26] used DO concentration and stream velocity to predict the presence of macroinvertebrates. According to Flecker and Feifarek [46] hydrodynamics (including stream velocity) play a crucial role in structuring tropic macroinvertebrate communities. Therefore, although only five physical-chemical variables were used in the models, explanatory variables embedded in the LRMs covered a wide spectrum of potential impacts.

Seasonal changes of environmental variables were taken into account [10,43]. Two samples in each year and at each sampling location (in the wet and dry season) were collected for two consecutive years for each river basin (Table S1; Figure S4–S8). By including these, potential seasonal uncertainties were integrated in our statistical analysis.

The outcome of the LRM per family was visualized as the estimated probability that the family was present as a function of a physical-chemical variable (e.g. Figure 3–5). Within the physical-chemical ranges that were investigated and for the studied macroinvertebrate families (see boxplots and colored zone of the response curves), nine out of twenty-five interaction effects were significant (Table 3). Hence, the corresponding variable-macroinvertebrate relationships were different between river basins and suggested different habitat preferences for the investigated families. For Libellulidae for instance, it was found that the effect of dissolved oxygen was different between river basins (Table 3; Figure 4B). Increasing DO concentrations were associated with a higher probability that Libellulidae were present in Ecuadorian river basin ($p = 0.02$), but in the Vietnamese and Ethiopian river basin this relationship was not significant ($p = 0.82$ and $p = 0.62$). In the Vietnamese and Ethiopian river basin, Libellulidae were less responsive to shifts in DO concentrations as the probability of

Table 5. Correlation coefficients of the physical-chemical variables.

	Stream velocity	Water temperature	pH	Conductivity	Dissolved oxygen
Stream velocity	/				
Water temperature	**−0.14 (<0.001)**	/			
pH	0.03 (0.571)	**−0.21 (<0.001)**	/		
Conductivity	**−0.39 (<0.001)**	**0.41 (<0.001)**	**0.15 (0.001)**	/	
Dissolved oxygen	**0.28 (<0.001)**	**−0.28 (<0.001)**	**0.23 (<0.001)**	**−0.42 (<0.001)**	/

Corresponding p-values are given between brackets; effects were tested at the 5% significance level. Significant relations are indicated in bold.

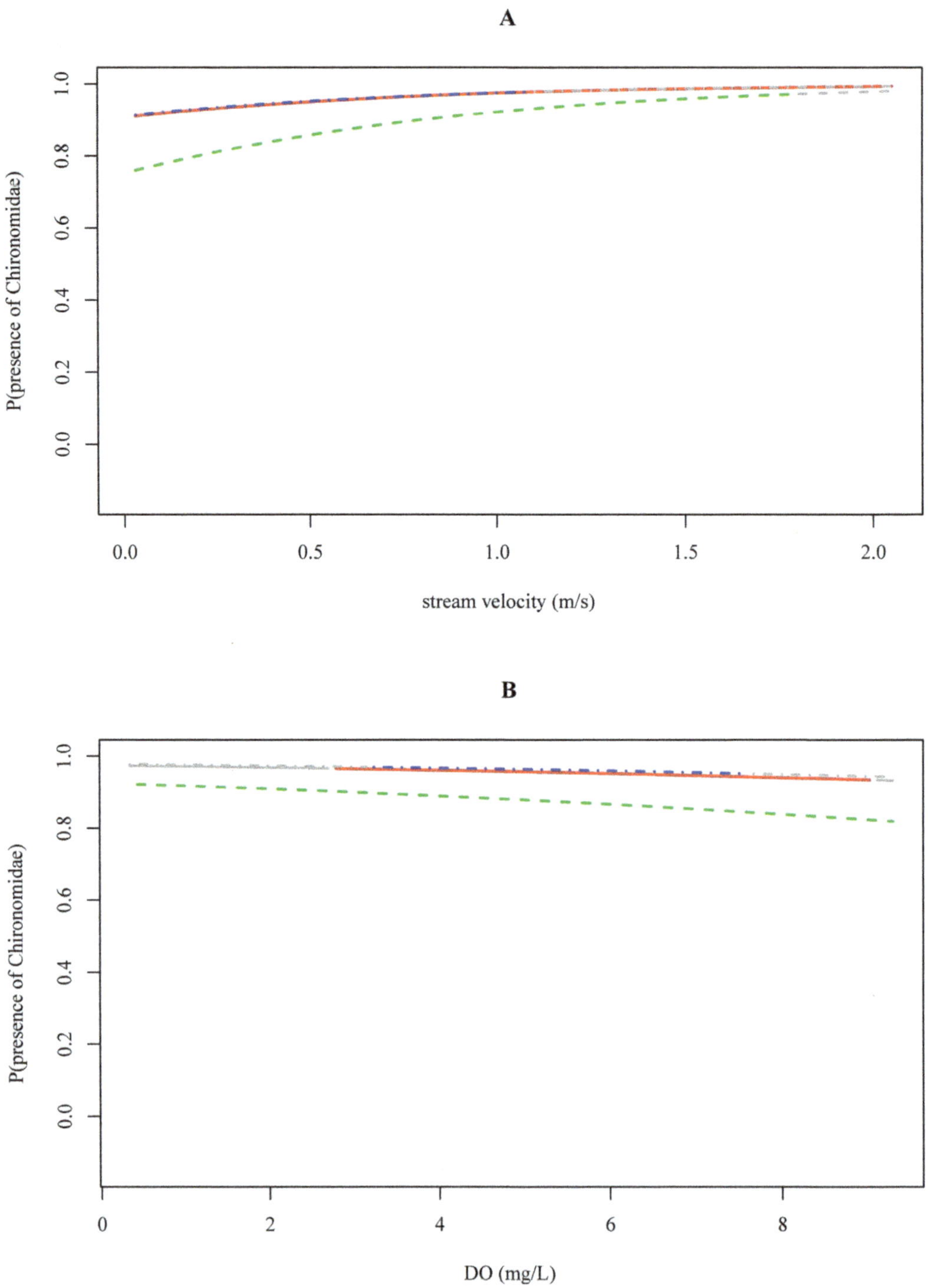

Figure 3. The probability of Chironomidae being present in relation to the stream velocity (A) and dissolved oxygen data (B) measured in Ecuador (red, solid), Ethiopia (green, dashed) and Vietnam (blue, dotdashed). The gray-colored ends of the response curves indicate extrapolation outside the observed physical-chemical range in the corresponding river basin.

occurrence hardly changed at varying DO concentrations (Table 4, Figure 4B). In the Vietnamese river basin, this can be explained by the relative few observed presences of some taxa, i.e. Libellulidae were found in only nine out of sixty samples (Table 2). In the Ethiopian river basin however, presence and absence data were more equally represented. Concerning the association between stream velocity and the occurrence of Libellulidae it was found that Libellulidae favored high currents in the Ecuadorian and Vietnamese river basin ($p = 0.01$ and $p = 0.02$, respectively). In the Ethiopian river basin, this relationship was not observed (Table 4; Figure 4A). However, a negative relationship did not mean that Libellulidae were absent in the Ethiopian river basin as there was still an estimated probability of 50% that Libellulidae were present within the range of observations (Figure 4A).

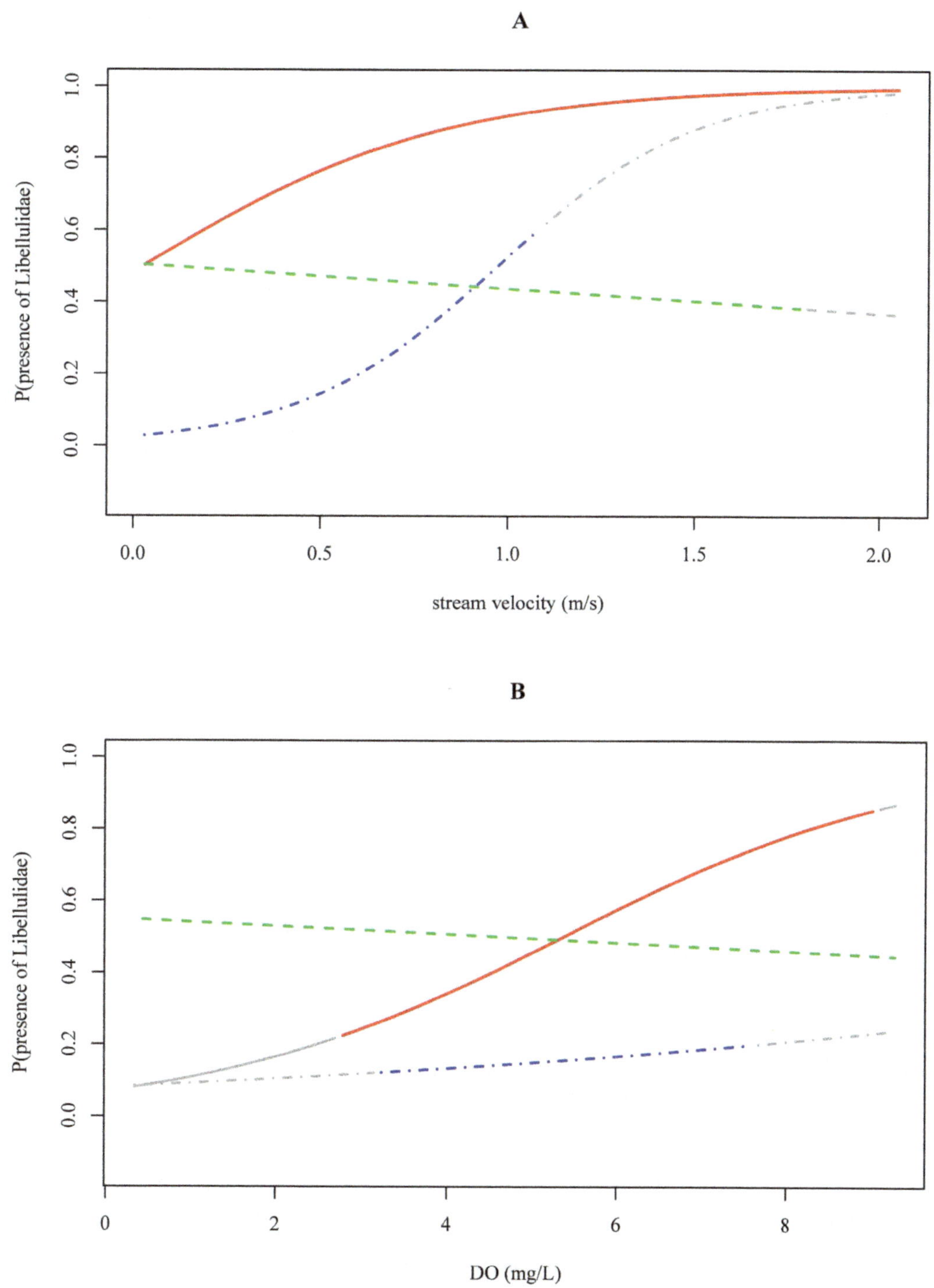

Figure 4. The probability of Libellulidae being present in relation to stream velocity (A) and dissolved oxygen data (B) measured in Ecuador (red, solid), Ethiopia (green, dashed) and Vietnam (blue, dotdashed). The gray-colored ends of the response curves indicate extrapolation outside the observed physical-chemical range in the corresponding river basin.

The fact that river basin dependent associations were found is not surprising. For instance, Bonada et al. [47] found that the response of macroinvertebrates to pollution was different between Mediterranean ecoregions. Also in the US, Zuellig and Schmidt [48] found dissimilar regional benthic invertebrate community compositions. Moreover, it was stated by Thorne and Williams [13] that due to untreated domestic and urban effluents in developing countries the relationship between individual physical-chemical variables and macroinvertebrates present can be extremely complex. Hence, this may lead to river basin specific associations between abiotic conditions and the biological communities.

Since the range of the observed physical-chemical conditions were not always equal between river basins, in some cases extrapolations outside the observed range were shown (Figure 4A). For instance, water temperature measurements significantly

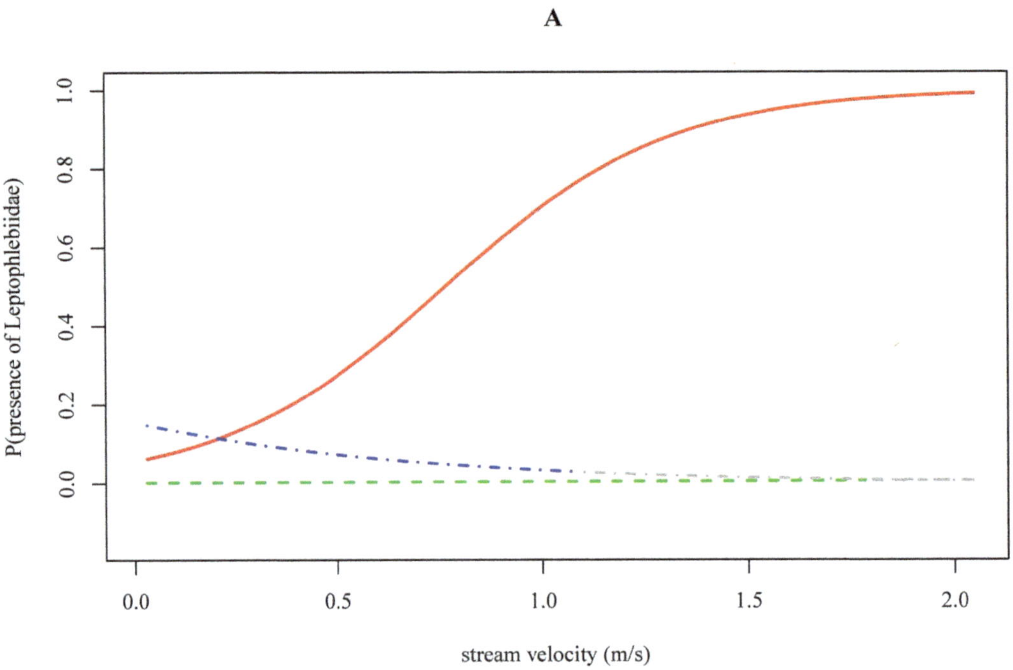

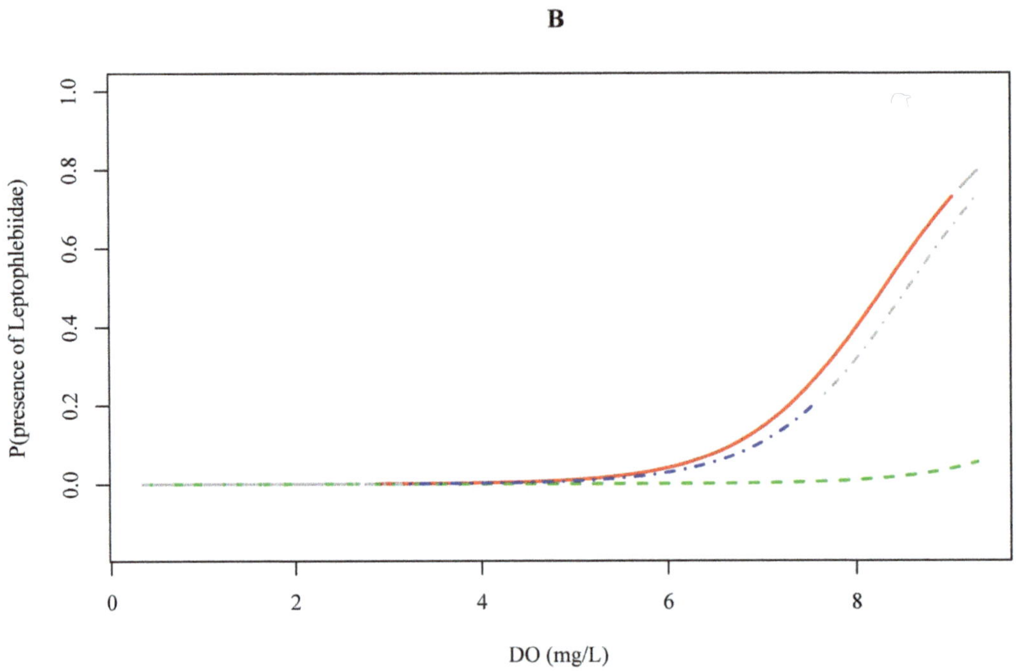

Figure 5. The probability of Leptophlebiidae being present in relation to stream velocity (A) and dissolved oxygen data (B) measured in Ecuador (red, solid), Ethiopia (green, dashed) and Vietnam (blue, dotdashed). The gray-colored ends of the response curves indicate extrapolation outside the observed physical-chemical range in the corresponding river basin.

differed between the three river basins (Table S2). Hence, it was found that the association between the occurrence of Libellulidae and water temperature was significantly different between river basins (Table 3). However, at the intersection of the three ranges, the estimated probabilities were similar (Figure S50). Therefore, major differences in the estimated probabilities across river basins are likely to be attributed to difference in the range of the water temperature between those basins.

No relation was found between the occurrence of Leptophlebiidae and stream velocity in the Ethiopian and the Vietnamese river basin. However, according to the data Leptophlebiidae occurred in both river basins in upstream sampling sites with a moderate to high stream velocity. The reason for the non-significant relation is the prevalence of the present-points. In the Vietnamese and Ethiopian river basin, Leptophlebiidae were only present in 18% and 2% of the samples, respectively. Consequently, the occurrence of Leptophlebiidae seems statistically not associated with stream velocity. Therefore, due to unbalanced dataset prevalence it is possible that not all models obtained represent true ecological relations [30]. As such, LRMs' outcomes should be carefully interpreted, but for those families and river basins where prevalence was in balance, the associations found, were ecologically relevant.

Presence and absence data are commonly used in macroinvertebrate research (e.g., [48–51]) as they provide a basic inventory to explore the species' ecology [52]. However, recently Howard et al. [53] concluded that using abundance data instead of presence and absence data could make models more informative since abundance data provide additional insight with regard to the population dynamics. Although it may result in a gain of information [54], including abundance data will also increase the variability of the response variable. In this perspective, Flecker and Feifarek [46] concluded that the abundance of macroinvertebrate families differed seasonally between one and four orders of magnitude. In this research the sample size (60–306) was relatively limited compared to the number of sampling locations (15–47) [55] and may not cover the entire seasonality in the macroinvertebrate abundance data. Therefore, it was more appropriate to analyze presence-absence data, which are more robust against seasonal changes. Moreover, logistic models have been frequently applied (e.g., [24–26]) as they are well suited to provide predictions of the probability of occurrence based on presence and absence data [52].

As an alternative to the conventional taxonomic division of the benthic invertebrates one can use functional traits to categorize macroinvertebrate assemblages (e.g., [18,56]). Such a trait-based approach is a promising alternative to taxonomy-based approaches for assessing the conditions of freshwater ecosystems [57]. However, trait-based studies have often a limited scope and further testing is needed to establish their reliability [48], especially in regions where knowledge on macroinvertebrates is poor.

Overall, LRMs based on GEE are a flexible way to model the probability of occurrence of macroinvertebrates as a function of environmental variables. We revealed similar as well as dissimilar abiotic preferences of macroinvertebrates between the three river basins, but these estimated probabilities are restricted to the observed range of the predictor within each river basin. In conclusion we found that associations between macroinvertebrates and abiotic conditions can be river basin-specific.

Supporting Information

Figure S1 Sampling sites in the Chaguana river basin in Ecuador.

Figure S2 Sampling sites in the Gilgel Gibe river basin in Ethiopia.

Figure S3 Sampling sites in the Cau river basin in Vietnam.

Figure S4 Boxplots showing the seasonality (dry (blue) and wet (red) season) of the stream velocity.

Figure S5 Boxplots showing the seasonality (dry (blue) and wet (red) season) of the water temperature.

Figure S6 Boxplots showing the seasonality (dry (blue) and wet (red) season) of the conductivity.

Figure S7 Boxplots showing the seasonality (dry (blue) and wet (red) season) of the pH.

Figure S8 Boxplots showing the seasonality (dry (blue) and wet (red) season) of the dissolved oxygen.

Figure S9 Boxplots indicate the observed stream velocity in Ecuador (red), Ethiopia (green) and Vietnam (blue) at which Chironomidae are found to be present (denoted by P on the left axis) and absent (denoted by A on the left axis). The sample sizes per boxplot are shown on the right axis.

Figure S10 Boxplots indicate the observed water temperature in Ecuador (red), Ethiopia (green) and Vietnam (blue) at which Chironomidae are found to be present (denoted by P on the left axis) and absent (denoted by A on the left axis). The sample sizes per boxplot are shown on the right axis.

Figure S11 Boxplots indicate the observed conductivity in Ecuador (red), Ethiopia (green) and Vietnam (blue) at which Chironomidae are found to be present (denoted by P on the left axis) and absent (denoted by A on the left axis). The sample sizes per boxplot are shown on the right axis.

Figure S12 Boxplots indicate the observed pH in Ecuador (red), Ethiopia (green) and Vietnam (blue) at which Chironomidae are found to be present (denoted by P on the left axis) and absent (denoted by A on the left axis). The sample sizes per boxplot are shown on the right axis.

Figure S13 Boxplots indicate the observed DO concentrations in Ecuador (red), Ethiopia (green) and Vietnam (blue) at which Chironomidae are found to be present (denoted by P on the left axis) and absent (denoted by A on the left axis). The sample sizes per boxplot are shown on the right axis.

Figure S14 Boxplots indicate the observed stream velocity in Ecuador (red), Ethiopia (green) and Vietnam (blue) at which Baetidae are found to be present (denoted by P on the left axis) and absent (denoted by A on the left axis). The sample sizes per boxplot are shown on the right axis.

Figure S15 Boxplots indicate the observed water temperature in Ecuador (red), Ethiopia (green) and Vietnam (blue) at which Baetidae are found to be present (denoted by P on the left axis) and absent (denoted by A on the left axis). The sample sizes per boxplot are shown on the right axis.

Figure S16 Boxplots indicate the observed conductivity in Ecuador (red), Ethiopia (green) and Vietnam (blue) at which Baetidae are found to be present (denoted by P on the left axis) and absent (denoted by A on the left axis). The sample sizes per boxplot are shown on the right axis.

Figure S17 Boxplots indicate the observed pH in Ecuador (red), Ethiopia (green) and Vietnam (blue) at which Baetidae are found to be present (denoted by P on the left axis) and absent (denoted by A on the left axis). The sample sizes per boxplot are shown on the right axis.

Figure S18 Boxplots indicate the observed DO concentrations in Ecuador (red), Ethiopia (green) and Vietnam (blue) at which Baetidae are found to be present (denoted by P on the left axis) and absent (denoted by A on the left axis). The sample sizes per boxplot are shown on the right axis.

Figure S19 Boxplots indicate the observed stream velocity in Ecuador (red), Ethiopia (green) and Vietnam (blue) at which Hydroptilidae are found to be present (denoted by P on the left axis) and absent (denoted by A on the left axis).

Figure S20 Boxplots indicate the observed water temperature in Ecuador (red), Ethiopia (green) and Vietnam (blue) at which Hydroptilidae are found to be present (denoted by P on the left axis) and absent (denoted by A on the left axis). The sample sizes per boxplot are shown on the right axis.

Figure S21 Boxplots indicate the observed conductivity in Ecuador (red), Ethiopia (green) and Vietnam (blue) at which Hydroptilidae are found to be present (denoted by P on the left axis) and absent (denoted by A on the left axis). The sample sizes per boxplot are shown on the right axis.

Figure S22 Bottom: Boxplots indicate the observed pH in Ecuador (red), Ethiopia (green) and Vietnam (blue) at which Hydroptilidae are found to be present (denoted by P on the left axis) and absent (denoted by A on the left axis).

Figure S23 Boxplots indicate the observed DO concentrations in Ecuador (red), Ethiopia (green) and Vietnam (blue) at which Hydroptilidae are found to be present (denoted by P on the left axis) and absent (denoted by A on the left axis). The sample sizes per boxplot are shown on the right axis.

Figure S24 Boxplots indicate the observed stream velocity in Ecuador (red), Ethiopia (green) and Vietnam (blue) at which Libellulidae are found to be present (denoted by P on the left axis) and absent (denoted by A on the left axis). The sample sizes per boxplot are shown on the right axis.

Figure S25 Boxplots indicate the observed water temperature in Ecuador (red), Ethiopia (green) and Vietnam (blue) at which Libellulidae are found to be present (denoted by P on the left axis) and absent (denoted by A on the left axis). The sample sizes per boxplot are shown on the right axis.

Figure S26 Boxplots indicate the observed conductivity in Ecuador (red), Ethiopia (green) and Vietnam (blue) at which Libellulidae are found to be present (denoted by P on the left axis) and absent (denoted by A on the left axis). The sample sizes per boxplot are shown on the right axis.

Figure S27 Boxplots indicate the observed pH in Ecuador (red), Ethiopia (green) and Vietnam (blue) at which Libellulidae are found to be present (denoted by P on the left axis) and absent (denoted by A on the left axis). The sample sizes per boxplot are shown on the right axis.

Figure S28 Boxplots indicate the observed DO concentrations in Ecuador (red), Ethiopia (green) and Vietnam (blue) at which Libellulidae are found to be present (denoted by P on the left axis) and absent (denoted by A on the left axis). The sample sizes per boxplot are shown on the right axis.

Figure S29 Boxplots indicate the observed stream velocity in Ecuador (red), Ethiopia (green) and Vietnam (blue) at which Leptophlebiidae are found to be present (denoted by P on the left axis) and absent (denoted by A on the left axis). The sample sizes per boxplot are shown on the right axis.

Figure S30 Boxplots indicate the observed water temperature in Ecuador (red), Ethiopia (green) and Vietnam (blue) at which Leptophlebiidae are found to be present (denoted by P on the left axis) and absent (denoted by A on the left axis). The sample sizes per boxplot are shown on the right axis.

Figure S31 Boxplots indicate the observed conductivity in Ecuador (red), Ethiopia (green) and Vietnam (blue) at which Leptophlebiidae are found to be present (denoted by P on the left axis) and absent (denoted by A on the left axis).

Figure S32 Boxplots indicate the observed pH in Ecuador (red), Ethiopia (green) and Vietnam (blue) at which Leptophlebiidae are found to be present (denoted by P on the left axis) and absent (denoted by A on the left axis). The sample sizes per boxplot are shown on the right axis.

Figure S33 Boxplots indicate the observed DO concentrations in Ecuador (red), Ethiopia (green) and Vietnam (blue) at which Leptophlebiidae are found to be present (denoted by P on the left axis) and absent (denoted by A on the left axis). The sample sizes per boxplot are shown on the right axis.

Figure S34 The probability of Chironomidae being present in relation to the stream velocity measured in Ecuador (red, solid), Ethiopia (green, dashed) and Vietnam (blue, dotdashed). The gray-colored ends of the response curves indicate extrapolation outside the observed physical-chemical range in the corresponding river basin.

Figure S35 The probability of Chironomidae being present in relation to the water temperature measured in Ecuador (red, solid), Ethiopia (green, dashed) and Vietnam (blue, dotdashed). The gray-colored ends of the response curves indicate extrapolation outside the observed physical-chemical range in the corresponding river basin.

Figure S36 The probability of Chironomidae being present in relation to the conductivity measured in Ecuador (red, solid), Ethiopia (green, dashed) and Vietnam (blue, dotdashed). The gray-colored ends of the response curves indicate extrapolation outside the observed physical-chemical range in the corresponding river basin.

Figure S37 The probability of Chironomidae being present in relation to the pH measured in Ecuador (red, solid), Ethiopia (green, dashed) and Vietnam (blue, dotdashed). The gray-colored ends of the response curves indicate extrapolation outside the observed physical-chemical range in the corresponding river basin.

Figure S38 The probability of Chironomidae being present in relation to dissolved oxygen (DO) concentration measured in Ecuador (red, solid), Ethiopia (green, dashed) and Vietnam (blue, dotdashed). The gray-colored ends of the response curves indicate extrapolation outside the observed physical-chemical range in the corresponding river basin.

Figure S39 The probability of Baetidae being present in relation to stream velocity measured in Ecuador (red, solid), Ethiopia (green, dashed) and Vietnam (blue, dotdashed). The gray-colored ends of the response curves indicate extrapolation outside the observed physical-chemical range in the corresponding river basin.

Figure S40 The probability of Baetidae being present in relation to water temperature measured in Ecuador (red, solid), Ethiopia (green, dashed) and Vietnam (blue, dotdashed). The gray-colored ends of the response curves indicate extrapolation outside the observed physical-chemical range in the corresponding river basin.

Figure S41 The probability of Baetidae being present in relation to conductivity measured in Ecuador (red, solid), Ethiopia (green, dashed) and Vietnam (blue, dotdashed). The gray-colored ends of the response curves indicate extrapolation outside the observed physical-chemical range in the corresponding river basin.

Figure S42 The probability of Baetidae being present in relation to pH measured in Ecuador (red, solid), Ethiopia (green, dashed) and Vietnam (blue, dotdashed). The gray-colored ends of the response curves indicate extrapolation outside the observed physical-chemical range in the corresponding river basin.

Figure S43 The probability of Baetidae being present in relation to dissolved oxygen (DO) concentration measured in Ecuador (red, solid), Ethiopia (green, dashed) and Vietnam (blue, dotdashed). The gray-colored ends of the response curves indicate extrapolation outside the observed physical-chemical range in the corresponding river basin.

Figure S44 The probability of Hydroptilidae being present in relation to stream velocity measured in Ecuador (red, solid), Ethiopia (green, dashed) and Vietnam (blue, dotdashed). The gray-colored ends of the response curves indicate extrapolation outside the observed physical-chemical range in the corresponding river basin.

Figure S45 The probability of Chironomidae being present in relation to the Hydroptilidae measured in Ecuador (red, solid), Ethiopia (green, dashed) and Vietnam (blue, dotdashed). The gray-colored ends of the response curves indicate extrapolation outside the observed physical-chemical range in the corresponding river basin.

Figure S46 The probability of Hydroptilidae being present in relation to the conductivity measured in Ecuador (red, solid), Ethiopia (green, dashed) and Vietnam (blue, dotdashed). The gray-colored ends of the response curves indicate extrapolation outside the observed physical-chemical range in the corresponding river basin.

Figure S47 The probability of Hydroptilidae being present in relation to pH measured in Ecuador (red, solid), Ethiopia (green, dashed) and Vietnam (blue, dotdashed). The gray-colored ends of the response curves indicate extrapolation outside the observed physical-chemical range in the corresponding river basin.

Figure S48 The probability of Hydroptilidae being present in relation to dissolved oxygen (DO) concentration measured in Ecuador (red, solid), Ethiopia (green, dashed) and Vietnam (blue, dotdashed). The gray-colored ends of the response curves indicate extrapolation outside the observed physical-chemical range in the corresponding river basin.

Figure S49 The probability of Libellulidae being present in relation to stream velocity measured in Ecuador (red, solid), Ethiopia (green, dashed) and Vietnam (blue, dotdashed). The gray-colored ends of the response curves indicate extrapolation outside the observed physical-chemical range in the corresponding river basin.

Figure S50 The probability of Libellulidae being present in relation to the water temperature measured in Ecuador (red, solid), Ethiopia (green, dashed) and Vietnam (blue, dotdashed). The gray-colored ends of the response curves indicate extrapolation outside the observed physical-chemical range in the corresponding river basin.

Figure S51 The probability of Libellulidae being present in relation to the conductivity measured in Ecuador (red, solid), Ethiopia (green, dashed) and Vietnam (blue, dotdashed). The gray-colored ends of the response curves indicate extrapolation outside the observed physical-chemical range in the corresponding river basin.

Figure S52 The probability of Libellulidae being present in relation to pH measured in Ecuador (red, solid), Ethiopia (green, dashed) and Vietnam (blue, dotdashed). The gray-colored ends of the response curves indicate extrapolation outside the observed physical-chemical range in the corresponding river basin.

Figure S53 The probability of Libellulidae being present in relation to dissolved oxygen (DO) concentration measured in Ecuador (red, solid), Ethiopia (green, dashed) and Vietnam (blue, dotdashed). The gray-colored ends of the response curves indicate extrapolation outside the observed physical-chemical range in the corresponding river basin.

Figure S54 The probability of Leptophlebiidae being present in relation to stream velocity measured in Ecuador (red, solid), Ethiopia (green, dashed) and Vietnam (blue, dotdashed). The gray-colored ends of the response curves indicate extrapolation outside the observed physical-chemical range in the corresponding river basin.

Figure S55 The probability of Leptophlebiidae being present in relation to the water temperature measured in Ecuador (red, solid), Ethiopia (green, dashed) and Vietnam (blue, dotdashed). The gray-colored ends of the response curves indicate extrapolation outside the observed physical-chemical range in the corresponding river basin.

Figure S56 The probability of Leptophlebiidae being present in relation to the conductivity measured in Ecuador (red, solid), Ethiopia (green, dashed) and Vietnam (blue, dotdashed). The gray-colored ends of the response curves indicate extrapolation outside the observed physical-chemical range in the corresponding river basin.

Figure S57 The probability of Leptophlebiidae being present in relation to pH measured in Ecuador (red, solid), Ethiopia (green, dashed) and Vietnam (blue, dotdashed). The gray-colored ends of the response curves indicate extrapolation outside the observed physical-chemical range in the corresponding river basin.

Figure S58 The probability of Leptophlebiidae being present in relation to dissolved oxygen (DO) concentration measured in Ecuador (red, solid), Ethiopia (green, dashed) and Vietnam (blue, dotdashed). The gray-colored ends of the response curves indicate extrapolation outside the observed physical-chemical range in the corresponding river basin.

Table S1 Seasonal differences per physical-chemical variable and per country. A p-value less than 0.05 demonstrates a significant difference between seasons for the mean physical-chemical variable that is considered for a specific country. Significant relations are indicated in bold.

Table S2 Country-wise differences per physical-chemical variable and per season. A p-value less than 0.05 demonstrates a significant difference between countries for the mean physical-chemical variable that is considered for a specific season. Significant relations are indicated in bold.

Acknowledgments

We would like to thank all people who contributed to the sampling campaigns.

Author Contributions

Conceived and designed the experiments: PG LDG AA THH. Performed the experiments: PB LDG STM AA THH. Analyzed the data: GE JDN OT. Contributed reagents/materials/analysis tools: PG LDG AA THH. Wrote the paper: GE JDN PB LDG STM AA THH PG OT.

References

1. Rosenberg DM, Resh VH (1993) Introduction to freshwater biomonitoring and benthic macroinvertebrates. In: Rosenberg DM, Resh VH, editors. Freshwater biomonitoring and benthic macroinvertebrates. Chapman and Hall, New York. pp. 1–9.
2. Kerans BL, Karr JR (1994) A benthic index of biotic integrity (B-IBI) for rivers of the Tennessee Valley. Ecol Appl 4: 768–785.
3. Cairns J, Pratt JR (1993) A history of biological monitoring using benthic macroinvertebrates. In: Rosenberg DM, Resh VH, editors. Freshwater Biomonitoring and Benthic Macroinvertebrates. Chapman and Hall, New York. pp. 10–27.
4. Molozzi J, Feio MJO, Salas F, Marques JOC, Callisto M (2013) Development and test of a statistical model for the ecological assessment of tropical reservoirs based on benthic macroinvertebrates. Ecol Indic 23: 155–165.
5. Chutter FM (1994) The rapid biological assessment of stream and river quality by means of the macroinvertebrate community in South Africa. In: Uys MC, editor. Classification of rivers, and environmental health indicators. Proceedings of a joint South African/Australian Workshop. Cape Town 7-11/2/1994. Africa. Water Research Commission Report No. TT 63/94, Pretoria, South Africa. pp. 217–234.
6. Gabriels W, Lock K, De Pauw N, Goethals PLM (2010) Multimetric Macroinvertebrate Index Flanders (MMIF) for biological assessment of rivers and lakes in Flanders (Belgium). Limnologica 40: 199–207.
7. Chandler JR (1970) A biological approach to water quality management. Wat Pollut Control 69: 415–422.
8. Armitage PD, Moss D, Wright JF, Furse MT (1983) The performance of a new biological water quality score system based on macroinvertebrates over a wide range of unpolluted running-water sites. Water Res 17: 333–347.
9. Alba-Tercedor J, Sanchez-Ortega A (1988) Um metodo rapido y simple para evaluar la calidade biologica de las aguas corrientes basado em el de Hellawell (1978). Limnética 4: 51–56.
10. Couceiro SRM, Hamada N, Forsberg BR, Pimentel TP, Luz SLB (2012) A macroinvertebrate multimetric index to evaluate the biological condition of streams in the Central Amazon region of Brazil. Biol Indic 18: 118–125.
11. Helson JE, Williams DD (2013) Development of a macroinvertebrate multimetric index for the assessment of low-land streams in the neotropics. Ecol Indic 29: 167–178.
12. Dudgeon D, Arthington AH, Gessner MO, Kawabata ZI, Knowler DJ, et al. (2006) Freshwater biodiversity: importance, threats, status and conservation challenges. Biol Rev 81: 163–182.
13. Marques M, Barbosa F (2001) Biological quality of waters from an impacted tropical watershed (middle Rio Doce basin, southeast Brazil), using benthic macroinvertebrate communities as an indicator. Hydrobiologia 457: 69–76.
14. Thorne R, Williams P (1997) The response of benthic macroinvertebrates to pollution in developing countries: a multimetric system of bioassessment. Freshw Biol 37: 671–686.
15. Moya N, Tomanova S, Oberdorff T (2007) Initial development of a multi-metric index based on aquatic macroinvertebrates to assess streams condition in the Upper Isiboro-Sécure Basin, Bolivian Amazon. Hydrobiologia 589: 107–116.
16. Tomanova S, Goitia E, Helešic J (2006) Trophic levels and functional feeding groups of macroinvertebrates in neotropical streams. Hydrobiologia 556: 251–264.
17. Dominguez-Granda L, Lock K, Goethals PLM (2011) Using multi-target clustering trees as a tool to predict biological water quality indices based on benthic macroinvertebrates and environmental parameters in the Chaguana watershed (Ecuador). Ecol Inform 6: 303–308.
18. Ambelu A, Lock K, Goethals PLM (2010) Comparison of modelling techniques to predict macroinvertebrate community composition in rivers of Ethiopia. Ecol Inform 5: 147–152.
19. Mereta ST, Boets P, Ambelu-Bayih A, Malu A, Ephrem Z, et al. (2012) Analysis of environmental factors determining the abundance and diversity of

macroinvertebrate taxa in natural wetlands of Southwest Ethiopia. Ecol Inform 7: 52–61.
20. Hoang TH, Lock K, Mouton A, Goethals PLM (2010) Application of classification trees and support vector machines to model the presence of macroinvertebrates in rivers in Vietnam. Ecol Inform 5: 140–146.
21. Nguyen HH, Everaert G, Gabriels W, Hoang TH, Goethals PLM (2014) A multimetric macroinvertebrate index for assessing the water quality of the Cau river basin in Vietnam. Limnologica 45: 16–23.
22. Ríos-Touma B, Encalada AC, Prat Fornells N (2011) Macroinvertebrate Assemblages of an Andean High-Altitude Tropical Stream: The Importance of Season and Flow. Int Rev Hydrobiol 96: 667–685.
23. R Core Team (2012) R: A language and environment for statistical computing. R Foundation for Statistical Computing, Vienna, Austria. ISBN 3-900051-07-0. Available: http://www.R-project.org/.
24. Zuur AF, Ieno EN, Walker NJ, Saveliev AA, Smith GM (2009) Mixed effects models and extensions in ecology with R. New York: Springer Science+Business Media, LLC 2009. 292p.
25. Pearce JL, Ferrier S (2000) Evaluating the predictive performance of habitat models developed using logistic regression. Ecol Model 133: 225–245.
26. Rushton SP, Ormerod SJ, Kerby G (2004) New paradigms for modelling species distributions? J App Ecol 41: 193–200.
27. Holguin-Gonzalez JE, Everaert G, Boets P, Galvis A, Goethals PLM (2013) Development and application of an integrated ecological modelling framework to analyze the impact of wastewater discharges on the ecological water quality of rivers. Environ Modell Softw 48: 27–36.
28. Liang KY, Zeger SL (1986) Longitudinal data analysis using generalized linear models. Biometrika 73: 13–22.
29. Horton NJ, Bebchuk JD, Jones CL, Lipsitz SR, Catalano PJ, et al. (1999) Goodness-of-fit for GEE: an example with mental health service utilization. Stat Med 18: 213–222.
30. Mouton AM, De Baets B, Van Broekhoven E, Goethals PLM (2012) Prevalence-adjusted optimisation of fuzzy models for species distribution. Ecol Model 220: 1776–1786.
31. Graham MH (2003) Confronting multicollinearity in ecological multiple regression. Ecology 84: 2809–2815.
32. Mereta ST, Boets P, De Meester L, Goethals PLM (2013) Development of a multimetric index based on benthic macroinvertebrates for the assessment of natural wetlands in Southwest Ethiopia. Ecol Indic 29: 510–521.
33. Armitage PD, Cranston P, Pinder LC (1995) The Chironomidae: The biology and Ecology of non-biting midges. London: Chapman and Hall. 572p.
34. Connolly NM, Crossland MR, Pearson RG (2004) Effect of low dissolved oxygen on survival, emergence, and drift of tropical stream macroinvertebrates. J N Am Benthol Soc 23: 251–270.
35. Tixier G, Felten V, Guérold F (2009) Life cycle strategies of Baetis species (Ephemeroptera, Baetidae) in acidified streams and implications for recovery. Fund Appl Limnol 174: 227–243.
36. Hynes HBN (1970) The ecology of running waters. Liverpool: Liverpool University Press. 555p.
37. D'Heygere T, Goethals PLM, De Pauw N (2003) Use of genetic algorithms to select input variables in decision tree models for the prediction of benthic macroinvertebrates. Ecol Model 160: 291–300.
38. Chapman LJ, Schneider KR, Apodaca C, Chapman CA (2004) Respiratory ecology of macroinvertebrates in a swamp-river system of East Africa. Biotrop 36: 572–585.
39. Dakou E, D'heygere T, Dedecker AP, Goethals PLM, Dimitriadou ML, et al. (2007) Decision tree models for prediction of macroinvertebrate taxa in the river Axios (Northern Greece). Aquat Ecol 41: 399–411.
40. Rawi CSMD, Al-Shami SA, Madrus MR, Ahmad AH (2013) Biological and ecological diversity of aquatic macroinvertebrates in response to hydrological and physicochemical parameters in tropical forest streams of Gunung Tebu, Malaysia: implications for ecohydrological assessment. Ecohydrology 7: 496–507.
41. Chapman G (1986) Ambient water quality criteria for dissolved oxygen. EPA 440/5-86-003. U.S. Environmental Protection Agency, Washington, DC.
42. Al-Shami SA, Heino J, Che Salmah MR, Abu Hassan A, Suhaila AH, et al. (2013) Drivers of beta diversity of macroinvertebrate communities in tropical forest streams. Freshw Biol 58: 1126–1137.
43. Blanchette M, Pearson R (2012) Macroinvertebrate assemblages in rivers of the Australian dry tropics are highly variable. Freshw Sci 31: 865–881.
44. Kasangaki A, Chapman LJ, Balirwa J (2008) Land use and the ecology of benthic macroinvertebrate assemblages of high-altitude rainforest streams in Uganda. Freshw Biol 53: 681–697.
45. Melo AS (2009) Explaining dissimilarities in macroinvertebrate assemblages among stream sites using environmental variables. Zoologia 26: 79–84.
46. Flecker AS, Feifarek B (1994) Disturbance and the temporal variability of invertebrate assemblages in two Andean streams. Freshw Biol 31: 131–142.
47. Bonada N, Rieradevall M, Dallas H, Davis J, Day J, et al. (2008) Multi-scale assessment of macroinvertebrate richness and composition in Mediterranean-climate rivers. Freshw Biol 53: 772–788.
48. Zuellig RE, Schmidt TS (2012) Characterizing invertebrate traits in wadeable streams of the contiguous US: differences among ecoregions and land uses. Freshw Sci 31: 1042–1056.
49. Turak E, Flack LK, Norris RH, Simpson J, Waddell N (1999) Assessment of river condition at a large spatial scale using predictive models. Freshw Biol 41: 283–298.
50. Everaert G, Boets P, Lock K, Dzeroski S, Goethals PLM (2011) Using classification trees to analyze the impact of exotic species on the ecological assessment of polder lakes in Flanders, Belgium. Ecol Model 2011: 2202–2212.
51. Everaert G, Pauwels IS, Boets P, Verduin E, de la Haye MAA, et al. (2013) Model-based evaluation of ecological bank design and management in the scope of the European Water Framework Directive. Ecol Eng 53: 144–152.
52. Pearce JL, Boyce MS (2006) Modelling distribution and abundance with presence-only data. J Appl Ecol 43: 405–412.
53. Howard C, Stephens PA, Pearce-Higgins JW, Gregory RD, Willis SG (2014) Improving species distribution models: the value of data on abundance. Methods Ecol Evol 5: 506–513.
54. Fukuda S, Mouton AM, De Baets B (2012) Abundance versus presence/absence data for modelling fish habitat preference with a genetic Takagi-Sugeno fuzzy system. Environ Monit Assess 184: 6159–6171.
55. Vaughan IP, Ormerod SJ (2005) The continuing challenges of testing species distribution models. J Appl Ecol 42: 720–730.
56. Boets P, Thas O, Van de Vijver E, Lock K, Töpke K (2013) Relating taxonomy-based traits of macroinvertebrates with river sediment quality based on basic and zero-inflated Poisson models. Ecol Inform 18: 49–60.
57. Menezes S, Baird DJ, Soares AMVM (2010) Beyond taxonomy: a review of macroinvertebrate trait-based community descriptors as tools for freshwater biomonitoring. J Appl Ecol 47: 711–719.
58. Dominguez-Granda L (2007) Indices based on macroinvertebrate communities for assessment of the quality of the Chaguana river in Ecuador. PhD dissertation, Ghent University, Gent, Belgium. 233p.
59. Matamoros D (2004) Predicting river concentrations of pesticides from banana plantations under data-poor conditions. PhD dissertation, Ghent University, Gent, Belgium. 206p.
60. MONRE (2006) National Strategy for Water Resources Up To 2020, Cultural and Information Publish House, Hanoi, Vietnam.
61. Demissie TA, Saathoff F, Sileshi YA, Gebissa Y (2013) Climate change impacts on the streamflow and simulated sediment flux to Gilgel Gibe 1 hydropower reservoir – Ethiopia. European International Journal of Science and Technology 2: 64–77.
62. Uhlenbrook S, Mohamed Y, Gragne AS (2010) Analyzing catchment behavior through catchment modelling in the Gilgel Abay, Upper Blue Nile River Basin, Ethiopia. Hydrol Earth Syst Sci 14: 2153–2165.
63. Broothaerts N, Kissi E, Poesen J, Van Rompaey A, Getahun K, et al. (2012) Spatial patterns, causes and consequences of landslides in the Gilgel Gibe catchment, SW Ethiopia. Catena 97: 127–136.
64. Colombo D, Maran S (2004) Gilgel Gibe II hydroelectric project environmental impact assessment main report. Report A4511403 by Centro Elettrotecnico Sperimentale Italiano (CESI), 135p.

15

Final Opportunity to Rehabilitate an Urban River as a Water Source for Mexico City

Marisa Mazari-Hiriart[1]*, Gustavo Pérez-Ortiz[1,2], María Teresa Orta-Ledesma[3], Felipe Armas-Vargas[4], Marco A. Tapia[2,5], Rosa Solano-Ortiz[5], Miguel A. Silva[6], Isaura Yañez-Noguez[3], Yolanda López-Vidal[7], Carlos Díaz-Ávalos[8]

1 Laboratorio Nacional de Ciencias de la Sostenibilidad, Departamento de Ecología de la Biodiversidad, Instituto de Ecología, Universidad Nacional Autónoma de México, México, D.F., México, **2** Departamento de Biología, Facultad de Ciencias, Universidad Nacional Autónoma de México, México, D.F., México, **3** Coordinación de Ingeniería Ambiental, Instituto de Ingeniería, Universidad Nacional Autónoma de México, México, D.F., México, **4** Posgrado en Ciencias de la Tierra, Instituto de Geología, Universidad Nacional Autónoma de México, México, D.F., México, **5** Posgrado en Ciencias Biológicas, Facultad de Ciencias-Instituto de Ecología, Universidad Nacional Autónoma de México, México, D.F., México, **6** Posgrado de Ciencias Bioquímicas, Departamento de Ingeniería Celular y Biocatálisis, Instituto de Biotecnología, Universidad Nacional Autónoma de México, Cuernavaca, Morelos, México, **7** Programa de Inmunología Molecular Microbiana, Departamento de Microbiología y Parasitología, Facultad de Medicina, Universidad Nacional Autónoma de México, México, D.F., México, **8** Departamento de Probabilidad y Estadística, Instituto de Investigaciones en Matemáticas Aplicadas y en Sistemas, Universidad Nacional Autónoma de México, México, D.F., México

Abstract

The aim of this study was to evaluate the amount and quality of water in the Magdalena-Eslava river system and to propose alternatives for sustainable water use. The system is the last urban river in the vicinity of Mexico City that supplies surface water to the urban area. Historical flow data were analyzed (1973–2010), along with the physicochemical and bacteriological attributes, documenting the evolution of these variables over the course of five years (2008–2012) in both dry and rainy seasons. The analyses show that the flow regime has been significantly altered. The physicochemical variables show significant differences between the natural area, where the river originates, and the urban area, where the river receives untreated wastewater. Nutrient and conductivity concentrations in the river were equivalent to domestic wastewater. Fecal pollution indicators and various pathogens were present in elevated densities, demonstrating a threat to the population living near the river. Estimates of the value of the water lost as a result of mixing clean and contaminated water are presented. This urban river should be rehabilitated as a sustainability practice, and if possible, these efforts should be replicated in other areas. Because of the public health issues and in view of the population exposure where the river flows through the city, the river should be improved aesthetically and should be treated to allow its ecosystem services to recover. This river represents an iconic case for Mexico City because it connects the natural and urban areas in a socio-ecological system that can potentially provide clean water for human consumption. Contaminated water could be treated and reused for irrigation in one of the green areas of the city. Wastewater treatment plants and the operation of the existing purification plants are urgent priorities that could lead to better, more sustainable water use practices in Mexico City.

Editor: Judi Hewitt, University of Waikato (National Institute of Water and Atmospheric Research), New Zealand

Funding: The authors would like to acknowledge the financial support of Gobierno del Distrito Federal, Secretaría de Medio Ambiente in conducting the project Plan Maestro de Manejo Integral y Aprovechamiento Sustentable de la Cuenca del Río Magdalena del Distrito Federal, through the Programa Universitario de Estudios sobre la Ciudad, Universidad Nacional Autónoma de México (2008–2009), and the project Sistema de Indicadores para el Rescate de los Ríos Magdalena y Eslava, Distrito Federal through the Programa Universitario de Medio Ambiente, UNAM (2009–2010). As well as the support of the Dirección General de Asuntos del Personal Académico through the Programa de Apoyo a Proyectos de Investigación e Innovación Tecnológica, PAPIIT IN219809. The funders had no role in study design, data collection and analysis, decision to publish, or preparation of the manuscript.

Competing Interests: The authors have declared that no competing interests exist.

* Email: mazari@unam.mx

Introduction

In megacities with populations exceeding ten million inhabitants, water is a scarce resource that is in high demand. The exploration and application of new, more sustainable water use and reuse approaches are urgently needed. Water managers must change the perspectives and policies that assume that water can only be used once by considering the benefits obtained from the natural environment surrounding the urban areas.

In emerging economies, cities grow before sufficient hydraulic infrastructure has been developed to handle the water supply and wastewater disposal requirements [1]. This lack of infrastructure significantly affects the surface and groundwater systems, making waterways vulnerable to pollution and leading to deleterious effects on water sources and human health. There is a need for more adequate approaches regarding processes that occur at the basin level, in addition to directly considering hydrological ecosystem services, such as water provision and water quality.

During the evolution of what is now the Mexico City Metropolitan Area (MCMA), lakes and rivers have been transformed into drains or converted into sewers in the vicinity of modern avenues, and freeways transit over piped waterways [2]. The hydraulic system of the Basin of Mexico, where the MCMA and its 22 million inhabitants are located, has been irreversibly altered. Of the 45 rivers in this region, most are piped to avoid flooding and unsanitary conditions.

In the Basin of Mexico, the current water demand is 77 m^3/s for first use, of which 71% is groundwater extracted through wells, 2% is derived from springs and surface water, 21% is from the Cutzamala surface water system, and 6% is from the Lerma groundwater system [3,4]. The subject of this research, the Magdalena-Eslava river system, represents the most important surface water source for human use in Mexico City.

Rivers flow from their origins in high areas to low-lying areas, exposing these water bodies to the integrated effects of human activities along their course and throughout the basin landscape. Anthropogenic impacts have impaired the ability of many river ecosystems to provide the goods and services upon which society depends. Watercourses may be degraded and rendered unable to support healthy aquatic communities such that natural processes are hampered and social value is lost.

Balancing the interactions between natural and constructed systems in urban areas is crucial for the future supply of water for large human settlements. Interest in river rehabilitation and restoration exists at the global level. River restoration is defined as an improvement in the integrity of a river through human intervention. The aim of river restoration is either to recover the hydrologic and ecological processes or to achieve rehabilitation, which means recovering a system that has not returned to its original state or is as healthy as it would have been if it had been fully restored [5].

The Magdalena-Eslava river system is one of the last non-piped rivers remaining in the MCMA. Given its contribution to the water supply in its upper reaches and the potential to recover some of the environmental services provided by the system, there is significant interest in the possibility of improving the water quality in the lower parts of its course. The aim of this study was to evaluate the Magdalena-Eslava river system in terms of the amount and quality of water, based on physicochemical and microbiological attributes that it offers the community. Accordingly, the study was performed based on historical and current data. As one of the last urban rivers, this study provides an opportunity to pose alternatives for more efficient water use.

Study area description

The study area, which is located in Mexico City, comprises the Magdalena and Eslava rivers. The river system covers two climatic regions. The lower area, which reaches to 3,050 masl, presents a temperate sub-humid climate with a summer rainy season and average temperatures that oscillate between 12 and 18°C. The upper basin has a semi-cold climate and a rainy season during the summer in areas higher than 3,050 masl, with an annual mean temperature between 5 and 12°C [6]. The rainy season ranges from May to October, with a monthly precipitation in excess of 250 mm in July [7]. The annual precipitation in most of the area is between 1,200 and 1,500 mm [8].

The Magdalena River originates in *Sierra de Las Cruces* from springs and runoff from the *La Palma*, *El Gavilán* and *El Muñeco* mountains in a location called *Cieneguillas*. The Magdalena River Basin is actually a sub-basin of the Basin of Mexico (hereafter called the Magdalena River Basin) and belongs to what is legally known as the *Zona Protectora Forestal Cañada de Contreras*, which was declared in 1932. The protected zone includes a fringe of 12 km from the river's origin, with a 500 m buffer on each side of the river, as declared in 1947 [9,10].

The Magdalena river basin extends from 2,570 to 3,870 masl [11], covering an area of 29.80 km^2. The Eslava River, which is the basin's largest tributary, flows through an area of 24.05 km^2, with an altitude descending to 2,510 masl at the urban limit.

The Magdalena River is 28.2 km long and flows 14.8 km through a natural forested area known as the Conservation Soil, with an additional 13.4 km crossing through the urban area. In the urban section, 8.8 km of river flows as an open canal, and 4.6 km is piped [12]. The Eslava River is 13.7 km long, with 10.96 km in the Conservation Soil and 2.74 km in urban areas [12]. The basins of the Magdalena and Eslava rivers are shown in Figure 1.

The Magdalena Basin, known as *Cañada de Contreras* in *Sierra de las Cruces*, is the only drainage of significant magnitude [13] within the Basin of Mexico and is primarily classified as temperate sub-humid. Despite its location in a peri-urban area, the Magdalena Basin serves as an important refuge for biodiversity and is home to 1,175 species, of which 212 are considered useful and 39 are listed in the at-risk category [14].

Three vegetation communities are present in the area, according to Ávila-Akerberg (2009) [8] and Nava (2003) [15]. The dominant vegetation types are *Abies religiosa* forest (46%) from 2,750 to 3,500 masl, *Pinus hartwegii* forest (29%) from 3,420 to 3,800 masl, *Quercus* (8.3%) and mixed forest (1.3%) from 2,600 to 3,370 masl, grassland (7.2%), and cloud forest (0.2%). Each type of forest produces different rates of runoff, as calculated by Jujnovsky et al. (2010) [16]. The *Abies* area produces 10,944,800 m^3 of water per year, the *Pinus* forest produces 6,878,000 m^3 of water per year, and the mixed oak forest generates 3,217,500 m^3 of water per year.

The natural vegetation cover has been altered as a result of human activities, resulting in an increase in erosion and the reduction of water and soil humidity retention. Therefore, water flows rapidly during flood events in the lower parts of the basin [13]. The forested area southwest of Mexico City represents the most important surface water source and groundwater recharge area. This area supplies water and regulates both flood events and the microclimate in the MCMA. Urban expansion represents a threat to the Conservation Soil area, which currently provides ecosystem services, such as fresh water.

At the request of the City, the *Gobierno del Distrito Federal* through its Environmental Ministry (*Secretaría del Medio Ambiente*), the two main universities located in Mexico City, the National Autonomous University of Mexico (UNAM) and the Metropolitan Autonomous University (UAM), generated a Master Plan to Rescue the Magdalena and Eslava Rivers [12]. This plan was developed by an interdisciplinary team that included engineers, architects, ecologists, biologists, architects, geographers, and sociologists, who proposed to rehabilitate the river as an urban area that could be used as a community recreation area while recovering the water quality and the watercourse running through the city [12]. A second project developed a system of indicators to measure the improvements resulting from the actions taken as a result of the Master Plan [17], which was not completely implemented.

The topic of river ecosystem recovery and rehabilitation has grown in relevance, receiving increasing attention in the last two decades. From an urban perspective, rivers are crucial for water supply, flood and drought regulation, storm-water management, water quality improvement, and erosion control; additionally, rivers serve as urban wildlife refuges. Overall, rivers can serve a range of purposes from supplying water to offering locations for social and recreational activities; hence, rivers often account for a significant fraction of the few recreational green areas in large cities [18,19].

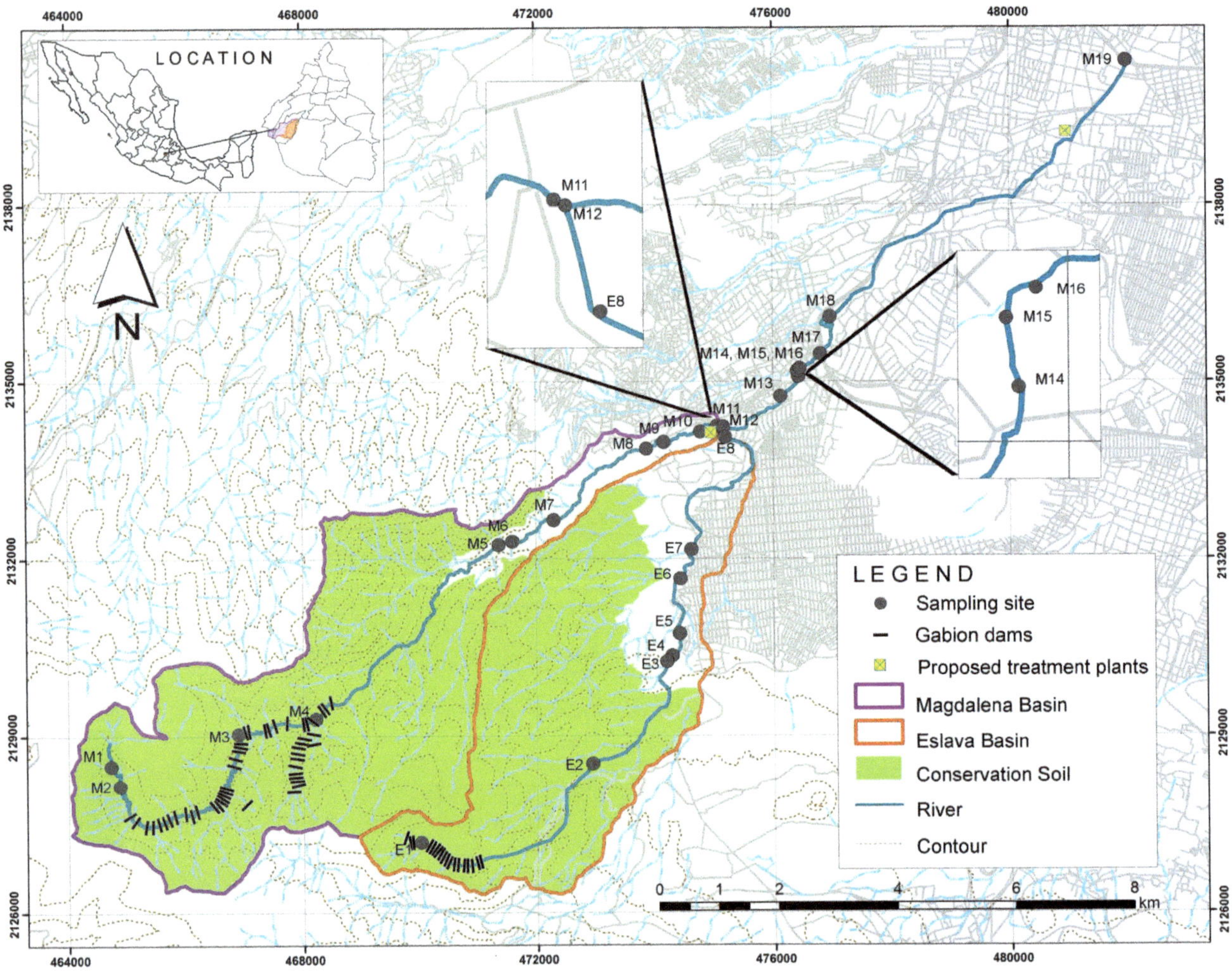

Figure 1. Study area, including the Magdalena and Eslava rivers within the Basin of Mexico and the Mexico City Metropolitan Area.

Methods

To determine the water flow, historical data were collected and analyzed (1973–2010); this approached allowed us to discern important variations due to seasonal or management effect in the river system. Based on current water quality measurements (2008–2012), we describe the actual condition of the river based on physicochemical and microbiological variables. The methods described in this section were applied.

Water flow

A flow analysis was conducted based on historical data recorded at the Santa Teresa hydrometric station (26440) to understand the water quantity and variability of flow at the only existing station of the National Water Commission. This hydrometric station is located after the junction of the Magdalena and Eslava rivers. As mentioned above, the Eslava is the main tributary of the Magdalena River, which also receives inflows from additional small creeks and natural run-off, combined with the discharge of wastewater from the surrounding urban area.

Daily information was obtained from the official flow surveys recorded at the Santa Teresa hydrometric station (26440) and was published in the National Data Bank of Surface Water [20] of the National Water Commission (*Comisión Nacional del Agua*). The data were analyzed for consistency and to detect and correct deviations from the historic series based on the double mass curve method [21,22]. The series of daily data spanning 38 years were then plotted, and two periods for the flow regime were identified.

We calculated the average monthly flow for each regime, i.e., the natural and the regulated areas, so we could compare the regulating effect of water infrastructure on the river. Finally, we plotted the average, minimum and maximum annual flows to observe the river's evolution over time.

Experimental design

After a general survey of the area, sampling sites for water quality evaluation were selected at places with little human or animal influence in the natural area and at others with inputs of urban runoff and wastewater in the urban area. Based on an initial diagnosis performed in 2008 and 2009 [12,17,23], the sampling locations were evaluated for redundancy in terms of the information provided. We computed a correlation matrix for the 19 sampling locations. After 2009, the sampling protocols included ten sampling sites. If two locations had a Pearson's correlation above 0.6 and were neighbors of each other, one was eliminated. This procedure allowed us to reduce the number of sampling locations from 19 to eight stations. We then added two sampling locations a few meters upriver of the intake of the water treatment plants that are located along the river. These ten stations were sampled from 2010 to 2012. The sampling covered the dry and rainy seasons, as shown in Table 1. The selected sampling sites, identified as representative of the Magdalena and Eslava rivers, are depicted in Figure 1. Fieldwork permission was given by the *Secretaría del Medio Ambiente* (SMA) and by the *Comisión de Recursos Naturales* (CORENA), the local authority that belongs to the *Gobierno del Distrito Federal* (GDF).

The behaviors of the two rivers were plotted independently, where the Magdalena River (main course) is presented with a five-year data set (2008–2012), while its tributary, the Eslava River, is shown with a two-year data set (rainy season 2010, 2011–2012) in Table 1.

Physicochemical analyses

As part of a long-term water quality monitoring program in the Magdalena Eslava river system, the attributes of temperature (°C), pH, electric conductivity (µS/cm), turbidity (NTU), total dissolved solids (TDS, mg/L), dissolved oxygen (DO, mg/L), DO saturation percentage, total phosphorus (total P, mg/L), orthophosphate (P-PO_4, mg/L), total nitrogen (total N, mg/L), ammonia (N-NH_3, mg/L), and nitrates (N-NO_3, mg/L) were measured.

The physicochemical variables of temperature, pH, electric conductivity, turbidity, total dissolved solids and dissolved oxygen were measured *in situ* over the course of five years. All measurements were made using a Multiparameter Water Quality Sonde YSI 6600-M (Yellow Springs, OH, USA). Five-hundred-milliliter water samples were taken and analyzed in triplicate for nutrients using a spectrophotometer HACH Model DR2400 (Loveland, CO, USA) following the HACH manual [24]. Total P was measured following the method of PhosVer3 by acid digestion USEPA (0.06 to 3.50 mg/L P) and the molybdovanadate method (1 to 100 mg/L PO_4^{3-}). Orthophosphate was measured using the amino acid method (0.23 to 30.00 mg/L PO_4^{3-}), and ascorbic acid was measured using PhosVer3 USEPA (0.02 to 2.50 mg/L PO_4^{3-}). Total N was measured using the persulfate digestion method (10 to 50 mg/L N) and (0.5 to 25.0 mg/L N). Ammonia was measured following the silicate method (0.02 to 2.50 mg/L NH_3-N), and nitrates were measured using the cadmium reduction method USEPA (0.01 to 0.50 mg/L NO_3-N) [24].

Microbiological analyses

Samples for microbiological analyses were taken at the same time and location as those collected for the physicochemical analyses.

Bacteria counts. One-liter samples for bacteriological analyses were collected from selected sites using polypropylene sterile flasks. Samples taken in triplicate were transported and stored at 4°C according to the American Public Health Association standard procedures [25,26]. Microbiological samples were processed within 24 h of collection. All samples were analyzed in triplicate, and the results were reported in colony-forming units (CFUs) [27,28].

Two bacteriological variables, including fecal coliform (FC, UFC/100 mL) and fecal enterococci (FE, UFC/100 mL), were measured constantly during the period from 2008–2012. Total coliforms (TC, UFC/100 mL) were determined only for the period 2008–2009. Positive samples were processed to identify the pathogenic and opportunistic bacteria.

Water samples were analyzed following the standard membrane filtration procedures to enumerate three bacterial types, namely, FC, FE, and TC. Membrane filters (0.45 µm cellulose acetate, Millipore MF type HA) were placed on M-FC agar for FC, KF Streptococcus agar for streptococci/enterococci, and m-Endo agar MF for total coliforms. Incubation was performed with a WTB Binder brand incubator at 35°C for 24 h for enterococci and total coliforms and at 44.5°C for 24 h for fecal coliforms, according to APHA [25,26]. The results are reported as colony forming units (CFUs).

Bacteria identification. Gram-stain and biochemical tests were performed to identify bacteria using a semi-automatic DADE MicroScan, AutoSCAN-4 (DADE International. West Sacramento, CA) [26]. Positive samples from water filtration isolates were selected based on morphology. Five colonies of each specific morphology were selected and identified by the MicroScan Auto SCAN-4 DADE (West Sacramento, CA).

Additional microbiological variables were analyzed during the period 2008–2009 and during the rainy season of 2010. We evaluated these variables because of their importance and relation to gastrointestinal diseases as a result of their prevalence in water in Mexico. These microorganisms included coliphage, enterovirus (EV), adenovirus F human serotypes 40 and 41 (AdV F), *Cryptosporidium parvum* oocysts and *Giardia lamblia* cysts.

Table 1. The sampling design for the Magdalena-Eslava river system.

Variable	Dry season	Rainy season
Magdalena River		
Physicochemical	2008, 2009, 2010, 2012	2008, 2009, 2010, 2011, 2012
Indicator bacteria	2008, 2009, 2010, 2012	2008, 2009, 2010, 2011, 2012
Pathogens	2008, 2009	2008, 2009, 2010
Coliphages	2008, 2009	2008, 2009
Virus	2008, 2009	2008, 2009
Protozoa	2008, 2009	2008, 2009
Eslava River		
Physicochemical	2011, 2012	2010, 2011, 2012
Indicator bacteria	2011, 2012	2010, 2011, 2012
Pathogens		2010

Coliphage. Coliphages were quantified in duplicate following the double-layer agar method based on the International Standard Water quality-Detection and enumeration of bacteriophages (ISO 10705-1, 1995) [29]. The plaques were quantified after the incubation period, and the results were reported in plaque forming units per milliliter (PFU/mL) [30].

The method detection limit was determined following the double-layer agar technique. Serial dilutions were performed using an MS2 phage with a known titer of 10^8 PFU/mL and *E. coli* K12– Hfr 3000 (ATCC 23631) as the host bacteria. The detection limit for this method was determined to be between 10 and 1 PFU [31].

The coliphage quantification with the *E. coli* Hs (pFamp) as host bacteria was performed in duplicate using the adaptation USEPA "Method 1601: Male-specific (F+) and Somatic Coliphage in Water by Two-step Enrichment Procedure)" [29,30].

Virus determination. For the detection of viruses and protozoa, 10-L water samples collected from the natural sites were concentrated by ultrafiltration following the method of Polaczyk et al. (2008) [32].

Enterovirus (EV) quantification was performed in duplicate on the water samples using RNA extraction techniques based on the RNA viral QIAamp (QIAGEN) mini kit. To quantify the genome number of EV, the basic protocol reported by Monpoeho et al. (2000) [33] was followed.

To obtain the AdV F genome number, the protocol by Xagoraraki et al. (2007) [34] was followed.

Protozoa quantification. *Cryptosporidium parvum* and *Giardia lamblia* were detected and quantified by immunofluorescence microscopy.

One milliliter of each sample (direct for wastewater or concentrated for clean samples) was analyzed in single samples using an indirect immunofluorescence technique in liquid phase according to Method 1623 [35], as modified by Rangel-Martínez [36] and Tapia-Palacios [37]. To count the (oo)cysts in the samples, an Axiostar Plus (Carl Zeiss, Gottingen, Germany) fluorescence microscope was used. Total scans of each slide were performed at 20X and 40X and, in special cases, at 100X for confirmation.

Statistical Analyses

The data gathered during the study were analyzed in a spatial and temporal context. Because the sampling locations were located along river courses, we assumed that the spatial coordinates were one-dimensional. In this coordinate system, we identified each site by a site code and by the distance from the first sampling location in the uppermost part of each river. We fitted generalized linear models (GLMs) with the adequate link function for each attribute [38]. For example, for each variable whose histogram clearly skewed to the left, we fitted GLMs with Gamma error and a logarithmic link function. These analyses were performed to explore the association between the physicochemical and biological variables, using site, season and year as covariates. Such analysis allowed us to detect significant effects of the covariates in each of the attributes considered in the analysis. We also created interaction plots to graphically depict the spatial behavior of the values of the variables along the downhill path of the Magdalena and Eslava rivers and at their confluence. Separating the courses in this way allowed us to compare the Magdalena and Eslava rivers before their merging point and to detect whether differences were present in the water quality variables relative to downriver points.

Results and Discussion

Water flow

Despite their relatively small size, the Magdalena River and its main tributary, the Eslava River, provide water to part of the southern area of Mexico City and are the main surface water source for the city, representing more than half of the surface water used. Several studies have been conducted to estimate the amount of water provided by the Magdalena River. In the 1950s, when water sources for the city were investigated, the Magdalena River registers showed a mean flow of 0.33 m^3/s or 10,400,000 m^3/yr [39]. The mean water volume for the Magdalena River was reported in 1975 as 6,970,000 m^3, equivalent to 0.22 m^3/s, and that for the Eslava River was 3,578,000 m^3, equivalent to 0.11 m^3/s [40]. These volumes total 0.33 m^3/s, with a total volume for potential use of 10,548,000 m^3/yr.

A first view of the hydrologic ecosystem services in terms of water quantity and quality with runoff estimates per environmental unit gives an annual production for 2002–2003 of 21,538,250 m^3, equivalent to 0.67 m^3/s, which is similar to the 0.58 m^3/s reported at the hydrometric station in 1999. From 1999–2001 [41], the average flow was 0.70 m^3/s, of which 81% corresponded to baseflow and 19% to surface runoff, providing water for approximately 118,237 inhabitants. In a complementary study, Jujnovsky et al. (2012) [42] assessed the water supply from the Magdalena River Watershed, determining the water balance based on the SWAT model, and obtained a water provision of 18,400,000 m^3/yr. The baseflow contribution accounted for 85%, while the surface runoff was approximately 15%, providing drinking water for approximately 78,500 inhabitants and being capable of supplying 153,203 potential beneficiaries.

Natural and regulated flow regimes. Our study shows the mean annual flow in the hydrometric station to be 0.78 m^3/s, with contributions of 0.45 m^3/s (57%) from the Magdalena River and 0.34 m^3/s (43%) from the Eslava River. The daily regimen is shown in Figure 2, based on historical data from 1973 to 2010, clearly depicting two periods. The first period is from 1973 to 1989, which represents the natural flow regime (NFR), and the second period is from 1990 to 2010, which represents a regulated flow regimen (RFR). This graph shows a decreasing flow and less seasonal variability in the maximum and minimum under the RFR.

The natural flow regime achieved rates of 0.90 m^3/s per year, surpassing the 2 m^3/s during 16 of the 17 years identified as NFR. Starting in 1990, the maximum flow rarely reached 2 m^3/s. The minimum regulated flow of the second period is three times higher than the natural minimum flow, varying from 0.13 to 0.37 m^3/s, that is, the minimum flow increased by 176%. In contrast, the maxima were reduced by 69% (Figure 2). Regulation structures, such as gabion dams, increase the minimum flow due to retention and water storage. At present, more than 60 regulation structures have been registered [12], as shown in Figure 1. These structures were constructed for different objectives, such as decreasing the flow velocity, creating micro wetlands and diminishing sediment volumes, among others [12]. These hydraulic structures are not recommended because they function as frontiers that break the biological continuity of lotic systems [17].

The hydrometric information is depicted in the curves of average monthly flow (Figure 3). The NFR includes the dry season (January to May), with a mean measurement of 0.29 m^3/s and a lower flow in April of 0.26 m^3/s. The rainy season shows an increase in flow from June to August and a decline in December (0.66, 1.72 and 0.46 m^3/s, respectively). At the end of the 1980s,

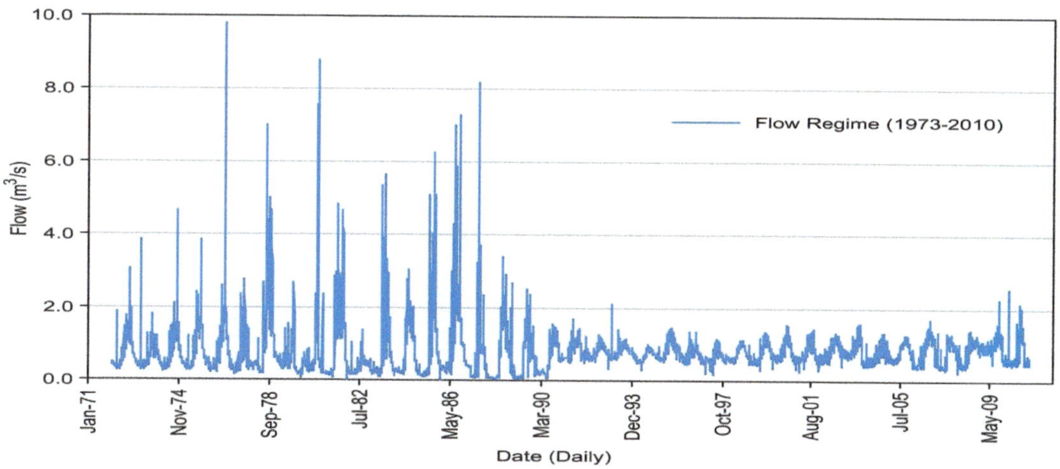

Figure 2. Daily volumetric flow regime (1973–2010) measured at the hydrometric station Santa Teresa (26440).

the basins of the Magdalena and Eslava rivers experienced flow variations, as observed in Figure 3. In the dry season, the RFR curve increased 128% compared with the natural regime (NFR). The RFR value for April was 0.60 m^3/s, compared with 0.26 m^3/s for the NFR in the same month. In the rainy season, the RFR curve diminished with respect to the NFR by 30% from July to October.

Figure 3 shows the changes experienced by the river before and after the hydraulic regulation in the periods mentioned previously, which is defined for four average monthly flow regimes. The inertial behavior of the minimum natural flow regime (mNFR), shows that the dry period (January-May) minimum flow was 0.06 m^3/s, except in March-April, when the minimum flow was 0.04 m^3/s. Starting in the rainy period, the minimum flow responded to the natural increase from June-August (a maximum flow of 0.57 m^3/s) until declining over the period of September to December from 0.51 to 0.09 m^3/s.

The minimum regulated flow regime (mRFR) differs from the natural condition (mNFR curve), with a greater effect observed during the dry season. In March and April, the minimum natural flows increased to 490% (from 0.04 to 0.24 m^3/s) and then increased 202% in January, February and May. From July to October, the mRFR curve increased 64% from its original condition (1973 to 1989); in November and December, the flows increased from 0.20 and 0.09 to 0.57 and 0.47 m^3/s, respectively. This increase can be attributed to the hydraulic structures and to external factors, such as wastewater discharge from irregular settlements in the lower drainage basins and surface runoff during the rainy season. The hydrological histogram (Figure 3) shows the discrepancy in hydrological conditions: under low-flow conditions, the mRFR curve corresponds to the NFR conditions. Therefore, the minimum regulated behavior mimics a natural condition.

Figure 4 depicts the annual flow over a 38-year period. In the first period (1973–1989), the maximum flow reached 10 m^3/s, with a mean annual maximum of 4.94 m^3/s. In the 1990s, the development of hydraulic infrastructure in the river caused a laminar flow effect. During the second period, from 1990 to 2010, the maximum flow decreased 69% (relative to 2.53 m^3/s as the maximum), with a mean annual maximum flow of 1.52 m^3/s.

The behavior of the mean annual flows (Figure 4) also reflects the effect of regulation. Although the mean flow in both periods is very similar, namely, 0.77 and 0.79 m^3/s, its volume has not changed significantly. Over time, extraction from the river has been balanced by wastewater inputs into the river, which has led

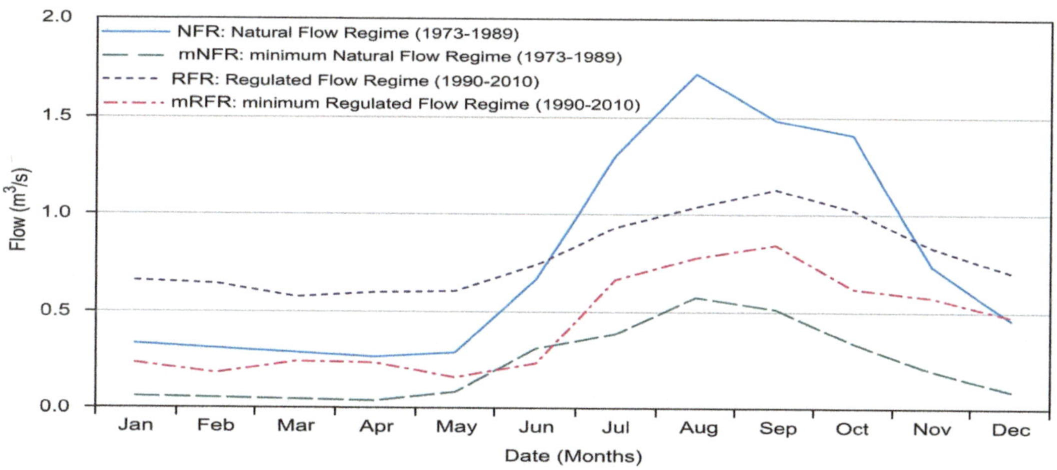

Figure 3. Interannual variation of the NFRs, RFRs, mNFRs and mRFRs measured at hydrometric station 26440.

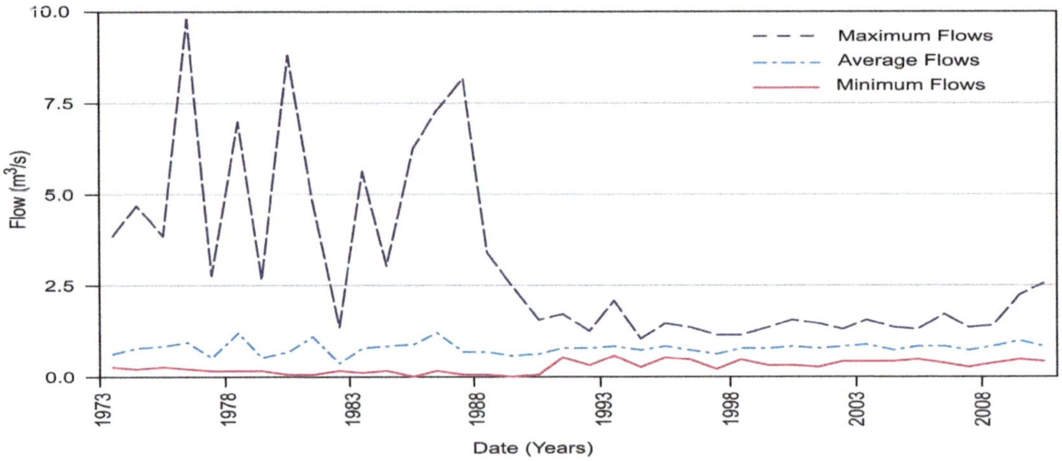

Figure 4. Annual volumetric flow regime at the outlet of the Magdalena and Eslava river basins. Average, maximum and minimum flows.

to an increase of 0.02 m^3/s. The exchange of clean water for untreated wastewater has significantly degraded the environmental condition of the river.

There are only two direct discharge points into the river over the course of the distance from its origin to the treatment plant; in contrast, after the treatment plant, 58 inputs are known [12]. Irregular settlements are problematic because they often lack a sewer system, thereby allowing wastewater to flow directly into the rock or soil and hence toward the lower discharge area where the river is located.

In Figure 5, a historical annual comparative analysis illustrates the natural flow from 1973 to 1989 in two flow-duration curves; during this time, the annual mean flow at the lower point of the Magdalena river was at least 0.54 m^3/s more than 90% of the time, which represents the probable flow that the river used to supply along the year. High flows with flows of approximately 1.14 m^3/s are present during a very short period (10% of the times). In the 1990s, the development of hydraulic infrastructure in the river caused a laminar flow effect; therefore, in the curve for the regulated period from 1990–2010, the mean flow at the river is at least 0.71 m^3/s for 90% of the time, which is an increase of 31% in comparison with the previous natural minimum; consequently, a 25% decrease in high flows is observed. It is also possible to observe changes to the slope of the natural flow curve (reference) with respect to the regulated flow curve, specifically, changing from 0.73 to 0.24%.

The behavior of the natural curve of collected frequencies (1973 to 1989) depicted in Figure 5 is representative of mountain rivers; therefore, based on the flow duration curve observed for the regulated period (1990 to 2010), a new behavior is observed. That is, the river is behaving if it were a floodplain river, where changes are less apparent. Generally, this response is due to changes to the flow of the Magdalena River as a result of anthropogenic activities, with an associated increase in the concentration of several water quality variables.

Figure 6 shows the distribution of water flux for the rainy and dry seasons for the years 2008–2010, starting with the dry season from January-April 2008, followed by the rainy season of 2008, the dry season from November 2008–April 2009, and so on. The figure demonstrates that the mean value is relatively constant for dry and rainy seasons, while the variability changed during those same years; hence, no pattern can be observed for the changes in variability. This is related to the high variability of the amount of rain falling in the upper parts of the basin, particularly during the rainy season of 2010. The low variability for the dry season from November–December 2010 is due in part to the small amount of data for this time lapse.

Water quality

Physicochemical and microbiological analyses. The spatial and temporal analyses for the physicochemical variables and nutrients are presented in Figures 7 and 8. These variables are those most representative of river water quality, such as electrical conductivity, dissolved oxygen, ammonia, total N, orthophosphates and total P, as well as fecal coliforms and fecal enterococci. These variables are considered representative because they have the highest absolute value in the factor loadings of the first three principal components. The factor loadings for these attributes account for more than 80% of the total variability of the Magdalena River system [43].

Although the number of samples at each site was not the same, it only affects the variance of the mean values of the measured water properties. This effect has been considered and corrected in all of our analyses by dividing the mean value estimates by $\sigma/\sqrt{n}$.

The GLM models fitted to the water quality variables [38] indicate statistically significant differences for all of the variables, except dissolved oxygen, in the upper and lower sections of both rivers. The point of change was the location at which human activities and settlements began on either or both river banks. Water quality variables sampled upstream of these urban areas fell within acceptable limits for use as a water supply. The values gradually fell to lower quality as the sampling sites became increasingly affected by the wastewater generated in the irregularly inhabited areas. The spatial trends along the river gradient for the eight most-relevant water quality attributes are shown as box plots in Figure 7 for the Magdalena River and in Figure 8 for the Eslava River.

The plot shows a change point at site M9, which is located at the border between the natural and urban areas: here, increasing concentrations of all values can be observed, except that of dissolved oxygen, which decreases. The highest concentrations were detected at site M15, where part of the water is piped to the city's drainage system. These values then decrease slightly at sites M17 and M18, where other effluents with residual water from the southern area of the city join the waterway.

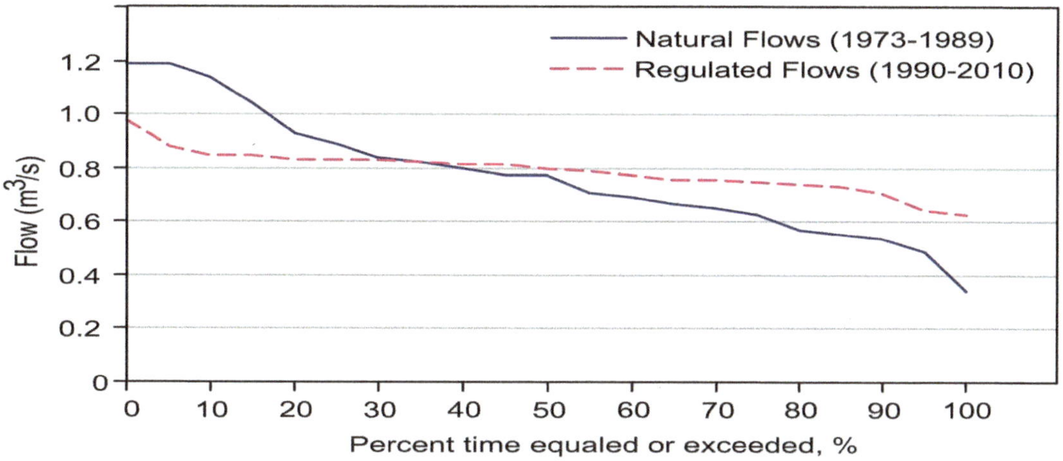

Figure 5. Duration curve of annual natural and regulated flows, Magdalena river basin, 1973–2010.

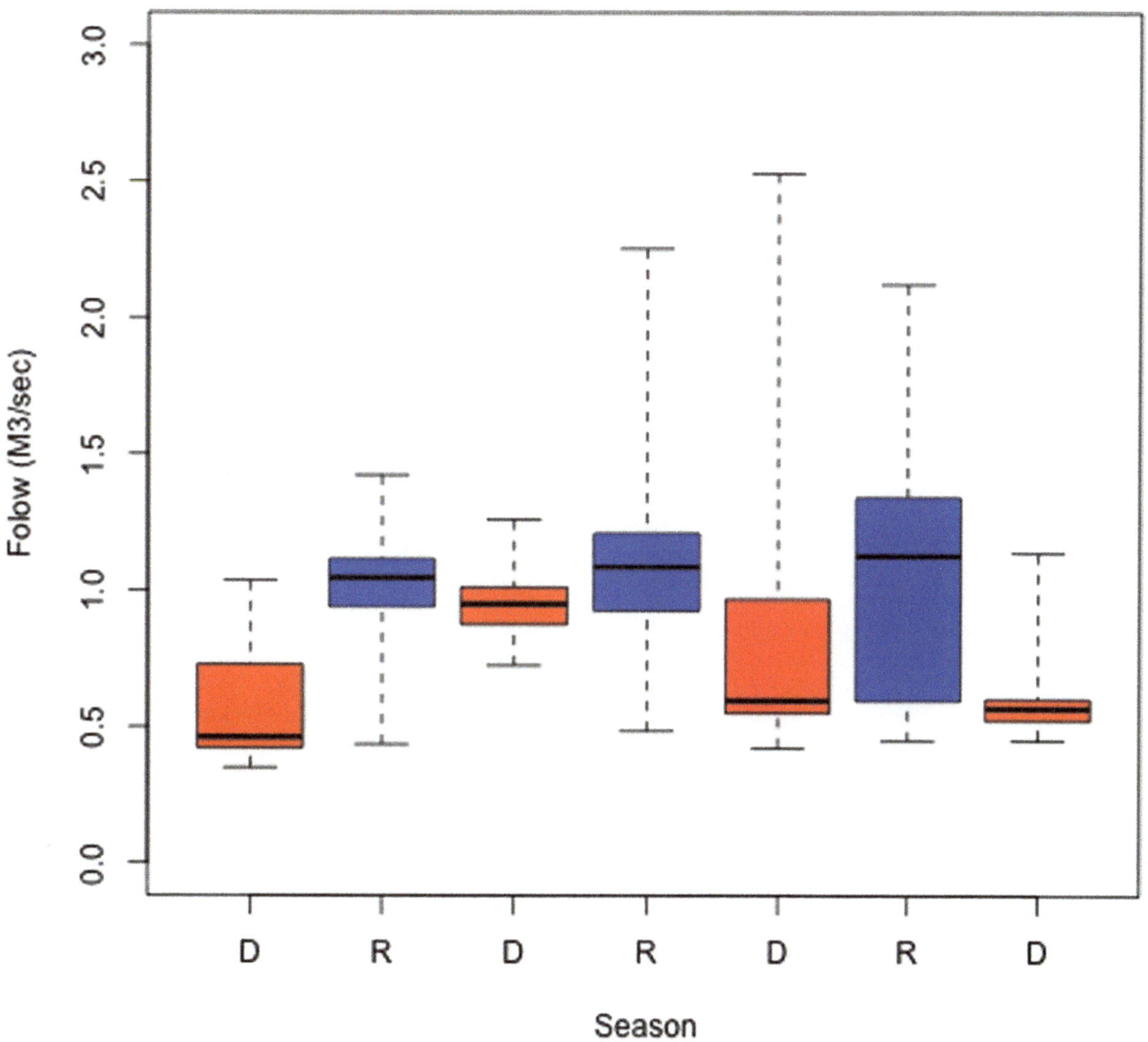

Figure 6. Water flux distribution from 2008–2010 grouped by season.

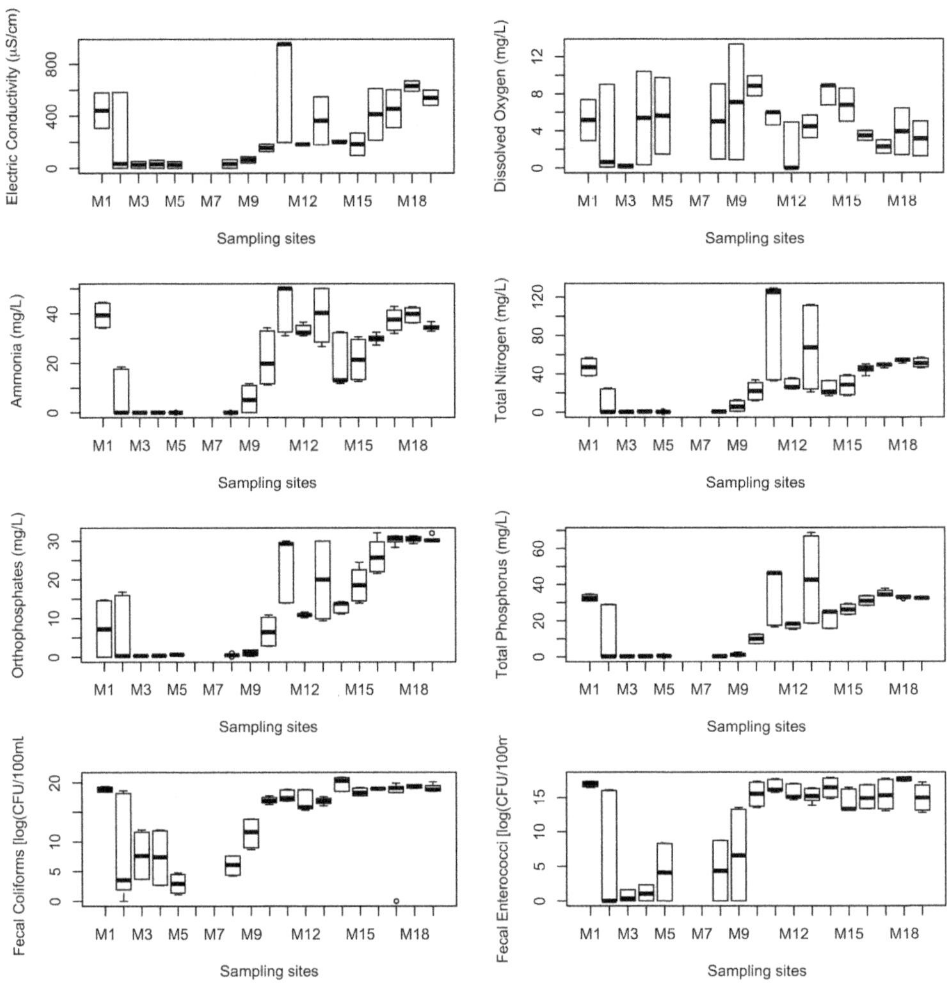

Figure 7. Spatial trend for the eight main water quality variables in the Magdalena River, based on 19 sampling stations.

In the Eslava River, all variables except dissolved oxygen are significantly different between the natural and urban areas, a change that occurs somewhere between E5 and E6. Further, all variables exhibit stable values until the confluence point. The Eslava River has lower water quality than the Magdalena River does at the confluence point, particularly in terms of electrical conductivity, the value of which is six times higher in the Eslava River; additionally, the value of total N is three times higher, while that of total P is twice the concentration of that in the Eslava River. The fecal enterococci values were three times as large on a log scale, which means that this measurement is 1,000 times greater. Please note this difference in the values of the axes on the graphs.

According to Mexican standards, several attributes surpass the NOM-001-SEMARNAT-1996 [44] limits for water discharged to Mexican water bodies. The total N exceeds these levels from M15 to M18 and is double the limit at E8 in the Eslava River. At E7, the values are slightly above the limit.

Because both rivers are used as surface water sources for human use and consumption, the LFDMA [45] limit is 1,000 FC per 100 mL. Taking this into account, the fecal coliforms surpass the limit and require treatment before distribution through the southern parts of the city. The standards are met at the Potabilization Plant La Magdalena, but there are many water intakes near settled populations in an irregular land ownership situation, which means that these populations are exposed to high levels of bacteria; this is an important risk factor that should be taken into further account.

Bacteria identification. The 164 bacterial isolates and identifications obtained from 516 water samples taken from the Magdalena River in the period 2008 to 2010 are presented in Table 2. These belong to nine different families and 18 genera [26,46]. The most abundant genera were *Enterococcus*, *Escherichia coli*, *Enterobacter* and *Klebsiella*. From the Eslava River, 27 bacterial isolates and identifications were obtained from 120 water samples taken during 2012. These belong to four different families and seven genera, as reported in Table 3.

Therefore, a variety of bacteria belonging to a total of 20 different genera were identified in the river system crossing the southern area of Mexico City. The most prevalent bacteria were from the Enterococcaceae family, *Enterococcus*, and from the Enterobacteriaceae family, *Escherichia*, *Enterobacter* and *Klebsiella*. Although these results serve as an initial approach to the identification of bacteria, it is important to note that all of these bacteria are common to the human gut and can be opportunistic pathogens that may cause illness to immune-compromised people who access this water when untreated.

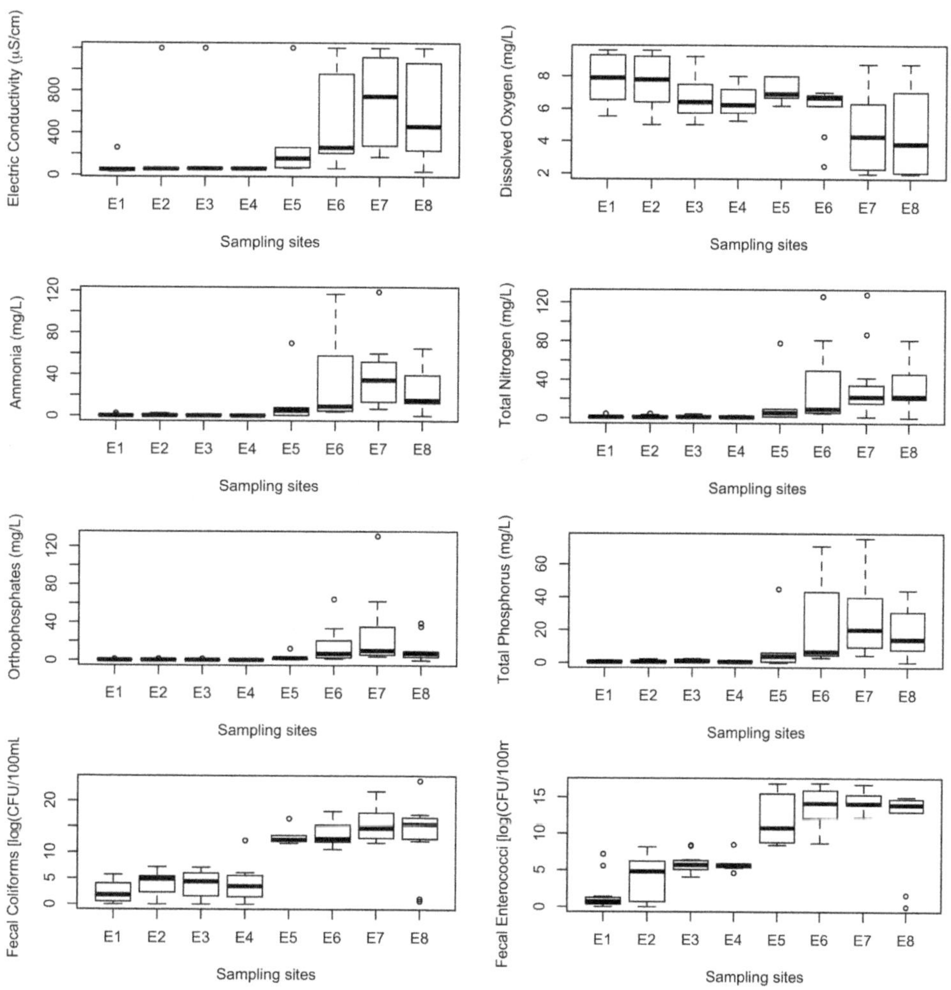

Figure 8. Spatial trend for the eight main water quality variables in the Eslava River, based on eight sampling stations.

Coliphage, virus and parasite analyses. The presence of coliphage was measured by bacterial hosts in samples from the urban zone; this finding is consistent with the inputs of raw and untreated wastewater to the Magdalena River (Table 4). As host bacteria, *E. coli* K12-Hfr registered the higher sensitivity, although larger counts were obtained for *E. coli* Hs (pFamp). It is important to note that there are two infection mechanisms of interest, either by adherence to the cell wall or infection via pilli, as in the coliphage F-specific [47,48,49,50]. Therefore, the host is more susceptible to infection and shows higher counts.

In the natural area sites, no coliphage presence was detected, which suggests that perturbation does not have a detectable impact using this method. Generally, the presence of a phage indicates fecal contamination by a virus; however, this level of analysis is not conclusive because it cannot establish the contaminants' origin, namely, whether human or animal. In urban areas, coliphages are present from the transition zone between the natural and urban zones, where the highest density was detected at site M19; hence, the number of phages increases with human perturbation and influence.

Both adenovirus (AdV F) and enterovirus (EV) exclusively affect humans; their presence indicates human fecal contamination [51,52]. Once these viral particles have been excreted into the environment, their number cannot increase because the presence of the cell host is required for their replication [53]. Therefore, the detection of AdV and EV reflects the fact that in the human population, the virus is circulating and replicating. This constitutes a threat that must be considered regarding the river because these viruses can cause infections to susceptible individuals, yielding viral gastroenteritis.

AdV F replicates quickly in the gastrointestinal track, so the amount of excreted viral particles is on average $1\times10^{10}/g$ in feces [53]. Such infections mainly occur in children under five years of age because adults who were infected as young children often acquire immunity [52]. Nevertheless, if adults are exposed to viral particles, they can be asymptomatically reinfected, with a lower amount of viral particles excreted in their feces. This can favor the dispersion of AdV F because these individuals will not be cautious in their behaviors because they are asymptomatic.

People can present infections and acute gastroenteritis as a result of AdV F [54]. It has been estimated that this virus is the cause of 9 to 12% of acute gastrointestinal cases in children under five years of age in Mexico; however, based on the lack of follow up regarding this type of pathogen and the lack of updated epidemiological information (few studies have been conducted since 1985), this may be an underestimation [55]. Due to the relevance of this virus as a gastroenteritis-causing agent, it is considered the second-most important virus after rotavirus

Table 2. The bacteria identified in the Magdalena River water samples in 2008, 2009 and 2010.

	Sites																			
Isolated microorganisms	**M1**	**M2**	**M3**	**M4**	**M5**	**M6**	**M7**	**M8**	**M9**	**M10**	**M11**	**M12**	**M13**	**M14**	**M15**	**M16**	**M17**	**M18**	**M19**	
n	**24**	**24**	**6**	**24**	**42**	**42**	**18**	**18**	**42**	**42**	**6**	**6**	**42**	**24**	**42**	**24**	**42**	**24**	**24**	**Total**
Moraxellaceae																				
Acinetobacter spp.																		1		1
Aeromonadaceae																				
Aeromonas sp.				1										1		1	1	1		5
Burkholderiaceae																				
Burkholderia sp.								1												1
Alcaligenaceae																				
Bordetella sp.								2												2
Enterobacteriaceae																				
Citrobacter spp.													1				1			2
Enterobacter spp.	1	1			1			3	1		2	1	3	2		1		1	2	19
Escherichia sp.	2	1			2			1	2	1	2	3	2	1	1	3	3	3		27
Hafnia sp.	1																			1
Klebsiella spp.								1	1		1	2	1		1	2	2	2		13
Proteus sp.					1															1
Providencia sp.			1	1																2
Salmonella sp.								1												1
Serratia spp.									2											2
Yersinia sp.						1		1												2
Leuconostocaceae																				
Leuconostoc sp.					1															1
Pseudomonadaceae																				
Pseudomonas sp.			2	3	1				1		1				1					9
Vibrionaceae																				
Vibrio sp.									1			2	1	2	1	1				8
Enterococcaceae																				
Enterococcus spp.	4	2		5	5	1	1	6	5	2	6	3	2	6	2	6	6	5		67
Total	8	4	3	10	11	2	1	16	13	3	12	11	10	12	6	14	13	13	2	164

Table 3. The bacteria identified in the Eslava River water samples in 2010.

	Sites								
Isolated microorganisms	**E1**	**E2**	**E3**	**E4**	**E5**	**E6**	**E7**	**E8**	**Total**
n	**15**	**15**	**15**	**15**	**15**	**15**	**15**	**15**	
Burkholderiaceae									
Ralstonia spp.				1					1
Bacillaceae									
Bacillus spp.	1								1
Enterobacteriaceae									
Escherichia spp.		1	1	1	1	1	1	2	8
Enterobacter spp.				1					1
Klebsiella spp.						1	1	2	4
Yersinia spp.								1	1
Enterococcaceae									
Enterococcus spp.	1	1	1	1	1	1	1	4	11
Total	2	2	2	4	2	3	3	9	27

Table 4. The geometric mean, standard deviation in log scale, maximum and minimum of Coliphage, Enterovirus, Adenovirus, *Cryptosporidium parvum* oocysts and *Giardia lamblia* cysts in the Magdalena River water from 2008 to 2009.

		Sites							
		M1	M5	M8	M11	M12	M17	M18	M19
Coliphage with host	Geometric mean	<1	<1		39	52	57	22	114
K12	Standard deviation	<1	<1		17	18	58	51	14
PFU/mL	Min-Max	<1	<1		29–65	27–65	9–140	7–110	97–127
	n	4	4		4	4	6	6	6
Coliphage with host	Geometric mean	<1	<1	<1	4	7	3	3	33
Hs	Standard deviation	<1	<1	<1	10	7	10	3	26
PFU/mL	Min-Max	<1	<1	<1	<1–18	<1–17	<1–20	<1–7	18–80
	n	4	4	2	4	4	6	6	6
Enterovirus Gen/mL	Geometric mean	<1	<1	<1	<1	<1	<1	<1	<1
	Standard deviation	<1	<1	<1	<1	<1	<1	<1	<1
	Min-Max	<1	<1	<1	<1	<1	<1	<1	<1
	n	3	5	6	4	4	6	4	4
Adenovirus Gen/mL	Geometric mean	10	<1	<1	12974	7299	98294	7305	97084
	Standard deviation	577	<1	<1	46297	28951	253772	10157	35432
	Min-Max	<1–1000	<1	<1	1130–930464	982–63008	24930–557800	1618–25186	71486–144840
	n	3	5	4		4	6	4	4
Oocysts	Geometric mean	<1	32	<1	6784	21	575	4244	<1
Cryptosporidium	Standard deviation	<1	707	<1	14849	5196	9504	2121	<1
parvum/L*	Min-Max	<1	0–1000	<1	2000–230002	<1–9000	<1–19000	3000–6000	<1
	n	2	2	1		3	3	2	2
Cysts *Giardia*	Geometric Mean	1000	71	<1	18	18	30	89	<1
lamblia/L*	Standard deviation	<1	3536	<1	2828	3464	16166	5657	<1
	Min-Max	1000–1000	<1–5000	<1	<1–4000	<1–6000	<1–28000	<1–8000	<1
	n	2	2	1	2	2	3	2	2

Cryptosporidium parvum and Giardia lamblia detection limit 1 (oo)cyst/mL [37]; Adenovirus (AdV) detection limit 10 genome copies [34]; Enterovirus (EV) detection limit 10 genome copies [74]; FRNA specific bacteriophage detection limit 1 PFU/mL [30], somatic bacteriophage detection limit 1 PFU/mL [30,34].

worldwide [56]; therefore, it is important to monitor the prevalence of this virus both in the human population and in residual water as an indicator of what is infecting a given population.

According to our results (Table 4), AdV F is present at both the natural and urban sites, indicating human fecal contamination in both locations. The presence of AdV F at the river origin (M1) could represent an isolated contamination event because it was only detected at a low density in the samples from this site. The presence of human fecal contamination coincides with the irregular settlements that discharge their wastewater directly to the river. Because the system once flowed continuously, the viral contamination issue is likely a recent phenomenon. Wastewater discharge is constant; unsurprisingly, the higher amounts of AdV F viral particles were detected in the urban zone due to wastewater discharge without proper treatment [54,57].

It is important to note that the number of AdV F particles can represent an important mechanism of exposure to the population, primarily for children under five, who may be more likely to come in direct contact with the water in urban areas; accordingly, this exposure can lead to moderate to severe gastroenteritis [53]. For the exposure dose, a point estimate was reported by Crabtree *et al.* [58], who considered AdV 4 as a surrogate because no AdV 40/41 dose-response has been published. Based on this finding, a 2 infectious viral particle can be considered the probability for infection based on an average dose of 1.66 [59].

Considering that the detection limit of the protocol applied for AdV 40/41 quantification was ten genome copies, as determined by serial dilutions from the positive control, it would not be possible to identify a hazard related to AdV 40/41. Nevertheless, the fact that the genome detection rate is commonly larger than the number of viral particles [60] clearly demonstrates that the detection capacity of 10 genome copies is adequate for the identification of adenovirus as a hazard.

Cryptosporidium parvum and *Giardia lamblia* are the most common intestinal protozoa in the environment, commonly infecting both humans and domestic and wild animals [61]. These protozoa are among the main disease-causing agents of diarrhea in the world [62]. Both species form resistance structures that allow them to survive outside the host; they can thus be transmitted through water [63]. Additionally, both species have a low infective dose (ID_{50}), specifically, 132 *Cryptosporidium* spp. oocysts [64] and 25–100 *Giardia lamblia* cysts [65]. Thus, a potential host consuming a low number of these microorganisms can readily develop an infection.

Based on these facts, the detection of *C. parvum* and *G. lamblia* in 50% and 44% of the Magdalena River samples, respectively (Table 4), means that the population and animals that use the water for diverse activities are exposed to parasites, which eventually could lead to an outbreak or epidemic. Children between 0 and 14 years of age and immunodeficient persons are susceptible.

For the Cuitzmala river, a rural area in Jalisco, Mexico, Tapia-Palacios (2012) [37] reported a range of 0–87 oocysts/L *C. parvum* and 0–2,655 cysts/L of *G. lamblia*, which in contrast with the results of this analysis for the urban area, though both species were detected in the two areas. The range in the natural zone was lower for both species (0–1,000 oocysts of *C. parvum* and 1,000–5,000 cysts/L of G. lamblia), which is in contrast with the urban areas that exhibited higher ranges (0–23,000 oocysts/L of *C. parvum, and* 0–28,000 cysts/L of *G. lamblia*). This can be explained by the input of wastewater from the urban area [12]. Untreated water could be increasing the number of oocysts present in the zone, leading to a potential population exposure risk because the population densities range from 1–2 log units more than the infective dose.

Because these rivers represent the only surface water source from within the Basin for Mexico City, particular concern should be held regarding the presence of certain microorganisms, such as the protozoa *Giardia* and *Cryptosporidium* and adenovirus, which are relatively resistant to the chlorination process in comparison to other microorganisms. Additionally, the presence of opportunistic and/or potentially pathogenic bacteria in a surface water source that is chlorinated and distributed to the population suggests that there should be additional funds to consider preventive measures, consistent with the practices carried out in other countries concerning water intended for human use.

Physical removal is critical to the control of *Cryptosporidium* because the organism is highly resistant to standard disinfection practices [66,67]. As mentioned by Betancourt and Rose (2004) [68], the effectiveness of conventional disinfection by chlorination and alternative disinfection procedures, such as chlorine dioxide, ozonation and ultraviolet irradiation (UV), toward inactivating *Cryptosporidium* has been the focus of current research. Recent USEPA criteria regarding disinfection processes for water containing this pathogen suggest disinfection with ultraviolet light or ozone and/or filtration [69]. The goals for systems that treat surface water must be disinfection and filtration to remove or inactivate 99.9% of *Giardia lamblia*, 99% of *Cryptosporidium* oocysts and 99.99% of viruses [69]. In Mexico and other developing countries, modern treatment processes should be applied, given the presence of these microorganisms in the surface waters.

One of the first steps when dealing with rivers is understanding their flow and any variations that occur throughout the year and across different years. In Mexico, monitoring practices were initiated but have since been abandoned for many rivers throughout the country. A sustainable approach to water management, especially in light of climate change, would be to monitor and report environmental and social conditions to provide a guide toward a more sustainable water management strategy.

The regimes of the Magdalena and Eslava rivers belong to homonymous basins that have been modified, and both exhibit two distinct historical periods. The periods are defined by the construction of excess hydraulic infrastructure works, such as gabion dams, 50 on the Magdalena and 18 on the Eslava (Figure 1), and marginal drains. The management of the flows in the Magdalena and Eslava rivers has caused significant fluctuations in flow magnitude (69%), has increased the frequency of maximum ordinary flows and has caused a significant increase (176%) in the minimum flow magnitude. This NFR alteration will certainly have consequences for the river communities in the natural area and for system rehabilitation.

These regime changes may contribute to the observed water quality degradation and contamination of the systems because the loss of velocity along a slope and other discontinuities in the river affect the inherent auto-depuration processes. As recorded in our fieldwork, the observed higher concentrations of the physico-chemical variables indicate the presence of higher levels of contaminants in the lower section of both basins, especially as the urban zone is entered.

Wastewater discharges also contribute to the degradation of water quality. Hence, even if discharge does not contribute significantly to flow volume, its effect on water quality is magnified due to the decrease in flow velocity. Further, wastewater discharge increases the concentrations of various water quality attributes.

In some rivers, contaminants are diluted through their tributaries. However, in this case, the Eslava River contributes its own contaminant load, which for some variables is orders of magnitude higher than the Magdalena's water, thereby hampering the auto-depuration process.

It is necessary to treat the residual wastewater dumped into the river and recover the fluvial processes and function of the rivers as ecosystems. This approach would provide several advantages. For example, there would be more water produced in the higher basin, which could be disinfected and distributed to the population. Additionally, a decrease in the amount of water that flows to the drainage system and that must be pumped out of the Basin of Mexico would be realized, as well as additional amounts of water once the wastewater has been treated and reused for irrigation near the river.

In this study, we demonstrate a practical and representative approach to achieving a detailed water quality program that could allow for the detection of change points. Additionally, we identify the problems at some locations and various methods for improving the respective environmental conditions. In this particular case, it is also of important that locations for the required wastewater treatment plants be suggested, as well as the type of plant and the installed capacity.

The water to be treated is considered domestic wastewater based on its Biochemical Oxygen Demand (DBO_5) and Chemical Oxygen Demand (COD), according to Metcalf and Eddy (1996) [70].

It is important to note the need for treatment plants to improve water quality. The wastewater discharge along the river has high concentrations of nitrogen and phosphorus forms, total dissolved solids, high conductivity, and a BOD_5 of approximately 130 mg/L to 240 mg/L (results not shown). These factors, combined with the high microorganism counts, argue in favor of the installation of several wastewater treatment plants (WWTPs) [71]. A general concept for viable plants and processing methods was presented by Orta de Velasquez (2008) [23], although treatability tests would be required, along with a specific design for each plant with an associated project feasibility study.

The wastewater treatment plants considered for this river, in all cases, included tertiary treatment. This treatment is implemented by biological secondary treatment, and the advanced treatment is implemented by physico-chemical treatment. The first treatment of activated sludge helps remove nutrients (N and P) and both soluble and colloidal biodegradable organic matter. Tertiary filtration is used to treat wastewater intended for reuse to remove any suspended matter that was not removed in the previous operations. Microbial standards can be met using both filtration (physical removal) and disinfection (inactivation). The removal and inactivation capabilities vary depending on the filtration technology or disinfectant applied. The first wastewater treatment plant constructed in the area of the Magdalena River was finished at the end of 2013, not clear if it is in operation at the moment. With the goal of removing both nutrients and pathogens, advanced treatment, which includes activated sludge, tertiary filtration and ultraviolet disinfection, was implemented. In this case, an ultraviolet light disinfection process was applied, which can effectively inactivate most viruses (111 mW*s/cm^2 for 6-log inactivation), *Giardia* (121 mW*s/cm^2 for 2-log inactivation) and *Cryptosporidium* (1,900 mW*s/cm^2 for 2-log inactivation) [72]. With this plant being operational, the river must have a basic flow level along its path to preserve some system functions.

At least two WWTP are proposed. The first could be placed near sites M11 and E8, which could treat low-quality water flowing from both rivers. A second WWTP could be placed before site M19, where the treated water could be used for irrigation in one of the few large green areas of the southern part of the city, known as *Viveros de Coyoacán*. The potential locations for these two plants are presented in Figure 1.

From the perspective of the ecosystem services provided by this lotic system, the amount of water provided by the Magdalena-Eslava river system accounts for the ecological flow in the natural area, after which the water goes through a potabilization process and is distributed for human use and consumption in the southern area of the city. The amount of water that is contaminated when clean water from the natural areas of both rivers combines with untreated wastewater is not accounted for, but it is necessary to consider the financial resources that are being wasted on these relatively small rivers.

The current average flow of the Magdalena river is 0.45 m^3/s, based on measured flow, and the cost per m^3 with the water subsidy is $8.17 (equivalent to 0.628 USD at an exchange rate of US $1 = $12.99 Mexican peso) Mexican pesos; without subsidy, the cost is $25.97 (equivalent to US $1.99) (SACM, Official Receipt. *Derechos por el suministro de agua. Magdalena Contreras. Manzana Baja*. September, 2013) [73]. Therefore, the amount of water lost accounts for 14,191,200 m^3 per year, which means a loss of $115,942,104 Mexican pesos (equivalent to US $8,925,489) at the subsidized cost, or $368,545,464 Mexican pesos (equivalent to US $28,371,475). In this example, we have only considered the Magdalena River, which has better water quality and more abundant flow. An equivalent of this at Mexico City prices for bottled water at $5.66 per liter ($5,660 per m^3) is an estimated $79,470,720,000 Mexican pesos (equivalent to US $6,117,838,337).

As part of any megacity water management strategy, there is a need to consider the amount of water lost and the financial capital that those losses represent. There is an urgent need to rehabilitate and recover the flow of the river and its water quality, which is an approach that could be replicated in more than 40 other rivers around Mexico City. The ecosystem service of clean water is being lost in the megacity, where water must be pumped at a very high cost from either groundwater systems or distant basins. These rivers serve as one example of similar situations that have arisen for other rivers in the Basin of Mexico, where water could be managed in a more sustainable way in the vicinity of this megacity. Doing so would yield enormous environmental and economic benefits in the long term.

From the perspective of providing water, water quality represents a fundamental issue. Better use of the existing water sources can be achieved in the MCMA. The rehabilitation of the Magdalena-Eslava river system could provide an example for other local and remote locations to follow.

Health issues are at stake, so action must be taken to ensure water quality in this particular system. The contaminated water flowing through the urban area should be improved to avoid future deleterious effects to the population.

Considering the treatment of wastewater that flows through the urban area, this water could also be used for irrigation in the green areas in the southern area of Mexico City.

Conclusions

The flow regimes of the Magdalena and Eslava rivers have been modified by the increase of hydraulic infrastructure constructed along the river, resulting in decreased maximum flows and increased minimum flows. These changes in flow regimes are directly related to the quality of the river water, due to the loss of velocity in the flow and the loss of continuity of flow.

The water quality of this system deteriorates as the river flows from a natural forested area to a semi-urban area, passing irregular settlements without basic water and drainage services before flowing then into an urbanized area.

The water quality is worst in the Eslava River at the confluence point, M12, after which either a dilution effect or an artifact of the complex drainage system arises, leading to an apparent improvement in the wastewater quality downstream. Organic matter, such as that found in the discharge, affects the oxygen balance and is usually accompanied by microbiological contamination.

The domestic and agricultural wastes contribute phosphates and ammonia nitrogen. Agriculture also contributes to the dissolved solid levels measured in the river.

The Magdalena Eslava river system could be used more efficiently. The river receives domestic discharge, and a portion of the flow passes through the *Delegación Magdalena Contreras*, where no industry is present.

The construction of at least two wastewater treatment plants is proposed, with the goal of restoring the river's water quality and maintaining the flow in the system. The suggested placement of these treatment plants is based on the low flow that occurs during the dry season and the sewer system conditions. By considering the landscape design to be a linear park, the treated wastewater flow was defined to rely on water flow and high water quality for recreational use. Additionally, the potential to reuse this water for green area irrigation and car washing exists. Treating the wastewater in this area of Mexico City would be relatively simple in comparison to treating the wastewater in other industrial areas of this metropolitan area.

Human activities, which have increased significantly in recent decades, have impacted the environment. The scale of urbanization, industrial operation and agricultural production has reached a point at which these activities impact the quality of the hydraulic resources. This is a global phenomenon to which Mexico City is not an exception.

Wastewater in the cities of the developing world transports a combination of liquid and solid residues from regular and irregular residential areas, public institutions, and industrial and commercial establishments. These residues eventually reach pluvial, surface and groundwater. Due to the diversity of compounds and microorganisms in wastewater, such as those observed in this study, it is not possible to provide details for all compounds but only to give an idea of the water technology required for improving water quality, with a view toward reuse and sustainability. Undoubtedly, there exist other compounds that were not measured as part of this evaluation, such as endocrine disruptors or organic compounds; these compounds must be evaluated, especially in the urban section of the lotic system, to develop a complete picture.

The amounts of several types of microorganisms in the urban zone, each with a different resistance to water and wastewater treatment, must be considered from the public health perspective because these microorganisms represent an important exposure factor for the local population.

Mexico City has lost the majority of its rivers due to mismanagement. The case of the Magdalena and Eslava rivers represents an opportunity to rehabilitate a major river system and could serve as a model for other waterways.

Acknowledgments

We thank Lucía Almeida Leñero for her enthusiastic collaboration throughout the years in the study area and the support received for this work. We thank Erick Hjort Colunga, Rosa Isabel Amieva, Ana Cecilia Espinosa, and María del Jazmín Aguilar for their invaluable assistance in the field and with laboratory work. We thank Alma de los Ríos for assistance in creating the figures.

Author Contributions

Conceived and designed the experiments: MMH MTOL CDA FAV. Performed the experiments: GPO MAT RSO MAS IYN. Analyzed the data: CDA MMH GPO TLV MTOL FAV. Contributed reagents/materials/analysis tools: MMH MTOL YLV CDA FAV. Wrote the paper: MMH GPO CDA FAV MATOL YLV MAT RSO. Performed the field work: GPO MAT MMH. Lab work: GPO MAT RSO YLV MAS.

References

1. Foster S, Lawrence A, Morris B (1998) Groundwater in urban development. Assessing management needs and formulating policy strategies. The World Bank. Washington, D.C.
2. Legorreta J (2009) Ríos, lagos y manantiales del Valle de México. Universidad Autónoma Metropolitana-Gobierno del Distrito Federal. Artes Impresas Eón, S.A. de C.V. México, D.F. 365 p.
3. Jiménez B (2009) Risks and integrated management of the urban water cycle in megacities of the developing world: Mexico City. In: Water and Urban Development Paradigms. Towards an integration of engineering, design and management approaches. Feyen J, Shannon K, Neville M (Eds.). CRC Press. Taylor & Francis Group. London. pp. 387–396.
4. Domínguez-Mora R (2011) Diagnóstico resumido de los problemas de abastecimiento de agua y de inundaciones en el Valle de México. Revista Digital Universitaria 12(2): 1067–6079.
5. Bradshaw AD (2002) Introduction and Philosophy. In: Perrow MR, Davy AJ. Handbook of Ecological Restoration. Principles of Restoration. Cambridge University Press. Cambridge, U.K.
6. García E (1988) Modificaciones al sistema de clasificación climática de Köppen. Instituto de Geografía, UNAM, México, D.F. 33 p.
7. Álvarez-Román KE (2000) Geografía de la Educación Ambiental: algunas propuestas de trabajo en el Bosque de Los Dinamos, Área de Conservación Ecológica de la Delegación Magdalena Contreras. Tesis Licenciatura en Geografía. Facultad de Filosofía y Letras, UNAM. México. 127 p.
8. Ávila-Akerberg V (2009) Forest quality in the southwest of Mexico City. Assessment towards ecological restoration of ecosystem services. Ph.D. Dissertation. Institute für Landespflege. Universität Freiburg. Germany. 138 p.
9. Secretaría de Medio Ambiente y Recursos Naturales, Instituto Nacional de Ecología (SEMARNAT-INE) (2003a) Acuerdo que declara Zona Protectora Forestal los bosques de La Cañada de Contreras, D.F. In: Áreas naturales protegidas de México con decretos federales. México, D.F. 302 p.
10. Secretaría de Medio Ambiente y Recursos Naturales, Instituto Nacional de Ecología (SEMARNAT-INE) (2003b) Declaración de Zona Protectora Forestal del Río Magdalena In: Áreas Naturales Protegidas de México con Decretos Federales. México, D.F. 317 p.
11. Ávila-Akerberg V (2004) Autenticidad de los bosques en la cuenca alta del río Magdalena. Diagnóstico hacia la restauración ecológica. Tesis Maestría. Facultad de Ciencias, UNAM. México. 112 p.
12. Programa Universitario de Estudios sobre la Ciudad-Gobierno del Distrito Federal (PUEC-GDF) (2008) Plan Maestro de Manejo Integral y Aprovechamiento Sustentable de la Cuenca del río Magdalena, Distrito Federal. Diagnóstico Integral. 90 p.: http://www.sma.df.gob.mx/sma/links/dowload/biblioteca/diagnostico_integral.pdf Accessed 2013 Aug 20.
13. Rzedowski GC de, Rzedowski J (2001) Flora fanerogámica del Valle de México. 2ª ed. Instituto de Ecología, A.C. y Comisión Nacional para el Conocimiento y Uso de la Biodiversidad. Pátzcuaro, Michoacán. 1406 p.
14. Cantoral-Uriza EA, Almeida-Leñero L, Cifuentes-Blanco L, Paniagua-León L, Luis-Martínez MA, et al. (2009) La diversidad de una cuenca en la Ciudad de México. Ciencias 94: 28–33.
15. Nava M (2003) Los bosques de la cuenca alta del río Magdalena, D.F., México. Un estudio de vegetación y fitodiversidad. Tesis Licenciatura Biología. Facultad de Ciencias. UNAM. México. 64 p.
16. Jujnovsky J, Almeida-Leñero L, Bojorge-García M, Monges YL, Cantoral-Uriza E,et al. (2010) Hydrologic ecosystem services: water quality and quantity in the Magdalena River, Mexico City. Hidrobiológica 20(2): 113–126.
17. Programa Universitario de Medio Ambiente-Gobierno Distrito Federal (PUMA-GDF) (2009) Sistema de Indicadores para el rescate de los ríos Magdalena y Eslava. México, D.F. 134 p. Available: http://www.sma.df.gob.mx/sma/links/download/archivos/sistema_indicadores.pdf. Accessed 2013 Aug 20.
18. Riley AL (1998) Restoring Streams in Cities. A Guide for Planners, Policymakers, and Citizens. Island Press. Washington, D.C.
19. González Reynoso AE, Hernández Muñoz L, Perló Cohen M, Zamora Saenz I (2010) Rescate de ríos urbanos. Propuestas conceptuales y metodológicas para la restauración de ríos. Universidad Nacional Autónoma de México. Programa de Estudios sobre la Ciudad. Ed. Color, S.A. de C.V. México, D.F. 109 p.

20. Banco Nacional de Datos de Aguas Superficiales (BANDAS) (2006) *National Data Bank Surface Waters*. Comisión Nacional del Agua. Available: <ftp://ftp.conagua.gob.mx/Bandas/Bases_Datos_Bandas> Accessed 2013 March 18.
21. Searcy JK, Hardison CH (1960) Double-mass curves. U.S. Geological Survey Water-Supply Paper 1541-B. USGS, Washington, D.C.
22. Yuefeng Y, Tijiu C, Xiaohua W, Mingfang Z, Cunyong J (2011) Effect of forest recovery on summer streamflow in small forested watersheds, Northeastern China. Hydrol Process 26 (8): 1208–1214. DOI: 10.1002/hyp.8204.
23. Orta de Velásquez MT, Yáñez Noguez I, Mazari Hiriart M, Ríos Soriano B, Velázquez Pedroza K, et al. (2008) Restauración de un rio urbano. Aspectos ambientales e hidrológicos. XXXI Congreso Interamericano de Ingeniería Sanitaria y Ambiental. Santiago de Chile, Chile. 12–15.
24. HACH (2002) Water analysis Handbook. 4th ed. HACH Co. CA. 1260 p.
25. American Public Health Association (APHA), American Water Works Association, Water Environmental Federation (2005) Standard Methods for Examination of Water and Wastewater. 21a ed. United Book Press. Washington. D.C.
26. Murray PR, Baron EJ, Pfaller MA, Tenover FC, Yolken RH (Eds.) (1995) Manual of Clinical Microbiology. American Society for Microbiology. Washington D. C. 1482 p.
27. Gerba CP (2000) Indicator microorganisms. In: Maier RM, Pepper IM, Gerba CP (Eds.) Environmental Microbiology. Academic Press. San Diego, California. pp. 491–503.
28. Toranzos GA, McFeters GA, Borrego JJ, Savill M (2007) Detection of microorganisms in environmental freshwaters and drinking water. In: Manual of Environmental Microbiology. Edited by Hurst CJ, Crawfortd RL, Garland JL, Lipson DA, Mills AL, et al. Washington D. C. ASM Press. pp. 249–264.
29. ISO 10705-1 (1995), Water quality -Detection and enumeration of bacteriophages-part 1: Enumeration of F-specific RNA bacteriophages. Geneva, Switzerland: International Organization for Standardization.
30. Solano-Ortiz R (2011) Avances metodológicos para la detección del virus de la hepatitis A y bacteriófagos en agua. Tesis Licenciatura, Biología. Facultad de Ciencias, UNAM. México. 68 p.
31. Fonseca-Salazar MA (2010) Bacteriófagos como indicadores de la calidad del agua en Ciudad Universitaria. Tesis Licenciatura, Biología. Facultad de Ciencias, UNAM. México, D.F. 62 p.
32. Polaczyk LA, Narayanan J, Cromeans LT, Hahn D, Roberts MJ, et al. (2008) Ultrafiltration-based techniques for rapid and simultaneous concentration of multiple microbe classes from 100-L tap water samples. J Microbiol Methods 73: 92–99.
33. Monpoeho S, Dehee A, Mignotte B, Schwartzbrod L, Marechal V, et al. (2000) Quantification of enterovirus RNA in sludge samples using single tube real-time RT-PCR. BioTechniques 29: 88–93.
34. Xagoraraki I, Kuo DH, Wong K, Wong M, Rose JB (2007) Occurrence of human adenoviruses at two recreational beaches of the Great Lakes. Appl Environ Microbiol 73: 7874–7881.
35. US Environmental Protection Agency (USEPA) (2005) Method 1623: *Cryptosporidium* and *Giardia* in Water by Filtration/IMS/FA, Cincinnati, OH, December 2005. 68 p.
36. Rangel-Martínez C (2010) Identificación de *Blastocytis hominis* en muestras de agua obtenidas de Ciudad Universitaria en la Ciudad de México. Tesis Licenciatura, Biología. Universidad Simón Bolívar. México D.F. 62 p.
37. Tapia Palacios MA (2012) Detección de *Cryptosporidium parvum* y *Giardia lamblia* en agua del río Cuitzmala, Jalisco. Tesis Licenciatura, Biología. Facultad de Ciencias, UNAM. México. 64 p.
38. McCullagh P, Nedler JA (1989) Generalized Linear Models. Monographs on Statistics and Probability, Vol. 37, Chapman and Hall, New York.
39. Hiriart F, Marsal RJ, Cruickshank G, Key F (1952) Contribución de la Comisión Federal de Electricidad a la Solución del Problema de Abastecimiento de Agua a la Ciudad de México. México. 132 p.
40. Departamento del Distrito Federal (DDF) (1975) Memoria de las Obras del Sistema de Drenaje Profundo del Distrito Federal. Tomo I. DDF-Talleres Gráficos de la Nación. México.
41. González-Martínez MT (2008) Modelación hidrológica como base para el pago por servicios ambientales en la microcuenca del río Magdalena, Distrito Federal. Tesis Maestría. Universidad Autónoma de Querétaro. México. 137 p.
42. Jujnovsky J, González-Martínez TM, Cantoral-Uriza EA, Almeida-Leñero L (2012) Assessment of Water Supply as an Ecosystem Service in a Rural-Urban Watershed in Southwestern Mexico City. Environ Manage 49 (3): 690–702. DOI 10.1007/s00267-011-9804-3.
43. Krzanowski WJ (1988) Principles of multivariate analysis: A user's perspective. Oxford University. Clarendon Press. New York. 563 p.
44. Diario Oficial de la Federación (DOF) (1997) NOM-001-SEMARNAT-1996. Que establece los límites máximos permisibles de contaminantes en las descargas de aguas residuales en aguas y bienes nacionales. 6 enero 1997.
45. Ley Federal de Derechos en Materia de Agua (LFDMA) (2012) Ley Federal en Materia de Agua. México.
46. Bergey's (2012) Bergey's Manual of Systematic Bacteriology (2001–2012), 2nd ed. Springer Verlag. New York.
47. Long CS, El-Khoury SS, Oudejans GS, Sobsey DM, Vinjé J (2005) Assessment of sources and diversity of male-specific coliphages for soured tracking. Environ Eng Sci 22(3): 367–377.
48. Schaper M, Jofre J (2000) Comparison of methods for detecting genotypes of F-specific RNA bacteriophages and fingerprinting the origin of feacal pollution in water samples. J Virol Methods 89: 1–10.
49. Havelaar HA, Hogeboom MW (1984) A method for the enumeration of male-specific bacteriophages in sewage. J Appl Bacteriol 56: 439–447.
50. Mooijman AK, Bahar M, Muniesa M, Havelaar HA (2002) Optimisation of ISO 10705-1 on enumeration of F-specific bacteriophages. Journal of Virological Methods 103: 129–136.
51. Muir P, Van Loon AM (1997) Enterovirus infections of the central nervous system. Intervirology 40: 153–66.
52. Wadell G (1984) Molecular Epidemiology of Human Adenoviruses. Curr Top Microbiol Immunol 110: 191–220.
53. Maier RM, Pepper IL, Gerba CP (2009) Environmental Microbiology. China. Academic Press. 598 p.
54. Hierholzer JC (1992) Adenoviruses in the immunocompromised host. Clin Microbiol Rev 5: 262–274.
55. Luna MM, Leboreiro JI, Zapata IB, Rendón-Macías ME (2013) Los adenovirus como causa de gastroenteritis aguda en los niños. Revista Mexicana de Pediatría 80: 98–104.
56. Enriquez CE, Hurst CJ, Gerba CP (1995) Survival of the enteric adenoviruses 40 and 41 in tap, sea, and waste water. Water Res 29: 2548–2553.
57. Ciruela P, Izquierdo C, Oviedo M, Domínguez A (2004) Rotavirus y adenovirus 40–41 como causantes de gastroenteritis infantil. Vacunas 5: 16–23.
58. Crabtree PA, Gerba CP, Rose JB, Hass CN (1997) Waterborne adenovirus a risk assessment. Water Science and Technology 35 (11–12): 1–6.
59. Kundu A, McBride G, Wuertz S (2013) Adenovirus-associated health risk for recreational activities in a multi-use coastal watershed based on site-specific quantitative microbial risk assessment. Water Res 47: 6309–6325.
60. El-Senousy WM, Osman GA, Melegy AA (2014) Survival of Adenovirus, Rotavirus, Hepatitis A Virus, Pathogenic Bacteria and Bacterial Indicators in Ground Water. World Applied Sciences Journal 29 (3): 337–348. DOI: 10.5829/idosi.wasj.2014.29.03.13849
61. Xiao L, Fayer R (2008) Molecular characterization of species and genotypes of Cryptosporidium and Giardia and assessment of zoonotic transmission. Int J Parasitol 38: 1239–1255.
62. Reynolds KA, Mena KD, Gerba CP (2008) Risk of waterborne illness via drinking water in the United States. Rev Environ Contam Toxicol 192: 117–158.
63. Plutzer J, Karanis P, Domokos K, Törökné A, Márialigeti K (2008) Detection and characterization of Giardia and Cryptosporidium in Hungarian raw, surface and sewage water samples by IFT, PCR and sequence analysis of the SSUrRNA and GDH genes. Int J Hyg Environ Health 211: 524–533.
64. Okhuysen PC, Chappell CL, Crabb JH, Sterling CR, Dupont HL (1999) Virulence of three distinct Cryptosporidium parvum isolates for healthy adults. J Infect Dis, 180: 1275–81.
65. Rendtorff RC (1979). The experimental transmission of Giardia lamblia among volunteer subjects. In:W. Jakubowski and J.C. Hoff (eds) Waterborne Transmission of Giardiasis. Report No. EPA-600/9-79-001, pp. 64-8. US Environmental Protection Agency. Environmental Research Centre, Cincinnati.
66. US Environmental Protection Agency (USEPA) (2001a) *Cryptosporidium*: Drinking Water Health Advisory. Office of Science and Technology EPA-822-R-01-009.Environmental Protection Office of Water: 31 p.
67. US Environmental Protection Agency (USEPA) (2001b) Drinking Water Criteria Document Addendum: *Cryptosporidium*. United States Environmental Protection Agency, Office of Water, Washington, DC.
68. Betancourt WQ, Rose JB (2004) Drinking water treatment processes for removal of *Cryptosporidium* and *Giardia*. Vet Parasitol 126 (1–2): 219–234.
69. US Environmental Protection Agency (USEPA) (2013) Basic Information about Pathogens and Indicators in Drinking Water. December 13, 2013. http://water.epa.gov/drink/contaminants/basicinformation/pathogens.cfm Accessed 2014 Jun 21.
70. Metcalf & Eddy Inc (1996) Ingeniería de Aguas Residuales Tratamiento, Vertido y Reutilización 1: 53–55, 145–46. México.
71. Velázquez PK (2009) Inventario de descargas y propuestas de tratamiento para el saneamiento del Río Magdalena. Tesis Ingeniería Química. FES Zaragoza. Universidad Nacional Autónoma de México.
72. US Environmental Protection Agency (USEPA 1999) *Alternative Disinfectants and Oxidants. Guidance Manual*. Ch. 8. Ultraviolet Radiation. Environmental Protection Office of Water EPA 815-R-99-014. USA.
73. Sistema de Aguas de la Ciudad de México (SACM) (2013) Official Receipt. Derechos por el suministro de agua. Magdalena Contreras. Manzana Baja. September, 2013.
74. Silva-Magaña MA (2013) Uso de indicadores virales para la evaluación de la calidad de agua potable y de reúso en Ciudad Universitaria, México. Tesis Licenciatura. Biología. Facultad de Ciencias, Universidad Nacional Autónoma de México. México: 70 pp.

16

A Preliminary Study on *Oxya fuscovittata* (Marschall) as an Alternative Nutrient Supplement in the Diets of *Poecillia sphenops* (Valenciennes)

Arijit Ganguly[1], Ranita Chakravorty[1], Angshuman Sarkar[2], Dipak K. Mandal[1], Parimalendu Haldar[1], Julieta Ramos-Elorduy[3], Jose Manuel Pino Moreno[3]*

1 Department of Zoology, Visva-Bharati University, Santiniketan, West Bengal, India, **2** Department of Statistics, Visva-Bharati University, Santiniketan, West Bengal, India, **3** Departamento de Zoología, Universidad Nacional Autonoma de Mexico, Mexico City, Distrito Federal, México

Abstract

Growth of the ornamental fish industry is being hindered by the scarcity of low cost feed; hence alternative protein supplements should be explored. In this context the present study aims to evaluate whether the grasshopper *Oxya fuscovittata* could be used as a supplement for fish meal in the diets of *Poecillia sphenops*, which is one of the most common ornamental fishes worldwide. The present work is divided into three phases: In the first phase proximate composition of the grasshopper is obtained and five diets are prepared where fish meal is gradually replaced by *Oxya* meal and named as control, D1, D2, D3 and D4. All the diets are formulated on iso-nitrogenous basis where the protein percentage is fixed at 400 g/kg. The second phase deals with feeding trial and in the third phase all the data of the feeding trial are subjected to a linear model. The feeding trial shows that the control, D1 and D2 fed fishes have almost similar results. The linear model proves that the variation in the indices are mainly due to replacement of fish meal by *Oxya* meal, not due to the variations of rice husk and mustard oil cake that are also used to formulate the diets of the present study. From the results two *Oxya* supplemented diets, i.e. D1 and D2 are proved to be almost equivalent to the control diet. Hence it is concluded that *Oxya* meal is able to replace 25% to 50% of fish meal from the diets of *P. sphenops*.

Editor: Alexander Chong Shu-Chien, Universiti Sains Malaysia, Malaysia

Funding: These authors have no support or funding to report.

Competing Interests: The authors have declared that no competing interests exist.

* Email: jpino@ib.unam.mx

Introduction

Fish farmers in general are facing difficulty in raising their fishes because of an alarming increase in the cost of feed which represents 60–70% of the total production cost [1]; and this is also true for ornamental fish industry. In concert with the high rising production cost, the market price of the ornamental fishes is also gaining height [2]. Moreover the conventional protein sources for fish industry i.e. fish meal, soy meal and other grains are being competed for by the human population as well [3]. Recently, with a global population boom the increasing demand and uncertain availability of these traditional protein sources resulted in the quest of other protein rich alternatives for inclusion into the fish diets [4,5]. In this context, insects being nutritionally rich could be a possible alternative [1]. Among the edible insects, acridids (i.e. short-horn grasshoppers of family Acrididae of the order Orthoptera) might have a great future to be established as a sustainable mini-livestock. Acridids are suitable insects for consumption because according to the report of Xiaoming et al. acridids have a high nutritional value and can be used to formulate good quality feed for livestock [6]. But for a constant supply to the feed developing companies there is a need to build acridid farms, for which a suitable species should be found out. In this context Anand et al. reports an estimation of probable annual biomass that could be produced in acridid farms [7]. According to their conclusions *Oxya fuscovittata* (Marschall) is most suitable because being tetra-voltine, they can complete four life cycles annually and could produce a huge biomass. Keeping this in mind the present study aims to utilize *O. fuscovittata* as a supplement of fish meal in the diets of *Poecilia sphenops* (Valenciennes).

Materials and Methods

The present work is divided into three phases; in the first phase analysis of nutrient composition of the ingredients are carried out, which is followed by feed formulation and diet quality assessment; in the second phase the feeding trial is conducted with the selected ornamental fish. During diet formulation fish meal is gradually replaced by *Oxya* meal; but to formulate diets on iso-nitrogenous basis, two other ingredients are needed to vary. Consequently the amounts of those ingredients differ from diet to diet. In this context it is hard to infer that the significant differences in the results of feeding trial is due to replacement of fish meal by *Oxya* meal, or due to the variations in the amounts of the other major ingredients. Hence, in the third phase all the data of the feeding trial are fitted to a linear model, to conclude which ingredient has greater effect on the variations of the observed indices during the feeding trial.

Table 1. Proximate composition of main ingredients.

Ingredients	CP(g/Kg±SD)	EE(g/Kg±SD)	C(g/Kg±SD)	TA(g/Kg±SD)
Fish Meal	483.1±07.2c	92.8±19.0b	381.7±10.1b	42.4±17.5a
Oxya Meal	640.7±02.3d	64.8±03.1a	241.8±04.7a	52.7±0.9a
Rice Husk	54.42±01.5a	49.0±08.0a	742.3±11.8d	154.3±11.1b
Mustard Oil Cake	306.3±12.8b	52.7±12.5a	606.5±12.2c	34.5±11.9a

Note: CP = crude protein, EE = ether extract (crude fat), C = carbohydrate, TA = total ash. Values with different letters within a column are significantly different (P<0.001) using Tukey's range test.

The present work has been conducted according to the rules of "Institutional Animal Ethics Committee" of Visva-Bharati University, and approved by the committee according to the Indian law.

Selection of model ornamental fish

To accomplish the present work we have chosen *P. sphenops* (more commonly "black molly") which could be easily reared, could complete life cycle within a short period and easy to breed. Moreover, it is one of the most common ornamental fishes to be seen in the aquariums.

Diet formulation

For the formulation of various supplementary diets *Oxya* meal, fish meal (dried *Harpadon nehereus* F. Hamilton 1822), mustard oil cake (MOC) and rice husk are used as major ingredients. Carboxymethyl cellulose (CMC) is used as a binder for the diets and calcium, table salt (NaCl) and vitamin-mineral mixture are used as additive. First of all proximate composition of the major ingredients are carried out according to AOAC methods [8]. Nitrogen is estimated using Kjeldahl method and crude protein is calculated by multiplying the amount of nitrogen with the factor 6.25 (988.05). Crude fat is estimated with soxhlet apparatus (920.39). Ash contents are determined by subjecting the samples to muffle furnace at 550°C for 6 hrs (942.05). The amount of carbohydrate content is calculated by difference method (100– sum of protein, fat and ash).

Five diets are prepared gradually replacing fish meal by *Oxya* meal with the ratio of 100:0, 75:25, 50:50, 25:75 and 0:100. The diet having "fish meal: *Oxya* meal" with the ratio of 100:0 is considered as control because there is no supplement of *Oxya* meal. The other diets are named D1, D2, D3 and D4 respectively. The amount of ingredients of all of the formulated diets are mixed in such a way so that each of them contain about 400 g/kg crude protein, because according to literature mollies need this much protein in their diets for optimal growth [9]. Maintaining the

Table 2. Ingredient composition and quality of formulated diets.

Ingredients	Control	D1	D2	D3	D4
Fish meal	600.00 g	450.00 g	300.00 g	150.00 g	0.00 g
Acridid meal	0.00 g	150.0 g	300.0 g	450.0 g	600.0 g
Mustard oil cake	339.00 g	267.80 g	173.70 g	80.10 g	12.70 g
Rice husk	11.00 g	82.20 g	176.30 g	269.90 g	337.30 g
Carboxy methyl cellulose	30.00 g	30.00 g	30.00 g	30.00 g	30.00 g
Vitamin-Mineral mixture[a]	7.00 g	7.00 g	7.00 g	7.00 g	7.00 g
NaCl	8.00 g	8.00 g	8.00 g	8.00 g	8.00 g
Calcium additive	5.00 g	5.00 g	5.00 g	5.00 g	5.00 g
Total	1 Kg	1 Kg	1 Kg	1 Kg	1 Kg
Formulated diet quality					
CP (g/Kg ± SD)	399.4±6.2a	400.1±5.3a	401.1±6.7a	399.9±8.1a	400.6±7.6a
EE (g/Kg ± SD)	63.1±11.3a	57.3±9.8a	52.7±12.3a	61.2±11.8a	52.3±8.7a
C (g/Kg ± SD)	498.4±7.7a	517.1±4.8b	522.1±10.1b	523.1±8.6b	523.3±4.4b
TA (g/Kg ± SD)	44.1±4.6b	27.5±3.1a	25.2±4.8a	21.8±3.6a	23.8±2.7a
Energy (kJ/g ± SD)	18.94±0.06a	19.15±0.04b	19.15±0.07b	19.39±0.06c	19.19±0.05b
P/E(mg protein per kJ ± SD)	20.82±0.19a	20.89±0.13a	21.14±0.11b	20.61±0.08a	21.18±0.19b
Striking time (in seconds)	6.50±1.22a	6.72±0.68a	7.06±1.57a	7.24±0.28a	6.36±0.16a

Note: CP = crude protein, EE = ether extract (crude fat), C = carbohydrate, TA = total ash, P/E = protein to energy ratio. Values with different letters within a column are significantly different (P<0.001) using Tukey's range test.

[a]Provided *per* Kg of diet: Vitamin A (as Acetate) 69000 IU, Vitamin D3 6900 IU, Vitamin E Acetate 172.5 mg, Ascorbic Acid 1035 mg, Thiamine Mononitrate 69 mg, Riboflavine 69 mg, Pyridoxine HCl 20.7 mg, Cyanocobalamine 103.5 mcg, Niacinamide 690 mg, Vitamin K 1.7 mg, Calcium Pentothenate 112.4 mg, Calcium Phosphate 890.1 mg, Magnesium Oxide 414 mg, Ferrous Fumerate 221.1 mg, Zinc (from zinc sulphate) 15.2 mg, Copper (from copper sulphate) 23.4 mg, Manganese (from manganese sulphate) 14 mg, Sodium Molybdate 1.7 mg, Sodium Borate 6.1 mg.

Table 3. Water quality of the experimental tanks.

Parameters	Range	Mean ± SD
Temperature (°C)	21.40–30.50	27.18±1.25
pH	7.34–7.71	7.52±0.48
Dissolve Oxygen (mg/L)	3.14–3.82	3.61±0.52
Hardness (mg/L)	162.40–167.60	165.67±2.20

amount of protein by adjusting the proportion of ingredients is done by the Pearson's square method. After formulation, diets are first made to dough forms that are subjected to a pelletizer (RJK-FFF-PM-52, Rajkumar Agro Engineers Pvt. Ltd, India), to make floating type pellets.

Assessment of diet quality

Crude protein, crude fat (ether extract), carbohydrate and total ash (TA) are estimated according to AOAC as already mentioned [8]. Energy contents of the diets are estimated by oxygen bomb calorimeter and expressed in kJ/g. Protein to energy (P/E) ratio is an important measure for the quality assessment of the formulated diets and it is expressed as "mg of protein/kJ of energy". Feed acceptability test is assessed as the time taken by the first fish to strike the first particle after the diet dropped into the tanks [10]. The time is recorded in seconds using a stop watch.

Water quality

Analysis of water of the experimental tanks has been done fortnightly in terms of temperature, pH, dissolved oxygen (DO) and hardness. Temperature and pH is measured using a combined pH meter and thermometer (ZPHI-9100), Zico India Ltd., but DO and hardness is measured by titration according to APHA [11].

Collection of experimental species

Same variety of twenty females and ten males of *P. sphenops* are bought from the local market of Bhubandanga, Santiniketan, West Bengal, India. They are reared in the laboratory using a *Spirulina* based diet till the females give birth to sufficient amount of hatchlings. One hundred and fifty (150) fingerlings of similar initial weight (about 0.01 g) are selected to start the feeding experiment, and are kept for 7 days of acclimation period prior to feeding.

Experimental setup

150 fingerlings are divided into 5 sets with 3 replications having 10 individuals in each group. They are placed in 15 experimental tanks of 26 L capacities. Half of the total volume of water is changed daily from each tank. The tanks are completely drained and thoroughly scrubbed once a week to prevent algal growth. Adult body weight and length are measured at the end of experiment after 176 days.

Feeding the fishes

The experimental fishes are fed with known amount of oven dried feed twice daily at 6.30 and 18.30 hrs for five minutes. The uneaten diets are siphoned off and oven dried. The amount of dried feed consumed is calculated by subtracting the left over dried feed from the known amount of dried feed offered.

Estimation of consumption, utilization, survival, growth and reproductive potential of *P. sphenops* when fed with the formulated diets

Growth is measured in terms of specific growth rate (SGR) and condition factor (K) according to De Silva & Anderson, and Moyle & Cech using the following equations [12,13].

$$SGR = [(\log W_{t2} - \log W_{t1}) \times 100]/T$$

$$K = 100 \times (weight/length^3)$$

Food consumption and utilization is measured in terms of protein efficiency ratio (PER), feed conversion ratio (FCR), and feed conversion efficiency (FCE) as suggested by De Silva and Anderson using the following equations [12].

Table 4. Results of Survival, SGR, PER, FCR, FCE, K and reproductive potential of *P. sphenops* fed with various formulated diets.

Response indices	Control	D1	D2	D3	D4
% Survival	84.19±2.36a	83.78±2.11a	83.93±1.82a	82.97±2.32a	83.31±1.97a
SGR	1.52±0.10c	1.52±0.08c	1.5±0.17c	1.26±0.06b	1.04±0.14a
PER	14.97±0.28d	15.97±0.24e	12.18±0.40c	3.16±0.26b	1.13±0.05a
FCR	1.16±0.03a	1.14±0.02a	1.18±0.05a	1.86±0.08b	2.21±0.11c
FCE	3.40±0.23d	3.62±0.14d	2.76±0.14c	0.97±0.13b	0.45±0.09a
K	1.66±0.0187c	1.55±0.192c	1.94±0.206c	1.01±0.097b	0.55±0.016a
Reproductive potential/female	14.33±1.53a	14.67±1.15a	15.67±0.58a	14.33±1.15a	15.33±0.58a

Note: SGR = Specific growth rate, PER = Protein efficiency ratio, FCR = Feed conversion ratio, FCE = Feed conversion efficiency, K = Condition factor. Values are means ± SD. Values with different letters within a row are significantly different ($P<0.001$) using Tukey's range test.

Table 5. Estimate of the regression coefficients of the effects of different factors used in the linear model on each response variable.

Response indices	Estimate of the regression coefficients			
	Intercept [P value]	Replacement [P value]	Rice husk [P value]	Mustard oil cake [P value]
SGR	0.0003 [0.4999]	0.0080 [0.4970]	0.0055 [0.4979]	0.0047 [0.4982]
PER	0.0004 [0.4998]	0.0074 [0.4972]	0.0107 [0.4960]	0.0044 [0.4983]
FCR	0.0002 [0.4999]	0.0089 [0.4967]	0.0016 [0.4994]	0.0051 [0.4981]
FCE	0.0004 [0.4999]	0.0070 [0.4974]	0.0090 [0.4966]	0.0042 [0.4984]
K	0.0003 [0.4999]	0.0077 [0.4971]	0.0067 [0.4975]	0.0045 [0.4983]

Note: SGR = Specific growth rate, PER = Protein efficiency ratio, FCR = Feed conversion ratio, FCE = Feed conversion efficiency, K = Condition factor, Intercept = constant effect, Replacement = *Oxya* meal that replaced fish meal, P value = Area under the t distribution above the observed t value of the student t-test that was conducted to know whether their corresponding factors had significant effect on the response.

$$PER = WWt/DWp$$

$$FCR = F/WWt$$

$$FCE = WWt/F$$

[Where W_{t1} = initial weight, W_{t2} = final weight, T = number of days, F = average dry weight of food ingested per individual, WWt = wet weight gained and DWp = dry weight of protein in feed].

Survival is measured in terms of percentage of individuals alive up to sexual maturity, whereas reproductive potential is measured in terms of total number of offspring born per female in the experimental sets.

Statistical analysis phase 1 and 2

All the data are presented as means ± SD. Each variable is first subjected to one way analysis of variance (ANOVA). Thereafter separation of mean values according to significance is done within each variable by Tuky's range test. All the statistical analyses are carried out using S Plus version 4.0 and PAST version 3.02.

Statistical analysis phase 3

To make sure whether the responses obtained in the feeding trial are due to the replacement of fish meal by *Oxya* meal or due to the variation of rice husk and mustard oil cake, the data of the response indices are fitted to the following linear model using MATLAB program (MATLAB 7.1)–

$$y = \beta_0 + \beta_1 x + \beta_2 (60 - x) + \beta_3 Z_1 + \beta_4 Z_2 + e$$

[Where, y = response indices, β_0 = effect of some constant factors that are identical for all the formulated diet fed sets, β_1 = effect due to x, x = *Oxya* meal, β_2 = effect due to (60-x), (60-x) = fish meal, β_3 = effect due to Z_1, Z_1 = rice husk, β_4 = effect due to Z_2, Z_2 = mustard oil cake, e = error].

$$\text{Or, } y = \beta_0 + 60\beta_2 + (\beta_1 - \beta_2)x + \beta_3 Z_1 + \beta_4 Z_2 + e$$

This equation could be simplified as–

$$y = \beta_0^* + \beta_1^* x + \beta_3 Z_1 + \beta_4 Z_2 + e$$

[Where, $\beta_0^* = (\beta_0 + 60\beta_2)$, and $\beta_1^* = (\beta_1 - \beta_2)$. In other words according to this model, β_0^*, β_1^*, β_3, and β_4 are the "estimate of the regression coefficients" of the effects due to their corresponding factors (i.e. intercept or the constant effect, *Oxya* meal that replaced fish meal, rice husk and mustard oil cake respectively) on the response indices].

In the next step, "estimate of the regression coefficients" are individually subjected to student t-test to know whether their corresponding factors have significant effect on the response indices. If the null hypothesis is accepted, one could conclude that the corresponding factor has no significant effect and if the null hypothesis is rejected, one could conclude that the corresponding factor has significant effect on the response.

Table 6. Comparison between the sum of squares of the effects due to Group1 and Group2 on the response indices.

Response Variables	Sum of squares due to Group1 [$\times 10^3$]	Sum of squares due to Group2 [$\times 10^3$]	P-value of the ratio of sum of squares
SGR	215.9000000	0.7788000	<0.001
PER	0.9156000	0.2053000	0.0883
FCR	543.9200000	26.7680000	<0.01
FCE	21.7300000	3.3322000	<0.01
K	135.9900000	3.4697000	<0.01

Note: SGR = Specific growth rate, PER = Protein efficiency ratio, FCR = Feed conversion ratio, FCE = Feed conversion efficiency, K = Condition factor, Group1 = replacement of fish meal by *Oxya* meal, Group2 = rice husk and mustard oil cake in combination.

After this test the four major factors of the formulated diets (i.e. *Oxya* meal, fish meal, rice husk and mustard oil cake) are divided into two groups–

replacement of fish meal by *Oxya* meal = Group1
rice husk + mustard oil cake = Group2

Then the effect of Group1 on the response indices are statistically compared with the same of Group2, to conclude which group has greater effect on the variation of the response indices. This is done by comparing the sum squares (SS) of the effect of Group1 and Group 2 because SS represents the sum of squared difference from the mean, and this difference gives rise to variance. This part is carried out using the same model already programmed in MATLAB.

Results

Diet formulation

Proximate compositions of the major ingredients i.e. rice husk, *Oxya* meal, MOC and fish meal are tabulated in table 1. It reveals that crude protein level is highest in *Oxya* meal (more than 640 g/kg) and crude fat is highest in fish meal (more than 90 g/kg). Contents of carbohydrate and total ash show higher values for rice husk. After getting the proximate composition of the major ingredients, their required proportions in the formulated diets are calculated. The final compositions of the diets are depicted in table 2.

Assessment of diet quality

Quality of the formulated diets is presented in table 2. Almost 400 g/kg of crude protein level is obtained in each of the diets with no significant difference ($p>0.05$, Tuky's range test). Similarly the values of crude fat (ether extract) again show statistically insignificant variation. The values of carbohydrate contents are also very similar; however, in the control diet it is slightly lower than the others ($p<0.05$, Tuky's range test). Total ash has a result which is reciprocal to the values of carbohydrate where the control diet shows statistically higher ($p<0.001$, Tuky's range test) results. Energy is quite similar, but the control and D3 has significantly lowest and highest mean values respectively ($p<0.001$, Tuky's range test). Protein to energy ratios (P/E) are almost the same (nearly 21 mg/kJ) for all the diets. Results of the acceptability test reveals that the time taken by the first fish to strike the first particle of the formulated diets is within about 6–7 seconds for each case. Hence no significant variation is obtained by Tuky's range test.

Assessment of water quality of the experimental tanks

The temperature, pH, dissolved oxygen and hardness of water during the feeding period are tabulated in table 3. Temperature of all the experimental tanks range from 21.4 to 30.5°C with a mean value of 27.18°C. The water is found to be slightly alkaline as the pH range falls between 7.34–7.71 with a mean value of 7.52. Dissolved oxygen has an average value of 3.61 mg/L and it ranges between 3.14–3.82 mg/L in all the experimental tanks. Assessment of hardness reveals that the water of the experimental tanks is hard with a mean value of 165.67 mg/L that ranges between 162.4–167.6 mg/L.

Consumption, utilization, survival, growth and reproductive potential of *P. sphenops* when fed with the formulated diets

The results of survival, SGR, PER, FCR, FCE, K and reproductive potential of *P. sphenops* are summarized in table 4. More than 80% survival is obtained for all the diet fed sets and they do not vary significantly. But for the other indices as a whole there is a tendency of the control diet, along with D1 and D2 fed *P. sphenops* to have similar results that are significantly better than D3 and D4 fed groups. In case of specific growth rate (%SGR) the values vary insignificantly ($p>0.05$, Tuky's range test) between the control, D1 and D2 fed sets and these values are higher than D3 and D4 fed ones. The values of feed conversion ratio (FCR) also show better results for the control, D1 and D2 diets as the results are significantly lower ($p<0.05$, Tuky's range test) than the groups fed with D3 and D4. On the contrary unlike SGR and FCR, the results of protein efficiency ratio (PER) and feed conversion efficiency (FCE) statistically vary between the control, D1 and D2 fed sets. For PER among these three diets D1 fed groups show highest values ($p<0.05$, Tuky's range test) followed by control and D2 fed groups. For FCE, though the values are insignificant between control and D1 fed groups, the sets fed with D2 shows a slightly lower value ($p<0.05$, Tuky's range test). Condition factor (K) also shows a similar trend, values vary insignificantly between control, D1 and D2 fed mollies and then it gradually decreases in D3 and D4 fed sets. Each female of all the experimental tanks are found to give birth to 13–16 offspring. On an average the values are insignificant ($p>0.05$, ANOVA, Tuky's range test) when data are compared between the sets fed on different formulated diets.

Linear model of response variables

The estimate of the regression coefficients of the effects of different factors used in the linear model on each response index are depicted in table 5. Survival and reproductive potential are excluded from this analysis as they vary insignificantly from diet to diet ($p>0.05$). Values of the student t-test that are conducted to know whether their corresponding factors have significant effect on the response indices are also presented in the same table. The insignificant p values ($p>0.05$, Tuky's range test) show that the null hypothesis is rejected for all. This means all the factors in question (i.e. replacement of fish meal by *Oxya* meal, rice husk and mustard oil cake) has certain amount of effect on the variation of the response data that are obtained during the feeding trial. Now, as it has been evident that all the factors in question are playing some role on the variation of the response data, in the next step it is essential to find out which factor is playing greater role. For this purpose the sum of squares (SS) of the effects due to replacement of fish meal by *Oxya* meal (Group1) is compared with the sum of squares (SS) of the effects due to rice husk and mustard oil cake in combination (Group 2). Results of the sum of squares along with their corresponding p values have been tabulated as table 6. This table has revealed that only excluding PER, for rest of the cases, Group1 has much greater effect on the variations in response data that are obtained during the feeding trial. This proves that the effect of replacement of fish meal by *Oxya* meal has greater influence on the variation of response than the effect of rice husk and mustard oil cake in combination.

Discussion

It is already known that mollies prefer hard water with a pH range of 7.5–8.2, and 18°C−28°C is the suitable temperature for them [14]. In the present study the result of water quality reveals that the water in the experimental tanks is hard with a temperature and pH within the optimum range. However, dissolved oxygen is a little lower than the optimum need (i.e. above 5 mg/L) for most of the fishes [15,16]. The main cause behind this lower DO may be because no aerator is used in the tanks. However, mollies have been reported to get adapted to hypoxic conditions as they could tolerate a level of DO as low as 1 mg/L [17]. In this context it could be stated that the water condition of the experimental tanks are ideal for rearing mollies.

Protein is one of the most important constituents in feed for fish growth. Before designing any experiment on the effect of formulated diets on growth of fishes, the dietary percentage of protein needed for their optimal growth should be obtained first [18]. Shim and Chua worked on poeciliids like *Poecilia reticulata* Peters 1859 and according to them 300–400 g/kg of protein in feed is optimum for their growth [19]. Parker also opines that 400 g/kg dietary protein is optimum for mollies [9]. Hence 400 g/kg of crude protein is fixed for all the formulated diets in the present study. According to some workers, 100–200 g/kg of lipid in fish diets gives optimum growth rates without producing an excessive fatty carcass [20]. In this context it may seem that the diets of the present study contain a little lower amount of fat (near about 50–60 g/kg). However, report of Fernando et al. states that poeciliids are fed with diets having 30–70 g/kg of fat in Singapore [21]. This report supports the view that the formulated diets of the present study contain sufficient amount of fat. It is well known that fish apparently do not have carbohydrate requirement, but it could be metabolized by many fishes and usually herbivorous fishes can metabolize carbohydrates better than carnivorous species [22,23]. According to the report of National Research Council warm water fish (such as mollies) are able to use high carbohydrate levels as an energy source [24]. This phenomenon is advantageous because carbohydrate sources are inexpensive as seen in the present case where all of the formulated diets contain a high amount of carbohydrate. Protein to energy (P/E) ratio is another important measure that gives us a clear picture whether the balance between protein and energy contents is maintained in the diets in proper amount. According to some authors for achieving maximum growth, the P/E ratio in fish diet should range from 19–22 mg/kJ [25–27]. These reports are encouraging because in the present case all of the formulated diets have P/E value of nearly 21 mg/kJ.

Before feeding trial it is necessary to be sure whether the experimental fish recognizes the diets as a palatable one. The results of low feed striking time with insignificant variation positively indicate about the palatability of all the formulated diets. The feeding trial reveals that the results of growth, food consumption and utilization are very similar when fed on control, D1 and D2. On the contrary the values when fed on D3 and D4 are of a lower grade. Among the very little information in literature on using insects as alternative protein supplement for fish, satisfactory results for *Heterobranchus longifilis* Valenciennes 1840 has been reported, where 50% fish meal is replaced by termite meal [28]. Martinez et al. also report satisfactory results when 30% cockroach meal is added to formulate diets for Japanese carp [29]. Along with food consumption, utilization and growth, the measurement of condition factor (K) is also important as it could express the final health of the adult individuals in terms of length and weight at the end of the feeding trial. Results of the condition factor follow a similar trend as observed in case of the other indices already mentioned, because here also the results of control, D1 and D2 diets show ideal results (more than 1.5 [30,31]). Unsatisfactory results are obtained in case of D3 and D4 fed groups might be because of mollies cannot tolerate a composition where more than 50% of fish meal is replaced by the *Oxya* meal. Or this could be an effect of variations in the amount of rice husk and mustard oil cake.

To clear the doubt whether the difference in response variables of feeding trial are due to replacement of fish meal by *Oxya* meal, or due to the effect of change in the amount of rice husk and mustard oil cake, a statistical explanation is important. In this regard results of the linear model quite clearly reveal that only excluding PER, variations of all the indices under consideration occur mainly due to the replacement of fish meal by *Oxya* meal. Here it should be kept in mind that PER is actually a measure of weight gain with respect to the amount of protein consumed through food. The statistical model shows that *Oxya* meal, fish meal, rice husk and mustard oil cake, all have certain effect on the response variables; it is quite clear that all these four ingredients contribute in the protein part of the formulated diets. In this context it could be explained that may be a significant amount of protein contribution by all the four major ingredients leads to an insignificant result when the effect of replacement of fish meal by *Oxya* meal is compared with the combined effect of rice husk and mustard oil cake in case of PER. Results of the present work is encouraging because it is quite clear from this study that at least up to 50% of fish meal could be effectively replaced by *Oxya* meal. This amount of replacement will have no negative effect on the growth and reproduction of *P. sphenops*. Moreover, it is also clear that the variations on the response indices are mainly due to this replacement and not due to the varied amount of rice husk and mustard oil cake.

The results of the present study as a whole support the view of rearing *O. fuscovittata* in "acridid farms" that will easily and constantly provide nutritionally rich animal protein supplement to the aqua-feed developers to formulate supplementary diets for ornamental fishes making the industry more viable. Moreover, if the concept of "acridids as alternative food source" is popularized it could lower the rate of overexploitation of fish meal. Consequently the "demand: supply" ratio of the fish meal could come down, resulting a lower market price of the fish meal also. However, similar studies with various proportions of *Oxya* supplement in the diets of other ornamental fishes are necessary in future. These works also should emphasize on obtaining the digestibility of the formulated diets, which is lacking in the present work. Then only one could be more assured of the suitability of *Oxya* meal as an effective alternative protein source.

Acknowledgments

The authors are thankful to the head of the Department of Zoology, Visva-Bharati University for providing necessary laboratory facilities. Mrs. Mousumi Das is specially acknowledged for her kind help regarding crude protein estimation.

Author Contributions

Conceived and designed the experiments: AG RC DM PH. Performed the experiments: AG RC AS DM PH JR JP. Analyzed the data: AS. Contributed reagents/materials/analysis tools: AG RC DM PH. Contributed to the writing of the manuscript: AG RC As DM PH JR JP.

References

1. Van Huis A, Itterbeeck JV, Klunder H, Mertens E, Halloran A, et al. (2013) Edible insects: Future prospects for food and feed security. FAO Forestry paper 171. Rome: Food and Agriculture Organization of the United Nations. 201 p.
2. Mohanta KN, Subramanian S (2011) Nutrient of common fresh water ornamental fishes. Technical Bulletin No. 27. Goa: ICAR Research Complex for Goa. 55 p.
3. Hossain MH, Ahammad MU, Howlider MAR (2003) Replacement of fish meal by broiler offal in broiler diet. Int J Poult Sci 2: 159–163.
4. Naylor RL, Goldburg RJ, Primavera JH, Kautsky N, Beveridge MCM, et al. (2000) Effect of aquaculture on world fish supplies. Nature 405: 1017–1024.

5. Nyirenda J, Mwabumba M, Kaunda E, Sales J (2000) Effect of substituting animal protein source with soybean meal in diets of *Oreochromis karongae* (Trewavas 1941). Naga, ICLARM q 23: 13–15.
6. Xiaoming C, Ying F, Hong Z, Zhiyong C (2010) Review of the nuritive value of edible insects. In: Durst PB, Johnson DV, Leslie RL, Shono K, editors. Forest insects as food: humans bite back. Proceedings of a workshop on Asia-Pacific resources and their potential for development. Bangkok: FAO Regional Office for Asia and the Pacific. 85–92.
7. Anand H, Das S, Ganguly A, Haldar P (2008) Biomass production of acridids as possible animal feed supplement. **J Environ &** Sociobiol 5: 181–190.
8. AOAC (Association of official analytical chemists) (1990) Official methods of analysis, 15 th edition, volume 1. Virginia: Association of Official Analytical Chemists. 737 p.
9. Parker R (2011) Aquaculture Science. Stamford, CT: Cengage Learning. 672 p.
10. Sogbesan OA, Ugwumba AAA (2006) Bionomics evaluation of garden snail (*Limicolaria aurora*, Jay, 1937; Gastropoda: Limicolaria) meat meal in the diet of *Clarias gariepinus* fingerlings (Burchell, 1822). Nig J Fisheries 2: 358–371.
11. APHA (American Public Health Association) (1985) Standard methods for the examination of water and waste water, 16 th edition. Washington DC: American Public Health Association. 1268 p.
12. De Silva SS, Anderson A (1998) Fish nutrition in aquaculture. London: Chapman and Hall. 319 p.
13. Moyle PB, Cech JJ (1996) Fishes: An introduction to ichthyology. Englewood Cliffs, NJ: Prentice Hall Inc. 590 p.
14. Bailey M, Sandford G (2001) The ultimate aquarium. London: Anness publishing Ltd. 256 p.
15. Lawrence C (2007) The husbandry of zebrafish (*Danio rerio*): A review. Aquaculture 269: 1–20.
16. Bahnasawy MH, El-Ghobashy AE, Abdel-Hakim NF (2009) Culture of the Nile tilapia (*Oreochromis niloticus*) in a recirculating water system using different protein levels. Egypt Aquat Biol Fish 13: 1–15.
17. Timmerman CM, Chapman LJ (2004) Hypoxia and interdemic variation in *Poecilia latipinna*. J Fish Biol 65: 635–650.
18. Guillaume J (1997) Protein and amino acids. In: D'Abramo LR, Conclin DE, Akiyama DM, editors. Crustacean nutrition, Advances in world aquaculture. Louisiana: World Aquaculture Society. 26–41.
19. Shim KF, Chua YL (1986) Some studies on the protein requirement of the guppy *Poecilia reticulata* (Peters). J Aquariculture Aquat Sci 4: 79–84.
20. Cowey CB, Sargent JR (1979) Nutrition. In: Hoar WS, Randall J, editors. Fish physiology. New York: Academic Press. 1–69.
21. Fernando AA, Phang VPE, Chan SY (1991) Diets and feeding regimes of poeciliid fishes in Singapore. Asian Fish Sci 4: 99–107.
22. Furuichi M, Yone Y (1981) Availability of carbohydrate in nutrition of carp and red sea breem. B Japan Soc Sci Fish 48: 945–948.
23. Kalita P, Mukhopadhyay PK, Mukherjee AK (2007) Evaluation of the nutritional quality of four unexplored aquatic weeds from North East India for the formulation of cost-effective fish feeds. Food Chem 103: 204–209.
24. National Research Council (NRC) (1983) Nutrient requirements of warm water fishes and shellfishes. Washington: National Academy Press.102 p.
25. Hasan MR, Moniruzzaman M, Farooque AMO (1990) Evaluation of leucaena and water hyacinth leaf meal as dietary protein sources for the fry of Indian major carp, Labeo rohita (Hamilton). In: Hirano R, Hanyu I, editors. Second Asian Fisheries Forum. Manila: Asian Fisheries Society. 275–278.
26. Akand AM, Hassan MR, Habib MAB (1991) Utilization of carbohydrate and lipid as dietary energy sources by stinging catfish H. fossilis (Bloch). In: DeSilva S, editor. Fish nutrition research in Asia. Proceedings of the fourth Asian fish nutrition workshop. Manila: Asian Fisheries Society. 93–100.
27. Arockiaraj AJ, Muruganandam M, Marimuthu K, Haniffa MA (1999) Utilization of carbohydrates as a dietary energy source by stripped murrel *Channa striatus* (Bloch) fingerlings. Acta Zool Taiwanica 10: 103–111.
28. Sogbesan O, Ugwumba AAA (2008) Nutritional evaluation of termite (*Macrotermes subhyalinus*) meal as animal protein supplements in the diets of *Heterobranchus longifilis* (Valenciennes, 1840) Fingerlings. Turk J Fish Aquat Sc 8: 149–157.
29. Martínez HH, Ramos-Elorduy J, Pino-Moreno JM, Acosta-Castaneda C (2008) Evaluation of diets including nymphs of *Periplaneta americana*. Cienc Pesq 16: 25–30. (In Spanish with English abstract).
30. Lagler KF (1956) Freshwater Fishery Biology. Iowa: W.C. Brown Company. 421 p.
31. Wade JW (1992) The relationship between temperature, food intake and growth of brown trout, *Salmon trutta* (L.) fed natural and artificial pelleted diet in earth ponds. J Aquat Sci 7: 59–71.

Deforestation and Benthic Indicators: How Much Vegetation Cover Is Needed to Sustain Healthy Andean Streams?

Carlos Iñiguez-Armijos[1,2]*, Adrián Leiva[1], Hans-Georg Frede[2], Henrietta Hampel[3], Lutz Breuer[2]

1 Departamento de Ciencias Naturales, Universidad Técnica Particular de Loja (UTPL), Sección de Ecología, Loja, Ecuador, **2** Institute for Landscape Ecology and Resources Management (ILR), Research Centre for Bio Systems, Land Use and Nutrition (IFZ), Justus-Liebig-University Giessen, Giessen, Germany, **3** Universidad de Cuenca, Quinta Balzaín, Cuenca, Ecuador

Abstract

Deforestation in the tropical Andes is affecting ecological conditions of streams, and determination of how much forest should be retained is a pressing task for conservation, restoration and management strategies. We calculated and analyzed eight benthic metrics (structural, compositional and water quality indices) and a physical-chemical composite index with gradients of vegetation cover to assess the effects of deforestation on macroinvertebrate communities and water quality of 23 streams in southern Ecuadorian Andes. Using a geographical information system (GIS), we quantified vegetation cover at three spatial scales: the entire catchment, the riparian buffer of 30 m width extending the entire stream length, and the local scale defined for a stream reach of 100 m in length and similar buffer width. Macroinvertebrate and water quality metrics had the strongest relationships with vegetation cover at catchment and riparian scales, while vegetation cover did not show any association with the macroinvertebrate metrics at local scale. At catchment scale, the water quality metrics indicate that ecological condition of Andean streams is good when vegetation cover is over 70%. Further, macroinvertebrate community assemblages were more diverse and related in catchments largely covered by native vegetation (>70%). Our results suggest that retaining an important quantity of native vegetation cover within the catchments and a linkage between headwater and riparian forests help to maintain and improve stream biodiversity and water quality in Andean streams affected by deforestation. This research proposes that a strong regulation focused to the management of riparian buffers can be successful when decision making is addressed to conservation/restoration of Andean catchments.

Editor: Tomoya Iwata, University of Yamanashi, Japan

Funding: This research was funded through a scholarship of the Secretaria Nacional de Educación Superior, Ciencia, Tecnología e Innovación of Ecuador (SENACYT), a DFG (Deutsche Forschungsgemeinschaft) grant (FR615/28-1/-2) and by the Universidad Técnica Partucular de Loja (UTPL). The support from Henrietta Hampel was thanks to the PROMETEO Program of SENESCYT. The funders had no role in study design, data collection and analysis, decision to publish, or preparation of the manuscript.

Competing Interests: The authors have declared that no competing interests exist.

* Email: cainiguez@utpl.edu.ec

Introduction

Tropical deforestation is one of the most important topics worldwide associated with biodiversity loss and ecosystem degradation [1–3]. However, the effects of deforestation on ecological condition of tropical streams have been rarely investigated despite evidence that forest loss negatively affects these streams [4–7]. In Ecuador, deforestation for agriculture and/ or urbanization is a critical threat for Andean ecosystems [8], reaching annual deforestation rates of 2.7% between 1989 and 2008 in the montane forests of southern Andes (Tapia Armijos, pers. communication). However many of these ecosystems provide essential environmental services including biodiversity conservation, carbon storage and water supply [9]. Several studies focused along the Ecuadorian Andean region have demonstrated that deforestation affects habitat structure and macroinvertebrate community assemblages in streams [10], litter breakdown rates [11] and hydrological processes [12]. Clearly the Ecuadorian montane streams are influenced by the surrounding landscape and several research groups have recently deployed considerable efforts to assess these relationships. Nonetheless, information gaps concerning the influence of land cover/use and spatial scale on stream condition remains, despite they affect water quality and aquatic diversity [13,14]. These data are crucial for management and conservation of developed catchments to insure long term sustainability [15].

In other regions, organizations such as the European Environment Agency (EEA), the United States Environmental Protection Agency (EPA) and national agencies in New Zealand have identified, and are currently considering within their monitoring programs, the impacts of land cover/use changes on stream ecological conditions based on responses of macroinvertebrate communities [16–19]. In addition, they have developed catchment and/or riparian rehabilitation projects in order to enhance stream condition [20]. In South and Central America such extensive monitoring or restoration programs by governmental initiatives are rare, despite their potential to reduce the impact of deforestation on stream condition [13]. In this regard, a Costa

Rican regulation has showed the effectiveness of protecting riparian buffers to reduce the impacts of deforestation on macroinvertebrate communities and sensitive taxa in tropical streams [4].

Several authors [10,21–23] have applied a variety of macroinvertebrate metrics (e.g. Fisher's index; EPT taxa) to assess the impact of land cover/use on the ecological condition of streams in South America focusing on the structure and composition of benthic communities. These studies have confirmed changes to aquatic biodiversity and water quality because of deforestation around the streams. Nevertheless, how does vegetation cover influences stream ecological condition? In streams running through forested catchments the macroinvertebrate communities are more heterogeneous, diversity of intolerant taxa to environmental stress is higher, and abundance and variety of habitats is greater [10,14,21]. Likewise, streams with forested riparian areas have low variation in water temperature and less nutrients and sediment loads along the entire length [4,13]. At the catchment scale, there is also an influence of vegetation cover on controlling hydrology and the entire diversity of the streams [12]. In this regard, estimating the extent of native vegetation cover in a catchment and the importance of riparian management and headwater forest protection to maintain and improve biodiversity and water quality of streams affected by deforestation is a pressing task for management and conservation [13,14,24].

Geographic Information Systems (GIS) allow researchers to explore correlations in spatial patterns with ecological features in the landscape. Allan [13] describes the ability to apply GIS tools with remote sensing data to assess the influence of land cover/use on stream ecological condition. Studies in temperate areas have described the influence of land cover and spatial scale on stream communities [25,26] and on stream water quality [14] by applying GIS based evaluation techniques to derive land cover data and to relate them with physical, chemical and biological data.

In this study, our objective was to investigate the influence of vegetation cover at three spatial scales (local, riparian buffer and catchment) on structure and composition of macroinvertebrate communities and on water quality of headwater streams in the southern Ecuadorian Andes. These streams are located in two contrasting regions of varying intensities of anthropogenic pressure (i.e. deforestation, population growth, and farming). The questions we addressed were: what spatial scale and how much vegetation cover are associated with healthy streams based on macroinvertebrate metrics? In order to address these questions, we calculated several diversity and water quality metrics from macroinvertebrate communities, derived spatial information on land cover from GIS, and related these metrics statistically to three spatial scales and stream parameters. Finally, we addressed management and conservation problems in order to help mitigating impacts of deforestation on stream ecosystems.

Methodology

For this study, the Ministry of Environment of Ecuador issued us a general research permit for every sampling site and to collect water samples and macroinvertebrate specimens. In the case of the Zamora upstream catchment (Zu), the Municipality of Loja granted a specific permission to work at sites Zu1–2, 4–5, 8 because these streams are used for the water supply of the city. To work at the San Francisco catchment (SF), the administration of the San Francisco Research Station and Nature and Culture International allowed us to access sites SF1–7. For all other sites we used the general research permit.

Most of macroinvertebrates collected in this study were identified at genus level and we do not know if any of these genera are endangered or protected, because of the lack of information for Ecuador.

Study area

We studied 23 streams of two headwater catchments of the Zamora River basin, located in the mountain forests in Ecuadorian southern Andes (Figure 1). The Zamora upstream catchment covers an area of 276.2 km^2 and ranges between 2,020 to 3,250 m a.s.l. It is the most populated catchment within the Zamora River basin with approximately 120, 000 inhabitants in the city of Loja. The San Francisco catchment is located in a sparsely populated area of the Zamora River basin with an area of 75.3 km^2 and an elevation range from 1,800 to 3,250 m a.s.l. The natural vegetation in both study catchments is evergreen mountain rain forests, humid shrub, cloud forest and páramo [27,28]. Anthropic activities in both catchments have substituted the original vegetation mainly to pasture, and to a lesser extent to cropland, pine (*Pinus*) and eucalyptus (*Eucalyptus*) plantations [29]. Urban development has only taken place in the Zamora upstream catchment and has spread along its main river. Both catchments have similar climatic conditions with a wet season from May to September and a dry season from October to April. The average annual precipitation and temperature in the Zamora upstream and San Francisco catchments are about 2,200 mm and 15.2°C respectively.

Stream measurements and macroinvertebrates sampling

We sampled 16 streams in the Zamora upstream catchment and 7 streams in the San Francisco catchment. Sampling focused on riffles with an altitudinal variation of 350 m among sites. Every site consisted of a 30–60 m reach placed at the outlet of each stream in order to capture the influences of land use upstream on macroinvertebrate communities. According to the sampling frequency applied in other studies with benthic macroinvertebrates both in Ecuador and in other Neotropical regions [4,10,11,30–34], we collected physical–chemical and macroinvertebrate data during both dry (November 2010; December 2012) and wet seasons (May 2011; April 2012) at the same stream reach. Physical–chemical parameters such as pH, temperature (*T*), specific conductance (SC), dissolved oxygen (DO), were assessed in–stream in each of the sampling periods (Table S1) with a portable YSI 556 Multi Probe System and were averaged from several measurements during the sampling campaigns. Grab samples were taken for measuring further water chemistry at the laboratory of the Universidad Técnica Particular de Loja (UTPL) using standard methods [35]: nitrate (NO_3^-), phosphate (PO_4^{3-}), biochemical oxygen demand during decomposition in a period of 5 days (BOD_5), total coliforms (TC) and turbidity (Table S1).

For sampling of macroinvertebrate communities several methods exist and each one provides different results, nevertheless a combination of sampling methods is preferable [30]. In this study, qualitative macroinvertebrates samples were taken along the stream reach applying two sampling methods: 1) collecting macroinvertebrates directly in their stream habitats from stones and leaves, and 2) using a D–frame net (500 µm mesh) for kick sampling at 3–6 points selected randomly. Collection method 1 was applied during 10 min along the reach to remove all visible macroinvertebrates from stones and collect leaves respectively. Selected stones collected were totally submerged and had an average diameter of 20 cm approximately. Leaves were stored in a plastic bag and rinsed with tap water to remove macroinvertebrates in the laboratory. For collection method 2, macroinverte-

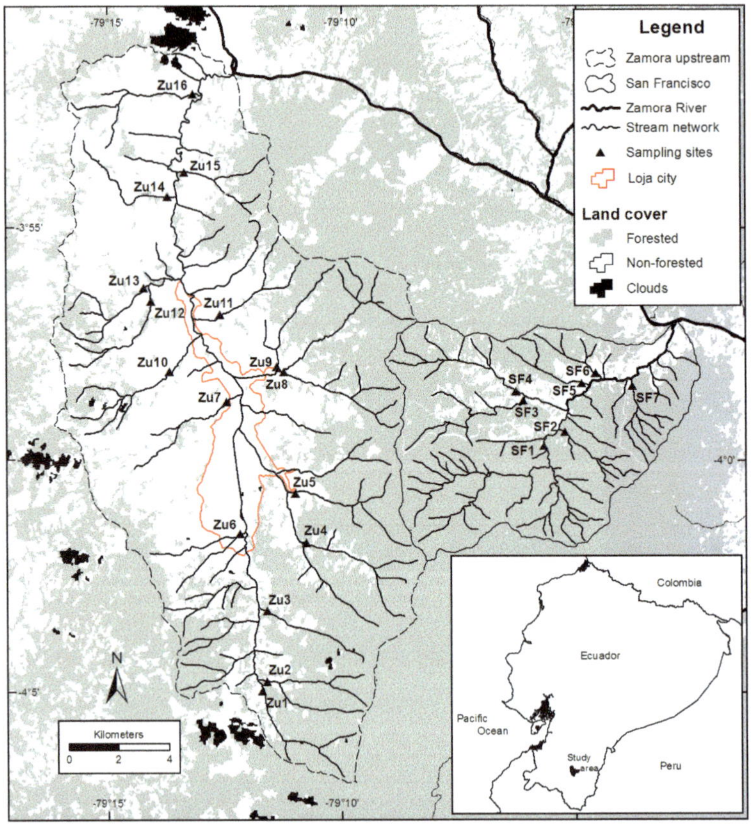

Figure 1. Study area and location of the sampling sites. Codes represent the sampling sites in the Zamora upstream (Zu1 to Zu16) and San Francisco (SF1 to SF7) catchments. Native vegetation cover was classified into forested and non-forested. Only the complete stream network for the study area is shown. Catchment characteristics are further shown in Table S1.

brates were sampled by kicking the substrate while holding the D-frame net downstream at each point. The application of this technique was based on its suitability for sampling in a variety of stream substrates and depths [36,37]. Both collections were combined into a single collection per site in order to avoid differences on macroinvertebrate communities due to seasonality of the Ecuadorian Andes [38,39]. All macroinvertebrates were preserved in 95% ethanol and brought to the laboratory to be sorted and identified to the lowest taxonomical level possible using identification keys for South American macroinvertebrates [40,41].

GIS data and vegetation cover

The Geographic Information System software ArcGIS 9.3 was used to evaluate spatial digital information. Catchments were delineated from a Digital Elevation Model (DEM) with a10 m resolution, available at the Research Unit 816 (RU 816) database (www.tropicalmountainforest.org/) of the Deutsche Forschungsgemeinschaft (DFG). Stream networks were obtained from the national cartography created by the National Mapping Agency (IGM; Instituto Geográfico Militar).

Based on Allan [13], the influence of vegetation cover (VC) on stream condition was evaluated at three spatial scales: 1) catchment scale representing the entire area upstream of the sampling site; 2) riparian scale, using a buffer of 30 m on each bank along the entire length of the stream; and 3) local scale, comprised by the same buffer width along a 100 m reach upstream of the sampling site. Buffer width chosen was based on regulations by local governments (called Municipal Ordinances) in Ecuador. These ordinances reflect several law articles of the Ecuadorian Government in terms of water resources protection, but it does not specify or recommend an amount of riparian vegetation.

A 2010 land use map provided by M.F. Tapia-Armijos (pers. communication) was used to quantify VC within each catchment. This map was generated for southern Ecuador from ASTER images (http://asterweb.jpl.nasa.gov/) with a minimum mapping unit of 1 ha and classified into eight main land use types: páramo, forest, shrub, pasture, crop, plantation, urban and un-vegetated. In Table 1 we show the land use and percentage area covered by each land use type within the 23 studied catchments, with forest and pasture classes as the most dominating ones. Urban land use was absent or very poorly represented. For this study we aggregated this classification to two land cover categories, i.e. forested and non-forested. Forest class included all types of native vegetation and non-forest class all types of disturbed areas (pasture, cropland, plantation, urban, un-vegetated). The forest class was used to determine the VC only at catchment scale, because resolution and accuracy may limit the interpretation of riparian land cover in mountainous areas [13,25,42]. In order to address the riparian and local scales, the VC within the 30 m buffer was digitized and interpreted from 2011 rectified colored orthophotographs (0.30 m resolution). These orthophotographs were obtained and provided by the Ecuadorian project SIG-TIERRAS (Ministerio de Agricultura, Ganadería, Acuacultura y

Table 1. Land use characteristics and area (%) of each land use type of the 23 studied catchments.

Catchment	Páramo	Forest	Shrub	Pasture	Plantation	Crop	Urban	Un-vegetated
Zu1	0.8	69.4	13.4	13.6	0.3	2.3	0.0	0.2
Zu2	9.3	61.1	14.4	12.5	0.5	2.1	0.0	0.0
Zu3	1.4	59.4	10.7	25.0	0.7	2.6	0.0	0.1
Zu4	2.8	46.9	19.3	25.2	0.5	5.3	0.0	0.2
Zu5	4.4	54.9	27.9	10.4	0.9	1.4	0.1	0.1
Zu6	0.3	8.8	17.6	66.0	2.2	2.9	1.5	0.7
Zu7	0.0	4.6	13.2	71.1	0.8	9.0	0.0	1.2
Zu8	2.7	39.0	25.6	30.9	0.7	1.0	0.0	0.2
Zu9	0.5	41.0	12.4	35.9	1.5	8.3	0.2	0.1
Zu10	0.4	8.0	28.0	47.1	2.5	13.4	0.0	0.6
Zu11	0.4	33.7	9.1	42.0	2.7	11.9	0.0	0.2
Zu12	0.2	17.6	4.7	57.3	2.9	16.3	0.0	0.9
Zu13	0.4	16.0	19.2	48.9	1.2	13.5	0.0	0.7
Zu14	0.0	6.5	0.1	73.2	1.9	15.0	0.0	3.3
Zu15	0.0	51.9	9.9	32.9	1.1	4.0	0.0	0.2
Zu16	0.0	15.6	0.2	72.4	2.5	7.0	0.0	2.3
SF1	0.0	82.3	5.3	5.6	0.0	0.0	0.0	6.8
SF2	2.0	93.5	4.0	0.0	0.0	0.0	0.0	0.5
SF3	0.0	78.0	5.0	15.8	0.3	0.0	0.0	0.9
SF4	1.9	87.6	3.7	6.3	0.1	0.0	0.0	0.3
SF5	0.4	81.8	3.0	14.1	0.4	0.0	0.0	0.3
SF6	0.4	76.3	2.8	18.9	1.2	0.0	0.0	0.4
SF7	2.0	92.0	3.9	2.0	0.0	0.0	0.0	0.1

Pesca; Proyecto Sistema Nacional de Información y Gestión de Tierras Rurales e Infraestructura Tecnológica).

Data analysis

We calculated nine indices (also referred as metrics) to describe ecological conditions of the streams. We calculated three structural, two compositional and four water quality indices (Table 2) that allowed us assessing stream conditions in response to increasing VC. The structural indices were richness, Fisher's index, and evenness (Pielou's index). For compositional indices we classified macroinvertebrates by functional feeding groups (FFG) according to [43,44], but we only used percentage of scrappers and shredders because they have an expected decreasing response to perturbation [36], in our case deforestation.

Water quality was assessed with the Water Quality Index (WQI), the EPT index, the Biological Monitoring Working Party index adapted to Colombia (BMWP/Col), and the %5 Dominant taxa. WQI is a single score composed of several measurements of various physical, chemical and bacteriological parameters, where low WQI values show high influence of environmental stress [45]. EPT index shows the taxa richness within the macroinvertebrate groups which are considered sensitive to stressors, and therefore should increase with increasing water quality [46]. BMWP/Col index provides a single value at the family level, representative of the organism's tolerance to stressors, where high tolerance to pollution corresponds to low BMWP scores [40]. %5 Dominant taxa is a measure of the abundance of the five most dominant taxa to the total number of organisms in the sample at each site [36]. When a community is dominated by few taxa it will have high dominance values, thus indicating the community is under the influence of environmental stress.

Two statistical approaches have been used to evaluate relationships between vegetation cover/land use and ecological indices; these are regression [14,42,47] and ordination analyses [4,10,48]. Prior to analyses, macroinvertebrate abundance data (x) were transformed by $\log_{10}(x)$ to reduce the effect of large numbers in single taxa (e.g. *Baetodes*). Rare taxa, i.e. taxa represented by three or less individuals in a sampling site, were excluded in order to avoid the influence of these taxa in the ordination analysis [49]. Because explanatory variables did not have a uniform scale, stream parameters were $\log_{10}(x+1)$ transformed in order to avoid the problem with 0 values and reduce the range of data. All of these variables were analyzed by pairs in scatterplot matrices, and through an exploratory Redundancy Analysis (RDA) to identify the least correlated and most informative variables (PO_4^{3-}, DO, TC and turbidity).

An important issue was to identify whether ecological condition of the streams was attributable to a land use type and/or a stream parameter. We used a stepwise multiple linear regression (SMLR) with calculated macroinvertebrate metrics as dependent variables against land use and PO_4^{3-}, DO, TC and turbidity to determine which of those independent variables affect our response variables. SMLR was applied with the 'stepAIC' function of 'MASS' package in R [50] using backward elimination of variables and Akaike Information Criterion for variables or predictors selection. In addition, we estimated the relative importance of predictors based on their contribution to R^2 using boostrap ('relaimpo' package). All p–values were adjusted with a Bonferroni correction for multiple comparisons during regression analyses. These analyses indicated that forest class (i.e. VC) was consistently more informative to explain macroinvertebrate metrics than other land use types or stream parameter (see Table S2 for SMLR results). Therefore, to examine the linear dependency of each of our response variables with increasing VC, macroinvertebrate metrics (dependent) were regressed against percentages of VC at each spatial scale (independent) using the method of least mean squares ('lm' function of 'stats' package).

We used a RDA ('rda' function of 'vegan' package) to identify macroinvertebrate community assemblages associated with percentages of VC, PO_4^{3-}, DO, TC and turbidity. RDA evaluates multivariate response data and is a powerful tool of constrained ordination in community ecology [51]. RDA works on a matrix Y of response data (e.g. macroinvertebrate data) and a matrix X of explanatory variables (e.g. water chemistry, VC). It identifies groups and shows the vector fitting of variables used in the analysis. While the vector fitting is commonly used for environmental interpretation in ordination diagrams, this implies a linear relationship between ordination scores and the environment

Table 2. Results of regression analysis between calculated macroinvertebrate metrics and percentage of vegetation cover at multiple spatial scales.

Spatial scale	**Catchment**		**Buffer**		**Local**	
Metric	R^2	p	R^2	p	R^2	p
Structural						
Richness	0.30	<0.01	0.34	<0.001	0.10	0.13
Fisher's index	0.50	<0.001	0.38	<0.001	0.01	0.30
Evenness	0.24	<0.01	0.11	0.11	0.00	0.93
Compositional						
% Scrappers	0.02	0.57	0.01	0.72	0.02	0.55
% Shredders	0.42	<0.001	0.46	<0.001	0.12	0.09
Water Quality						
WQI†	0.56	<0.001	0.50	<0.001	0.12	0.11
EPT index	0.53	<0.001	0.59	<0.001	0.03	0.39
BMWP/Col	0.44	<0.001	0.46	<0.001	0.17	0.05
%5 Dominant taxa	0.85	<0.001	0.60	<0.001	0.02	0.50

†WQI is the unique metric derived from water chemistry data showed in Table S1.

variables and may not always be appropriate if it is desired to understand the ordination of studied sites or species in relation to one or more variables. Thus, we fitted the surface of the best predictor variable, identified with an ANOVA of the RDA results, to detect how macroinvertebrate communities are grouped and spread in relation to this variable. We examined whether macroinvertebrate communities were different between those groups using an Analysis of Similarities (ANOSIM; 'anosim' function of 'vegan' package). ANOSIM is a nonparametric procedure that provides a test of whether there is a significant difference between two or more groups of sampling units.

Results

Stream condition and vegetation cover

Relationships between VC and all metrics are generally stronger at catchment and buffer scale than at local scale where associations are not apparent (Table 2). Significant correlations are found on catchment and buffer scale in 90% and 80% of all cases. Within the structural metrics, only evenness does not show significant relationship with the percentage of VC at buffer scale ($R^2 = 0.11$; $p = 0.11$). In general, correlations decrease with smaller spatial scales, disappearing completely at local scale.

As for compositional metrics, only one FFG shows a strong relationship with VC at the two larger scales ($p<0.001$). The percentage of shredders is more related to VC at catchment and buffer scales (Table 2). Scrappers show weaker regressions at all spatial scales ($R^2<0.02$). At local scale no significant correlation existed at all.

All water quality metrics have stronger relationships at catchment and buffer scale than at local scale ($R^2>0.44$ and $p<0.001$, Table 2). Within water quality metrics, the %5 Dominant taxa depicts the highest correlation across all scales with 0.85 on catchment scale. Regression slopes indicate significant differences in water quality when VC increases at both spatial scales. However, there are some streams with a large area covered by forest at catchment and buffer scale which present low values in water quality indices (Figure 2B–C; F–G). For example, in both spatial scales the EPT index increase with VC ($R^2 = 0.53$; $p<0.001$ and $R^2 = 0.59$; $p<0.001$), but there is one site with approximately 70% of forest cover whose index value is less than 50, indicating fairly poor water quality. In addition, the strongest relationship is inversely correlated between the %5 Dominant taxa and the percentage of VC in the catchment ($R^2 = 0.85$; $p = 0.001$). As a general pattern most of the macroinvertebrate metrics indicate good water quality when the percentage of native VC is higher than 70% within the catchment and the riparian buffer (see Figure 2).

Macroinvertebrate communities and vegetation cover

All structural indices are related with VC at catchment scale rather than at other spatial scales (Table 2). Thus, we decided to further analyze the community assemblage patterns related to catchment forest cover. A correlation analysis run for those variables showed scatterplot matrices and correlation values from 0 to 1.We defined a decision threshold of 0.50 to remove those variables which are correlated with more than one variable. Remaining variables include VC, DO, PO_4^{3-}, TC and turbidity. RDA reveal that 51% of the variation in the data can be explained by these five selected variables and the ANOVA identified that catchment VC is the best predictor for macroinvertebrate composition in the ordination ($p<0.001$). In Figure 3A, most variables are spread along negative axis 1 and negative axis 2, with VC, PO_4^{3-} and turbidity being the most influential factors for axis 1, while DO and TC are more associated with axis 2. Figure 3B depicts the fitted surface of the best predictor variable (i.e. VC) to detect how macroinvertebrate communities are grouped and spread in relation to VC. Thus the ordination plot illustrates two visibly distinguishable groups; one group associated to catchments with a forest cover above 70% which is even more aggregated than the other group related to lower VC. The ANOSIM indicates that both groups of macroinvertebrate assemblages were strongly dissimilar between both ranges of VC ($R = 0.55$; $p = 0.001$). Both RDA plot and Bray–Curtis distance of the ANOSIM suggest that macroinvertebrate assemblages differ in taxonomical composition when there is a decline in the percentage of VC within the catchments.

Overall, macroinvertebrate communities in the 23 study streams present specific patterns. They are composed of a total of 13,512 individuals belonging to 11 orders, 33 families and 53 genera (data available at PANGAEA – Data Publisher for Earth & Environmental Science; doi:10.1594/PANGAEA.831687). Only two taxa (taxa–1 = Empididae; taxa–2 = Pyralidae) were not identified to genera level.

Comparing occurrence and abundance patterns between both groups of catchments (i.e. less than 70% and greater than 70% VC) we detected several singularities. In catchments with a VC between 70 to 100% we collected 4,061 individuals of 11 orders. In contrast 9,451 individuals of nine orders were found in catchments with less VC (Basommatophora and Lepidoptera were absent). We found 33 families, all of them present in catchments highly forested; whilst 10 families were absent in catchments with less than 70% forest cover (Calamoceratidae, Ceratopogonidae, Coenagrionidae, Dolichopodidae, Gyrinidae, Odontoceridae, Physidae, Psephenidae, Ptilodactylidae, and Pyralidae). With regard to the 53 genera found, only *Scirtes* (Coleptera) and *Molophilus* (Diptera) were absent in catchments above 70% forest cover, and 25 genera were absent in catchments with lower forest cover showing the greatest difference between both groups of catchments at this taxonomic level. Here genera absent at <70% catchment VC were: *Physa* (Basommatophora); *Noelmis*, *Phanacerus*, *Dineutus*, *Anchytarsus*, *Psephenops*, *Elodes* (Coleoptera); *Probezzia*, *Aphrosytus*, *Hexatoma* (Diptera); *Andesiops*, *Moribaetis*, *Haplohyphes*, *Tricorythodes*, *Farrodes* (Ephemeroptera); taxa–2 (Lepidoptera); *Argia* (Odonata); *Phylloicus*, *Leptonema*, *Macrostemum*, *Grumichella*, *Nectopsyche*, *Triplectes*, *Marilia*, *Cernotina* (Tricoptera).

A trend can be observed here as well, confirmed by the ANOSIM, as there are differences in macroinvertebrate compositions between both groups of catchments (Table 3). We detect an increase for almost all metrics of the macroinvertebrate communities with an increase of the percentage of VC in the catchment, except for evenness where no changes are observed.

Discussion

Stream condition and vegetation cover

Our results show a stronger relationship between VC and the macroinvertebrate metrics at catchment and buffer scales. This suggests an important linkage between headwater and riparian forests to conserve diversity and water quality of the streams in Ecuadorian Andean catchments. We observe that stream condition begins to improve when native VC within the catchment is over 70% (see Figure 2). Therefore our suggestion is to retain at least 70% native VC within the catchment in order to keep sufficient ecological condition of the streams. Such a percentage of VC can contribute significantly to riparian management for protection or restoration of watersheds where forest has been

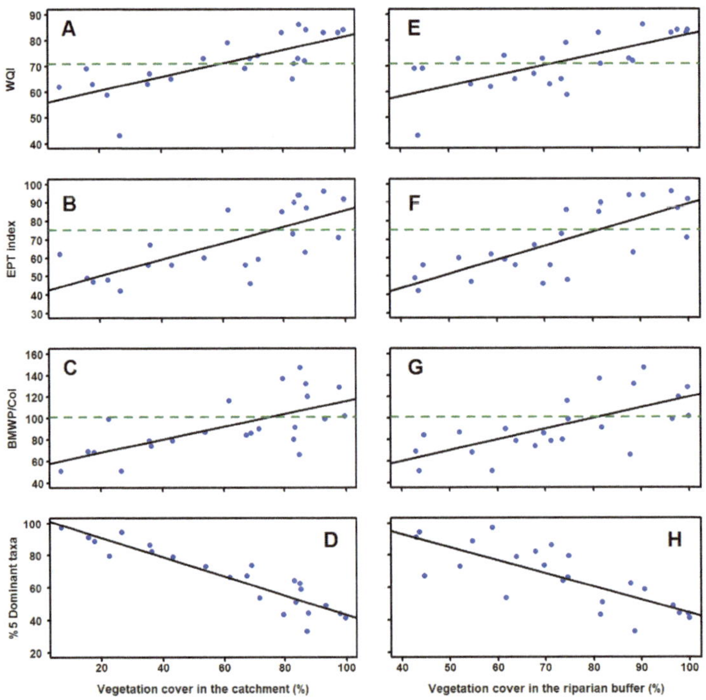

Figure 2. Water quality indices plotted against vegetation cover (%) at catchment and buffer scales. Horizontal green dashed lines represent thresholds of good water quality defined for each metric in the literature. Only the %5 Dominant index does not have ranges of tolerance, but high percentages indicate stress in the communities. For both spatial scales R^2 values are over 0.44 with a p–value <0.001, see Table 2 for specific values of each regression.

replaced by grasslands. Further, our findings show that local scale strategies (e.g. conserving small forest patches in riparian areas) do not improve water quality and are unlikely to improve the overall ecological condition of streams. According to [13,47], the maintenance of ecological conditions may require conservation of more than only a few native forest patches, and efforts should

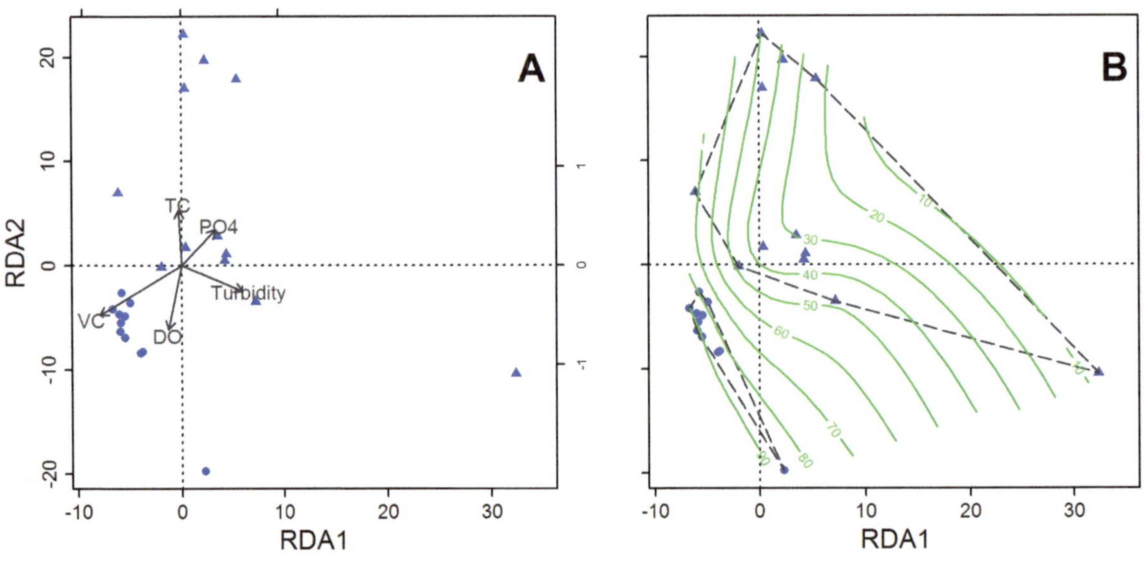

Figure 3. Redundancy Analysis (RDA) of macroinvertebrate community assemblages from catchments with different vegetation covers (VC). This ordination analysis detected two groups of macroinvertebrate assemblages: solid circles represent macroinvertebrate communities in catchments with 70–95% in VC and triangles indicate macroinvertebrate communities in catchments with less than 70% in VC. Each symbol represents the community of a single catchment, i.e. macroinvertebrate community at catchment SF3. A Shows the RDA ordination with vector fitting of five variables selected during analysis. DO = dissolved oxygen, TC = total coliforms, PO_4 = phosphate. B Displays in green curves the fitted surface only for the percentage of VC, which is the most explicative variable of the ordination (65%). Dashed lines enclose and separate the detected groups.

Table 3. Diversity patterns of macroinvertebrate communities in both groups of catchments identified in RDA ordination plot (Figure 3B). VC = catchment vegetation cover.

Vegetation cover	<70% VC		>70% VC		Response to increasing VC
Metric	**Mean**	**SD**	**Mean**	**SD**	
Structural					
Richness	14.3	2.1	20.1	6.2	Increase
Fisher's index	2.5	0.4	4.9	1.3	Increase
Evenness	0.7	0.1	0.7	0.1	Variable
Compositional					
% Scrappers	26.0	16.3	30.6	17.3	Increase
% Shredders	0.3	0.8	3.1	2.6	Increase

focus on considering larger spatial patterns that have more important consequences for stream conservation.

Macroinvertebrates diversity is more related to native VC at catchment and buffer scale than at a local scale. Many studies have documented the influence of upstream vegetation on stream ecological condition downstream, especially on water chemistry and habitat conditions [13,14,52]. For example, DO increases as the stream enters the forest [53]. Likewise, the similar relationships shown between water quality metrics and VC at catchment and buffer scale is not surprising given the strong longitudinal connectivity in a riverine landscape [13,54]. Riparian forests can influence water chemistry and habitat quality, which are directly linked to the macroinvertebrate metrics and help to minimize the effects of catchment deforestation on streams [4,42,55].

In this study, there were two particular sampling points in streams of catchments Zu4 and Zu8 (see Figure 1 and Table S1). Both had low values in water quality indices but a relatively high VC (60–68%). We assume that livestock and minor roads which are located parallel to the streams influence these metrics. For example, in [4] the authors found a stream affected by inputs of sediments and nutrients derived from roads and livestock, despite having a wide riparian buffer. We believe those elements are affecting the macroinvertebrate composition of very well forested catchments as discussed below.

Macroinvertebrate communities, indicators and vegetation cover

Analyzing the composition of macroinvertebrate communities of 23 Andean streams, we are able to identify direct relationships with the coverage of the catchment's native forest. The difference between macroinvertebrate assemblages increases when decreasing the percentage of native VC. We find a group with a more aggregated distribution when VC is large in the RDA plot. This pattern was also observed by [14] in a larger sample of streams in New Zealand. It supports the premise that deforestation is affecting the macroinvertebrate community assemblages in Ecuadorian Andean streams as in other Neotropical streams [4,21,56].

The difference between both groups of macroinvertebrate communities detected in the RDA is also revealed in the ANOSIM. Bücker et al. [10] found similar patterns in the San Francisco catchment having less variation at structural and compositional level and higher water quality metric values in forested than in deforested streams. Comparable results were also found in other Andean streams [11] and Amazon streams [21] of northern Ecuador. Nonetheless, among all metrics used here and mostly applied in similar studies, %5 Dominant taxa reveals a superior fitting to the regression slope despite it has not been used to reflect deforestation effects on streams before. In [36], authors denoted that high values of this metric indicate a community under environmental stress.

In terms of the FFG composition, we observe a direct response to increasing VC in both groups of macroinvertebrates (Table 3). In average, the abundance of shredders present in streams with > 70% VC in catchment is always higher than in less forested catchments. Shredders abundance is typically linked to leaf litter from riparian vegetation and it is sensitive to riparian zone impacts [43,57,58]. Shedders could be considered as an indicator of deforestation in Andean streams, despite proportion of shredders in macroinvertebrate communities in our catchments are always small (see Table 3). According to [4,5] this seems to be a general pattern in tropical streams. On average, scrapers abundance is slightly higher in more forested catchments in contrast to other Neotropical streams where their abundance is higher in less forested streams [4]. Additionally no association between scrappers and VC was detected.

Management and conservation

Allan et al. [13,47] have recommended that it is preferable to retain as much forest cover as possible in a catchment to preserve natural stream condition. However, in tropical regions like Ecuadorian Andes, deforestation has been increasing throughout the years affecting a large part of catchments along the Andes [29,59,60]. Deforestation has stopped in the upper Andean catchments because of topography and accessibility, preserving headwater forests and/or páramo [29]. However, this native vegetation in the headwaters is not enough to sustain healthy streams. In Andean countries, human activities are especially concentrated on midstream sections and valleys of catchments affecting water/habitat quality downstream and other rivers after the confluence. For example, within Guayas river basin in Ecuador there are several cities and in combination with intensive agricultural land use they are affecting the entire river network and its water quality [61]. Our results suggest that retaining over 70% native forest cover in Andean catchments now dominated by grasslands and keeping the riparian buffers, the water quality and stream biodiversity could be enhanced or at least preserved. However, this task is not so easy to achieve if there is not a strong regulation at national level.

The Ecuadorian constitution defines the right to water and promotes the protection, restoration and management of basins and water resources; but it does not mention how much forest needs to be retained in a catchment or a recommended width for riparian buffers. Best conservation actions that intend maintaining as much forest as possible have been focused on catchments for drinking water supply from initiatives of local governments (Municipalities) supported by non–governmental organizations, through the establishment of trust funds for catchment conservation. Also, local governments through Municipal Ordinances have decided to protect riparian strips with different buffer widths (15–30 m), depending if the stream is running in the city or outside the urban boundary. Riparian buffers could considerably reduce the effects of deforestation on benthic communities and water quality, but there is need for a proper management of riparian areas [4]. Unfortunately, Municipal Ordinances in Ecuador do not include the management of riparian buffers. Here, we assessed the effect of a 30 m buffer width on stream condition showing strong associations with the water quality metrics. Therefore, we also recommend protection and an effective management of native riparian forest buffers in Andean streams to assist headwater forests to sustain ecological condition of the streams.

We did not evaluate any management approaches of riparian areas, but we would like to recommend some strategies that governments can include in national regulations. We believe that the most important problems that managers will face in Andean streams are livestock density, residential areas and parallel roads along riparian zones which affect water quality and macroinvertebrate communities in forested catchments. Therefore, these problems must be addressed; for example, livestock exclusion jointly with a long–term research program to assess its effectiveness [62]; or an in–depth monitoring to further study road effects on stream and riparian vegetation [63]. Another important issue to be considered is the restoration of riparian vegetation, mainly because it enhances water quality, as our study indicates. Restoration positively affects habitat and macroinvertebrate community heterogeneity [48] and can be ecologically successful by using native plant species and avoiding the use of invasive plant species, which can affect structure and functionally of streams [64]. Finally, as mentioned by [4] we would like to emphasize that "riparian forest buffers are not a replacement for continuous forest, but can promote biodiversity and stream ecosystem integrity in tropical catchments affected by deforestation".

Supporting Information

Table S1 List of environmental variables considered in the 23 studied Andean streams. Data of water parameters are means obtained from several measurements in two sampling campaigns. Altitude was determined once at the sampling sites. Catchment area and percentage of vegetation cover (VC) at three spatial scales were calculated in ArcGIS 9.3 Geographic Information System software. T = Temperature, SC = Specific Conductance, DO = Dissolved Oxygen, NO_3^- = Nitrate, PO_4^{3-} = Phosphate, BOD_5 = Biochemical Oxygen Demand in 5 days. Turbidity in Nephelometric Turbidity Units or NTU and Total Coliforms (TC) as Colony Forming Units or CFU.

Table S2 Stepwise multiple linear regression analysis between macroinvertebrate metrics, land use and stream parameters for 23 Andean streams. Degrees of freedom for all regressions were 1, 20. AIC = Akaike Information Criterion used for selection of predictor variables. DO = Dissolved Oxygen, PO_4^{3-} = Phosphate, TC = Total Coliforms.

Acknowledgments

Special thanks are due to the personnel of the San Francisco Research Station, Nature and Culture International, Municipality of Loja and the Ministry of Environment for the provision of research facilities. Many thanks are due to C. Prieto to provide us additional information on stream parameters and macroinvertebrates of the Zamora catchment; and to L.M. Cun for his support with field sampling and stream measurements. The help of C. Romero, S. Aguilar and M. Villa with water analyses is appreciated. We would like to thank S. Bourret for proof reading, C. Espinosa for his assistance with R functions and routines, and P.F. Quintana–Ascencio and two anonymous reviewers for constructive comments on previous drafts of the manuscript.

Author Contributions

Conceived and designed the experiments: CIA HGF LB. Performed the experiments: CIA AL. Analyzed the data: CIA LB HH. Contributed reagents/materials/analysis tools: CIA HGF LB. Wrote the paper: CIA HH LB.

References

1. Laurance WF (1999) Reflections on the tropical deforestation crisis. Biol Conserv 91: 109–117.
2. Achard F, Eva HD, Stibig H-J, Mayaux P, Gallego J, et al. (2002) Determination of deforestation rates of the world's humid tropical forests. Science (80-) 297: 999–1002.
3. Lambin EF, Geist HJ, Lepers E (2003) Dynamics of l and-use and land -cover change in tropical regions. Annu Rev Environ Resour 28: 205–241.
4. Lorion CM, Kennedy BP (2009) Relationships between deforestation, riparian forest buffers and benthic macroinvertebrates in neotropical headwater streams. Freshw Biol 54: 165–180.
5. Wantzen KM, Wagner R (2006) Detritus processing by invertebrate shredders: a neotropical – temperate comparison. J North Am Benthol Soc 25: 216–232.
6. Iwata T, Nakano S, Inoue M (2003) Impacts of past riparian deforestation on stream communities in a tropical rain forest in Borneo. Ecol Appl 13: 461–473.
7. Ordóñez MV (2011) Influencia del uso del suelo y la cobertura vegetal natural en la integridad ecológica de los ríos altoandinos al noreste del Ecuador Universidad San Francisco de Quito.
8. Sierra R, Campos F, Chamberlin J (2002) Assessing biodiversity conservation priorities: ecosystem risk and representativeness in continental Ecuador. Landsc Urban Plan 59: 95–110.
9. Buytaert W, Cuesta-Camacho F, Tobón C (2011) Potential impacts of climate change on the environmental services of humid tropical alpine regions. Glob Ecol Biogeogr 20: 19–33.
10. Bücker A, Sondermann M, Frede H-G, Breuer L (2010) The influence of land-use on macroinvertebrate communities in montane tropical streams – a case study from Ecuador. Fundam Appl Limnol 177: 267–282.
11. Encalada AC, Calles J, Ferreira V, Canhoto CM, Graça MAS (2010) Riparian land use and the relationship between the benthos and litter decomposition in tropical montane streams. Freshw Biol 55: 1719–1733.
12. Crespo PJ, Feyen J, Buytaert W, Bücker A, Breuer L, et al. (2011) Identifying controls of the rainfall–runoff response of small catchments in the tropical Andes (Ecuador). J Hydrol 407: 164–174.
13. Allan JD (2004) Landscapes and riverscapes: the influence of land use on stream ecosystems. Annu Rev Ecol Evol Syst 35: 257–284.
14. Death RG, Collier KJ (2009) Measuring stream macroinvertebrate responses to gradients of vegetation cover: when is enough enough? Freshw Biol 55: 1447–1464.
15. Van Sickle J, Baker J, Herlihy A, Bayley P, Gregory S, et al. (2004) Projecting the biological condition of streams under alternative scenarios of human land use. Ecol Appl 14: 368–380.
16. Collier KJ, Hamer M (2012) The ecological condition of Waikato wadeable streams based on the Regional Ecological Monitoring of Streams (REMS) Programme. Waikato Regional Council, Technical Report, Waikato.
17. European Environment Agency (2012) European waters - assessment of status and pressures. EEA Report No 8. Copenhagen.
18. Environmental Protection Agency (2013) National Rivers and Streams Assessment 2008–2009: A Collaborative Survey. Draft February 28. Washington.
19. Perrie A, Morar S, Milne JR, Greenfield S (2012) River and stream water quality and ecology in the Wellington region: State and trends. Greater Wellington Regional Council, Publication No. GW/EMI-T-12/143, Wellington.
20. Parkyn SM, Davies-Colley RJ, Halliday NJ, Costley KJ, Croker GF (2003) Planted Riparian Buffer Zones in New Zealand: Do They Live Up to Expectations? Restor Ecol 11: 436–447.
21. Bojsen BH, Jacobsen D (2003) Effects of deforestation on macroinvertebrate diversity and assemblage structure in Ecuadorian Amazon streams. Arch für Hydrobiol 158: 317–342.

22. Miserendino ML, Casaux R, Archangelsky M, Di Prinzio CY, Brand C, et al. (2011) Assessing land-use effects on water quality, in-stream habitat, riparian ecosystems and biodiversity in Patagonian northwest streams. Sci Total Environ 409: 612–624.
23. Villamarín C, Rieradevall M, Paul MJ, Barbour MT, Prat N (2013) A tool to assess the ecological condition of tropical high Andean streams in Ecuador and Peru: The IMEERA index. Ecol Indic 29: 79–92.
24. Gergel SE, Turner MG, Miller JR, Melack JM, Stanley EH (2002) Landscape indicators of human impacts to riverine systems. Aquat Sci 64: 118–128.
25. Stewart JS, Wang L, Lyons J, Horwatich JA, Bannerman R (2001) Influences of watershed, riparian-cprridor and reach-scale characteristics on aquatic biota in agricultural watersheds. J Am Water Resour Assoc 37: 1475–1487.
26. Sutherland AB, Meyer JL, Gardiner EP (2002) Effects of land cover on sediment regime and fish assemblage structure in four southern Appalachian streams. Freshw Biol 47: 1791–1805.
27. Göttlicher D, Obregón A, Homeier J, Rollenbeck R, Nauss T, et al. (2009) Land-cover classification in the Andes of southern Ecuador using Landsat ETM+ data as a basis for SVAT modelling. Int J Remote Sens 30: 1867–1886.
28. Thies B, Meyer H, Nauss T, Bendix J (2014) Projecting land-use and land-cover changes in a tropical mountain forest of Southern Ecuador. J Land Use Sci 9: 1–33.
29. Torrachi S, Tapia MF, Escudero A, de la Cruz M (2013) Deforestación en una región montañosa megadiversa en los Andes: Dinámica del paisaje en el sur de Ecuador. In: de la Cruz M, Maestre FT, editors. Avances en el Análisis Espacial de Datos Ecológicos: Aspectos Metodológicos y Aplicados. Móstoles: ECESPA–Asociación Española de Ecología Terrestre. pp. 275–289.
30. Stein H, Springer M, Kohlmann B (2008) Comparison of two sampling methods for biomonitoring using aquatic macroinvertebrates in the Dos Novillos River, Costa Rica. Ecol Eng 34: 267–275.
31. De Cáceres M, Legendre P, Moretti M (2010) Improving indicator species analysis by combining groups of sites. Oikos 119: 1674–1684.
32. Sites RW, Willig MR, Linit MJ (2003) Macroecology of Aquatic Insects: A Quantitative Analysis of Taxonomic Richness and Composition in the Andes Mountains of Northern Ecuador. Biotropica 35: 226–239.
33. Jacobsen D (2004) Contrasting patterns in local and zonal family richness of stream invertebrates along an Andean altitudinal gradient. Freshw Biol 49: 1293–1305.
34. Tarras-Wahlberg N., Flachier A, Lane S., Sangfors O (2001) Environmental impacts and metal exposure of aquatic ecosystems in rivers contaminated by small scale gold mining: the Puyango River basin, southern Ecuador. Sci Total Environ 278: 239–261.
35. American Public Health Association (1995) Standard Methods for the Examination of Water and Wastewater. 19th ed. American Public Health Association, editor Washington DC.
36. Barbour MT, Gerritsen J, Snyder BD, Stribling JB (1999) Rapid Bioassessment Protocols for Use in Streams and Wadeable Rivers: Periphyton, Benthic Macroinvertebrates, and Fish. 2nd ed. Environmental Protection Agency.
37. Hauer FR, Resh VH (2007) Macroinvertebrates. In: Hauer FR, Lamberti GA, editors. Methods in Stream Ecology. San Diego CA: Academic Press. pp. 435–454.
38. Ríos-Touma B, Encalada AC, Prat Fornells N (2011) Macroinvertebrate Assemblages of an Andean High-Altitude Tropical Stream: The Importance of Season and Flow. Int Rev Hydrobiol 96: 667–685.
39. Jacobsen D, Encalada AC (1998) The Macroinvertebrate fauna of Ecuadorian high–land streams in the wet and dry season. Arch für Hydrobiol 142: 53–70.
40. Roldán Pérez G (2003) Bioindicación de la calidad del agua en Colombia. Uso del método BMWP/Col. 1st ed. Medellín: Editorial Universidad de Antioquia.
41. Fernádez HR, Domínguez E (2001) Guía para la determinación de los artrópodos bentónicos sudamericanos. 1st ed. Tucumán: Universidad Nacional de Tucumán.
42. Sponseller R, Benfield E, Valett H (2001) Relationships between land use, spatial scale and stream macroinvertebrate communities. Freshw Biol.
43. Cummins KW, Merritt RW, Andrade PCN (2005) The use of invertebrate functional groups to characterize ecosystem attributes in selected streams and rivers in south Brazil. Stud Neotrop Fauna Environ 40: 69–89.
44. Tomanova S, Goitia E, Helešic J (2006) Trophic levels and functional feeding groups of macroinvertebrates in neotropical streams. Hydrobiologia 556: 251–264.
45. Brown R (1970) A Water Quality Index – Do We Dare? Water Sew Work 11: 339–343.
46. Lenat DR (1988) Water quality assessment of streams using a qualitative collection method for benthic macroinvertebrates. J North Am Benthol Soc 7: 222–233.
47. Allan JD, Erickson DL, Fay J (1997) The influence of catchment land use on stream integrity across multiple spatial scales. Freshw Biol 37: 149–161.
48. Harding JS, Benfield EF, Bolstad P V, Helfman GS, Jones III EBD (1998) Stream biodiversity: The ghost of land use past. Proc Natl Acad Sci U S A 95: 14843–14847.
49. Legendre P, Gallagher E (2001) Ecologically meaningful transformations for ordination of species data. Oecologia 129: 271–280.
50. R Development Core Team (2011) R: A language and environment for statistical computing. R Foundation for Statistical Computing, Vienna. http://www.r-project.org/.
51. Bocard D, Gillet F, Legendre P (2011) Numerical Ecology with R. 1st ed. New York: Springer.
52. Harding JS, Claassen K, Evers N (2006) Can forest fragments reset physical and water quality conditions in agricultural catchments and act as refugia for forest stream invertebrates? Hydrobiologia 568: 391–402.
53. Suga CM, Tanaka MO (2012) Influence of a forest remnant on macroinvertebrate communities in a degraded tropical stream. Hydrobiologia 703: 203–213.
54. Shandas V, Alberti M (2009) Exploring the role of vegetation fragmentation on aquatic conditions: Linking upland with riparian areas in Puget Sound lowland streams. Landsc Urban Plan 90: 66–75.
55. Rios SL, Bailey RC (2006) Relationship between Riparian Vegetation and Stream Benthic Communities at Three Spatial Scales. Hydrobiologia 553: 153–160.
56. Couceiro SRM, Hamada N, Luz SLB, Forsberg BR, Pimentel TP (2006) Deforestation and sewage effects on aquatic macroinvertebrates in urban streams in Manaus, Amazonas, Brazil. Hydrobiologia 575: 271–284.
57. Lecerf A, Dobson M, Dang CK, Chauvet E (2005) Riparian plant species loss alters trophic dynamics in detritus-based stream ecosystems. Oecologia 146: 432–442.
58. Lecerf A, Richardson JS (2010) Litter decomposition can detect effects of high and moderate levels of forest disturbance on stream condition. For Ecol Manage 259: 2433–2443.
59. James LA, Marcus WA, Harden CP (2006) Human impacts on headwater fluvial systems in the northern and central Andes. Geomorphology 79: 249–263.
60. Molina A, Vanacker V, Balthazar V, Mora D, Govers G (2012) Complex land cover change, water and sediment yield in a degraded Andean environment. J Hydrol 472: 25–35.
61. Borbor-Córdova MJ, Boyer EW, McDowell WH, Hall CA (2006) Nitrogen and phosphorus budgets for a tropical watershed impacted by agricultural land use: Guayas, Ecuador. Biogeochemistry 79: 135–161.
62. Sarr D (2002) Riparian livestock exclosure research in the Western United States: A critique and some recommendations. Environ Manage 30: 516–526.
63. Jones JA, Swanson FJ, Wemple BC, Snyder KU (2000) Effects of roads on hydrology, geomorphology, and disturbance patches in stream networks. Conserv Biol 14: 76–85.
64. Hladyz S, Kajsa A, Giller PS, Woodward G (2011) Impacts of an aggressive riparian invader on community structure and ecosystem functioning in stream food webs. J Appl Ecol 48: 443–452.

18

Effects of Land Use, Topography and Socio-Economic Factors on River Water Quality in a Mountainous Watershed with Intensive Agricultural Production in East China

Jiabo Chen[1], **Jun Lu**[1,2]*

1 Department of Natural Resources, College of Environment and Natural Resources, Zhejiang University, Hangzhou, Zhejiang Province, China, **2** China Ministry of Education Key Lab of Environment Remediation and Ecological Health, Zhejiang University, Hangzhou, Zhejiang Province, China

Abstract

Understanding the primary effects of anthropogenic activities and natural factors on river water quality is important in the study and efficient management of water resources. In this study, analysis of Variance (ANOVA), Principal component analysis (PCA), Pearson correlations, Multiple regression analysis (MRA) and Redundancy analysis (RDA) were applied as an integrated approach in a GIS environment to explore the temporal and spatial variations in river water quality and to estimate the influence of watershed land use, topography and socio-economic factors on river water quality based on 3 years of water quality monitoring data for the Cao-E River system. The statistical analysis revealed that TN, pH and temperature were generally higher in the rainy season, whereas BOD_5, DO and turbidity were higher in the dry season. Spatial variations in river water quality were related to numerous anthropogenic and natural factors. Urban land use was found to be the most important explanatory variable for BOD_5, COD_{Mn}, TN, DN, NH_4^+-N, NO_3^--N, DO, pH and TP. The animal husbandry output per capita was an important predictor of TP and turbidity, and the gross domestic product per capita largely determined spatial variations in EC. The remaining unexplained variance was related to other factors, such as topography. Our results suggested that pollution control of animal waste discharge in rural settlements, agricultural runoff in cropland, industrial production pollution and domestic pollution in urban and industrial areas were important within the Cao-E River basin. Moreover, the percentage of the total overall river water quality variance explained by an individual variable and/or all environmental variables (according to RDA) can assist in quantitatively identifying the primary factors that control pollution at the watershed scale.

Editor: Miguel Lopez-Ferber, Ecole des Mines d'Alès, France

Funding: Chinese National Key Technology R&D Program (Grant No. 2012BAC17B01). The funders had no role in study design, data collection and analysis, decision to publish, or preparation of the manuscript.

Competing Interests: The authors have declared that no competing interests exist.

* Email: jlu@zju.edu.cn

Introduction

The deterioration of river water quality has become a primary environmental concern due to unsustainable anthropogenic activities; the demand for freshwater has been rapidly increasing in many developing countries, especially in China [1–3]. River water quality is controlled by complex anthropogenic activities and natural factors at both the river and watershed scales [1], [7–9]. Understanding the temporal and spatial variations in river water quality and estimating the primary regional factors that affect water quality can assist researchers in establishing priorities for sustainable water management [2], [10–12]. The relationships between water quality parameters and land use/cover, population density and point source discharge have been frequently studied [1], [7], [9], [13–15]. Simultaneously, topography and animal waste discharge are considered important factors that affect watershed river water quality [7], [13–16]. Particularly, several countries, e.g., China, contain large mountainous areas [17] and extensive animal production that lack strict management techniques [16], [18,19].

Recently, Pearson correlations have been widely employed to determine the relationships between environmental variables and river water quality [1], [7]. This method is simple and provides quantitative information; however, Pearson correlations lack visualization. Multiple regression analysis (MRA) is a useful tool that is commonly used to determine that watershed characteristics that best explain the spatial variability of an individual river water quality variable [7], [20]. This method lacks general information regarding pollution types. Moreover, principal component analysis (PCA) has become a widely accepted method in river water quality assessment and source apportionment studies in the last decade [2]. This method is commonly used to obtain the types of specific pollution sources without explanatory variables; however, PCA can only provide preliminary information regarding pollution types. Redundancy analysis (RDA), a multivariate statistical analysis method, has been proved useful for qualitative analysis of the interactions between river water quality and watershed characteristics in highly complex systems [13], [21,22]. RDA can reveal the influences of environmental factors on overall water quality, not only on single water quality variables [20]. Several commonly used statistical methods for pollution source identification are complementary (i.e., they have their own benefits and limits). Summary of commonly used statistical methods on pollution source identification in recent years is listed in **Table S1**. However, comprehensive applications of different statistical methods to evaluate the effects of environmental variables on water quality have not been fully explored in river studies in China. Furthermore, few studies have used RDA analysis to quantitatively evaluate the effects of watershed characteristics on the overall river water quality.

The Cao-E River is located upstream of the Qiantang Estuary, which is a major riverine system in Zhejiang Province, China, and is the primary source of industrial, agricultural and domestic water supplies [10]. Furthermore, the blue algae bloom that occurred in July 2004 occurred in the Qiantang Estuary [23,24].

In this study, ANOVA, Pearson correlation analysis, MRA, PCA and RDA were applied to investigate the effects of sub-watershed land use, topography and socio–economic factors (including animal waste discharge) on river water quality at the watershed scale for the Cao-E River system in a GIS environment. The objectives of this study were threefold: (1) to examine the temporal and spatial variations in water contamination within the study area, (2) to investigate the relationships between the river water quality parameters and land use, topography and socio-economic factors, and (3) to identify the primary pollution sources to estimate the possible sources that affect the water quality parameters in the Cao-E River basin. The results can be helpful for water conservation in the Cao-E River basin, and they provide a valuable tool for water quality agencies to develop assessment strategies for effective water quality management and rapid solutions for water pollution problems at the watershed scale.

Materials and Methods

1 Ethics statement

No specific permits were required for the described field studies; the sampling did not cause any disturbance to the environment or to the protected species at the sampling sites.

2 Study area

The Cao-E River (29°08′-30°15′ N, 120°30′-121°15′ E) is located in Zhejiang Province, East China. The Cao-E River is one of the main tributaries of the Qiantang Estuary, which flows into the East China Sea. The river system contains a main stream and several major tributaries (i.e., the CT River, the XC River, the CL River, the HZ River, the XS River and the YT River; **Figure 1**). The main stream is ~197 km long with an average slope of 3.0‰. The Cao-E River basin has a typical subtropical monsoon climate with an average annual rainfall of 1500 mm and an average ambient temperature of 16.2°C [25]. The watershed is within a mountainous region. Specifically, mountains and hills cover 2/3 of the area; the remaining area is covered by plains with intensive agricultural production [26].

3 Data Sources

Water quality data. The stream water samples (the number of samples used for statistical analysis is 36) were collected from July 2003 to June 2006 at 20 sampling sites throughout the watershed (Figure 1). Specific sampling stations that were largely influenced by point source pollution were excluded from further analysis. These specific sampling sites should meet the following conditions: (1) there was an obvious outlet of point pollution discharge near the sampling site, and (2) The standard deviation of the water quality parameter (BOD_5) at the specific sampling sites were triple more than the standard deviation for total sampling sites. The samples were collected once a month between 9:00 am and 16:00 pm. Moreover, the water samples were collected at approximately 0.3 m below the water surface in the central stream, placed into plastic bottles (2.5 L), transported to the laboratory and stored at 0~4°C for subsequent chemical analysis. The chemical measurements were performed in the laboratory within 24 h after collecting the water samples. Moreover, dissolved oxygen (DO), water temperature (T), pH and electrical conductivity (EC) were measured when the water samples were collected with a hand-held multi-parameter instrument (Multi 340i SETs, The Merck Co. Ltd., Germany). The turbidity was measured with a hand-held Turbiquant 1000IR (Multi 340i, The Merck Co. Ltd., Germany). The following chemical and biological water quality parameters were measured: chemical oxygen demand (COD_{Mn}), 5-day biochemical oxygen demand (BOD_5), total nitrogen (TN), total phosphorus (TP) and dissolved phosphorus (DP), which were measured using standard methods [27]. The ammonium nitrogen (NH_4^+-N) content was measured using an Astoria analyzer system (AAS, Brown Rupee CO. Ltd., Germany) after filtration through a 0.45-μm filter (Hailing Medicine, Zhejiang Province, China). The nitrate nitrogen (NO_3^--N) content was measured from the absorbance of the sample at 220 nm (A220) and 275 nm (A275) using a visible and ultraviolet spectrophotometer (Shanghai Spectrum Instruments, Shanghai, China) after filtration through a 0.45-μm filter; the results were determined by the following relationship: A220 - 2×A275 [27].

Land use/cover, topography and socio-economic data and GIS analysis. Latitude and longitude were measured using a hand-held GPS. The catchment boundaries were delineated on a 30-m spatial resolution digital elevation model (DEM); the stream network was represented using the Soil and Water Assessment Tool (SWAT) model. ArcGIS 9.3 Desktop GIS software was used to calculate the watershed characteristics for each sampling site. All datasets were converted to a common digital format (WGS 1984) and a common coordinate system (Albers Equal Area). Hydrologic units were not solely defined by the watershed above an individual point [28]; a hydrologic unit on the Cao-E River may cover 2/3 or more of the entire watershed above that sampling station (most hydrologic units on tributaries encompass the entire watershed above a sampling station).

The socioeconomic data, including land use change, the gross domestic product, animal husbandry output and human population in the watershed, were obtained from the Shengzhou (SZ), Xinchang (XC), Shangyu (SY) and Shaoxing (SX) County Statistical Yearbooks for the period 2003–2006 and were adjusted with some typical investigation in several towns within the watershed. The temporal variations in land use were considered to be minimal. The annual changes in forest areas and water areas were less than 0.05%, the annual decrease rate in cropland areas was less than 1% and the annual increase rate in urban areas was less than 2.5%., respectively, for the entire watershed and throughout the study period. Thus, Landsat Thematic Mapper imagery from 2006 was used to map the land cover in the study area [29]. The land cover was categorized into the following 4 classes: forest, water, cropland and urban land. The annual increases in human population density, gross domestic product per capita and animal husbandry output per capita were less than 1.5%, 10% and 2%, respectively, for the entire watershed and throughout the study period. Therefore, the averages of the socio-economic data for the entire study period were selected in this study. The slope map was derived from the DEM using a GIS spatial analyst tool (ESRI, 2006). The land use, slope, DEM and

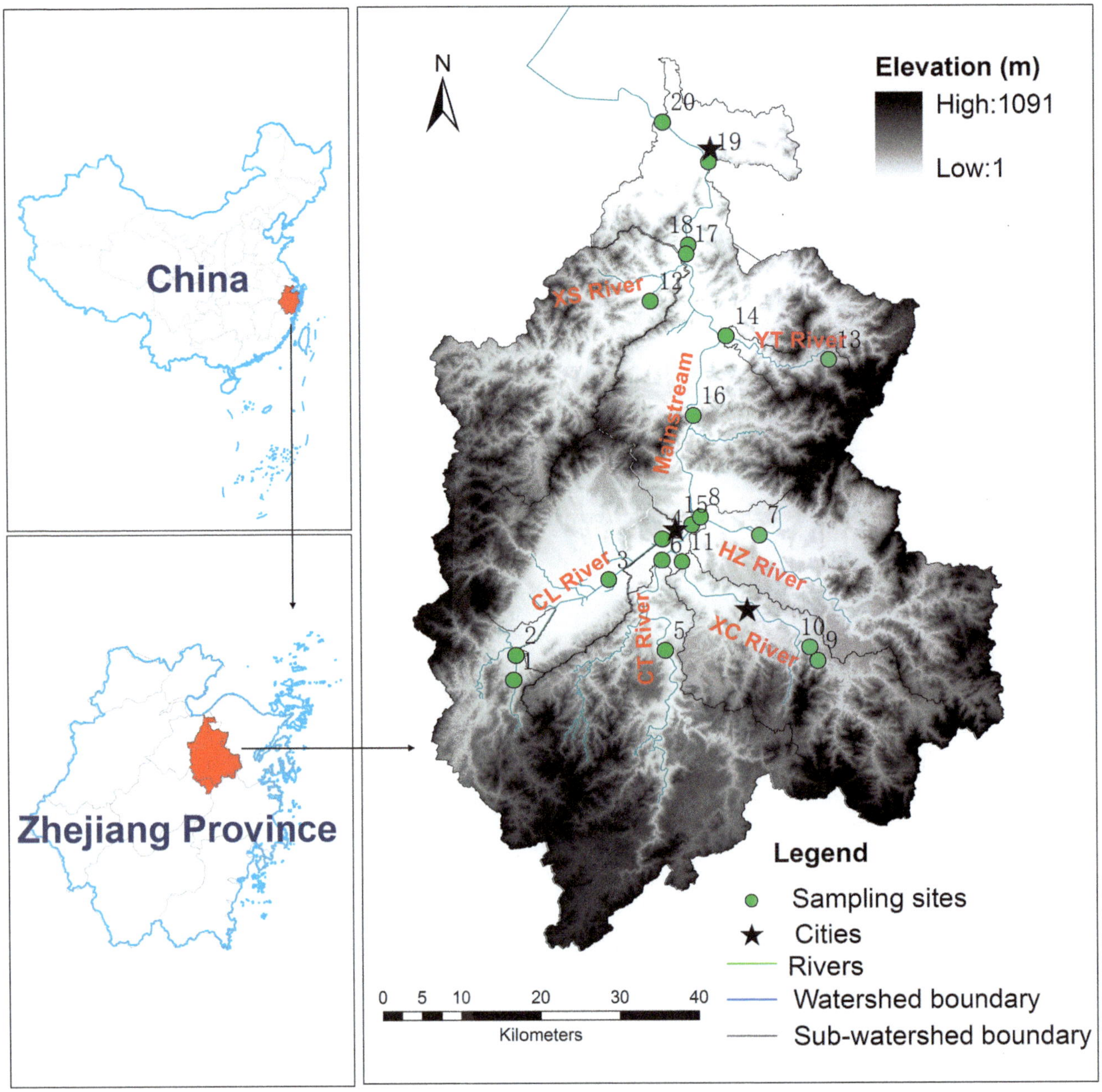

Figure 1. The Geographical Location Of The Sampling Sites In The Cao-E River System. (Sampling Stations 1-Cl1, 2- Cl2, 3-Cl3, 4-Cl4, 5- Ct 1,6-Ct2,7- Hz1,8- Hz2,9- Xc1,10- Xc2,11- Xc3,12- Xs1,13- Yt1,14- Yt2,15- M1,16- M2,17- M3,18- M4,19- M5, 20- M6).

administrative map (with attribute data, including the gross domestic product, animal husbandry output and human population for every town) were used to calculate the respective land use area (forest, water, urban land and cropland), mean elevation, mean slope, area, human population density, gross domestic product per capita and animal husbandry output per capita within each sub-watershed using a GIS spatial analyst tool. The abbreviations and statistics of the sub-watershed characteristics in the Cao-E River basin are summarized in **Table 1**.

4 Data analysis

Analysis of variance (ANOVA) was used to compare river water quality variations between different seasons (i.e., rainy and dry) via the Student-Newman-Keuls multiple-range test ($P<0.05$). Relationships among the considered variables were tested using Pearson's coefficient with statistical significance set at $P<0.05$. Major gradients and principal patterns in the water quality data between wet and dry seasons were detected using principal component analysis (PCA). PCA transforms a dataset consisting of p variables (analytical constituents) that are interrelated or correlated to various degrees into a new dataset containing p new orthogonal and uncorrelated variables; these variables are called the principal components (PCs) [30], [35]. The PCs are linear functions of the original variables in which the sum of their variances is equal to that of the original variables [30]. The PCs following a descending ordered according to their variances with PC1 corresponding to the variable with the largest variance. Algebraically, for p original variables (i.e., $x_1, x_2, \ldots, x_p$). Additional

Table 1. Abbreviations And Basic Statistical Information Of The Sub-Watershed Characteristics In The Cao-E River Basin.

	Parameters	Abbreviation	Range	Mean ± Sd
Land Use	Forest Percentage (%)	Frt	25.72–74.50	50.31±12.36
	Water Percentage (%)	Wt	0.13–3.39	1.87±0.87
	Urban Percentage (%)	Ub	0.45–14.44	4.09±3.52
	Cropland Percentage (%)	Crp	21.58–57.73	43.85±10.18
Topography	Mean Elevation (M)	Elv	125.12–556.54	306.17±116.17
	Mean Catchment Slope (°)	Slp_{mean}	10.20–17.72	13.74±2.29
	Catchment Area (Km^2)	Ar	38.39–2286.21	764.53±674.58
Socio-Economic Factors	Human Population Density (Persons/Km^2)	Hp_d	123.01–576.68	316.11±119.12
	Gross Domestic Product Per Capita (10^5 Chinese Yuan)	Gdp_{pc}	135.51–2151.16	582.38±427.88
	Animal Husbandry Output Per Capita (10^5 Chinese Yuan)	Aho_{pc}	8.07–39.40	20.56±8.61

PCs up to p can be formulated in a similar manner. The variances of the PCs are the eigenvalues; the coefficients or weights, are the eigenvectors extracted from the covariance or correlation matrix. In the example PCA, the correlation matrix was used in all cases. The goal of PCA is that the first k PCs (where $k<<p$) retain most of the information in all of the p original variables, which effectively reduces the practical dimensionality of the dataset. More specifically, if the correlations are high among many of the original variables, the first few PCs contain (or explain) a large percentage of the total variance and may be used to describe multivariate patterns or water quality variations across the watershed nearly as well as using the complete set of p original variables. These patterns are often related to specific sources of contamination [30].

The stepwise regression analysis was performed to determine the environmental or socio-economic factors (TP, $P<0.10$ and the others, $P<0.05$) that best explained the spatial variability of an individual river water quality variable [20]. Variables that were strongly intercorrelated with others (variance inflation factor >10) in the initial analysis, were removed and a further analysis was carried out with the remaining environmental variables for MRA and RDA [20], [31,32]. Variance inflation factor (VIF) is a common way for detecting multicollinearity, a general rule is that the Variance Inflation Factor should not exceed 10 [49]. RDA was performed using Canoco 4.5 software to evaluate the relationships between river water quality and the environmental variables [33]. RDA was chosen because previous inspection of the data revealed a linear response rather than a unimodal response in the primary river water quality variables [34]. A Monte Carlo permutation test was used to verify the significance of the models [29]. Two series of matrices (i.e., for river water quality and environmental variables) from the RDA results were visualized as ordination diagrams using the CanoDraw software package for Windows. The river water quality parameters and the environmental/socio-economic variables were represented with arrows. Correlations between the watershed characteristics and/or the river water quality parameters were obtained by projecting the watershed characteristics onto each river water quality parameter in which higher values indicated higher correlations [34]. Application summary of 5 statistical methods in this study was listed in **Table S2**. The statistical analyses, except RDA, were completed using the SAS 9.1 software for Windows.

Results

1 Physico-chemical water quality in the Cao-E River basin

ANOVA was used to compare river water quality variations between the different seasons (i.e., rainy and dry). Temporal and spatial variations in the physico-chemical parameters are shown in **Table 2** and **Figure 2**. More specifically, BOD_5, TN, DO, pH, T and turbidity exhibited significant temporal variations (**Table 2**, $p<0.10$). Moreover, TN, pH and T were generally higher in the rainy season (April to September), whereas higher values for BOD_5, DO and turbidity occurred in the dry season (October to March). Most of the water quality parameters, except pH and T, exhibited considerable spatial variations. BOD_5, COD_{Mn} and NH_4^+-N were higher in urban areas compared with surrounding rural areas, whereas DO exhibited a reverse response to the extent of urbanization. TN, DN, NO_3^--N, TP and DP were higher in the lower part of the basin where more land has been developed and were lower in the upper-eastern part of the Cao-E River basin. Turbidity and EC were higher in the lower part of the basin (**Figure 2**) where the reach is often affected by tides and/or point source pollution [36,37].

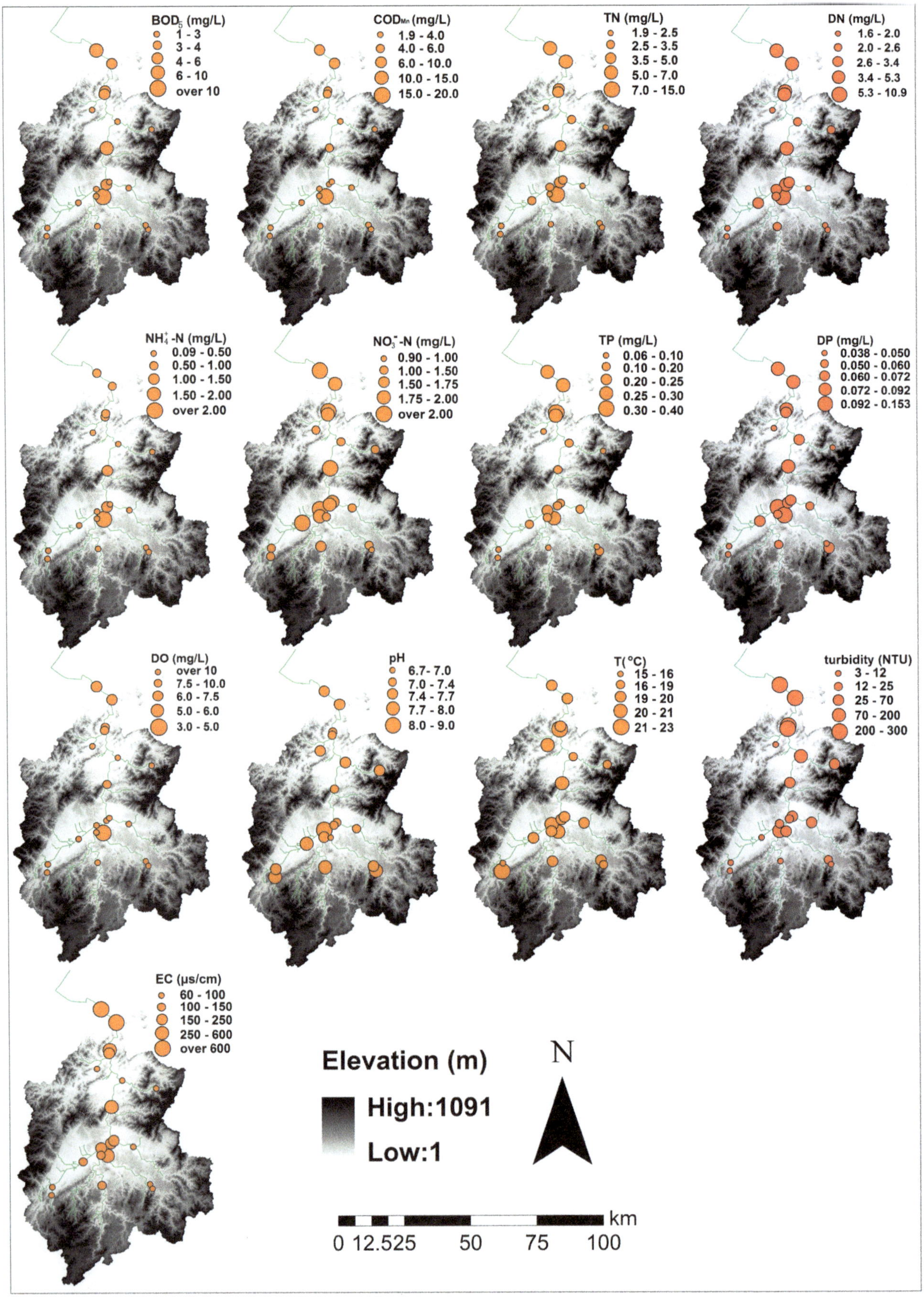

Figure 2. Spatial Variations In Bod5, Codmn, Tn, Dn, Nh_4^+-N, No_3^--N, Tp, Dp, Do, Ph, T, Turbidity And Ec In The Study Area.

Table 2. Seasonal Averages, Coefficients Of Variation [Cv (%)] And River Water Quality Parameter Ranges.

Water Quality Parameters	Rainy Season		Dry Season		
	Average [Cv (%)]	Range	Average [Cv (%)]	Range	*P-Value*
Bod_5 (Mg/L)	5.912 [183]	0.05–173.5	7.850 [181]	0–255.00	0.052
Cod_{mn} (Mg/L)	3.845 [87]	0.39–43.87	4.0361 [99]	0.49–47.75	0.557
Tn (Mg/L)	4.793 [103]	0.29–45.85	4.208 [73]	0.56–31.09	0.049
Dn (Mg/L)	3.549 [90]	0.06–22.53	3.295 [80]	0.03–30.07	0.219
Nh_4^+-N (Mg/L)	0.605 [195]	0–15.87	0.759 [2099]	0–19.58	0.156
No_3^--N (Mg/L)	1.677 [41]	0.01–4.30	1.606 [69]	0.08–15.08	0.283
Tp (Mg/L)	0.150 [108]	0–1.88	0.168 [113]	0–3.86	0.317
Dp (Mg/L)	0.061 [91]	0–0.49	0.076 [253]	0–3.49	0.163
Do (Mg/L)	7.164 [33]	0–19.55	8.045 [42]	0–17.86	<0.001
Ph	7.538 [7]	5.33–9.34	7.449 [8]	5.45–10.12	0.055
T (°C)	24.967 [24]	8.40–37.20	14.262 [64]	4.70–105.00	<0.001
Turbidity (Ntu)	70.011 [105]	0.83–586.10	92.939 [113]	0.91–1111	0.002
Ec (μs/Cm)	306.38 [347]	14–14420	348.56 [282]	14–11230	0.557

The Probabilities Associated With The Student-Newman-Keuls Multiple-Range Test Are Also Provided.

Major gradients and principal patterns in the water quality data between the wet and dry seasons were detected using PCA. The PCA based on only the dry season water quality data indicated that 4 significant factors (i.e., PCs with an eigenvalue >1) explained 87.8% of the total variance (**Table 3**). The first factor accounted for 50.2% of the total variance and had a strong positive correlation with BOD_5, COD_{Mn}, TN, DN, NH_4^+-N, and T. The PCA of the wet season water quality data showed that the eigenvalues for the 2 most significant factors were 8.442 and 1.817, accounting for 78.9% of the total variance. Factor 1 accounted for 64.9% of the total variance and was positively correlated with BOD_5, COD_{Mn}, TN, DN, TP and DP (**Table 3**).

2 Relationships between watershed characteristics and river water quality

Relationships between the river water quality variables and their corresponding sub-watershed land use, topography and socio-economic factors were explored using a Pearson correlation test, MRA and RDA. All environmental variables, including land use, topography and socio-economic factors, were utilized to

Table 3. River Water Quality Variables (13) On Rotated Pcs For Datasets During The Dry And Wet Seasons.

Water Quality Parameters	Dry Season				Wet Season	
	Pc1	Pc2	Pc3	Pc4	Pc1	Pc2
Bod_5 (Mg/L)	0.354	−0.110	0.008	0.076	0.312	−0.041
Cod_{mn} (Mg/L)	0.350	0.200	−0.187	0.128	0.310	0.263
Tn (Mg/L)	0.380	0.001	0,077	0.083	0.333	−0.054
Dn (Mg/L)	0.362	−0.044	0.147	0.127	0.311	−0.197
Nh_4^+-N (Mg/L)	0.345	−0.215	0.019	0.084	0.267	−0.346
No_3^--N (Mg/L)	0.275	−0.020	0.425	−0.079	0.283	0.174
Tp (Mg/L)	−0.221	0.409	0.077	−0.476	0.314	0.200
Dp (Mg/L)	0.097	0.329	0.609	−0.090	0.300	−0.018
Do (Mg/L)	−0.270	0.361	0.091	−0.204	0.272	0.349
Ph	−0.146	0.328	0.268	0.614	−0.184	0.546
T (°C)	0.418	−0.420	0.119	−0.212	0.010	0.246
Turbidity (Ntu)	0.222	0.340	−0.416	−0.312	0.283	0.246
Ec (μs/Cm)	0.241	0.308	−0.335	0.382	0.247	0.395
Eigenvalues	6.522	2.344	1.518	1.028	8.442	1.817
Percentage Of Variance	0.502	0.180	0.117	0.079	0.649	0.140
Cumulative Percentage Of Variance	0.502	0.682	0.799	0.878	0.649	0.789

Pcs Refer To Principal Components. The Significant Factors (I.E., Pcs With An Eigenvalue >1) Are List In The Table.

Table 4. The Pearson Correlation Coefficients For The River Water Quality Parameters Compared With The Corresponding Sub-Watershed Characteristics.

	Frt	Wt	Ub	Crp	Elv	Slp_{mean}	Ar	Hp_d	Gdp_{pc}	Aho_{pc}
Bod_5	−0.52*	0.11	0.76**	0.35	−0.31	−0.40	−0.08	0.59**	0.32	0.09
Cod_{mn}	−0.61**	0.31	0.90**	0.40	−0.58	−0.64**	0.10	0.78**	0.60**	0.42
Tn	−0.72**	0.27	0.96**	0.51*	−0.61**	−0.74**	0.37	0.87**	0.60**	0.48*
Dn	−0.70**	0.17	0.91**	0.51*	−0.53*	−0.68**	0.34	0.81**	0.48*	0.40
Nh_4^+-N	−0.60**	0.15	0.84**	0.42	−0.39	−0.51*	0.14	0.69**	0.38	0.19
No_3^--N	−0.54*	0.13	0.41	0.50*	−0.52*	−0.76**	0.70**	0.60**	0.45*	0.70**
Tp	−0.63**	0.37	0.77**	0.46*	−0.82**	−0.76**	0.46**	0.77**	0.66**	0.80**
Dp	−0.60**	0.08	0.61**	0.50*	−0.48*	−0.63**	0.30	0.64**	0.45*	0.48*
Do	0.48*	−0.43	−0.75**	−0.29	0.45*	0.59**	−0.63**	−0.71**	−0.44	−0.36
Ph	0.53*	0.01	−0.66**	−0.42	0.37	0.33	−0.44	−0.53*	−0.16	−0.15
T	−0.43	0.18	0.44	0.36	−0.44	−0.49*	0.46	0.53*	0.23	0.33
Turbidity	−0.28	0.40	0.42	0.15	−0.71**	−0.51*	0.36	0.45*	0.58**	0.81**
Ec	−0.21	0.61**	0.52*	0.02	−0.64**	−0.60**	0.20	0.57**	0.88**	0.78**

Abbreviations Are Provided In Table 1. The Figures In The Table Are Correlation Coefficients.
*Indicates Significance At The 0.05 Probability Level.
**Indicates Significance At The 0.01 Probability Level.

evaluate the response of river water quality to environmental gradients. The correlation analysis indicated that CRP was positively correlated with TN, DN, NO_3^--N, TP and DP ($r > 0.45, p < 0.05$; r refers to the correlation coefficient). Moreover, UB was positively correlated with BOD_5, COD_{Mn}, TN, DN, NH_4^+-N, TP, DP and EC ($r > 0.50, p < 0.05$) and negatively correlated with DO and pH ($r < -0.65, p < 0.01$). FRT was positively correlated with DO and pH ($r > 0.47, p < 0.05$) and exhibited a negative correlation with BOD, COD, TN, DN, NH_4^+-N, NO_3^--N TP, and DP ($r < -0.50, p < 0.05$). Furthermore, SLP_{mean} was positively correlated with DO ($r = 0.59, p < 0.01$) and negatively correlated with COD_{Mn}, TN, DN, NH_4^+-N, NO_3^--N, TP, DP, T, turbidity and EC ($r < -0.48, p < 0.05$). The elevation was positively correlated with DO ($r = 0.45, p < 0.05$) and negatively correlated with COD_{Mn}, TN, DN, NO_3^--N, TP, DP, turbidity and EC ($r < -0.47, p < 0.05$). Additionally, GDP_{pc} exhibited a positive correlation with COD, TN, DN, NO_3^--N, TP, DP, turbidity, and EC ($r > 0.44, p < 0.05$). AHO_{pc} was positively correlated with TN, NO_3^--N, TP, DP, turbidity and EC ($r > 0.47, p < 0.05$). Lastly, HP_d exhibited a negative correlation with DO and pH ($r < -0.52, p < 0.05$) and a positive correlation to all other river water quality parameters ($r > 0.44, p < 0.05$) (**Table 4**).

MRA demonstrated that no individual environmental factor was able to describe the overall water quality; however, most of the physico-chemical water parameters could be sufficiently predicted using 1 to 3 environmental factors (**Table 5**). Specifically, BOD_5 could be predicted using UB, AR and GDP_{pc}; COD_{Mn} with UB and AR; TN, DN, pH and DO with UB; NH_4^+-N with UB and HP_d; DP with HP_d; turbidity with AHO_{pc}; EC with GDP_{pc}; TP with UB and AHO_{pc}; and NO_3^--N with UB and SLP_{mean}.

The RDA of the overall water quality as the dependent variable suggested that FRT, UB, CRP, ELV, SLP_{mean}, AR, HP_d, GDP_{pc} and AHO_{pc} explained 15.7–50.8% of the river water quality spatial variation, and a combination of the topography, land use and socio-economic factors explained 86.1% of the variation in overall water quality (**Table 6**). The ordination diagram of overall water quality and topography, land use and socio-economic factors indicated that the first RDA axis displayed a pollution gradient (e.g., COD_{Mn} and TN increased along the axis), which was positively correlated with UB and HP_d and negatively correlated with FRT and SLP_{mean} (**Figure 3**), accounting for 52.4% of the total variance in water quality (**Table 7**). The second axis was related to AHO_{pc} and only explained 16.5% of the total variance (**Table 7**).

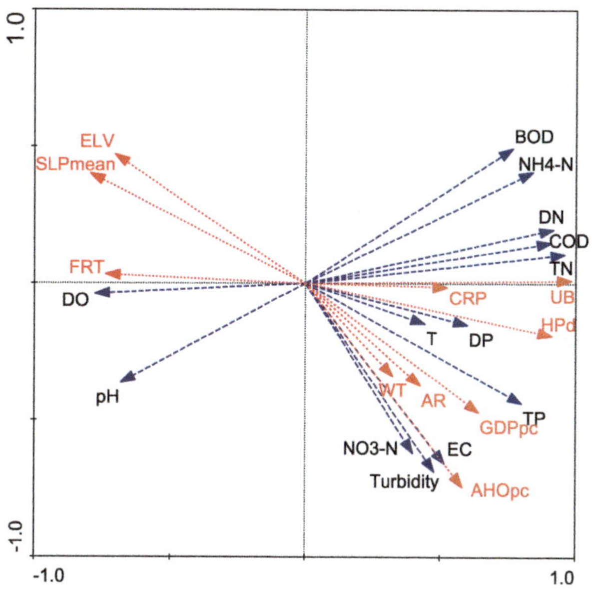

Figure 3. Biplots Of The River Water Quality Parameters And Watershed Characteristics In The Cao-E River Basin According To The Redundancy Analysis.

Table 5. Stepwise Multiple Regression Models For The River Water Quality Parameters And Watershed Characteristics In The Cao-E River Basin In Eastern China.

Parameters	Independent Variables	Regression Equation	R^2	Adjusted R^2	Significance
Bod_5	Ub, Ar, Gdp_{pc}	$1.053+566.924ub-0.012ar-0.015gdp_{pc}$	0.856	0.830	**
Cod_{mn}	Ub, Ar	1.451+88.727ub-0.002ar	0.903	0.892	**
Tn	Ub	1.668+69.168ub	0.912	0.907	**
Dn	Ub	1.231+53.629ub	0.821	0.811	**
Nh_4^+-N	Ub, Hp_d	$1.407+61.854ub-1.030hp_d$	0.805	0.782	**
No_3^--N	Slp_{mean}, Ub	$5.161012-0.233slp_{mean}-7.675ub$	0.720	0.670	**
Tp	Ub, Aho_{pc}	$0.012+1.093ub+0.050aho_{pc}$	0.794	0.770	*
Dp	Hp_d	$0.022+0.015hp_d$	0.406	0.373	**
Do	Ub	9.802-50.534ub	0.562	0.538	**
Ph	Ub	7.724012-5.814ub	0.439	0.408	**
Turbidity	Aho_{pc}	$-115.264+96.209aho_{pc}$	0.662	0.643	**
Ec	Gdp_{pc}	$-261.761+1.009gdp_{pc}$	0.778	0.766	**

Abbreviations Are Provided In Table 1. The Parameters Without Regression Models Are Not Listed.
*Indicates Significance At The 0.05 Probability Level.
**Indicates Significance At The 0.01 Probability Level.

Table 6. The Rda Results For The Percentage Of The Overall River Water Quality Variance Explained By An Individual Variable And All The Explanatory Variables.

Explanatory Variables	Variance Explained%	*P* Value
Land Use		
Frt	29.6	0.002
Wt	8.8	0.16
Ub	50.8	0.002
Crp	16.2	0.016
Topography		
Elv	29.8	0.004
Slp_{mean}	36.3	0.002
Ar	15.7	0.032
Socio-Economic Factors		
Hp_d	44.6	0.002
Gdp_{pc}	26.2	0.032
Aho_{pc}	26.9	0.002
Land Use + Topography + Social-Economic Factors	86.1	0.002

Abbreviations Are Provided In Table 1.

Discussion

1 Seasonal effects

Seasonal variations in the flow caused by the subtropical monsoon climate can partially explain the river water quality temporal variations [7], [38]. The relatively high TN concentrations in the rainy season were attributed to a flushing effect [36]. BOD_5, which corresponds to point source pollution, were lower in the rainy season due to the dilution by precipitation [2]. Relatively high T in the rainy season resulted from the specific climatic conditions during this period (T is higher in the rainy season than in the dry season; see **Table 2**). The lower values of DO and higher values of pH in the rainy season may related to some environmental factors variations, such as flow rate, water temperature, aquatic plant growth and so on [50], [39–41]. The reasons are complex and the further research is needed in the future. The turbidity was higher in the dry season, which may be primarily because the reduced flow during this period caused the river water to be easily influenced by tides. Moreover, Odokuma and Okpokwasili [40] reported that the other parameters such as phosphate showed significantly higher values in the rainy season than in the dry season in the New Calabar River, Nigeria. These different results compared with our study were mainly due to the regional climate difference. However, the absence of a significant difference in the other physico-chemical parameters between the dry and wet seasons indicated mixed and irregular influences, e.g., point sources and diffusion may have played an important role [1], [42].

2 Land cover effects

In the study area, land-use types could play important roles in affecting major river water quality parameter distributions, as reflected by their considerable variability (**Figure 2**). This result is corroborated by the strong positive correlations between urban land and BOD_5, COD_{Mn}, TN, DN, NH_4^+-N and TP (**Table 5**) and the strong negative correlations with DO and pH. Moreover, most of the nutrient variables exhibited lower concentrations in the forest-dominated region, whereas the cropland-dominated region had high nutrient concentrations. The region downstream of the urban area exhibited higher concentrations of BOD_5, COD, NH_4^+-N, TP and DP and lower pH and DO (**Figure 2**); this finding agrees with many related studies [21], [43,44]. Studies have also shown that certain river water quality parameters (e.g., TN and TP) are primarily determined by agricultural land use at

Table 7. The Rda Results Showing The Eigenvalues And Percentage Of The Overall River Water Quality Parameter Variance Explained By The First Two Rda Axes And All The Rda Axes.

	Axis 1	Axis 2	All Axes
Eigenvalues	0.524	0.165	0.861
Percentage Of Water Quality Variance Explained	52.4	16.5	86.1
Dominant Environmental Factors	Ub(0.967)	Aho_{pc}(−0.699)	
	Hp_d (0.897)		
	Slp_{mean}(−0.780)		
	Frt (−0.725)		

Abbreviations Are Provided In Table 1. The Number In Parentheses Indicates The Canonical Coefficients Of The Environmental Variables With The First Two Rda Axes.

the watershed scale in many parts of the world [1], [20]. However, this study found that agricultural land use in watersheds plays a minor role in explaining the spatial variations in river water quality (explaining 16.2% of the total overall water quality variance) in the Cao-E River basin. The different result is primarily due to the enhanced effects of urban runoff and incompletely treated industrial waste and domestic pollution on river water quality compared to agricultural runoff. In addition, vegetated buffers adjacent to the croplands can substantially mitigate agricultural runoff of nutrients and other contaminants via deposition, absorption, and denitrification [20].

Urban areas are primarily located along river networks in the Cao-E River basin. Therefore, urban effects on the river water quality were expected. Urban land use comprised a much smaller percentage of the Cao-E River basin than the cropland in this study. However, as observed from the regression model (BOD_5, COD_{Mn}, TN, DN, NH_4^+-N, NO_3^--N, TP, DO and pH) and the RDA analysis (explaining 50.8% of the total overall water quality variance) results, urban areas play a pivotal role in influencing water quality (**Tables 5 and 6**). This observation is most likely due to two factors: (1) the deficient capacity of urban sewage treatment plants leads to domestic pollution and industrial wastewater inputs into the Cao-E River system, and (2) the high percentage of impervious surfaces and over fertilization of grassy areas can increase discharge rates, sedimentation and pollutant runoff to streams [44].

3 Topography effects

Topographical factors played important roles in explaining spatial variations in river water quality within the Cao-E River basin (SLP_{mean} was 36.3% and ELV was 29.8%). Topography largely regulated the river water quality parameters. Most related studies have shown that higher SLP_{mean} and/or ELV lead to higher erosion rates, which subsequently increase the rate at which particulate matter enters a water body [13], [21], [45]. However, our results indicated an inverse relationship. Here, SLP_{mean} and ELV were both negatively correlated with TN, DN, NO_3^--N, TP and DP; SLP_{mean} exhibited a highly significant negative correlation with NO_3^--N values (**Figure 3 and Table 5**). The slope effects on water chemistry varied. Watershed physical characteristics, such as soil properties (soil texture and soil drainage), morphological variables (drainage density and elongation) and particularly surficial debris, largely affected water chemistry in river waters [37], [46]. In addition, these negative correlations existed primarily because ELV and SLP_{mean} were strongly correlated with land cover (**Table 8**). High elevations and/or steep catchments are primarily occupied by forest land cover; however, cropland is primarily located in regions in which the topography is relatively flat and has a low ELV. Therefore, the forested mountain area, which has higher SLP_{mean} and/or ELV, may export fewer nutrients than flat land (e.g., cropland). Similar results were published regarding water quality spatial variations in the Han River basin in South Korea [7] and non-point source effects on stream nutrient concentrations in the Seattle region of the USA [47].

4 The effects of socio-economic factors

The socio-economic factors commonly have different extent effects on river water quality, especially in developing countries such as China. Our study shows that socioeconomic factors have significant effects on river water quality during the study period (**Table 5**) and that HP_d, AHO_{pc}, and GDP_{pc} explain 44.6%, 26.9% and 26.2% of the total overall river water quality variance, respectively (**Table 6**). Furthermore, HP_d is a fundamental

Table 8. The Pearson Correlation Matrix For The Watershed Characteristics Of The Cao-E River Basin In Eastern China.

	Frt	Wt	Ub	Crp	Elv	Slp_{mean}	Ar	Hp_d	Gdp_{pc}	Aho_{pc}
Frt	1.00									
Wt	0.09	1.00								
Ub	−0.79**	0.31	1.00							
Crp	−0.96**	−0.31	0.59**	1.00						
Elv	0.57**	−0.48*	−0.71**	−0.41	1.00					
Slp_{mean}	0.83**	−0.33	−0.82**	−0.70**	0.74**	1.00				
Ar	−0.44	0.17	0.42	0.37	−0.44	−0.56*	1.00			
Hpd	−0.79**	0.37	0.95**	0.60**	−0.79**	−0.91**	0.55*	1.00		
Gdp_{pc}	−0.38	0.49*	0.69**	0.18	−0.65**	−0.67**	0.30	0.76**	1.00	
Aho_{pc}	−0.47*	0.20	0.55*	0.36	−0.69**	−0.72**	0.47*	0.64**	0.72**	1.00

*Indicates That The Correlation Was Significant At The 0.05 Probability Level.
**Indicates That The Correlation Was Significant At The 0.01 Probability Level.

parameter for predicting DP and AHO_{pc} for TP and turbidity (**Table 5**). This result can be attributed to high population density, the massive animal production and the lack of strict management regulations for animal waste in the study area. The population density exceeds 380 people km^{-2}, the annual animal output value exceeds 2.0 billion Yuan and the annual hog production exceeds 1.0 million heads for the entire basin [48]. Most human and animal waste with incomplete treatment is discharged into the surrounding water bodies. Moreover, EC was positively correlated with GDPpc, which was most likely related to point source pollution [37].

In addition, some researches [4–6] point out that climate condition is also an important factor of influencing river water quality. Thus, we will further study it in our future work.

Conclusions

The primary results of this study can be summarized as follows:

1) The in-stream water quality in the Cao-E River basin streams suggested that TN, pH and T were generally higher in the rainy season, whereas BOD_5, DO and turbidity were higher in the dry season.
2) Spatial variations in river water quality are typically associated with several anthropogenic and natural factors. Urban land cover was determined to be the most important explanatory variable for BOD_5, COD_{Mn}, TN, DN, NH_4^+-N, NO_3^--N, TP, DO and pH. Moreover, Animal husbandry output per capita was an important predictor for TP and turbidity, and Gross domestic product per capita largely determined the spatial variations in EC. The remaining unexplained variance resulted from other factors, such as topography.
3) 3) Pollution control for animal waste discharge in rural settlements was important in the study area. Moreover, agricultural runoff, industrial pollution and domestic pollution in urban and industrial areas were also important factors within the Cao E River basin.
4) The percentage of total overall river water quality variance explained by individual variables and/or all environmental variables in the study area (as determined using RDA) can assist in quantitatively identifying primary pollution control factors at the watershed scale.

This study improved our understanding of the anthropogenic activities and natural factors that affect river water quality and can assist in the design of efficient strategies for controlling river water pollution at the watershed scale. Moderate-resolution remote sensing data were adopted in this study; future investigations will require high-resolution DEM maps and additional land use classes to better evaluate the effects of specific environmental variables on overall river water quality.

Supporting Information

Table S1 Summary of commonly used statistical methods on pollution source identification in recent years.

Table S2 Application summary of 5 statistical methods in this study.

Acknowledgments

We would like to acknowledge the members of the Zhejiang University soil physical laboratory for assistance in the field and laboratory.

Author Contributions

Conceived and designed the experiments: JL JC. Performed the experiments: JC. Analyzed the data: JC JL. Wrote the paper: JL JC.

References

1. Li S, Gu S, Liu W, Han H, Zhang Q (2008) Water Quality In Relation To Land Use And Land Cover In The Upper Han River Basin, China. Catena 75: 216–222.
2. Zhang Y, Guo F, Meng W, Wang X (2009) Water Quality Assessment And Source Identification Of Daliao River Basin Using Multivariate Statistical Methods. Environ Monit Assess 152: 105–121.
3. Zhou N, Westrich B, Jiang S, Wang Y (2011) A Coupling Simulation Based On A Hydrodynamics And Water Quality Model Of The Pearl River Delta, China. J Hydrol 396: 267–276.
4. Astaraie-Imani M, Kapelan Z, Fu G, Butler D (2012) Assessing The Combined Effects Of Urbanisation And Climate Change On The River Water Quality In An Integrated Urban Wastewater System In The Uk. J Environ Manage 112: 1–9.
5. Costa-Cabral M, Coats R, Reuter J, Riverson J, Sahoo G, et al. (2013) Climate Variability And Change In Mountain Environments: Some Implications For Water Resources And Water Quality In The Sierra Nevada (Usa). Clim Change 116: 1–14.
6. Aubert Ah, Gascuel-Odoux C, Merot P (2013) Annual Hysteresis Of Water Quality: A Method To Analyse The Effect Of Intra-And Inter-Annual Climatic Conditions. J Hydrol 478: 29–39.
7. Chang H (2008) Spatial Analysis Of Water Quality Trends In The Han River Basin, South Korea. Water Res 42: 3285–3304.
8. Hoorman J, Hone T, Sudman Jr T, Dirksen T, Iles J, et al. (2008) Agricultural Impacts On Lake And Stream Water Quality In Grand Lake St. Marys, Western Ohio. Water Air Soil Pollut Int 193: 309–322.
9. Ahearn Ds, Sheibley Rw, Dahlgren Ra, Anderson M, Johnson J, et al. (2005) Land Use And Land Cover Influence On Water Quality In The Last Free-Flowing River Draining The Western Sierra Nevada, California. J Hydrol 313: 234–247.
10. Huang F, Wang X, Lou L, Zhou Z, Wu J (2010) Spatial Variation And Source Apportionment Of Water Pollution In Qiantang River (China) Using Statistical Techniques. Water Res 44: 1562–1572.
11. Chang N (2005) Sustainable Water Resources Management Under Uncertainty. Stoch Environ Res Risk Assess 19: 97–98.
12. Kolovos A, Christakos G, Serre Ml, Miller Ct (2002) Computational Bayesian Maximum Entropy Solution Of A Stochastic Advection - Reaction Equation In The Light Of Site - Specific Information. Water Resour Res 38: 51–54.
13. Sliva L, Dudley Williams D (2001) Buffer Zone Versus Whole Catchment Approaches To Studying Land Use Impact On River Water Quality. Water Res 35: 3462–3472.
14. Heathcote Aj, Filstrup Ct, Downing Ja (2013) Watershed Sediment Losses To Lakes Accelerating Despite Agricultural Soil Conservation Efforts. Plos One 8: E53554.
15. Keatley Be, Bennett Em, Macdonald Gk, Taranu Ze, Gregory-Eaves I (2011) Land-Use Legacies Are Important Determinants Of Lake Eutrophication In The Anthropocene. Plos One 6: E15913.
16. Wei R, Ge F, Huang S, Chen M, Wang R (2011) Occurrence Of Veterinary Antibiotics In Animal Wastewater And Surface Water Around Farms In Jiangsu Province, China. Chemosphere 82: 1408–1414.
17. Xianghao Z (2008) Study Of Protection And Construction Of Mountain Ecological Security Barrier In China. J Mount Sci (In Chinese) 1: 4–5.
18. Jiang X, Sommer Sg, Christensen Kv (2011) A Review Of The Biogas Industry In China. Energy Policy 39: 6073–6081.
19. Tong L, Li P, Wang Y, Zhu K (2009) Analysis Of Veterinary Antibiotic Residues In Swine Wastewater And Environmental Water Samples Using Optimized Spe-Lc/Ms/Ms. Chemosphere 74: 1090–1097.
20. Liu W, Li S, Bu H, Zhang Q, Liu G (2012) Eutrophication In The Yunnan Plateau Lakes: The Influence Of Lake Morphology, Watershed Land Use, And Socioeconomic Factors. Environ Sci Pollut Res Int 19: 858–870.
21. Ye L, Cai Q, Liu R, Cao M (2009) The Influence Of Topography And Land Use On Water Quality Of Xiangxi River In Three Gorges Reservoir Region. Environ Geol 58: 937–942.
22. Tian C, Lu X, Pei H, Hu W, Xie J (2013) Seasonal Dynamics Of Phytoplankton And Its Relationship With The Environmental Factors In Dongping Lake, China. Environ Sci Pollut Res Int 185: 2627–2645.
23. Yang X, Wu X, Hao H, He Z (2008) Mechanisms And Assessment Of Water Eutrophication. J Zhejiang Univ Sci B 9: 197–209.

24. Chen D, Lu J, Shen Y, Gong D, Deng O (2011) Spatio-Temporal Variations Of Nitrogen In An Agricultural Watershed In Eastern China: Catchment Export, Stream Attenuation And Discharge. Environ Pollut 159: 2989–2995.
25. Jin S, Lu J, Chen D, Shen Y, Shi Y (2009) Relationship Between Catchment Characteristics And Nitrogen Forms In Cao-E River Basin, Eastern China. J Environ Sci (China) 21: 429–433.
26. Shen Y, L U J, Chen D, Shi Y (2011) Response Of Stream Pollution Characteristics To Catchment Land Cover In Cao-E River Basin, China. Pedosphere 21: 115–123.
27. Sepbc (2002) Water And Wastewater Analysis Method. Prepared And Published By: State Environment Protection Bureau Of China (Sepbc). Beijing: China Environmental Science Press. 784 P.
28. Dodds Wk, Oakes Rm (2004) A Technique For Establishing Reference Nutrient Concentrations Across Watersheds Affected By Humans. Limnol Oceanogr Methods 2: 333–341.
29. Chen J, Lu J (2013) Establishment Of Reference Conditions For Nutrients In An Intensive Agricultural Watershed, Eastern China. Environ Sci Pollut Res Int 4: 2496–2505.
30. Olsen Rl, Chappell Rw, Loftis Jc (2012) Water Quality Sample Collection, Data Treatment And Results Presentation For Principal Components Analysis–Literature Review And Illinois River Watershed Case Study. Water Res 46: 3110–3122.
31. Myers Rh (1990) Classical And Modern Regression With Applications: Duxbury Press Belmont, Ca. 350p.
32. Miserendino Mial, Casaux R, Archangelsky M, Di Prinzio Cy, Brand C, et al. (2011) Assessing Land-Use Effects On Water Quality, In-Stream Habitat, Riparian Ecosystems And Biodiversity In Patagonian Northwest Streams. Sci Total Environ 409: 612–624.
33. Braak Ct, Šmilauer P (2002) Canoco Reference Manual And Canodraw For Windows User's Guide: Software For Canonical Community Ordination (Version 4.5). Section On Permutation Methods. Microcomputer Power, Ithaca, New York. 300p.
34. Lepš J, Šmilauer P (2003) Multivariate Analysis Of Ecological Data Using Canoco. New York: Cambridge University Press. 283 P.
35. Johnson Ra, Wichern Dw (2002) Applied Multivariate Statistical Analysis: Prentice Hall Upper Saddle River, Nj. 755p.
36. Chen D, Lu J, Yuan S, Jin S, Shen Y (2006) Spatial And Temporal Variations Of Water Quality In Cao-E River Of Eastern China. J Environ Sci (China) 18: 680–688.
37. Li S, Xia X, Tan X, Zhang Q (2013) Effects Of Catchment And Riparian Landscape Setting On Water Chemistry And Seasonal Evolution Of Water Quality In The Upper Han River Basin, China. Plos One 8: E53163.
38. Shah Vg, Dunstan Rh, Geary Pm, Coombes P, Roberts Tk, et al. (2007) Comparisons Of Water Quality Parameters From Diverse Catchments During Dry Periods And Following Rain Events. Water Res 41: 3655–3666.
39. Karaer F, Küçükballi A (2006) Monitoring Of Water Quality And Assessment Of Organic Pollution Load In The Nilüfer Stream, Turkey. Environ Monit Assess 114: 391–417.
40. Odokuma Lo, Okpokwasili Gc (1997) Seasonal Influences Of The Organic Pollution Monitoring Of The New Calabar River, Nigeria. Environ Monit Assess 45: 43–56.
41. Kannel Pr, Lee S, Lee Y (2008) Assessment Of Spatial–Temporal Patterns Of Surface And Ground Water Qualities And Factors Influencing Management Strategy Of Groundwater System In An Urban River Corridor Of Nepal. J Environ Manage 86: 595–604.
42. Chen Js, Gao Xm, He Dw, Xia Xh (2000) Nitrogen Contamination In The Yangtze River System, China. J Hazard Mater 73: 107–113.
43. Pratt B, Chang H (2012) Effects Of Land Cover, Topography, And Built Structure On Seasonal Water Quality At Multiple Spatial Scales. J Hazard Mater 209: 48–58.
44. Mehaffey Mh, Nash Ms, Wade Tg, Ebert Dw, Jones Kb, et al. (2005) Linking Land Cover And Water Quality In New York City's Water Supply Watersheds. Environ Monit Assess 107: 29–44.
45. Xiao B, Wang Q, Wu J, Huang C, Yu D (2010) Protective Function Of Narrow Grass Hedges On Soil And Water Loss On Sloping Croplands In Northern China. Agric Ecosyst Environ 139: 653–664.
46. Lindell L, Aa Str O M M, O Berg T (2010) Land-Use Change Versus Natural Controls On Stream Water Chemistry In The Subandean Amazon, Peru. Appl Geochem 25: 485–495.
47. Brett Mt, Arhonditsis Gb, Mueller Se, Hartley Dm, Frodge Jd, et al. (2005) Non-Point-Source Impacts On Stream Nutrient Concentrations Along A Forest To Urban Gradient. Environ Manage 35: 330–342.
48. Shaoxing City Statistical Bureau (2005) Shaoxing Statistical Yearbook. Shaoxing: Shaoxing City Press. 300p.
49. Belsey Da, Kuh E, Welsch Re (1980) Regression Diagnostics: Identifying Influential Data And Sources Of Collinearity: John Wiley. 320p.
50. Zhao J, Fu G, Lei K, Li Y (2011) Multivariate Analysis Of Surface Water Quality In The Three Gorges Area Of China And Implications For Water Management. J Environ Sci (China) 23: 1460–1471.

Hypoxia, Blackwater and Fish Kills: Experimental Lethal Oxygen Thresholds in Juvenile Predatory Lowland River Fishes

Kade Small[1], R. Keller Kopf[2]*, Robyn J. Watts[1,2], Julia Howitt[2,3]

1 School of Environmental Sciences, Charles Sturt University, Albury, New South Wales, Australia, **2** Institute for Land, Water and Society, Charles Sturt University, Albury, New South Wales, Australia, **3** School of Agricultural and Wine Sciences, Charles Sturt University, Wagga Wagga, New South Wales, Australia

Abstract

Hypoxia represents a growing threat to biodiversity in freshwater ecosystems. Here, aquatic surface respiration (ASR) and oxygen thresholds required for survival in freshwater and simulated blackwater are evaluated for four lowland river fishes native to the Murray-Darling Basin (MDB), Australia. Juvenile stages of predatory species including golden perch *Macquaria ambigua*, silver perch *Bidyanus bidyanus*, Murray cod *Maccullochella peelii*, and eel-tailed catfish *Tandanus tandanus* were exposed to experimental conditions of nitrogen-induced hypoxia in freshwater and hypoxic blackwater simulations using dried river red gum *Eucalyptus camaldulensis* leaf litter. Australia's largest freshwater fish, *M. peelii*, was the most sensitive to hypoxia but given that we evaluated tolerances of juveniles (0.99±0.04 g; mean mass ±SE), the low tolerance of this species could not be attributed to its large maximum attainable body mass (>100,000 g). Concentrations of dissolved oxygen causing 50% mortality (LC_{50}) in freshwater ranged from 0.25±0.06 mg l^{-1} in *T. tandanus* to 1.58±0.01 mg l^{-1} in *M. peelii* over 48 h at 25–26°C. Logistic models predicted that first mortalities may start at oxygen concentrations ranging from 2.4 mg l^{-1} to 3.1 mg l^{-1} in *T. tandanus* and *M. peelii* respectively within blackwater simulations. Aquatic surface respiration preceded mortality and this behaviour is documented here for the first time in juveniles of all four species. Despite the natural occurrence of hypoxia and blackwater events in lowland rivers of the MDB, juvenile stages of these large-bodied predators are vulnerable to mortality induced by low oxygen concentration and water chemistry changes associated with the decomposition of organic material. Given the extent of natural flow regime alteration and climate change predictions of rising temperatures and more severe drought and flooding, acute episodes of hypoxia may represent an underappreciated risk to riverine fish communities.

Editor: Erik V. Thuesen, The Evergreen State College, United States of America

Funding: Support for this research came from a CSU honours scholarship (K. Small) and from the Ecological Responses to Altered Flow Regimes Research Cluster which represents a collaboration between the CSIRO Water for a Healthy Country Flagship, Griffith University, the University of New South Wales, Monash University, Charles Sturt University, La Trobe University, and the Arthur Rylah Institute of the Victorian Department of Environment and Primary Industries. The funders had no role in study design, data collection and analysis, or preparation of the manuscript.

Competing Interests: The authors have declared that no competing interests exist.

* E-mail: rkopf@csu.edu.au

Introduction

Hypoxia and blackwater events occur in lowland rivers and wetlands world-wide and can be a natural phenomenon associated with rising temperature, floodplain inundation and the subsequent decomposition of organic material [1,2,3]. Excess loads of allocthonous organic material, nutrients or rising temperature can stimulate acute episodes of hypoxia and water chemistry changes. Hypoxic blackwater events are usually characterized by elevated dissolved organic carbon (DOC) concentrations, tea coloured water, low dissolved oxygen concentration [2,3] and reduced pH associated with the release of organic acids and polyphenols from terrestrial organic matter; some of which may be toxic [4]. Blackwater rivers, as opposed to sporadic episodes of hypoxic blackwater, are the result of a consistent annual cycle of floodplain inundation [5].

Fish inhabiting rivers and wetlands where hypoxia occurs may be adapted to tolerate low dissolved oxygen concentrations and variable water quality using a variety of mechanisms ranging from air-breathing to behavioural avoidance [6]. Using behaviour referred to as aquatic surface respiration (ASR), many species of fish will rise to the surface when exposed to hypoxia, in an attempt to extract oxygen from the thin surface layer of water that is contact with the atmosphere (Figure 1). Severe, widespread, or prolonged low-oxygen events may, however, lead to sub-lethal effects and mortalities [2,4,7,8]. In the United States hypoxia is estimated to be the proximate cause of 5.3% of fish kills [7], although low oxygen may be the ultimate cause of death in some mortalities ascribed to eutrophication, pollution or extreme climate events such as flooding and drought. Developing an understanding of low dissolved oxygen tolerances and sub-lethal effects on aquatic animals will be essential in developing natural resource management guidelines to mitigate the effects of human-induced hypoxia.

Given the natural occurrence of drought, flooding and blackwater events in lowland rivers of the Murray-Darling Basin (MDB) in south-eastern Australia [1,2,3], it is reasonable to assume that native fish in this ecosystem would be well adapted to

Figure 1. A severe hypoxic blackwater event resulted in extensive fish kills in the Murray-Darling Basin, Australia in 2010/11. Photograph shows a Murray cod *Maccullochella peelii* (Mitchell) appearing to perform aquatic surface respiration (ASR) during the hypoxic blackwater event in the Edward River near Moulamein, New South Wales. Printed under a CC BY license with permission from L. Pearce, 2011.

acute hypoxia and high levels of DOC. However, after 10 years of drought, since the year 2000, the MDB experienced large-scale floods followed by hypoxic blackwater events which caused widespread fish kills [2,8]. Estimates of the total number of fish mortalities and species composition following this event have not been published but anecdotal and scientific reports on fish kills in the MDB have tended to focus on large-bodied recreationally important species [7,8,9]. It remains unclear whether large-bodied species are disproportionately reported in fish kills because of low tolerances to hypoxic blackwater or because small-bodied species are under-reported. Previous research on Australian freshwater fishes suggests a relatively high degree of hypoxia, DOC and toxic leachate tolerance amongst small-bodied species [10,11,12] but very little research has been undertaken on the hypoxic blackwater tolerances of large-bodied lowland fishes [4,13,14] most commonly reported in fish kills [7,9].

This study aimed to quantify low dissolved oxygen (DO) blackwater tolerances in juvenile stages of four lowland river fishes native to the MDB, Australia. The species examined included juvenile stages of: Murray cod *Maccullochella peelii* (Mitchell), golden perch *Macquaria ambigua* (Richardson), silver perch *Bidyanus bidyanus* (Mitchell) and eel-tailed catfish *Tandanus tandanus* (Mitchell). As adults, these species are top predators and consumers in lowland river ecosystems of the MDB and are important to recreational fisheries. The objectives of this research were to: 1) determine if there are species-specific differences in lethal oxygen concentrations and ASR responses and 2) determine whether lethal oxygen concentrations and ASR responses are influenced by increasing DOC concentrations within simulated blackwater.

Materials and Methods

Experimental design

Between two-hundred and two-hundred and fifty juveniles each of four species (Table 1) were sourced from Urarah Fisheries (*M. peelii*), New South Wales Department of Primary Industries – Narrandera (*M. ambigua*), Murray Cod Hatcheries (*T. tandanus*) and Silverwater Fisheries (*B. bidyanus*). All hatcheries used wild broodstock. Juveniles were used to develop low oxygen thresholds because of conservation restrictions concerning the death of large numbers of threatened wild-caught adults and number of years required for hatchery fish to reach maturity. All fish were in the juvenile stage of development and approximately 6 to 14 months of age, ranging from 30 mm to 88 mm standard length (Table 1). Differences in hatch date resulted in inter-specific variation in body size but there were no significant differences in mean mass or length within species' treatments. Fish were transported within one day from the hatcheries to a temperature controlled room maintained at 25–26°C for the duration of experimentation.

Each species was acclimated to a constant temperature of 25–26°C for between 9 to 15 days prior to experimentation. Lights were programmed for a 12 h light and dark cycle and fish were fed blood worms twice daily except during experimentation. Species were acclimated separately in 750 l holding tanks equipped with a UV light filter and a recirculating bio-ball filtration system that maintained oxygen levels above 7 mg l^{-1}. All experiments were undertaken in glass aquaria, 23 cm (L)×23 cm (W)×45 cm (H), custom built to achieve minimal surface area and limit oxygen diffusion with the atmosphere. A static flow during all experiments was utilized to reduce variation in DO concentration. Prior to each experiment oxygen sensors (YSI rapid pulse electrochemical sensor; 6-series) were calibrated against temperature-specific oxygen saturated distilled water.

Hypoxia tolerance of each species was quantified using median lethal oxygen concentration (LC_{50}) and aquatic surface respiration (ASR_{50}) experiments conducted in accordance with international guidelines for acute toxicity testing [15] in: 1) hypoxic freshwater (no dissolved organic carbon/leaf litter) and 2) simulated hypoxic blackwater created using dried leaf litter. Experiments were carried out with permission from a Charles Sturt University Animal Care and Ethics (ACEC) committee protocol; number 11/018. The acute toxicity guidelines and ACEC protocol aimed to minimize the number of animals used whilst generating LC_{50} and ASR_{50} estimates under hypoxic blackwater and hypoxic freshwater experimental conditions for each species.

Hypoxic freshwater

Seven oxygen treatments, each containing 10 fish replicated per aquaria, were maintained during hypoxic freshwater LC_{50} and ASR_{50} experiments for each species. Treatments consisted of DO concentrations of 0.1 mg l^{-1}, 0.7 mg l^{-1}, 1.3 mg l^{-1}, 1.9 mg l^{-1}, 2.5 mg l^{-1}, 3.1 mg l^{-1} and a control (8 mg l^{-1}) that were maintained at 25–26°C for 48 h. The experimental temperature of 25–26°C was warmer than the majority of hypoxic blackwater events documented in the MDB [1–3], although it does represent the mid-point of the range of temperatures recorded during the 2005 blackwater event in the Barmah Forest [3]. Given that oxygen metabolism in fish increases with temperature, therefore, the hypoxia tolerances estimated here were intended to provide a precautionary estimate for application by natural resource managers. Previous studies assessing hypoxia on small-bodied fish species of the MDB were conducted at 25°C, as it represents a temperature commonly reached in summer in billabongs in the MDB [12] and the main river channel of the Murray River at Barmah [3]. Dissolved oxygen was regulated using nitrogen (N_2) or atmospheric air. Hypoxic freshwater experiments contained no leaf litter and therefore little or no DOC. Atmospheric air and N_2 were regulated via pressure valves and separate air stones placed at the bottom of experimental tanks filled with 17.5 l of holding tank water. Experimental DO concentrations were maintained at all times and treatments to within 0.3 mg l^{-1} of the pre-determined oxygen concentration or were excluded from the analyses.

Ten fish within each aquaria served as replicates for each species-oxygen treatment combination and one randomly selected DO treatment was repeated. The experimental design resulted in 20 replicates for the repeated treatment and 10 replicates for other

Table 1. Mean body size measurements and water quality parameters for hypoxia experiments conducted on Murray-Darling Basin fishes.

Species	Hypoxic freshwater						Simulated hypoxic blackwater					
	n	Length (mm)	Mass (g)	Temp. (°C)	pH range	DO (mg l^{-1}) range	*n*	Length (mm)	Mass (g)	Temp. (°C)	pH range	DO (mg l^{-1}) range
Murray cod *Maccullochella peelii*	80	50.42 (0.68)	0.99 (0.04)	25.63 (0.15)	7.69–8.65	0.16–8.24	60	51.34 (0.80)	1.01 (0.02)	25.02 (0.07)	6.61–7.80	0.17–8.26
Golden perch *Macquaria ambigua*	80	36.06 (0.88)	0.45 (0.03)	25.49 (0.09)	7.67–8.06	0.07–8.14	60	37.38 (0.43)	0.48 (0.02)	24.51 (0.11)	6.77–7.51	0.06–8.2
Eel-tailed catfish *Tandanus tandanus*	80	83.90 (1.64)	3.67 (0.19)	26.35 (0.22)	7.23–8.74	0.09–8.68	60	83.45 (0.92)	3.61 (0.10)	25.73 (0.12)	6.49–7.11	0.08–8.5
Silver perch *Bidyanus bidyanus*	70	35.4 (0.69)	0.40 (0.02)	26.22 (0.14)	7.41–8.55	0.09–8.67	50	35.98 (0.47)	0.43 (0.02)	26.71 (0.09)	6.05–7.22	0.09–8.3

Fish were exposed to 48 h lethal dissolved oxygen (DO) in hypoxic freshwater and simulated hypoxic blackwater. Mean ±(SE).

species-oxygen treatment combinations. Repeated treatments included DO concentrations of 3.1 mg l^{-1} for *M. ambigua*, 0.7 mg l^{-1} for *T. tandanus* and 2.5 mg l^{-1} for *M. peelii*. A more rigorous experimental design would have replicated each aquaria of 10 fish at least three times, although the total number of fish used here ultimately represented trade-offs among statistical power, animal ethics and limitations concerning species conservation status and project resources. The present design provided sufficient statistically power for all four species but individual oxygen-species treatments (n = 10) were largely unreplicated at the tank level.

The highest DO treatment concentration, used for all species, was pre-determined from range finding trials conducted in accordance with the international guidelines for acute toxicity testing [15]. In range finding trials oxygen levels were reduced in 10 increments from 8 mg l^{-1} down to 0.5 mg l^{-1} over a two hour period. This period of time allows for comparisons with other similar studies and avoids issues with a decline that is too rapid or too gradual [12]. The oxygen concentration two increments above the first signs of hypoxia stress (a change in ventilation rate, aquatic surface respiration (ASR) or loss of equilibrium) was selected as the upper boundary concentration (3.1 mg l^{-1}). The oxygen treatments, 0.1 mg l^{-1} and 0.7 mg l^{-1}, required 2.0 g and 0.5 g per 17.5 l treatment tank respectively of dissolved sodium sulphite in order to reach low DO levels. Sodium sulphite was added to treatment tanks as a solid and stirred prior to immersing fish within aquaria. The use of sodium sulphite in developing low DO for experimentation on fish is common [11,16,17] and although salt is not directly lethal to fish at the concentrations used here [18], the potential for compounding effects of increasing salinity and other toxic effects of the compound on hypoxia tolerance may warrant further investigation.

Experiments started by netting 10 fish and transferring them from 750 l holding tanks directly into an experimental treatment tank set to a pre-determined DO concentration. Immersing fish directly within experimental tanks set to a pre-determined DO concentration avoided bias associated with differences in time required to reduce oxygen levels among tanks but may have contributed to elevated rates of mortality at low oxygen levels. Water from holding tanks was used in all experimental and control tanks to avoid additional stress during transfer. Dissolved oxygen, temperature and pH were measured in each experimental and control tank at intervals of 0 h, 1 h, 4 h, 8 h, 16 h, 24 h, 32 h, 40 h, 48 h and DO concentrations were regulated with bubbled N_2 gas or atmospheric air between intervals if necessary, to ensure treatment DO levels were maintained.

Mortality assessments were made during water quality testing intervals and death was classified as a loss of equilibrium and the cessation of gill ventilation. A binomial classification of dead (1) or alive (0) at the end of the 48 h experimental period was used in statistical analyses, although all mortalities were removed immediately from treatments and standard length (mm) and mass (±0.01 g) were measured. Aquatic surface respiration was quantified every five minutes for the first two hours of each experiment and then at four-hourly intervals up to 48 h. Aquatic surface respiration measurement periods typically lasted less than five seconds and classified fish as performing ASR (1) or alive but not performing ASR (0). Individuals ventilating with their mouths touching the top 1 mm of the water surface layer were recorded as performing ASR [12], whereas fish simply swimming near the surface were excluded. A viewing curtain prevented disturbing fish when ASR behaviour was recorded. Binomial ASR observations over the 48 hr experimental period were used in statistical analyses.

Simulated hypoxic blackwater

A gradient of five simulated blackwater concentrations across aquaria was developed by immersing different masses of naturally abscised river red gum *Eucalyptus camaldulensis* leaf litter within experimental tanks. Oxygen concentrations were reduced via natural leaf litter decomposition and fish respiration within aquaria and did not require the use of nitrogen or sodium sulphite. Lethal oxygen concentration (LC_{50}) and ASR_{50} measurements were otherwise conducted in accordance with hypoxic freshwater experiments. Leaf litter was collected from a fallen branch on private property along the Murray River between Albury and Howlong, NSW. A private land owner granted permission to collect samples and the NSW Office of Environment and Heritage confirmed that, within this region, the species is not threatened, or part of a threatened ecological community, and that no permit is required for scientific collection.

Prior to experimentation, leaf litter and twigs, ≤5 mm diameter were oven-dried at 60°C for 24 to 48 hrs. Five geometrically increasing masses of dried leaf litter containing 0 g (control), 3 g, 9 g, 27 g, and 81 g were placed in nylon stockings weighted with 250 g of cobble. These masses resulted in effective leaf litter concentrations ranging from 0 g l^{-1} to 4.59 g l^{-1} among aquaria which resulted in an oxygen concentration gradient ranging from 8.4±0.4 mg l^{-1} down to 0.46±0.28 mg l^{-1} respectively without the use of nitrogen or sodium sulfite. Controls consisted of cobble in nylon stockings but no leaf litter.

Leaf litter bags were allowed to soak for 24 hrs prior to experimentation and they were then removed and 10 fish were immersed in aquaria. Fish mortality, ASR assessments, and water quality measurements of DO, pH and temperature were carried out using the same methodology as described in hypoxic freshwater experiments. In addition, two 60 ml water samples were taken from each simulated blackwater treatment at the 0 h and 48 h interval. The mean concentration of DOC for each leaf litter treatment was used to examine its effect on fish mortality and ASR responses. Dissolved organic carbon concentrations were determined at a National Association of Testing Authorities (NATA) Australia accredited lab where an international standard method, 5310 C [19], was applied in which total organic carbon is considered all non-purgeable carbon.

Natural blackwater event

Water quality conditions of experimental treatments were compared to a naturally occurring hypoxic blackwater event sampled in the MDB from the Edward River (Natural 1) and Wakool River (Natural 2) in December, 2010. John and Anthony Conallin from the Murray CMA and Luke Pearce from the NSW Department of Primary Industries confirmed the presence or absence of fish species mortalities in each river following the event. Water quality measurements were recorded during the hypoxic black water conditions, while fish mortalities in the field were verified during and after events. Water quality measurements of DO, pH and temperature were collected in flowing water within the main river channel during both events and water samples of less than one litre were collected for measurement of DOC concentration. Collection of water samples was made with permission from private land owners and the NSW Office of Water.

Statistical analyses

Generalised Linear Mixed Models (GLMM's) [20] fit by maximum likelihood were used to test hypotheses and evaluate which combination of variables (DO, species, DOC, temperature, pH) best predicted mortality and ASR responses. Significance was evaluated at a 5% alpha level of type I error and results are presented as means ±1 SE unless otherwise stated. Observations of individual fish mortalities and ASR responses to low dissolved oxygen experiments were fitted to separate GLMM's using a binomial logit function in the 'lme4' package developed for R [20]. Individual responses of fish were nested randomly within tanks and we tested two null hypotheses: 1) there are no species-specific differences in lethal oxygen concentrations and ASR responses and 2) lethal oxygen concentrations and ASR responses are not influenced by increasing DOC in simulated blackwater.

Null hypotheses one and two were tested using the GLMM formula: Mortality or ASR~DO+ species +DOC+ temperature +pH+(1|tank) where mortality = dead/1 or alive/0; ASR = performing ASR/1 or alive but not performing ASR/0; DO = mean DO within tank; species = species; DOC = mean dissolved organic carbon concentration; temperature = mean temperature; pH = mean pH and tank = experimental treatment aquaria set to a pre-determined DO concentration or simulated blackwater treatment. Null hypothesis one was rejected if species was a significant factor influencing mortality or ASR responses and null hypothesis two was rejected if DOC significantly affected either response. All possible combinations of factors were modelled and the Akaiki Information Criterion (AIC) was used to determine which model provided the most parsimonious fit to mortality and ASR responses. If correlation coefficients between fixed terms exceeded +/−0.70 then the term explaining the least amount of variation was excluded from the model. Residuals were checked visually for random dispersion and normality [20].

Lethal DO concentrations ($LC_{50's}$) and aquatic surface respiration ($ASR_{50's}$) values were estimated by fitting logistic regressions to individual responses of fish predicted by the most parsimonious GLMM for mortality and ASR data:

$$y = a + \frac{b-a}{1+\left(\frac{x}{ASR_{50} or LC_{50}}\right)^{-Hillslope}}$$

where the fitted parameter *a* is the minimum *y*-axis value, *b* is the asymptote, LC_{50} or ASR_{50} represent the median response and *hillslope* is the slope at its midpoint [21]. Logistic functions were restricted to data that met the criteria of normality and constant variance and parameters were reported only for models that provided significant ($P<0.05$) fits to the GLMM predicted values. Logistic regressions were fitted to the mean observed DO concentration measured within tanks rather than predetermined oxygen targets.

Results

Lethal oxygen concentrations

There were significant species-specific differences in mortality responses to low DO (Table 2) and therefore null hypothesis one was rejected. The GLMM factors: DO, species and DOC provided the most parsimonious (AIC = 168) fit to mortality responses and estimated values from this model were fitted significantly by logistic regression in all species (Figure 2; Table 3). Species-specific differences in sensitivity to low DO were similar in both hypoxic freshwater and simulated blackwater experiments (Table 3). The most tolerant species in hypoxic freshwater and simulated hypoxic blackwater was *T. tandanus* and the most sensitive was *M. peelii* (Figure 2). The oxygen concentration which caused 50% mortality in *M. peelii* within hypoxic freshwater experiments resulted in 4%, 9% and 4% percent mortality in *M. ambigua*, *B. bidyanus* and *T. tandanus* respectively. A repeated

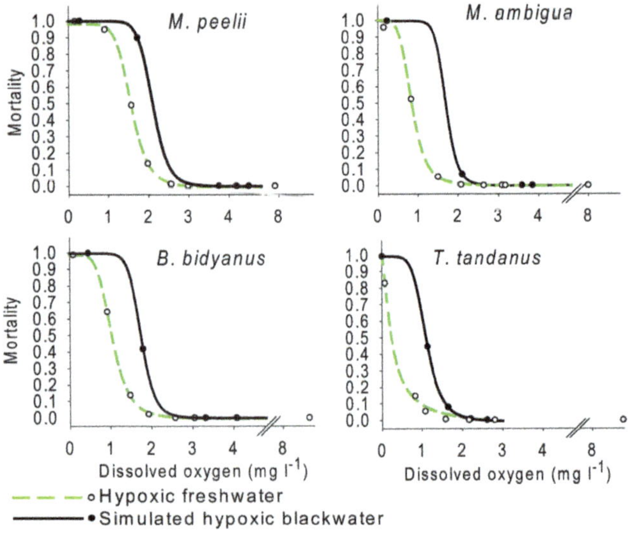

Figure 2. Lethal oxygen concentrations of four Murray-Darling Basin fishes. Proportion of juvenile Murray cod *Maccullochella peelii* (Mitchell), golden perch *Macquaria ambigua* (Richardson), silver perch *Bidyanus bidyanus* (Mitchell) and eel-tailed catfish *Tandanus tandanus* (Mitchell) dead at different dissolved oxygen concentrations at 25–26°C. Solid black and dashed green lines represent significant ($P<0.05$) logistic regressions fitted to simulated hypoxic blackwater and hypoxic freshwater experiments respectively. Circles represent model predicted values.

treatment for *B. bidyanus* could not be conducted due to a limited number of fish available although the standard error of LC_{50} (Table 3) estimates for this species was the same or lower than other species in hypoxic freshwater and simulated blackwater experiments.

Increasing DOC concentrations from simulated blackwater significantly affected mortality (Table 2) and therefore null hypothesis two rejected. Reductions in pH also significantly affected mortality but due to autocorrelation with DOC, its effect reported in Table 2 was modelled as an independent factor. Estimates of $LC_{50's}$ (Table 3) for all species occurred at higher oxygen concentrations in the presence of elevated DOC and lower pH's that occurred in simulated blackwater experiments (Table 1; 4). The oxygen concentration which caused more than 50% mortality in all species under simulated blackwater conditions resulted in less than 10% mortality of all species within hypoxic freshwater (Figure 2).

Time to mortality increased with increasing oxygen concentration from 15.39±1.68 min. (mean ±*SE*) at 0.1± mg l^{-1} DO to 128.60±32.61 min. at 0.7 mg l^{-1}. Logistic regressions in simulated blackwater experiments (Figure 2) predicted that first mortalities may start at oxygen concentrations as high as 2.4 mg l^{-1} to 3.1 mg l^{-1} in *T. tandanus* and *M. peelii* respectively. First mortalities of *B. bidyanus* and *M. ambigua* in hypoxic blackwater were predicted to start at 2.8 mg l^{-1} and 2.6 mg l^{-1} respectively. Several deaths of *B. bidyanus* and *M. peelii* were attributed to aggressive behaviour recorded in one control and the 0.17 g l^{-1} leaf litter treatment. Evaluating aggressive behaviour in relation to DO was beyond the scope of this experiment but it is interesting to note that this activity was limited to oxygen concentrations exceeding 4.45 mg l^{-1} and was most common above 8 mg l^{-1}. The effect of aggressive behaviour on mortality responses was minimized by the application of a GLMM which predicted that no mortalities could be attributed statistically to oxygen limitation at such high concentrations.

Natural blackwater events resulted in verified deaths of adult *M. peelli, M. ambigua* and *B. bidyanus* in the Edward-Wakool River system (Table 4). The natural blackwater events occurred at lower water temperatures but were associated with more acidic pH's than the simulated blackwater treatments in the present study. The dissolved oxygen concentration, pH and DOC of natural blackwater events fell between the conditions of the 1.53 g l^{-1} and 4.59 g l^{-1} simulated blackwater treatments in the present study. Percent mortality of *M. peelii* was the highest of all species and ranged from 90% in the 1.53 g l^{-1} to 99% in the 4.59 g l^{-1} simulated blackwater treatments (Table 4). Natural blackwater events resulted in confirmed mortalities of all species except for *T. tandanus* which was historically present in the Edward-Wakool system but has not been sampled in recent unpublished fisheries surveys (J. Conallin personal communication). *Tandanus tandanus* also had the highest tolerance to low DO in simulated blackwater treatments (Table 3) and field measurements of DO during the

Table 2. Generalised Linear Mixed Model (GLMM) estimates ±SE combining all factors to predict mortality and aquatic surface respiration (ASR) responses of four Murray-Darling Basin fishes exposed to hypoxic freshwater and simulated blackwater.

	Mortality				**ASR**			
Factors	**Standardized estimate**	**SE**	***z* value**	***P*-value**	**Standardized estimate**	**SE**	***z* value**	***P*-value**
Temperature (°C)	0.31	0.21	1.45	NS	−0.04	0.22	−0.18	NS
pH	0.56	0.22	2.57	<0.05	0.99	0.25	4.01	<0.001
Dissolved organic carbon	0.28	0.06	4.78	<0.001	−0.06	0.02	−4.72	<0.001
Dissolved oxygen	−4.22	0.53	−7.87	<0.001	−1.51	0.18	−8.47	<0.001
Murray cod *Maccullochella peelii*	4.8	0.74	6.45	<0.001	1.50	0.41	3.63	<0.001
Golden perch *Macquaria ambigua*	1.72	0.63	2.71	<0.05	1.34	0.42	3.16	<0.05
Silver perch *Bidyanus bidyanus*	2.65	0.6	4.44	<0.001	1.24	0.41	3.06	<0.05
Eel-tailed catfish *Tandanus tandanus*	1.85	0.53	3.48	<0.001	0.40	0.38	1.04	NS
AIC	177				364			

P-values listed as NS represent no significant effect. The most parsimonious GLMM models according to the Akaiki Information Criterion (AIC) included: Dissolved Oxygen, Species and Dissolved Organic Carbon for mortality (AIC = 168) and ASR (AIC = 360) responses.

Table 3. Logistic function parameters predicting lethal concentrations (LC) of dissolved oxygen (DO) in four Murray-Darling Basin fishes.

	Hypoxic freshwater					Simulated hypoxic blackwater				
Species	***LC*$_{50\%}$ DO (mg l^{-1})**	***a***	***b***	***Hill slope***	***P-value***	***LC*$_{50\%}$ DO (mg l^{-1})**	***a***	***b***	***Hill slope***	***P-value***
Murray cod *Maccullochella peelii*	1.58 (<0.01)	<−0.01 (<0.01)	0.98 (<0.01)	7.65 (0.13)	P<0.001	2.11 (<0.01)	<−0.01 (<0.01)	0.99 (<0.01)	11.41 (0.01)	P<0.001
Golden perch *Macquaria ambigua*	0.85 (<0.01)	<−0.01 (<0.01)	1.00 (0.01)	4.99 (0.19)	P<0.001	1.70 (<0.01)	<−0.01 (0.03)	0.99 (<0.01)	12.28 (0.03)	P<0.001
Eel-tailed catfish *Tandanus tandanus*	0.25 (0.06)	−0.03 (<0.01)	1.00 (0.10)	1.44 (0.21)	P<0.001	1.11 (<0.01)	<−0.01 (<0.01)	0.99 (<0.01)	5.99 (0.02)	P<0.001
Silver perch *Bidyanus bidyanus*	1.04 (<0.01)	<−0.01 (<0.01)	0.99 (<0.01)	5.32 (0.04)	P<0.001	1.75 (<0.01)	<−0.01 (<0.01)	1.00 (<0.01)	10.91 (0.01)	P<0.001

Experiments were conducted over 48 hrs at 25–26°C in hypoxic freshwater and simulated hypoxic blackwater. Parameters were estimated ± (SE) for significant ($P<0.05$) logistic regressions.

natural blackwater events did not reach the LC_{50} estimates for this species.

Aquatic surface respiration

Aquatic surface respiration was reported for the first time in juveniles of all four species and this behaviour was performed in hypoxic freshwater and simulated hypoxic blackwater experiments (Figure 3). Field observations of adult *M. peelii* swimming with their body positioned upward and their mouth at the water surface (Figure 1) were recorded in the Edward River during a hypoxic blackwater event. Null hypothesis one was rejected on the basis of significant differences in ASR responses among species (Table 3). The GLMM factors: DO, species and DOC provided the most parsimonious (AIC = 360) fit to ASR data, although pH was also a significant factor influencing the response (Table 3). Estimated values from the most parsimonious GLMM were fitted by a logistic regression for all species except *M. peelii* in the blackwater simulation and both experiments on *T. tandanus* (Table 5). In both species where the model did not fit, the logistic regression failed to predict ASR responses because a majority of individuals did not perform this behaviour at the lowest oxygen concentrations (Figure 3).

Null hypothesis two was rejected because ASR responses were influenced significantly by the presence DOC (Table 3) under simulated blackwater conditions. Proportionally fewer individuals of all species performed ASR under blackwater simulations when compared to hypoxic freshwater experiments (Figure 3). First observations of ASR behaviour (Figure 3) usually preceded first observations of mortality within species (Figure 2). Logistic regressions in hypoxic freshwater predicted that ASR may start at oxygen concentrations ranging from 2.59 mg l^{-1} to 4.96 mg l^{-1} in *B. bidyanus* and *M. peelii*, although first observations in these species occurred at 1.93 mg l^{-1} and 2.99 mg l^{-1} respectively. The first observation of ASR in *T. tandanus* and *M. ambigua* occurred in hypoxic freshwater at 2.81 mg l^{-1} and 3.17 mg l^{-1} respectively.

The proportion of fish performing ASR increased in all species and experiments with declining dissolved oxygen concentration. All species performed ASR (Figure 3) at levels of dissolved oxygen where no mortality was recorded (Figure 2) and at more extreme levels of hypoxia where mortality ensued. It is notable that a single *M. ambigua* survived the entire 48 h period at 0.15 mg l^{-1} DO and continuously utilised ASR behaviour. At the same oxygen concentration total mortality was observed across all species and other individuals of the same species within less than 60 min.

Discussion

Hypoxia thresholds

The fishes examined in this study were unexceptional in their ability to survive in low dissolved oxygen freshwater and in simulated blackwater. Significant species-specific differences were apparent, although rates of mortality increased following a logistic function when oxygen concentrations fell below a threshold of approximately 2–3 mg l^{-1} at 25–26°C. The range of hypoxia tolerances reported for juvenile lowland river fishes here may be considered intermediate on a global-scale comparison with other freshwater fishes. On the extreme end of the hypoxia tolerance spectrum, the crucian carp *Carassius carassius* (L.) and goldfish *Carassius auratus* (L.) can withstand anoxia at temperatures of 20–25°C for a few hours or days [22], whereas exposing some cold-adapted fishes to oxygen saturated (>8 mg l^{-1}) water at 25°C will result in mortality [23].

The hypoxia tolerances of the freshwater fishes examined here were comparable to other temperate stream fishes world-wide [24–26], whereby *T. tandanus* was the most tolerant and *M. peelii* amongst the most sensitive. The median lethal oxygen concentrations observed here ranged from 0.25 mg l^{-1} to 1.58 mg l^{-1} at 25–26°C in freshwater and 1.11 mg l^{-1} to 2.11 mg l^{-1} in simulated blackwater. At 25°C in freshwater the $LC_{50's}$ of 35 fishes in the mid-western United States ranged from 0.49 mg l^{-1} to 1.59 mg l^{-1} [24]. At the same temperature, the mean dissolved oxygen concentration resulting in a loss of equilibrium ranged from 0.95 mg l^{-1} to 2.66 mg l^{-1} in other prairie stream fishes in the United States [25]. The $LC_{50's}$ of four native New Zealand fishes and rainbow trout, *Oncorhynchus mykiss* (Walbaum) ranged from 0.54 mg l^{-1} to 2.65 mg l^{-1} at 15°C [26].

It is not appropriate to extend our experimental estimates to predict the magnitude of hypoxic blackwater fish kills in the wild, although our study provides some indication as to whether water quality conditions are approaching lethal thresholds. Notwithstanding the considerable physiological importance of cooler water conditions on hypoxia tolerance [23], which prevailed during the natural blackwater events in 2010/11 [2]. Our results suggest that low oxygen levels of less than 2.0 mg l−1, in addition to the high DOC (>10 mg l^{-1}) and low pH (4.2–6.3), which occurred in the

Table 4. Mortality (%) of four Murray-Darling Basin fishes exposed to simulated blackwater experiments and confirmed mortalities associated with natural hypoxic blackwater events in the Edward-Wakool river system in 2010/11.

Blackwater sample	DO (mg l^{-1})	pH	Temp. (°C)	DOC (mg l^{-1})	Mortality (%) Murray cod *Maccullochella peelii*	Golden perch *Macquaria ambigua*	Eel-tailed catfish *Tandanus tandanus* [a]	Silver perch *Bidyanus bidyanus*
Natural 1	1.07 (0.08)	4.18 (0.04)	19.4 (0.02)	17.7 (0.08)	Confirmed	Confirmed	-	Confirmed
Natural 2	1.98 (0.11)	6.25 (0.09)	18.7 (0.05)	16.1 (0.07)	Confirmed	Confirmed	-	Confirmed
Control (0 g l^{-1})	8.4 (0.4)	7.41 (0.05)	25.62 (0.03)	2.7 (0.05)	0	0	0	0
Simulated blackwater (0.17 g l^{-1})	4.83 (0.38)	7.22 (0.04)	26.26 (0.08)	3.58 (0.38)	0	0	0	0
Simulated blackwater (0.51 g l^{-1})	4.0 (0.3)	7.05 (0.01)	26.09 (0.06)	4.93 (0.13)	0	0	0	0
Simulated blackwater (1.53 g l^{-1})	1.8 (0.44)	6.76 (0.03)	25.93 (0.06)	9.45 (0.55)	90	7	8	45
Simulated blackwater (4.59 g l^{-1})	0.46 (0.28)	6.05 (0.01)	26.41 (0.05)	30.0 (0.7)	>99	>99	>99	>99

Water quality and dissolved organic carbon (DOC) measurements of simulated blackwater treatments (mass of leaf litter g l^{-1}) were aggregated across species for comparison with two natural blackwater events. Mean ±SE.
a. Historically present in the Edward-Wakool system but not sampled in recent unpublished fisheries surveys.

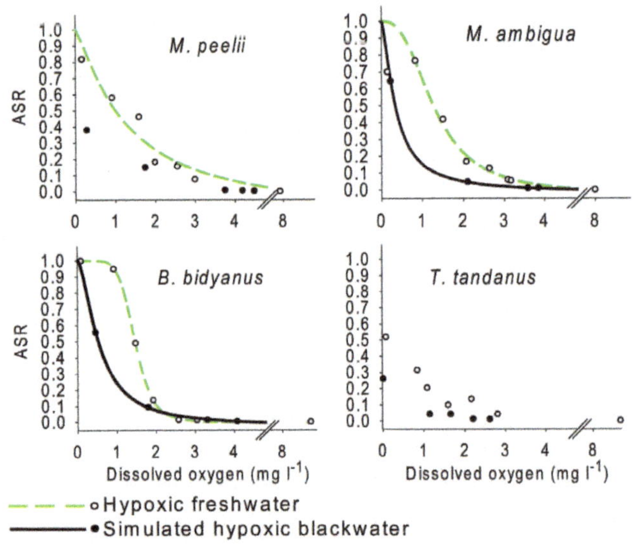

Figure 3. Aquatic Surface Respiration (ASR) in four Murray-Darling Basin fishes. Proportion of juvenile Murray cod *Maccullochella peelii* (Mitchell), golden perch *Macquaria ambigua* (Richardson), silver perch *Bidyanus bidyanus* (Mitchell) and eel-tailed catfish *Tandanus tandanus* (Mitchell) performing ASR at different dissolved oxygen concentrations at 25–26°C. Solid black and dashed green lines represent significant ($P<0.05$) logistic regressions fitted to simulated hypoxic blackwater and hypoxic freshwater experiments respectively. Circles represent model predicted values.

Edward-Wakool system and across much of the southern basin for several weeks or months during these events [2], must have approached or exceeded lethal threshold limits for all species except for *T. tandanus*. Confirmed field mortalities of the remaining three species in the Edward-Wakool system indeed corroborate this assumption. It is important to consider, however, that fish in the wild may be able to avoid hypoxic blackwater by seeking refuge habitat [6], while fish in our study were confined within experimental treatment tanks that may have exaggerated estimates of mortality.

Our estimates may have been exaggerated by not gradually acclimating fish to hypoxia. Lethal hypoxia experiments on freshwater fish in the United States suggest that metabolic rates require 24 hours to return to standard levels after handling [27]. We suggest, therefore, that future studies focusing on lethal hypoxia thresholds in fish utilize a gradual reduction in oxygen levels following a minimum recovery period of 24 hours after handling. Conversely, our lethal oxygen concentrations may have underestimated rates of mortality in the wild given that our experiments were brief (48 hrs) compared to the weeks or months [2] of hypoxic blackwater conditions that fish may be exposed to in the wild.

Hypoxia is a natural phenomenon in aquatic ecosystems, but conservation and management mitigation measures may be required if the frequency, duration, spatial extent or potency of these events has changed due to anthropogenic pressures [2]. Low oxygen survival and sub-lethal thresholds, including but not limited to those reported here, could be developed into hypoxia monitoring and warning systems that catalyse fish kill mitigation approaches and reporting [7]. The thresholds reported here provide a starting point to begin developing hypoxia mitigation strategies but must be carefully interpreted in light of potential sub-lethal effects [6], ambient and antecedent environmental conditions. It is essential to consider that hypoxia thresholds will

Table 5. Logistic function parameters predicting the proportion of live fish performing aquatic surface respiration (ASR) over 48 hrs at different dissolved oxygen (DO) concentrations.

	Hypoxic freshwater					Simulated hypoxic blackwater				
Species	*$ASR_{50\%}$* **DO (mg l^{-1})**	*a*	*b*	*Hill slope*	*P-value*	*$ASR_{50\%}$* **DO (mg l^{-1})**	*a*	*b*	*Hill slope*	*P-value*
Murray cod *Maccullochella peelii*	1.25 (0.11)	−0.15 (0.06)	1.00 (0.05)	1.25 (0.11)	P<0.001	-	-	-	-	NS
Golden perch *Macquaria ambigua*	1.31 (0.07)	−0.03 (0.04)	1.00 (0.04)	2.67 (0.34)	P<0.001	0.35 (0.10)	−0.02 (<0.01)	1.00 (0.17)	1.53 (0.16)	P<0.001
Eel-tailed catfish *Tandanus tandanus*	-	-	-	-	NS	-	-	-	-	NS
Silver perch *Bidyanus bidyanus*	1.47 (<0.01)	<−0.01 (<0.01)	1.00 (<0.01)	6.66 (0.06)	P<0.001	0.54 (0.13)	−0.03 (0.01)	1.00 (0.20)	1.68 (0.20)	P<0.001

Experiments were conducted at 25–26°C in hypoxic freshwater and simulated hypoxic blackwater. Parameters were estimated ± (SE) for significant ($P<0.05$) logistic regressions.

change according to temperature, pH, DOC [4,10] and other water chemistry parameters. Consequently, a better understanding how water quality parameters interact with fish physiology will be essential in developing future hypoxia mitigation plans. Elevated rates of mortality in blackwater simulations in the present study, when compared to low oxygen freshwater conditions, demonstrate that small reductions in pH and elevated DOC can exacerbate hypoxia driven mortality. Reductions in pH decrease the oxygen affinity of gills and it therefore directly affects hypoxia tolerance. The reduced pH of natural blackwater events and blackwater simulations is most likely attributable to increasing concentrations of dissolved organic acids, carbon dioxide or toxic polyphenols [4]. Polyphenols make up a large component of DOC in Australian rivers and at high concentrations some of these compounds are lethal to fish [4,10].

Future research focusing on temperature effects on hypoxia-driven mortality and sub-lethal effects on MDB fishes is warranted. Since hypoxia-driven mortality declines exponentially with decreasing temperature [23,28], the oxygen thresholds reported in this study could be used as precautionary management reference points under similar water quality conditions and temperatures equal to or below 25–26°C. Rising temperature directly increases metabolic demands in ectotherms [29] and therefore it is reasonable to assume for our study species that lethal oxygen concentrations will increase, or decrease, according to the ambient thermal conditions. Resting oxygen metabolism in species of fish generally increase by a factor of approximately 2.4 for every 10°C (Q_{10}) temperature interval between 0°C and 30°C [29]. The corresponding Q_{10} for hypoxia tolerance has been estimated to be 2.09 for marine fish [28], although comparable hypoxia Q_{10} estimates for freshwater fish assemblages do not exist. Temperature was held constant in the present study to control for its effect, although future research on hypoxia tolerances of MDB fishes conducted at a range of temperatures is needed.

All four species performed ASR which is an adaptation that many fishes exhibit when exposed to low dissolved oxygen [6,16]. It is notable that ASR has not previously been documented in the peer-reviewed literature for any of the species examined here. The $ASR_{50's}$ reported here were within the upper range of 0.48 to 1.98 mg l^{-1} [12] at 25°C for small-bodied native MDB fishes including Australian smelt *Retropinna semoni* (Weber), flat-headed galaxias *Galaxias rostratus* (Klunzinger), carp gudgeon compex *Hypseleotris spp.* and southern pygmy perch *Nannoperca australis* (Gunther). An unexpected observation from our experiment was that ASR was performed less frequently under simulated blackwater conditions than in hypoxic freshwater conditions. It is possible that a reduction in ASR behaviour within blackwater simulations may have resulted as a consequence of fish experiencing other physiological stressors, including toxic polyphenols [4] or reduced pH, beyond the direct effects of hypoxia. Nonetheless, field observations of *M. peelii* performing ASR (Figure 1) under more acidic pH's and higher DOC concentrations than were simulated here (Table 4) suggest that this behaviour is utilized by adults when exposed to hypoxic blackwater conditions in the wild.

Life-history and hypoxia tolerance

The species-specific differences in hypoxia tolerance observed here were broadly consistent with expectations based on known life-history strategies and evolutionary lineages of the species examined. For example, a species of catfish (*T. tandanus*) that typically occupies benthic slow flowing habitats, including wetlands and floodplains, and is relatively sedentary [30] compared to the other fishes studied was the most hypoxia and blackwater tolerant fish examined. The high tolerance of this species is typical of Siluriform fishes and, although not observed in *T. tandanus*, some species in this order are capable of aerial respiration [31]. Although the most hypoxia tolerant species (*T. tandanus*) also had the greatest mean body mass and length within our experiments, species-specific differences in hypoxia tolerance among the remaining species followed no discernable pattern associated with body mass. *Macquaria ambigua* (Percichthyidae) and *B. bidyanus* (Terapontidae) which are more mobile but spawn in potentially low oxygen floodplain environments [30] were the second most tolerant group to low dissolved oxygen and blackwater.

One outstanding question remaining from the present study is whether or not the hypoxia tolerances of the juveniles examined here are representative of adults that are more commonly reported in fish kills in the MDB [8,9]. A review on this subject across multiple species of marine fish concluded that body size itself plays little or no direct role in determining hypoxia tolerance and that the tolerance of a given species is usually size-independent after the larval stage [32]. The authors of this study purport that if there are size-specific differences in hypoxia tolerance, larger individuals of a given species would generally be expected to survive longer periods of time under hypoxic conditions compared to smaller individuals. However, there are examples of size, trophic status and developmental stage-dependent hypoxia tolerances within and among species of fish [23,33,34]. Therefore, further research focusing on hypoxia tolerances of different developmental stages, body sizes and life-histories of MDB fishes would be useful. Mortality of *M. ambigua* larvae was identified by [4] in response to dissolved oxygen concentrations as high as 4.3 mg l^{-1}at 24.5°C which suggests that larvae may be more sensitive than the juveniles examined here. Our results do not preclude the possibility that adults may be more, or less, sensitive to hypoxia than juveniles but emphasizes that *M. peelii* appears to be less hypoxia tolerant than other fishes in the MDB and elsewhere world-wide [24–26]. The relatively high hypoxia sensitivity of *M. peelii* may also explain, in part, why this species is disproportionately reported in blackwater fish kills in the MDB [7,8,9].

In contrast to popular anecdote, it appears that the hypoxia sensitivity of *M. peelii* may not be entirely related to its large maximum attainable body mass (>100,000 g). The *M. peelii* examined here were juveniles (<1.2 g) and not the largest species yet they were still the most sensitive examined. The comparatively low hypoxia tolerance of *M. peelii* appears to be species-specific or, at least, not entirely size-specific. We suggest that Australia's largest freshwater fish species is sensitive to hypoxia, not because it a large-bodied apex predator *per se*, but because it has evolved a life-history strategy that primarily utilizes lotic river channel environments and habitats [35,36], which are less prone to severe hypoxic blackwater conditions than floodplains, wetlands or backwater areas commonly occupied by other species [32]. It is notable, however, that juveniles of all four of the species examined here were historically stranded in large numbers within isolated floodplain pools following spring over bank floods [37]. Similar scale floodplain strandings of juveniles of large-bodied species have rarely, if ever, been reported in recent years within the MDB. Small-bodied and invasive species are now commonly found in isolated floodplain pools of the MDB and these fishes are seemingly unaffected by extremely high DOC (16–50 mg l^{-1}) and low dissolved oxygen (0.4–6.8 mg l^{-1}) at temperatures ranging from 15–18°C [10].

The moderate hypoxia survival thresholds reported for the species examined here are in stark contrast to the prevalence of unique adaptations, such as air-breathing and other physiological mechanisms, that facilitate long-term survival of fish in blackwater rivers [38,39]. Adaptations to survive prolonged and widespread hypoxia are obligatory in many blackwater rivers, while fishes of the MDB may have historically been able to withstand blackwater events that were more localized and may have occurred under cooler winter conditions [2], without anthropogenic barriers that could limit access to refuge habitat. For comparison, floodplain forests surrounding Amazonian blackwater rivers are inundated each year for three to eleven months, covering an area of approximately 70,000 km^2 [39], whereas blackwater events in the MDB are typically localized at the river or wetland scale and are characterised by extreme variability in area, timing and duration [1,3]. There is evidence, however, that the severity of hypoxic blackwater events may be increasing due to anthropogenic pressures [2] and results from our study suggest that fish which have life-histories adapted to in-channel environments [35], such as *M. peelii*, may be most vulnerable.

Acknowledgments

We thank the CSU Faculty of Science laboratory staff for assistance in maintaining fish and the Narrandera Fisheries Centre for providing fish free of charge. J. and A. Conallin from the Murray Catchment Management Authority and L. Pearce from NSW Department of Primary Industries provided photos, confirmed field mortalities and aquatic surface respiration responses following the Edward-Wakool fish kill. This work was carried out with permission from the CSU Animal Care and Ethics committee; protocol number 11/018.

Author Contributions

Conceived and designed the experiments: KS RK RW JH. Performed the experiments: KS. Analyzed the data: RK KS. Contributed reagents/materials/analysis tools: JH. Wrote the paper: RK KS.

References

1. Hladyz S, Watkins SC, Whitworth KL, Baldwin DS (2011) Flows and hypoxic blackwater events in managed ephemeral river channels. Journal of Hydrology 401: 117–125.
2. Whitworth KL, Baldwin DS, Kerr JL (2012) Droughts, floods and water quality: Drivers of a severe hypoxic blackwater event in a major river system (the southern Murray-Darling Basin, Australia). Journal of Hydrology: 190–198.
3. Howitt JA, Baldwin DS, Rees GN, Williams JL (2007) Modelling blackwater: predicting water quality during flooding of lowland river forests. Ecological Modelling 203: 229–242.
4. Gehrke PC, Revell MB, Philbey AW (1993) Effects of river red gum, *Eucalyptus camaldulensis*, litter on golden perch, *Macquaria ambigua*. Journal of Fish Biology 43: 265–279.
5. Meyer JL (1990) A blackwater perspective on riverine ecosystems. Bioscience 40: 643–651.
6. Chapman LJ, Mckenzie DJ (2009) Behavioral responses and ecological consequences. In: Richards, J. G., A. P. . Farrell, & C. J. . Brauner (editors). Hypoxia in Fishes. Elsevier, San Diego. Fish Physiology 27: 26–77.
7. La VT, Cooke SJ (2011) Advancing the science and practice of fish kill investigations. Reviews in Fisheries Science 19: 21–33.
8. King AJ, Tonkin Z, Lieshcke J (2012) Short-term effects of a prolonged blackwater event on aquatic fauna in the Murray River, Australia: considerations for future events. Marine and Freshwater Research 63: 576–586.
9. Koehn JD (2004) The loss of valuable Murray cod in fish kills: a science and management perspective. In: Lintermans, M., and B. . Phillips, Eds. Management of Murray Cod in the Murray–Darling Basin. Canberra: Murray–Darling Basin Commission.
10. McMaster D, Bond N (2008) A field and experimental study on the tolerances of fish to *Eucalyptus camaldulensis* leachate and low dissolved oxygen concentrations. Marine and Freshwater Research 59: 177–185.
11. Gee JH, Gee PA (1991) Reactions of Gobioid Fishes to Hypoxia: Buoyancy Control and Aquatic Surface Respiration. Copeia 1: 17–28.
12. McNeil DG, Closs GP (2007) Behavioural responses of a south east Australian floodplain fish community to gradual hypoxia. Freshwater Biology 52: 412–420.
13. Gehrke PC (1988) Response surface analysis of teleost cardio-respiratory responsesto temperature and dissolved oxygen. Comparitive Biochemistry Physiology. A 89: 587–592.
14. Gehrke PC (1991) Avoidance of inundated floodplain habitat by larvae of golden perch (*Macquaria ambigua* Richardson): Influence of water quality or food distribution? Marine and Freshwater Research 42(6): 707–719.
15. Anon (1992) *Guideline for Testing of Chemicals 203 Fish, Acute Toxicity Test.* Paris, France: Organisation for Economic Cooperation and Development. Available: http://www.oecd.org/chemicalsafety/assessmentofchemicals/1948241.pdf.
16. Kramer DL, McClure M (1982) Aquatic surface respiration, a widespread adaptation to hypoxia in tropical freshwater fishes. Environmental Biology of Fishes 7(1): 47–55.
17. Lewis WM (1970) Morphological adaptations of Cyprinodontoids for inhabiting oxygen deficient waters. Copeia 1970(2): 319–326.
18. James KR, Cant B, Ryan T (2003) Responses of freshwater biota to rising salinity levels and implications for saline water management: a review. Australian Journal of Botany 51: 703–713.
19. Anon (1998) *Standard Methods for the Examination of Water and Waste Water*, 20th Edition. American Public Health Association, Washington DC.
20. Zuur A, Leno EN, Walker N, Saveliev AA, Smith GM (2009) Mixed effects models and extensions in ecology with R. Springer, New York.
21. Anon (2006) *SigmaPlot for Windows Version 10.0.* Systat Software, Inc. Available: http://www.sigmaplot.com. Accessed 2014 Mar 20.
22. Nilsson GE, Renshaw GMC (2004) Hypoxic survival strategies in two fishes: extreme anoxia tolerance in the North European crucian carp and natural hypoxic preconditioning in a coral-reef shark. Journal of Experimental Biology 207: 3131–3139.
23. Jacobson PC, Jones TS, Rivers P, Pereira DL (2008) Field estimation of a lethal oxythermal niche boundary for adult Ciscoes in Minnesota lakes. Transactions of the American Fisheries Society 137: 1464–1474.
24. Smale MA, Rabeni CF (1995) Hypoxia and hyperthermia tolerances of headwater stream fishes. Transactions of the American Fisheries Society 124: 698–710.

25. Ostrand KG, Wilde GR (2004) Temperature, dissolved oxygen, and salinity tolerances of five prairie stream fishes and their role in explaining fish assemblage patterns. Transactions of the American Fisheries Society 130: 742–749.
26. Landman MJ, Van den Heuvel MR, Ling N (2005) Relative sensitivities of common freshwater fish and invertebrates to acute hypoxia. New Zealand Journal of Marine and Freshwater Research 39: 1061–1067.
27. Moss DD, Scott DC (1961) Dissolved-oxygen requirements of three species of fish. Transactions of the American Fisheries Society 90: 377–393.
28. Vaquer-Sunyer R, Duarte CM (2011) Temperature effects on oxygen thresholds for hypoxia in marine benthic organisms. Global Change Biology 17: 1788–1797.
29. Clarke A, Johnston NM (1999) Scaling of metabolic rate with body mass and temperature in teleosts fish. Journal of Animal Ecology 68: 893–905.
30. Lintermans M (2007) Fishes of the Murray-Darling Basin. An Introductory Guide. Canberra, Australia: Murray-Darling Basin Authority.
31. Cala P (1987) Aerial respiration in the catfish, *Eremophilus mutisii* (Trichomycteridae, Siluriformes), in the Rio Bogota Basin, Colombia. Journal of Fish Biology 31: 301–303.
32. Nilsson GE, Ostlund-Nilsson S (2008) Does size matter for hypoxia tolerance in fish? Biological Reviews 83: 173–189.
33. Kalinin AL, Rantin FT, Glass ML (1993) Dependence on body size of respiratory function in *Hoplias malabaricus* during graded hypoxia. Fish Physiology and Biochemistry 12: 47–51.
34. Anjos MB, De Oliveira RR, Zuanon J (2008) Hypoxic environments as refuge against predatory fish in the Amazonian floodplains. Brazilian Journal of Biology 68: 45–50.
35. Humphries P, King AJ, Koehn JD (1999) Fish, flows and flood plains: links between freshwater fishes and their environment in the Murray-Darling River system, Australia. Environmental Biology of Fishes 56: 129–151.
36. Jones MJ, Stuart IG (2007) Movements and habitat use of common carp (*Cyprinus carpio*) and Murray cod (*Maccullochella peelii*) juveniles in a large lowland Australian river. Ecology of Freshwater Fish 16: 210–220.
37. Anderson HK (1919) Rescue operation on the Murrumbidgee river. Australian Zoologist 1: 157–160.
38. Kramer DL, Lindsey CC, Moodie GEE, Stevens ED (1978) The fishes and the aquatic environment of the central Amazon basin, with particular reference to respiratory patterns. Canadian Journal of Zoology 56: 717–729.
39. Saint-Paul U, Zuanon J, Villacorta Correa MA, Garcia M, Noemi Fabre N, et al. (2000) Fish communities in central Amazonian white and blackwater floodplains. Environmental Biology of Fishes 57: 235–250.

20

Integrating Bacterial and Viral Water Quality Assessment to Predict Swimming-Associated Illness at a Freshwater Beach: A Cohort Study

Jason W. Marion[1¤a], Cheonghoon Lee[1¤b], Chang Soo Lee[1¤c], Qiuhong Wang[2], Stanley Lemeshow[3], Timothy J. Buckley[1¤d], Linda J. Saif[2], Jiyoung Lee[1,4]*

1 Division of Environmental Health Sciences, College of Public Health, The Ohio State University, Columbus, Ohio, United States of America, **2** Food Animal Health Research Program, Ohio Agricultural Research and Development Center, Department of Veterinary Preventive Medicine, The Ohio State University, Wooster, Ohio, United States of America, **3** Division of Biostatistics, College of Public Health, The Ohio State University, Columbus, Ohio, United States of America, **4** Department of Food Science and Technology, The Ohio State University, Columbus, Ohio, United States of America

Abstract

Background & Objective: Recreational waters impacted by fecal contamination have been linked to gastrointestinal illness in swimmer populations. To date, few epidemiologic studies examine the risk for swimming-related illnesses based upon simultaneous exposure to more than one microbial surrogate (e.g. culturable *E. coli* densities, genetic markers). We addressed this research gap by investigating the association between swimming-related illness frequency and water quality determined from multiple bacterial and viral genetic markers.

Methods: Viral and bacterial genetic marker densities were determined from beach water samples collected over 23 weekend days and were quantified using quantitative polymerase chain reaction (qPCR). These genetic marker data were paired with previously determined human exposure data gathered as part of a cohort study carried out among beach users at East Fork Lake in Ohio, USA in 2009. Using previously unavailable genetic marker data in logistic regression models, single- and multi-marker/multi-water quality indicator approaches for predicting swimming-related illness were evaluated for associations with swimming-associated gastrointestinal illness.

Results: Data pertaining to genetic marker exposure and 8- or 9-day health outcomes were available for a total of 600 healthy susceptible swimmers, and with this population we observed a significant positive association between human adenovirus (HAdV) exposure and diarrhea (odds ratio = 1.6; 95% confidence interval: 1.1–2.3) as well as gastrointestinal illness (OR = 1.5; 95% CI: 1.0–2.2) upon adjusting for culturable *E. coli* densities in multivariable models. No significant associations between bacterial genetic markers and swimming-associated illness were observed.

Conclusions: This study provides evidence that a combined measure of recreational water quality that simultaneously considers both bacterial and viral densities, particularly HAdV, may improve prediction of disease risk than a measure of a single agent in a beach environment likely influenced by nonpoint source human fecal contamination.

Editor: Andrew C. Singer, NERC Centre for Ecology & Hydrology, United Kingdom

Funding: This study was funded by the Public Health Preparedness for Infectious Diseases program (http://phpid.osu.edu) and The Ohio State University Targeted Investment in Excellence program sponsored by the Offices of Academic Affairs, the President, and Research. The human health and exposure data collection effort was funded by the Ohio Water Development Authority. The funders had no role in study design, data collection and analysis, decision to publish, or preparation of the manuscript.

Competing Interests: The authors have declared that no competing interests exist.

* Email: lee.3598@osu.edu

¤a Current address: Department of Environmental Health Science, Eastern Kentucky University, Richmond, Kentucky, United States of America
¤b Current address: Department of Environmental Health and Institute of Health and Environment, Graduate School of Public Health, Seoul National University, Seoul, Korea
¤c Current address: Environmental Biotechnology Research Center, Korea Research Institute of Bioscience and Biotechnology, Yuseong-gu, Daejeon, Korea,
¤d Current address: U.S. Environmental Protection Agency, National Exposure Research Laboratory, Research Triangle Park, North Carolina, United States of America

Introduction

Beach water advisories are issued to discourage human contact (e.g., swimming, wading, etc.) when water is potentially harmful to human health. In the United States, many of these advisories are issued at freshwater beaches when densities of fecal indicator bacteria (*E. coli* or enterococci) are observed or predicted to be in excess of single-day maximum criteria. Prior to November 26,

2012, the U.S. Environmental Protection Agency (USEPA) had established single-day maximum criteria, which were derived from several large epidemiological studies performed in the early 1980s [1]. In 2003, the 1986 *E. coli* criteria were supported by a comprehensive meta-analysis demonstrating *E. coli* was the most consistent indicator for predicting gastrointestinal illness in freshwater [2]. This same study indicated that methods for quantifying *E. coli* were problematic for same-day water quality advisories. The method required 18–24 h incubation and therefore was untimely for communication of risk to susceptible beachgoers [2].

Since the 2003 meta-analysis, several epidemiological studies evaluating associations between illness and more rapid measures of fecal indicators have been performed with an emphasis on genetic markers of bacterial or viral contamination, which have demonstrated the effectiveness of rapidly measured *Enterococcus* via quantitative polymerase chain reaction (qPCR) for predicting human illness at sewage-impacted Great Lakes and marine beaches [3,4]. However, in epidemiology studies on swimmers at non-point source-impacted beaches, associations between qPCR-based bacterial markers and gastrointestinal (GI) illness were not observed [5,6]. A potential reason for the lack of association in both studies could have related to the source of fecal contamination, which was likely dominated by avian sources [7], potentially presenting less risk for swimming-associated gastrointestinal (GI) illness than equivalent amounts of human-associated fecal contamination [8]. The diffuse nature of the source(s) of fecal contamination has been proposed as a possible explanation for the lack of association between bacterial genetic markers and human illness [9–11].

To date, few epidemiological studies on beachgoers have been performed at areas dominated by human-associated non-point source contamination, particularly in freshwater environments. Several peer-reviewed epidemiological studies have evaluated illness associations with human enteric viruses in European recreational waters [12–14] and U.S. waters [6,15,16]. Epidemiological studies pertaining to viruses are highly relevant as viruses are known to have a broad distribution in the aquatic environment [17], and are the most observed etiological agents in disease outbreaks associated with untreated U.S. recreational waters [18]. Recent epidemiological studies on swimmers [9–11] have suggested that viruses were the primary etiologic agents associated with recreational water illnesses based upon relatively short incubation periods following swimming exposure [10,19]. Similarly with regards to incubation periods, among limited-contact recreational users of freshwater in Chicago, Illinois (U.S.A.), most gastrointestinal illnesses generally developed within three days of water exposure [19].

To date, the majority of recreational water-related epidemiological studies have been performed to identify effective single indicators, often bacterial, for practical application among those making beach management decisions. Multiple fecal indicators and multiple genetic markers have been evaluated in recreational waters with respect to their association with each other as well as GI illness frequency among swimmers. There is some inconsistency in the findings of studies that have studied associations (or the lack thereof) between fecal indicator densities and measured pathogen and/or viral levels in freshwater [20] and coastal nvironments [21,22], and legitimate concerns remain regarding the use of single bacterial indicators for predicting and communicating swimming-associated illness risks to the general public [23–25].

To address these knowledge gaps, this study examined previously undescribed samples which were archived during our prospective cohort study [26] that gathered culture-based *E. coli* and human exposure data. In our 2010 study [26], new GI illnesses among swimmers were associated with increasing culture-based *E. coli* densities at East Fork Lake (Ohio, USA); however, no assessment of associations between genetic marker densities and human health occurred. In 2012, the U.S. Environmental Protection Agency (EPA) emphasized a need for additional evaluation of rapid molecular methods for the timely determination of water quality to protect recreational water users [27]. Building upon this need, we examined associations between GI illness incidence among swimmers exposed to multiple genetic markers targeting both viruses and bacteria in this non-point source human-impacted beach environment. This study presented a unique opportunity to successfully use our historical exposure and health data coupled with our previously unexplored genetic material to do a timely assessment of the effectiveness of viral and bacterial molecular markers for predicting GI illness risk among swimmers who used the East Fork beach in 2009.

Materials and Methods

Overall Approach

The exposure and health data were collected using the prospective cohort study approach adopted by Wade et al. in their epidemiological studies related to recreational water [3,4]. This study builds upon our epidemiological investigations of recreational water-related illnesses by evaluating human illness associations with our recently qPCR-determined densities of bacteria and viruses. In brief, beach water samples, beach water quality data, and beach user exposure/behavior data were collected at East Fork State Park (39.0198° N; 84.1432° W) near Cincinnati, Ohio, U.S.A. Approvals to use this location for water sample collection and beachgoer interviews were kindly provided orally and in writing by Chris Dauner (Park Manager, East Fork State Park), and Dan West (Chief, Ohio Department of Natural Resources, Division of Parks and Recreation). Complete health and exposure data were collected from 891 beachgoers from 278 households. As described in our 2010 study [26], participants reported their health status via a telephone questionnaire after an eight- to nine-day follow-up period. More recently we quantified densities of four viral (human adenovirus (HAdV), human enterovirus (HEntV), and human norovirus genogroup I (HNoV GI) and genogroup II (HNoV GII)) and four bacterial (*E. coli* by *uidA* and 23S, Enterococci by 23S, *Bacteroides-Prevotella* by HuBac) markers from 23 samples collected over 23 weekend days from 2009 using qPCR and paired them with human exposure data.

Ethics Statement

Approvals for the study design, questionnaires, verbal consent procedures, and related materials were obtained from the Institutional Review Board (IRB) at The Ohio State University (IRB Protocol #2009H0107). For this study, adult participants (18 years of age or older) capable of providing verbal consent were welcomed into the study by investigators or key personnel trained and authorized by the university to obtain consent. The written consent waiver permitting verbal consent was granted given the low study risk and because written consent would have required surname information and the gathering of consent documentation in a periodically windy and wet beach environment. Thus, written consent procedures would have increased the likelihood of losing or damaging documentation which would have then included participant surnames that could potentially be matched with telephone numbers and health data. Verbal consent was recorded on the datasheet for each household interviewed. In addition, participants were provided with a card sharing contact information for the principal investigator and the ethics coordinator in the Office of Responsible Research Practices at the university.

Enrollment and Exposure Survey

The East Fork Beach Study [26] recruited beach user households into the study at the beach and interviewed them upon their departure to ascertain information regarding their activities at the beach on that same day. For each household in the study, an adult spokesperson provided investigators with information requested from a standard script. The obtained information was used for classifying exposure status and obtaining information related to potential confounding variables (age, gender, ethnicity, food consumption, source of food/drink, number of beach visits per year, duration of time at the beach, hand hygiene, extent of sand exposure, distance traveled, and number of other beach users). More specifically, individuals were asked whether any household members had: (1) no water contact; (2) waded, played or swam in the water; and/or (3) immersed their head in the water. Questions pertaining to any recent or on-going illnesses were also asked. Individual beach users were dichotomously classified as having or not having each of the above-mentioned exposures as well as any prevalent symptoms of gastrointestinal illness. These exposure data were recently paired with date-specific *E. coli* genetic marker density data to further establish our exposure classification of each beach user.

Health Outcomes Survey

Health outcome data were obtained by Marion et al. [26] in which enrolled households were contacted eight to nine days after their beach visit via telephone. Only adult household members who were at the beach at the time of enrollment were eligible for interview participation. Interviews were conducted using a standard questionnaire inquiring about illness among household members at the beach at the time of enrollment. Reported illnesses were recorded as a yes/no response for a variety of symptoms (e.g., stomach cramps, nausea, diarrhea, vomiting, headache, fever, etc.) for each person who visited the beach. For comparing with other studies, persons were further classified by gastrointestinal illness. Similar to the Wade et al. studies [3,4], persons were defined to have experienced "GI illness" if they were reported to have had any one of the following: nausea, stomach ache or stomach cramps, diarrhea (three or more loose or watery stools in a 24-hour period), or vomiting. With respect to nausea, all persons in this study who reported nausea were considered positive for GI illness. This GI illness definition is slightly different than the Wade et al. studies [3,4], which only included nausea cases into their GI illness definition when the condition interfered with daily activities. In our study, we were unable to make this distinction as we did not ascertain if the nausea interfered with daily activities. Beyond the "GI Illness" definition, using the highly credible gastrointestinal illness (HCGI) definition employed by Colford [5] as "HCGI-1", we likewise coded individuals as positive for HCGI if they were reported to have experienced any of the following: (1) vomiting; (2) diarrhea and fever, (3) stomach-ache and fever, or (4) nausea and fever.

Sample Collection and Water Analysis

Beach water samples used for obtaining genetic marker data were collected on the same day as enrollment and administration of the exposure questionnaire on summer weekends during the time of day generally used for swimming. In total, 91% of samples were collected at the median time of 13:45±3 h. Two samples (9%) were collected after 16:00 on days when daily swimming attendance was low (n = 4, n = 28) in anticipation of more swimmers late in the day. For each day, 3 L was collected into multiple autoclaved 500 mL bottles (Nalgene, Rochester, NY, USA) and sterile Whirl-Pak bags (Nasco, Fort Atkinson, WI, USA) over 23 weekend days. Samples were collected near the beach center in water with an approximate depth of one meter by sweeping containers 30 cm below the water surface. Within 30 minutes after collection, a 1 L sample aliquot was used for membrane filtration for culturable *E. coli*, a 1 L sample aliquot was temporarily stored (24–48 h) at 4°C and then immediately filtered for bacterial marker analysis, and a 1 L aliquot was stored at −20°C for virus analysis.

E. coli densities were quantified for three separate filtration volumes (20, 50, and 100 mL) for each sample using the culture-based method described in EPA Method 1603 [27] and Marion et al. [26]. Other water quality parameters including temperature, pH, dissolved oxygen, and turbidity were also measured using a YSI 600XL data sonde (Yellow Springs Instruments, Yellow Springs, Ohio, U.S.A.) and a Hach 2100P Turbidimeter (Loveland, Colorado, U.S.A.) as described in Marion et al. [26]. UV index data were obtained from the National Oceanic and Atmospheric Administration [28]. Rainfall, lake stage (surface elevation), and lake inflow data were obtained from the U.S. Army Corps of Engineers [29].

A detailed description of sample preparation, water concentration, and quantification of the bacterial and viral marker densities by qPCR is provided as (**Text S1**), including a description of the primers and probes used in this study (**Table S1**). In brief, samples were filtered to capture bacteria for quantifying the following bacterial markers: *uid*A genes (uidA) and 23S rRNA genes (23S *E. coli*) of *E. coli*, 16S rRNA genes of *Bacteroides-Prevotella* (HuBac), and 23S rRNA genes of *Enterococcus* spp. (23S *Enterococcus*). For concentrating viruses (HEntV, HAdV, HNoV GI, and HNoV GII), membrane filtration was performed using cation (Al^{3+})-coated membranes. DNA or RNA was then extracted from the microbes captured on the membranes using a QIAmp DNA Stool Kit or the RNeasy Mini Kit (Qiagen, Valencia, CA, USA), respectively. RNA extracts for HEntV were amplified by reverse transcription (RT)-PCR and quantified using TaqMan real-time RT-PCR (RT-qPCR) with the QIAGEN OneStep RT-PCR Kit (Qiagen, Valencia, CA, USA). DNA extracts for HAdV and all bacterial markers were amplified by PCR and quantified using a TaqMan-based real-time qPCR system. All RT-qPCR and qPCR reactions except HNoV GI and HNoV GII were carried out in an ABI 48-well StepOne Real Time System (Applied Biosystems, Foster City, CA). RT-qPCR assays for HNoV GI and HNoV GII were carried out with an Eppendorf Mastercycler RealPlex^2 (Eppendorf, Germany). Quantification of the markers was then determined using standard curves generated by plotting Ct values (y) versus the log gene copy numbers of the target microbes (x) and each limit of detection (LOD) was determined based on the lowest gene copy number from which above 90% of the replicates could be amplified in each qPCR assay. Additionally, with regards to these samples, the presence/absence of qPCR inhibition was previously evaluated for East Fork samples [20] using the Sketa22 qPCR assay as described in Haugland et al. [30].

Data Analysis

The analysis focused on healthy susceptible individuals, and persons who were classified as prevalent or existing cases based upon responses to the enrollment questionnaire were excluded in analyses using incident cases. Logistic regression was employed to evaluate crude associations between genetic marker exposure and new illness among swimmers reporting head immersion. The data for the microbiological terms, including *E. coli* density data, were skewed, accordingly they were log (base 10)-transformed. Prior to log-transforming *E. coli* results, samples with 0 CFU/100 mL were assigned a value of 1 CFU/100 mL. In scatterplots and models containing genetic marker densities, for markers not detected after 45 cycles, densities were assigned a value of half the detection limit in log CFUs per 100 mL or log gene equivalents (GE) per 100 mL for bacterial and viral markers, respectively.

Multivariable logistic regression was used for estimating odds ratios for any GI illness, diarrhea, and HCGI among swimmers and non-swimmers (negative controls) as recommended in observational studies [31] and done by Colford et al. [9] in their beach user health study. The swimmer population was defined as beach users who reported immersing their head in the beach water. Non-swimmers were defined as beach users reporting no water contact or limited water contact that did not include head immersion in the beach water. All swimmers were assumed to have been exposed to the same density of *E. coli* and the various genetic markers observed from the single sample collection for that day. Densities of *E. coli* and the various genetic markers were determined from the single mid-day water sample collection.

The multivariable logistic regression models considered potential confounders and/or modifying influences, including covariates related to demographic, exposure, meteorological, and water quality factors. Since data were obtained from households, for the purpose of statistical analysis, the data were treated as clustered by household [32]. For constructing multivariable models, a backward selection approach was employed whereby covariates were dropped stepwise based on the highest p-value until arriving at the most parsimonious model [32]. The rule of ten events (cases) per covariate was used in logistic regression modeling [33]. Due to the low frequency of HCGI cases, HCGI model construction was limited to only two microbiological terms by relaxing the rule of ten events per covariate [34].

Multivariable logistic regression models were evaluated for fit via the Hosmer-Lemeshow Goodness-of-Fit Test [32] and for model discrimination via the area under the receiver-operating-characteristic curve (AUC) [32]. The modeling efforts and assessments of logistic model fit and discrimination were performed using Stata 11 (Stata Corporation, College Station, TX, USA). A time-series plot was generated with Minitab 16 (Minitab Inc., State College, PA, USA).

Results

Water Quality

Among the 23 weekend water samples collected for qPCR analysis, bacterial markers were more readily detected than viruses (Table 1). HEntV and HAdV were detected in 22% (5/23) and 35% (8/23) of all samples, respectively, whereas, HNoV GI and GII were not detected (0/23). As previously reported [20], qPCR inhibition was not observed in any of the 23 samples. Culturable *E. coli* densities greater than 0 CFU/100 mL were observed in 91% (21/23) of all samples. The time-series plot of culturable *E. coli* and HAdV densities by day (Figure 1) demonstrate the lack of association between the two water quality parameters, which was confirmed by correlation analysis (Spearman's $\rho = 0.147$; $p = 0.503$). Overall, culturable *E. coli* densities exceeded Ohio single-day maximum criteria (>235 CFU/100 mL) in 8.6% (2/23) of our samples; however, no beach advisories were issued by local or state authorities due to their adherence to the one sample per week guideline used by the Ohio Department of Health for inland waters. Other water quality parameters are summarized in Marion et al. [26].

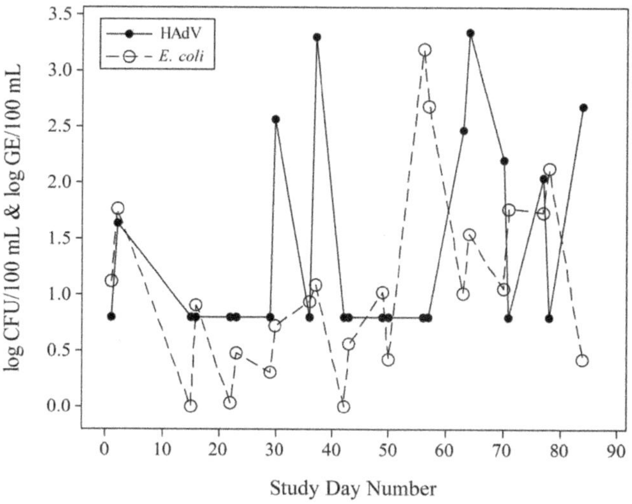

Figure 1. Time series plot for HAdV and *E. coli* densities measured during the 2009 swimming season at the study beach (East Fork Lake, Ohio).

Study Population

A total of 891 individuals were included in this study (from 278 households). The study population was 93% white, 2.5% black, 1% Hispanic and 3.5% other. The average study participant age was 24 years (median =21 years). Young children (≤5 years) represented 13% of the participants on whom data were available. Females were more represented (56%) than males (44%). The population described in this report is a subset of the 965 individuals from 300 households in our previous study [26], which was conducted over 26 weekend days. Genetic marker data were only obtained for 23 of 26 days of the 2009 East Fork Beach Study [26], and accordingly, this study reports the results of the investigation using genetic marker data. The collection of sufficient water for obtaining qPCR data was an additional task added to the previous study on the third sample day.

Among the participants in this study, 618 (69%) individuals reported head immersion in East Fork beach waters. Prevalent GI illness was reported for 18 (2.9%) of these swimmers at enrollment (Table 2). Chronic GI problems were self-reported at enrollment in 9 (50%) of these 18 individuals. All 18 individuals with prevalent GI illness (including chronic GI problems) were excluded from the exposure-related illness models. The prevalence of fever, nausea, and stomach cramping among participants was not ascertained at enrollment.

Water Quality and Human Illness

The qPCR-based densities for bacteria and viruses do not show any significant associations ($p<0.05$) with HCGI, GI, and diarrhea in single marker models using univariable and multivariable logistic regression when treating marker densities as continuous terms or binary terms in models (Table 3). Among all the continuous terms, HAdV had the strongest association with HCGI (crude OR (cOR) =1.6; 95% CI 0.90–2.9) and diarrhea (adjusted OR (aOR) =1.5; 95% CI 0.98–2.4); however, these associations were not significant but were noteworthy ($p = 0.105$; $p = 0.066$, respectively). When evaluating the association of either HEntV or HAdV detection with HCGI, GI, and diarrhea

Table 1. Microbial water quality measured by various genetic markers and *E. coli* at East Fork Lake, Ohio (N = 23).

Genetic Marker or Fecal Indicator	No. Samples Above LOD[a]	Detection Limit	Median	Range
HAdV (log gene equivalents/100 mL)	8	1.6	BD[b]	BD-3.34
HEntV (log gene equivalents/100 mL)	5	1.9	BD	BD-2.13
uidA (log copies/100 mL)	19	1.5	2.08	1.50–2.49
23S *E. coli* (log copies/100 mL)	19	2.46	2.94	2.46–3.19
HuBac (log copies/100 mL)	18	2.05	3.04	2.05–3.39
23S *Enterococcus* (log copies/100 mL)	17	1.93	2.19	1.93–2.88
E. coli (log CFU/100 mL)	21	ND[c]	1.01	LOD-3.19

[a]Limit of Detection or 0 CFU/100 mL for *E. coli*.
[b]Below Detection.
[c]Not Determined.

incidence, no association was observed ($p>0.05$). The covariates used in the illness models are summarized in **Table S2** and **Table S3**, respectively. The most significant covariate, food consumption at the beach (see **Table S2** and **Table S3**), was used in all genetic marker and *E. coli* models except the HCGI model, as food consumption at the beach was common to all HCGI cases. Beyond food consumption at the beach, specific conductivity and previous 72-hour UV results were the most significant covariates in all of the GI illness models. With respect to the diarrhea illness model, food consumption and previous 72-hour UV data were used as covariates, as they were identified as most significant through the backward elimination approach. In evaluating HCGI models relying on two microbiological terms, the only two-term model achieving significance was a combined model using HAdV and culturable *E. coli* ($p = 0.003$; likelihood ratio test). None of the bacterial markers were significant predictors of HCGI in models using an additional marker.

Using this same bacteria and virus approach for GI illness, we observed a similar association. Since more GI cases were detected than HCGI, the model was able to contain an additional two covariates. In this model, significant ($p<0.05$) or near significant odds ratios ($p = 0.056$) were observed for all covariates in this model (Table 4), including food consumption, specific conductivity, *E. coli*, and HAdV. Increases in water conductivity were associated with protective effects (aOR<1); whereas, food consumption, increasing HAdV, and increasing *E. coli* were identified as risk factors for GI (aOR>1). The use of two microbial terms was supported in model building as the removal of the HAdV term resulted in significantly lower log-likelihood values for HCGI, GI, and diarrheal models ($p<0.05$), and the two microbial terms had the low p-values throughout the entire backward selection procedure allowing their retention in the model. The final four-term model was evaluated using a negative control group (the non-swimmer group), and no significant associations were observed for any model terms, suggesting no day-effects, and implying a water quality-related association in our exposed group. Lastly, all possible models for predicting GI illness using a single bacterial marker in tandem with either culturable *E. coli* or viral marker levels were developed, and no bacterial markers were associated with illness.

Models constructed for predicting diarrhea using *E. coli* and HAdV preliminarily indicate that HAdV is associated with diarrheal disease incidence as the aOR for HAdV was significant ($p = 0.023$) when adjusting *E. coli* density by including *E. coli* density in the model. Compared to the HCGI and GI models, the diarrheal model provided poor discrimination (AUC <0.70) with

Table 2. Health status among study participants at enrollment and follow-up.

Reported Health Outcomes	Swimmers (n = 618)		Non-Swimmers (n = 274)	
	Health Status at Beach Enrollment (No. (%))	Health Status at Follow-up[a] (No. (%))	Health Status at Beach Enrollment (No. (%)	Health Status at Follow-up[a] (No. (%))
No Reported GI Problems	600 (97)	562 (94)	259 (95)	251 (97)
Any GI Illness	18 (2.9)	38 (6.3)	15 (5.5)	8 (3.1)
Chronic GI Problems	9 (1.5)	ND[b]	8 (2.9)	ND[b]
Diarrhea	6 (0.97)	28 (4.6)	5 (1.8)	8 (3.1)
Fever	ND	10 (1.6)	ND	2 (0.73)
Stomach Cramps	ND	8 (1.2)	ND	8 (1.2)
Nausea + Other GI Illness	ND	7 (1.2)	ND	1 (0.36)
Nausea Only	ND	3 (0.50)	ND	0 (0.00)
Vomiting	6 (0.97)	10 (1.6)	3 (1.1)	1 (0.36)
HCGI	ND	13 (2.2)	267 (97)	1 (0.37)

[a]Health status at telephone follow-up excluding positive cases at enrollment.
Not determined as information was not collected by the questionnaire.

Table 3. Odds ratios for HCGI, GI, and diarrhea associated with exposure to varying levels of various molecular genetic markers and/or fecal indicators at East Fork Lake, Ohio.

Molecular Marker	HCGI		Any GI Illness		Diarrhea	
	cOR[a] (95% CI)	Wald (*p*)	aOR[b] (95% CI)	Wald (*p*)	aOR (95% CI)	Wald (*p*)
HEntV (+)[c]	1.6 (0.34–7.6)	0.552	0.76 (0.21–2.7)	0.674	0.17 (0.02–1.3)	0.081
HAdV (+)	2.1 (0.65–7.1)	0.212	1.3 (0.45–3.5)	0.654	2.2 (0.65–7.6)	0.202
uidA *E. coli* (+)	1.1 (0.22–5.4)	0.892	0.78 (0.27–2.2)	0.637	0.88 (0.24–3.2)	0.843
23S *E. coli* (+)	0.69 (0.18–2.7)	0.595	0.78 (0.30–2.0)	0.599	1.1 (0.31–3.8)	0.905
HuBac (+)	1.5 (0.32–7.1)	0.604	1.1 (0.38–3.2)	0.846	0.86 (0.25–2.9)	0.815
23S *Enterococcus* (+)	0.70 (0.21–2.4)	0.586	0.77 (0.27–2.2)	0.613	0.58 (0.18–1.9)	0.363
Log HAdV	1.6 (0.90–2.9)	0.105	1.2 (0.77–1.9)	0.399	1.5 (0.97–2.4)	0.066
Log uidA *E. coli*	1.0 (0.38–2.9)	0.927	0.76 (0.38–1.5)	0.442	0.98 (0.34–2.9)	0.974
Log 23S *E. coli*	0.78 (0.38–1.6)	0.489	0.85 (0.50–1.4)	0.526	1.2 (0.48–3.2)	0.648
Log HuBac	1.5 (0.63–3.7)	0.355	1.1 (0.67–1.8)	0.7	0.78 (0.49–1.2)	0.289
Log 23S *Enterococcus*	0.96 (0.38–2.4)	0.921	0.89 (0.49–1.6)	0.714	0.67 (0.37–1.2)	0.207

[a]Crude odds ratio.
[b]Adjusted odds ratios, see Table S2 and S3, to see the covariates used for adjustment.
[c](+), Positive detection by qPCR, binary term.

respect to the ability of the model to properly classify diarrhea cases (Table 5). The Goodness-of-Fit test results (Table 5) also demonstrate the GI and HCGI models have better fits, as the p-values for those models are higher, indicating that the modeled results are not significantly different than the observed GI and HCGI results from the health survey.

For considering possible interaction effects, a water quality index was constructed based upon the detection limit of HAdV and the median *E. coli* density. Table 6 presents the four water quality index values, and the respective *E. coli* and HAdV densities represented by each water quality index level. When evaluating associations for this model, the reference was set as the

Table 4. Multivariable logistic regression models for predicting new HCGI[a], GI, and diarrhea among swimmers immersing their head and beachgoers not immersing their head in beach water (East Fork Lake, Ohio).

Model & Exposure	Covariate	β	SE_{β}	Wald (*p*)	AOR (95% CI)
HCGI Swimmers	Log HAdV (gene equivalents/100 mL)	0.8191	0.2864	0.005	2.3 (1.3–3.9)
	Log *E. coli* (CFU/100 mL)	1.101	0.3571	0.002	3.0 (1.5–6.1)
	Constant Term	−7.003			
GI Swimmers	Log HAdV (gene equivalents/100 mL)	0.421	0.1978	0.034	1.5 (1.0–2.2)
	Log *E. coli* (CFU/100 mL)	0.6515	0.2473	0.009	1.9 (1.2–3.1)
	Consumed Food at the Beach	1.543	0.7019	0.029	4.7 (1.2–19)
	Specific Conductivity (μS)	−0.092	0.0477	0.056	0.91 (0.83–1.0)
	Constant Term	19.37			
GI Non-Swimmers	Log HAdV (gene equivalents/100 mL)	0.0788	0.5481	0.886	1.1 (0.37–3.2)
	Log *E. coli* (CFU/100 mL)	0.2675	0.3346	0.425	1.3 (0.67–2.5)
	Consumed Food at the Beach	0.4254	0.8889	0.633	1.5 (0.26–8.9)
	Specific Conductivity (μS)	0.04	0.0279	0.154	1.0 (0.98–1.1)
	Constant Term	−15.33			
Diarrhea Swimmers	Log HAdV (gene equivalents/100 mL)	0.453	0.1977	0.023	1.6 (1.1–2.3)
	Log *E. coli* (CFU/100 mL)	0.3275	0.2466	0.186	1.4 (0.85–2.3)
	Constant Term	−4.226			
Diarrhea Non-Swimmers	Log HAdV (gene equivalents/100 mL)	0.0034	0.4905	0.994	1.0 (0.38–2.6)
	Log *E. coli* (CFU/100 mL)	0.2309	0.336	0.493	1.3 (0.65–2.4)
	Constant Term	−3.735			

[a]A model could not be constructed for HCGI among non-swimmers due to the low sample size resulting in an insufficient number of HCGI cases for model development.

Table 5. Model diagnostics for models from Table 4, pertaining to discrimination (AUC) and model calibration (goodness-of-fit).

Model Name for Swimmers	AUC[a]	Goodness-of-Fit[b] (*p*)
Multivariable HCGI Model	0.75	0.895
Multivariable GI Model	0.753	0.958
Multivariable Diarrhea Model	0.64	0.206

[a]Area under the receiver-operator-characteristic curve (AUC).
[b]Hosmer-Lemeshow Goodness-of-Fit Test [32].

group of days in which *E. coli* densities were lowest and HAdV was not detected. Here we observe multiple associations with increasing viral and/or bacteria densities and both GI and diarrheal illness incidence. Swimmers with the presumed greatest exposure to *E. coli* and HAdV had the greatest odds of reporting new GI illness (OR = 6.1; 95% CI 1.5–25) compared to swimmers with the least exposure to *E. coli* and HAdV. Exposure to the highest levels of *E. coli* when HAdV levels were low also presented increased GI illness risk (OR = 5.4; 95% CI 1.5–19) compared to those swimmers with low *E. coli* and low HAdV exposure. In the diarrheal illness model, significantly increased odds of diarrhea were only observed in the group who swam on days with elevated *E. coli* and HAdV compared to the reference group (OR = 5.2; 1.3–22). Model discrimination for each model (GI and diarrhea) was evaluated and the AUC values were 0.63 and 0.65, respectively.

Discussion

This prospective cohort study from an inland U.S. beach demonstrates the predictive potential of an integrative, multi-microbial approach for estimating recreational waterborne disease risk from viral and bacterial indicators. The term 'indicator' used here does not imply fecal indicator, but instead refers to 'health-relevant' indicators. Our results showed a positive association between increasing densities of individual health-relevant indicators (HAdV and *E. coli*) and increased odds of GI and HCGI among swimmers at the studied beach. More importantly, the two-indicator model (HAdV and *E. coli*) represented a significant improvement over single health indicator approaches with this data. Although speculative, particularly due to the small sample size of the study, the findings of this two-indicator approach are etiologically plausible as a combined viral + bacterial (ViBac) approach may account for GI and HCGI illness associations with viral and bacterial pathogens. Based upon our cohort study results, we have some reason to speculate that a multiple health-indicator approach may be beneficial for natural recreational waters such as the East Fork beach, which is part of a watershed comprised of mixed land uses and a high density of septic systems, all of which may promote a more diffuse type of fecal contamination than observed in point-source impacted waters. Within the beach watershed of interest, two municipal wastewater treatment plants (Williamsburg and Bethel WWTPs) and several smaller WWTPs operate, while malfunctioning septic systems are presumed to be a very significant source of contamination [35]. In the East Fork of the Little Miami River watershed, which includes the study beach, the septic system density in sub-catchments has been directly linked to human-associated fecal contamination through the use of human-associated genetic markers [35]. The only Harsha Lake (East Fork Lake) tributary investigated by Peed et al. [35] had the highest mean human-specific marker (HF183) density and the second highest density of septic systems ($41/km^2$) among tributaries investigated suggesting diffuse septic system-associated fecal contamination impacts this beach.

In health studies of swimmers in waters impacted by nonpoint source and/or diffuse fecal contamination, the lack of association between bacteria-based genetic markers and human illness has been observed [5,6,10,16]. Similarly, we did not observe associations between qPCR-based bacterial densities and human illness at our inland freshwater beach. The association of culturable *E. coli* with HCGI and GI was already observed [26], but the lack of association of GI and HCGI with human-specific genetic markers (e.g., HuBac) was unknown in this study population. Two possible reasons for our findings are as follows: (1) transport and fate varies considerably between bacterial genetic markers and viable waterborne pathogens, including viruses; and/or (2) our sample size was not big enough to detect significant associations.

With respect to viruses at the beaches with diffuse contamination, male-specific coliphage density has been associated with HCGI and GI illness in marine studies [5,16]. Similarly, our study observed an association between a virus (HAdV) and GI illness; however, unlike the marine studies, our study did not evaluate

Table 6. Odds ratios for GI and diarrhea associated with exposure to varying levels of HAdV and *E. coli* at East Fork Lake, Ohio.

Water Quality Index	Any GI Illness			Diarrhea		
	Cases/*n* (%)	OR (95% CI)	Wald (*p*)	Cases/*n* (%)	OR (95% CI)	Wald (*p*)
Group 1: Low HAdV & Low *E. coli* (HAdV <DL[a], log *E. coli* <1.05)	3/187 (1.6)	Referent		3/189 (1.6)	Referent	
Group 2: Low HAdV & High *E. coli* (HAdV <DL, log *E. coli*>1.05)	16/199 (8.0)	**5.4 (1.5–19)**	**0.011**	9/199 (4.5)	2.9 (0.74–12)	0.126
Group 3: High HAdV & Low *E. coli* (HAdV> DL, log *E. coli* <1.05)	11/126 (8.7)	5.9 (0.88–39)	0.067	9/127 (7.1)	4.7 (0.56–40)	0.153
Group 4: High HAdV & High *E. coli* (HAdV> DL, log *E. coli*>1.05)	8/88 (9.0)	**6.1 (1.5–25)**	**0.011**	7/90 (7.8)	**5.2 (1.3–22)**	**0.023**

[a]Detection Limit.

coliphages, focusing solely on viral pathogens. Like the associations between coliphages and illness [5,16], our finding of HAdV associations with HCGI and GI incidence in the adjusted models [Table 4] preliminarily suggests a possible benefit of using a viral indicator for evaluating recreational water illness risk in certain environments.

Similar to the Colford et al. study [5], human noroviruses (HNoV GI and HNoV GII) were not detected; however, human adenoviruses (HAdV) were detected from eight (35%) samples at East Fork versus one detection in the Colford et al. study [5]. The infrequent detection of HEntV (22%) in our study is not entirely unexpected as enteroviruses have been observed at densities below 1 PFU/100 mL in European waters [36] and have been infrequently detected (9% of samples) elsewhere [37]. These findings are similar to earlier findings from Fleisher et al. [14] that determined positive enterovirus detection (by cell culture) was less likely than negative detection in United Kingdom waters, which was also described as a limitation of the utility of the enterovirus assay for swimming advisory determination. The low detection frequency of viruses coupled with practicality of measurement presents challenges for establishing their use unless monitored in tandem with other indicators of water quality. Their use in a monitoring scheme may complement fecal indicator bacteria which have not effectively predicted viral densities in freshwater and marine environments [17,21,38,39]. Among viruses, adenovirus has shown promise as a potential human-associated marker of fecal contamination, whereby adenovirus presence has been linked to human-associated fecal problems as reflected by human-associated bacterial markers [40,41].

Despite a lack of association between viruses and bacteria, swimming advisories are primarily bacteria-based for practical reasons. Furthermore, there appears to be growing consensus among experts suggesting viruses are responsible for many water-related illnesses [18,19,42]. It is becoming more widely understood that enteric viruses (e.g., noroviruses, rotaviruses) are responsible for a substantial number of GI illnesses among U.S. beach users [18]. With viruses gaining attention and being more frequently measured, approaches for monitoring viruses to support integrative health risk models for water-associated illnesses are meaningful. Viral indicators of GI illness, particularly enteroviruses and bacteriophages, were described as promising predictors in 2003 by Wade et al. [2], but were also described as being limited by the difficulty and time for cultivation and enumeration. Now, with increasing use of molecular approaches such as qPCR, viral detection and enumeration is becoming less difficult and more rapid. In less contaminated or nonpoint source impacted waters, the use of a more commonly detected virus, like HAdV, may serve as an effective viral marker over noroviruses or enteroviruses for beach monitoring, since the detection frequency of adenovirus (36%) has been observed to be much greater than norovirus detection (9%) in European water samples [41,43].

The low detection frequency of norovirus in our samples and elsewhere may involve RNA virus instability, unstable DNA amplification potential from single-stranded RNA, and virus seasonality since the most prevalent human norovirus (GII.4) has clear winter peak seasonality [44] and therefore is less likely to be observed in summer surface waters. Unlike HNoV and HEntV, HAdV is a double-stranded DNA virus presumed to have better stability in environments and greater opportunities for successful detects. Furthermore, adenoviruses are believed to be up to 60 times more resistant to damage from ultraviolet irradiation than RNA viruses [45] likely affording HAdV particles greater integrity and viability leading to more undamaged infectious particles [46]. Given detection frequencies, Wyn-Jones et al. [41] encourages the consideration of adenovirus over norovirus as a recreational water quality indicator

Accordingly, water quality advisories should be protective for preventing human illness from all potential infectious agents as deemed practical with the available methods. The need for bacteria-based indicators of health risk are warranted, as the historical paradigm supports the notion that individual fecal indicators like *E. coli* and enterococci as well as human-specific markers explain the potential pathogenic bacterial genera (*Shigella, Camplyobacter, Salmonella*, etc.) as well as gastrointestinal illness among swimmers. However, the need for data attempting to address and/or quantify viral-associated human health risks is significant, and some attempt has recently been made using a site-specific quantitative microbial risk assessment (QMRA) for adenovirus illness [47]. Although speculative, virus monitoring may be beneficial in recreational waters where the contamination sources of the aquatic system are complex and impacted by a combination of non-point source human-associated contamination as well as agricultural runoff, as in Kundu et al. [47]. The concept of 'know your beach' appears to be particularly important as it relates to water quality when fecal indicator bacteria are low, and some consideration to broadening the paradigm to include multiple microbiological terms may be warranted with further research, with a particular emphasis on viruses capable of predicting recreational water illnesses more efficiently. Similar approaches demonstrating the value of a holistic design, like our combined virus-bacteria model, for enhancing risk assessments pertaining to water are already gaining attention in microbial source tracking studies using toolbox approaches [22]. Beyond water quality analysis, additional assessments of other known exposures presenting health risks, such as sand quality [48] and hydrology [49] are also recommended to be considered for more holistic beach condition modeling [25]. In this study, our integrative risk model using two microbiological terms, a fecal indicator bacteria and a viral genetic marker, provided a better assessment of human health risk than any single indicator/marker approach with limited data. Additional epidemiological studies using HAdV or other indicators of viral contamination that control for illnesses explained by bacteria are warranted for improving our predictive capability of swimming-related illness as well as our characterization of water quality at inland beaches.

Study Limitations

Future studies with larger populations are needed for drawing less speculative conclusions. The total swimmer population after exclusion is relatively small (n = 600) for rigorous analysis and may be leading us to spurious conclusions. Given the low number of cases presented, model construction was limited to two to four covariates, which therefore limited our ability to adequately assess multiple confounders, interaction effects, and linearity. The cohort study approach used here and in other U.S. studies [1,3,4] also presents several methodological limitations [50] related to potential misclassification of exposure and disease among study participants. We acknowledge a variety of confounders (assessed and not assessed) may be responsible for GI illness cases beyond adverse water quality conditions, such as food consumption at the beach. For example, personal risk perceptions were not measured in this study, which have been linked to self-reported illness among swimmers in other studies [15]. To reduce self-selection bias among swimmers, this study performed modeling comparing swimmers in various exposure groups to a low exposure swimmer group. Exposure misclassification is also relevant, particularly if beach water quality varies significantly throughout each study day which would limit reliable exposure estimates from a single daily sampling effort. Lastly, illness misclassification relying on self-reported information was more likely in this study than in studies using direct follow-up examinations.

Despite this sample size limitation, several constructed models did provide acceptable discrimination (AUC>0.70; Table 5) and acceptable model fit, particularly the HCGI and GI models from Table 4. Although the study presents several limitations, the study likely represents the only prospective epidemiological investigation exploring HAdV associations with recreational water-associated illness at this important beach type (inland lake recreational water). Future studies with larger samples sizes are warranted, particularly in beach environments where fecal contamination is human-associated and from dispersed sources or a mixture of point and dispersed sources.

Conclusions

- No association was observed between exposure to various qPCR-measured bacterial markers and recreational water-associated illness among study participants.
- Increasing levels of qPCR-measured HAdV were associated with increased odds of recreational water-associated gastrointestinal illness among the swimmers.
- Combined measurement of HAdV and culturable *E. coli* densities potentially enhances recreational water illness prediction among swimmers.
- Future studies with larger sample sizes enabling adjustment for additional confounding variables are needed to permit more robust investigations of human illness associations with exposures to HAdV and other viral markers in recreational water environments.

Supporting Information

Table S1 Primers and probes used in this study.

Table S2 Adjusted odds ratios for covariates used in GI illness models from Table 3.

Table S3 Adjusted odds ratios for covariates used in diarrheal illness models from Table 3.

Text S1 Detailed methods for sample processing and qPCR procedures.

Acknowledgments

The authors are grateful to the East Fork beach users, as well as Scott Fletcher, Chris Dauner, Bob Bowman, and Dan West from the Ohio Division of Parks and Recreation. The authors thank James Rosenblum, Michael Lofton, Matthew Farley, Sara Zolinski, and Kaedra Wetzel for assisting in the administration of the beach user questionnaire. The authors are thankful to Mary K. Estes at the Baylor College of Medicine for kindly providing the RNA transcript of Hu/NoV/GI/Norwalk.

Author Contributions

Conceived and designed the experiments: JL JWM. Performed the experiments: JWM CSL CL. Analyzed the data: JWM JL SL CL CSL QW. Contributed reagents/materials/analysis tools: JL LJS QW. Wrote the paper: JWM JL TJB SL CL CSL QW LJS.

References

1. Dufour A (1984) Health Effects Criteria for Fresh Recreational Waters. U.S. Environmental Protection Agency Report EPA-600-1-84-2004. Cincinnati, OH: U.S. Environmental Protection Agency, Office of Research and Development Available: http://www.epa.gov/nerlcwww/documents/frc.pdf. Accessed 2014 June 19.
2. Wade TJ, Pai N, Eisenberg JN, Colford JMJ (2003) Do U.S. Environmental Protection Agency water quality guidelines for recreational waters prevent gastrointestinal illness? A systematic review and meta-analysis. Environ Health Perspect 111: 1102–1109.
3. Wade TJ, Calderon RL, Sams E, Beach M, Brenner KP, et al. (2006) Rapidly measured indicators of recreational water quality are predictive of swimming-associated gastrointestinal illness. Environ Health Perspect 114 (1): 24–28. doi: 10.1289%2Fehp.8273.
4. Wade TJ, Sams E, Brenner KP, Haugland R, Chem E, et al. (2010) Rapidly measured indicators of recreational water quality and swimming-associated illness at marine beaches: a prospective cohort study. Environmental Health 9:66. doi: 10.1186/1476-069X-9-66.
5. Colford JM, Wade TJ, Schiff KC, Wright CC, Griffith JF, et al. (2007) Water quality indicators and the risk of illness at beaches with nonpoint sources of fecal contamination. Epidemiology . 18 (1), 27–35. doi: 10.1097/01.ede.0000249425.32990.b9.
6. Sinigalliano CD, Fleisher JM, Gidley ML, Solo-Gabriele HM, Shibata T, et al. (2010) Traditional and molecular analyses for fecal indicator bacteria in non-point source recreational marine waters. Water Res 44 (13): 3763–3772. doi: dx.doi.org/10.1016/j.watres.2010.04.026.
7. Gruber SJ, Kay LM, Kolb R, Henry K (2005) Mission Bay bacterial source identification study. Stormwater 6: 40–51.
8. Soller JA, Schoen ME, Bartrand T, Ravenscroft J, Ashbolt NJ (2010) Estimated human health risks from exposure to recreational waters impacted by human and non-human sources of faecal contamination. Water Res 44 (16): 4674–4691. doi: 10.1016/j.watres.2010.06.049.
9. Colford JM, Schiff KC, Griffith JF, Yau V, Arnold BF, et al. (2012) Using rapid indicators for Enterococcus to assess the risk of illness after exposure to urban runoff contaminated marine water. Water Res 46 (7): 2176–2186. doi: 10.1016/j.watres.2012.01.033.
10. Arnold BF, Schiff KC, Griffith JF, Gruber JS, Yau V, et al. (2013) Swimmer illness associated with marine water exposure and water quality indicators: impact of widely used assumptions. Epidemiology 24 (6): 845–853. doi: 10.1097/01.ede.0000434431.06765.4a.
11. Yau VM, Schiff KC, Arnold BF, Griffith JF, Gruber JS, et al. (2014) Effect of submarine discharge on bacterial indicators and swimmer health at Avalon Beach, CA, USA. Water Res 59: 23–36. doi: 10.1016/j.watres.2014.03.050.
12. Fewtrell L, Jones F, Tech B, Kay D, Wyer MD, et al. (1992) Health effects of white-water canoeing. Lancet 339 (8809): 1587–1589. doi:10.1016/0140-6736(92)91843-W.
13. Pike E (1994) Health Effects of Sea Swimming, Phase III – Final Report to the Department of the Environment. WRC Report: DoE 3412/2. Medmenham, U.K. UK: Water Research Center. pp 9–24.
14. Fleisher JM, Kay D, Salmon RL, Jones F, Wyer MD, et al. (1996) Marine waters contaminated with domestic sewage: nonenteric illnesses associated with swimmer exposure in the United Kingdom. Am J Public Health 86 (9): 1228–1234.
15. Fleisher JM, Fleming LE, Solo-Gabriele HM, Kish JK, Sinigalliano D, et al. (2010) The BEACHES Study: health effects and exposures from non-point source microbial contaminants in subtropical recreational marine waters. Int J Epidemiol 39 (5): 1291–1298. doi: 10.1093/ije/dyq084.
16. Abdelzaher AM, Wright ME, Ortega C, Hasan AR, Shibata T, et al. (2011) Daily measures of microbes and human health at a non-point source marine beach. J Water Health 9 (3): 443–457. DOI:10.2166/wh.2011.146Griffen DW, Donaldson KA, Paul JH, Rose JB (2003) Pathogenic human viruses in coastal waters. Clin Microbiol Rev 16 (1): 129–143. doi: 10.1128/CMR.16.1.129-143.2003.
17. Hlavsa MC, Roberts VA, Anderson AR, Hill VR, Kahler AM, et al. (2011) Surveillance for Waterborne Disease Outbreaks and Other Health Events Associated with Recreational Water -United States, 2007-2008. MMWR 60 (ss12): 1–32.
18. Soller JA, Bartrand T, Ashbolt NJ, Ravenscroft J, Wade TJ (2010). Estimating the primary etiologic agents in recreational freshwaters impacted by human sources of fecal contamination. Water Res 44 (16): 4736–4747. doi: 10.1016/j.watres.2010.07.064.
19. Dorevitch S, Pratap P, Wroblewski M, Hryhorczuk DO, Li H, et al. (2012) Health risks of limited-contact water recreation. Environ Health Perspect 120 (2): 192–197. doi: 10.1289/ehp.1103934.
20. Lee CS, Lee C, Marion JW, Wang Q, Saif L, et al. (2014) Occurrence of human enteric viruses at freshwater beaches during swimming season and its link to water inflow. Sci Total Environ 472: 757–766. doi: 10.1016/j.scitotenv.2013.11.088.
21. Noble RT, Fuhrman JA (2001) Enteroviruses detected by reverse transcriptase polymerase chain reaction from the coastal waters of Santa Monica Bay,

California: low correlation to bacterial indicator levels. Hydrobiologia 460: 175–184. doi: 10.1023/A:1013121416891.

22. McQuaig S, Griffith J, Harwood VJ (2012) Association of fecal indicator bacteria with human viruses and microbial source tracking markers at coastal beaches impacted by nonpoint source pollution. Appl Environ Microbiol 78 (18): 6423–6432. doi: 10.1128/AEM.00024-12.
23. Ashbolt NJ, Grabow WOK, Snozzi M (2001) Indicators of microbial water quality. In: Fewtrell L, Bartram J, editors. Water Quality: Guidelines, Standards and Health. London: World Health Organization: pp.289–315.
24. Boehm AB, Ashbolt NJ, Colford JM, Dunbar LE, et al. (2009) A sea of change ahead for recreational water quality criteria. J Water Health 7 (1): 9–20. doi: 10.2166/wh.2009.122.
25. Abdelzaher AM, Solo-Gabriele HM, Phillips MC, Elmir SM, Fleming LE (2013) An alternative approach to water regulations for public health protection at swimming beaches. J. Environ Public Health 2013: 138521. DOI: 10.1155/2013/138521.
26. Marion JW, Lee J, Lemeshow S, Buckley TJ (2010) Association of gastrointestinal illness and recreational water exposure at an inland U.S. beach. Water Res 44 (16): 4796–4804. doi: 10.1016/j.watres.2010.07.065.
27. U.S. Environmental Protection Agency (2002) Method 1603: *Escherichia coli* (*E. coli*) in Water by Membrane Filtration Using Modified membrane-Thermotolerant *Escherichia coli* Agar (Modified mTEC). Office of Water Report EPA-821-R-02-023. Washington D.C. Available: http://www.epa.gov/nerlcwww/documents/1603sp02.pdf. Accessed 2014 June 19.
28. National Oceanic and Atmospheric Administration (2013) Database: UV Index Bulletins Archives. http://www.cpc.ncep.noaa.gov/products/stratosphere/uv_index/uv_archive.shtml. Accessed 2014 June 19.
29. U.S. Army Corps of Engineers, Louisville District (2013) Database: Yearly Lake Reports. http://www.lrl.usace.army.mil/wcds/dlbrpt.htm. Accessed 2014 June 19.
30. Haugland RA, Siefring SC, Wymer LJ, Brenner KP, Dufour AP (2005) Comparison of *Enterococcus* measurements in freshwater at two recreational beaches by quantitative polymerase chain reaction and membrane filter culture analysis. Water Res 39 (4): 559–568. doi: 10.1016/j.watres.2004.11.011.
31. Lipsitch M, Tchetgen ET, Cohen T (2010). Negative controls: a tool for detecting confounding and bias in observational studies. Epidemiology 21 (3): 383–388. doi: 10.1097/EDE.0b013e3181d61eeb.
32. Hosmer DW, Lemeshow S (2000) Applied logistic regression, 2nd ed. New York: Wiley and Sons. 375 p.
33. Peduzzi P, Concato J, Kemper E, Holford TR, Feinstein AR (1996). A simulation of the number of events per variable in logistic regression analysis. J Clin Epidemiol 49 (12): 1373–1379.
34. Vittinghoff E, McCulloch CE (2007). Relaxing the rule of ten events per variable in logistic and cox regression. Am J Epidemiol 165 (6): 710–718. doi: 10.1093/aje/kwk052.
35. Peed LA, Nietch CT, Kelty CA, Meckes M, Mooney T, et al. (2011) Combining land use information and small stream sampling with PCR-based methods for better characterization of diffuse sources of human fecal pollution. Environ Sci Technol 45 (13): 5652–5659. doi: 10.1021/es2003167.
36. Lodder WJ, de Roda Husman AM (2005) Presence of noroviruses and other enteric viruses in sewage and surface waters in the Netherlands. Appl Environ Microbiol 71 (3): 1453–1461. doi: 10.1128/AEM.71.3.1453-1461.2005.
37. Haramoto E, Katayama H, Oguma K, Ohgaki S (2005) Application of cation-coated filter method to detection of noroviruses, enteroviruses, adenoviruses, and torque teno viruses in the Tamagawa River in Japan. Appl Environ Microbiology 71 (5): 2403–2411. doi: 10.1128/AEM.71.5.2403-2411.2005.
38. Jiang S, Noble R, Chu W (2001) Human adenoviruses and coliphages in urban-runoff impacted coastal waters of Southern California. Appl Environ Microbiol 67 (1): 179–184. doi: 10.1128/AEM.67.1.179-184.2001.
39. Payment P, Lemieux M, Trudel M (1982) Bacteriological and virological analysis of water from four fresh water beaches. Water Res 16 (6): 939–943. doi: 10.1016/0043-1354(82)90026-4.
40. Newton RJ, VandeWalle JL, Borchardt MA, Gorelick MH, McLellan SL (2011) Lachnospiraceae and Bacteroidales alternative fecal indicators reveal chronic human sewage contamination in an urban harbor. Appl Environ Microbiol 77 (19): 6972–6981. doi: 10.1128/AEM.05480-11.
41. Wyn-Jones AP, Carducci A, Cook N, D'Agostino M, Divizia M, et al. (2011) Surveillance of adenoviruses and noroviruses in European recreational waters. Water Res 45 (3): 1025–1038. doi: 10.1016/j.watres.2010.10.015.
42. Sinclair RG, Jones EL, Gerba CP (2009) Viruses in recreational water-borne disease outbreaks: a review. J Appl Microbiol 107 (6): 1769–1780. doi: 10.1111/j.1365-2672.2009.04367.x.
43. Albinana-Gimenez N, Miagostovich MP, Calgua B, Huguet JM, Matia L, et al. 2009. Analysis of adenoviruses and polyomaviruses quantified by qPCR as indicators of water quality in source and drinking-water treatment plants. Water Res 43 (7): 2011–2019. DOI: 10.1016/j.watres.2009.01.025.
44. Leshem E, Wikswo M, Barclay L, Brandt E, Storm W, et al. (2013) Effects and clinical significance of GII.4 Sydney Norovirus, United States, 2012–2013. Emerg Infect Dis 19 (8): 1231–1238. doi: 10.3201/eid1908.130458.
45. Gerba CP, Gramos DM, Nwachuku N (2002) Comparative inactivation of enteroviruses and adenovirus 2 by UV light. Appl Environ Microbiol 68: 5167–5169. doi: 10.1128/AEM.68.10.5167-5169.2002.
46. Fongaro G, Nascimento MA, Rigotto C, Ritterbusch G, da Silva AD, et al. (2013) Evaluation and molecular characterization of human adenovirus in drinking water supplies: viral integrity and viability assays. J Virol 10: 166. doi: 10.1186/1743-422X-10-166.
47. Kundu A, McBride G, Wuertz S (2013) Adenovirus-associated health risks for recreational activities in a multi-use coastal watershed based on site-specific quantitative microbial risk assessment. Water Res 47 (16): 6309–6325. doi: 10.1016/j.watres.2013.08.002.
48. Heaney CD, Sams E, Dufour AP, Brenner KP, Haugland RA, et al. (2012) Fecal indicators in sand, sand contact, and risk of enteric illness among beachgoers. Epidemiol 23 (1): 95–106.
49. Ge Z, Whitman RL, Nevers MB, Phanikumar MS (2012) Wave-induced mass transport affects daily *Escherichia coli* fluctuations in nearshore water. Environ Sci Technol 46 (4): 2204–2211. doi: 10.1021/es203847n.
50. Fleisher JM, Jones J, Kay D, Stanwell-Smith R, Wyer M, et al. (1993). Water and non-water related risk factors for gastroenteritis among swimmers exposed to sewage-contaminated marine waters. Int J Epidemiol 22 (4): 698–708. doi: 10.1093/ije/22.4.698.
51. Lee C, Agidi S, Marion JW, Lee J (2012) *Arcobacter* in Lake Erie beach waters: an emerging gastrointestinal pathogen linked with human-associated fecal contamination. Appl and Environ Microbiol 78 (16): 5511–5519. doi: 10.1128/AEM.08009-11.
52. Lee J, Deininger RA (2004) Detection of *E. coli* in beach water within 1 hour using immunomagnetic separation and ATP bioluminescence. Luminescence 19 (1): 31–36. doi: 10.1002/bio.753.
53. Haramoto E, Katayama H, Oguma K, Ohgaki S (2005) Application of cation-coated filter method to detection of noroviruses, enteroviruses, adenoviruses, and torque teno viruses in the Tamagawa River in Japan. Appl Environ Microbiol 71 (5): 2403–2411. doi: 10.1128%2FAEM.71.5.2403-2411.2005.
54. Monpoeho S, Dehee A, Mignotte B, Schwartzbrod L, Marechal V, et al. (2000) Quantification of enterovirus RNA in sludge samples using single tube real-time RT-PCR. Biotechniques 29 (1): 88–93.
55. Kageyama T, Kojima S, Shinohara M, Uchida K, Fukushi S, et al. (2003) Broadly reactive and highly sensitive assay for Norwalk-like viruses based on real-time quantitative reverse transcription-PCR. J Clin Microbiol 41 (4): 1548–1557. doi: 10.1128/JCM.41.4.1548-1557.2003.
56. Heim A, Ebnet C, Harste G, Pring-Akerblom P (2003) Rapid and quantitative detection of human adenovirus DNA by real-time PCR. J Med Virol 70 (2): 228–239. doi: 10.1002/jmv.10382.
57. Chern EC, Brenner K, Wymer L, Haugland RA (2009) Comparison of fecal indicator bacteria densities in marine recreational waters by QPCR. Water Qual Expo Healt 1 (3–4): 203–214.
58. Chern EC, Siefring S, Paar J, Doolittle M, Haugland RA (2011) Comparison of quantitative PCR assays for *Escherichia coli* targeting ribosomal RNA and single copy genes. Lett Appl Microbiol 52 (3): 298–306. doi: 10.1111/j.1472-765X.2010.03001.x.
59. Bernhard AE, Field KG (2000) Identification of nonpoint sources of fecal pollution in coastal waters by using host-specific 16S ribosomal DNA genetic markers from fecal anaerobes. Appl Environ Microbiol 66 (4): 1587–1594. doi: 10.1128/AEM.66.4.1587-1594.2000.
60. Okabe S, Okayama N, Savichtcheva O, Ito T (2007) Quantification of host-specific *Bacteroides-Prevotella* 16S rRNA genetic markers for assessment of fecal pollution in freshwater. Appl Microbiol Biotechnol 74 (4): 890–901. doi: 10.1007/s00253-006-0714-x.
61. Ludwig W, Schleifer KH (2000) How quantitative is quantitative PCR with respect to cell counts? Syst Appl Microbiol 23 (4): 556–562. doi: 10.1016/S0723-2020(00)80030-2.
62. U.S. Environmental Protection Agency (2010) Method A: Enterococci in Water by TaqMan Quantitative Polymerase Chain Reaction (qPCR) Assay. EPA-821-R-10-004. Washington, D.C. Available: http://water.epa.gov/scitech/methods/cwa/bioindicators/upload/rapid1.pdf. Accessed 2014 June 20.
63. Lee CS, Lee J, (2010) Evaluation of new gyrB-based real-time PCR system for the detection of *B. fragilis* as an indicator of human-specific fecal contamination. J Microbiol Methods 82 (3): 311–318. doi: 10.1016/j.mimet.2010.07.012.

Abundances of Tetracycline, Sulphonamide and Beta-Lactam Antibiotic Resistance Genes in Conventional Wastewater Treatment Plants (WWTPs) with Different Waste Load

Mailis Laht[1¶], Antti Karkman[2¶], Veiko Voolaid[1], Christian Ritz[3], Tanel Tenson[1], Marko Virta[2], Veljo Kisand[1]*

1 Institute of Technology, University of Tartu, Tartu, Estonia, **2** Department of Food and Environmental Sciences, University of Helsinki, Helsinki, Finland, **3** Department of Nutrition, Exercise and Sports, Faculty of Science, University of Copenhagen, Copenhagen, Denmark

Abstract

Antibiotics and antibiotic resistant bacteria enter wastewater treatment plants (WWTPs), an environment where resistance genes can potentially spread and exchange between microbes. Several antibiotic resistance genes (ARGs) were quantified using qPCR in three WWTPs of decreasing capacity located in Helsinki, Tallinn, and Tartu, respectively: sulphonamide resistance genes (*sul1* and *sul2*), tetracycline resistance genes (*tetM* and *tetC*), and resistance genes for extended spectrum beta-lactams (bla_{oxa-58}, bla_{shv-34}, and $bla_{ctx-m-32}$). To avoid inconsistencies among qPCR assays we normalised the ARG abundances with 16S rRNA gene abundances while assessing if the respective genes increased or decreased during treatment. ARGs were detected in most samples; *sul1, sul2,* and *tetM* were detected in all samples. Statistically significant differences (adjusted $p<0.01$) between the inflow and effluent were detected in only four cases. Effluent values for bla_{oxa-58} and *tetC* decreased in the two larger plants while *tetM* decreased in the medium-sized plant. Only bla_{shv-34} increased in the effluent from the medium-sized plant. In all other cases the purification process caused no significant change in the relative abundance of resistance genes, while the raw abundances fell by several orders of magnitude. Standard water quality variables (biological oxygen demand, total phosphorus and nitrogen, etc.) were weakly related or unrelated to the relative abundance of resistance genes. Based on our results we conclude that there is neither considerable enrichment nor purification of antibiotic resistance genes in studied conventional WWTPs.

Editor: Jorge M. O. Fernandes, University of Nordland, Norway

Funding: The research was funded by the Academy of Finland and EnSTe graduate school, and the European Regional Development Fund through the Centre of Excellence in Chemical Biology in Estonia. Additional funding was provided by COST Action TD0803. The funders had no role in study design, data collection and analysis, decision to publish, or preparation of the manuscript.

Competing Interests: The authors have declared that no competing interests exist.

* Email: kisand@ut.ee

¶ These authors contributed equally to this work.

Introduction

Antibiotic resistance (AR) has become a worldwide problem, making infectious diseases more resilient thus making treatment more difficult and costly [1]. AR is not confined to hospital environments, and is able to spread between both human dominated and natural environments. Increased amounts of antibiotic residues [2–4] and antibiotic resistant bacteria (ARB) are found in human-related environments such as agricultural settings (e.g. farms, soil etc.) [5,6] and surface-, drinking- and wastewaters [7–11]. Wastewater treatment plants (WWTPs) receive sewage from various sources, including hospitals and households which are both important sources of antibiotics and their residues [12–14]. and antibiotic resistant bacteria (ARB) [15–17]. The presence of antibiotics and antibiotic residues [18]. ARB, and antibiotic resistance genes (ARG) have been confirmed in many WWTPs [19–23]. Bacteria from various environments, including human, soil, and activated sludge, are mixed in WWTPs and therefore these facilities are considered to be important "hot-spots" for AR and spread of resistance genes [18,24–26]. The presence of antibiotics, ARB, and ARG in the same setting creates an environment that selects for AR and provides an opportunity for genetic material housing ARGs to transfer between bacterial species via horizontal gene transfer [13,21,24,25,27,28]. In a metagenomic study of plasmids it was shown that numerous medically relevant ARG can be found in WWTPs (140 ARGs in a single WWTP and 123 in the effluent water) [29]. Therefore, there is a concern that resistance genes will spread in the bacterial population and further into more natural environments less impacted by human activity [30–33].

Many studies of antibiotic resistant bacteria and resistance genes have used culture-based assays, which are biased towards specific cultivable pathogenic or environmental species [17,19,21,23,34]. Culture-dependent data does not reflect the real variability or the actual amount of resistance genes present in a given WWTP, so these studies normally characterize only a small subset of the total population [35,36]. Surprisingly, only a limited number of studies have used quantitative methods to investigate resistance genes in the total communities in effluent waters [21,33,37–40]. The small number of quantitative studies could be one reason why the overall impact of WWTPs in spreading resistance to the environment has not yet been properly evaluated.

The current study was designed to investigate the role of three conventional WWTPs in the distribution of ARGs using culture-independent quantitative methods. We thus focus on quantitative whole community level measurements on the water-phase of the inflow and effluent of WWTPs which has not been studied using quantitative methods in sufficient detail to assess the impact of WWTPs on the distribution of ARGs. To quantify the number of ARGs we analysed the total DNA from wastewater and effluent samples using quantitative real-time PCR (qPCR). Our original hypothesis was that conventional WWTPs increase the relative abundance of ARGs during processing because they do not employ technologies that target the removal of genetic elements.

Materials and Methods

1.1. Sample collection

Both raw wastewater and final effluent water samples were collected from three city WWTPs of decreasing capacity located in the Baltic Sea catchment area: large (Helsinki, Finland), medium (Tallinn, Estonia), and small (Tartu, Estonia). No specific permissions were required for sampling these locations, and the sampling was carried out in collaboration with each WWTP staff. The wastewater treatment technology employed in these three WWTPs is similar and typical of other facilities located in both Nordic (Finland, Sweden, Norway) and Eastern European (Estonia, Latvia, Lithuania, Poland) countries (EEA 2013). The WWTP in Helsinki is both the largest wastewater treatment plant in Finland and all Nordic countries; with about 0.8 million residents in the Helsinki metropolitan area. The majority of wastewater in Estonia is produced in Tallinn (~350 000 residents) and Tartu (~100 000). To abide by European law (91/271/EEC), the main steps of treatment are: primary treatment - mechanical treatment steps (sand, grit, fat and grease removal, pre-sedimentation); secondary treatment - biological treatment (activated sludge); followed by tertiary treatment - final deep purification using a combination of methods (chemical, mechanical and biological as in secondary sedimentation, bio-filters etc.). The WWTP effluent in both Helsinki and Tallinn is directed into the Baltic Sea while the WWTP effluent in Tartu flows into the Emajõgi River which forms part of the Baltic Sea catchment area. The main steps in the WWTP process and water-phase treatment are presented in Figure S1 in File S1.

Samples were collected over a one year period from December 2010 to December 2011 at five different time points, each representing a different season (four seasons; winter was sampled twice, Table S1A in File S1). In each sampling period three consecutive samples were taken on separate days at 1–3 day intervals; the exact sampling dates and monitored variables are given in Table S1 in File S1. Composite samples were taken, collected over 24 h periods, except in Tartu in the 2011 winter season when a grab sample was taken owing to technical problems (the automatic sampler was frozen because of extremely low temperatures).

1.2. Collection of total microbial community

The water samples were stored at 4°C pending filtration (within a couple of hours). Ten ml of influent water and 100 ml of effluent water were filtered through polycarbonate filters (pore size 0.22 µm, diameter 47 mm, GE Water & Process Technologies). For the last two time points at the Tallinn WWTP 200 ml of effluent water was filtered because a new treatment step (bio-filter) was added in September 2011.

1.3. DNA extraction

For the first three time points (Dec 2010; March 2011; June 2011) from the Helsinki WWTP, DNA was extracted from the samples using a MoBio PowerWater DNA isolation kit (MoBio Laboratories, Inc., CA, USA). For all other samples, the DNA was extracted using the modified bead beating and silica-membrane method (nucleic acid binding on to silica particles [41]. Method in brief: for lysis: 400 µl TE+50 µl lysozyme (50 mg/ml from egg yolk) was added to the filter and incubated at 37°C for 15 min. Fifty µl of 10% SDS and 500 µl lysis buffer BQ1 (NucleoSpin, Macherey-Nagel)+zirconium beads (0.1 mm diameter; burned at 500°C) were added and the samples were processed by a 5 min beating on a bead beater (Biospec products, Minibead beater). Thereafter, proteinase K was added followed by 15 min at 65°C with constant shaking. Five hundred µl of 96% ethanol was added and the whole volume was applied to commercial silica-membrane columns (NucleoSpin Macherey-Nagel). Finally, the total DNA was recovered in 50 µl of elution buffer BE (NucleoSpin Macherey-Nagel, 5 mM Tris/HCl pH 8.5). The concentration of extracted DNA was measured using a NanoDrop Spectrophotometer ND-1000 (absorption readings at 260 nm). The extracted DNA was stored at −20°C pending further analysis.

1.4. Detection and quantification of ARG copy number by qPCR

Seven resistance genes were surveyed: *sul1, sul2, tetM, tetC,* $bla_{shv\text{-}34}$, $bla_{ctx\text{-}m\text{-}32}$, and $bla_{oxa\text{-}58}$. For the first three time point samples (Dec 2010; March 2011; June 2011) from the Helsinki WWTP (Assay 1), qPCR was performed using a Dynamo Flash SYBR Green qPCR kit (Thermo Scientific, Lithuania) and a 7300 Real-Time PCR system (Applied Biosystems, Foster City, CA, USA). The thermal cycling conditions were as follows: 95°C for 7 min, 40 cycles at 95°C for 10 s and Tm for 30 s. A melting curve was obtained to confirm specificity of amplification. Reactions were conducted in 10 µl volumes on 96-well plates containing 1×Dynamo Flash SYBR Green master mix, 0.3 µM of each primer and 1×ROX passive reference dye. Template DNA was used in qPCR reactions in the range 2–12 ng DNA per reaction; a fixed dilution of raw DNA exctract was used. In parallel with the ARGs, the16S rRNA gene copy numbers were quantified.

The qPCR for detecting 16S rRNA gene and ARGs for all the other samples (Assay 2) used the 7900HT Fast Real-Time PCR System (Applied Biosystems). Reactions were conducted in 10 µl volumes on 384-well plates containing 1×Maxima SYBR Green/ROX qPCR Master Mix (Thermo Scientific), 0.4 µM of each primer. The two-step thermal cycling conditions for detecting 16S rRNA gene were as follows: 95°C for 10 min, 40 cycles of 95°C 15 s, 60°C 1 min. For ARGs the thermal cycling conditions were as follows: 95°C for 10 min, 40 cycles at 95°C for 15 s, Tm °C for 30 s and 72°C for 30 s. Melting curves were obtained to confirm specificity of amplification. Template DNA was used in the qPCR reaction in the range 3×10^{-4} to 22.4 ng DNA per reaction. This was obtained by a wider range of dilutions (to avoid inhibition) of the raw DNA extraction product, three levels of 10 fold dilutions

Table 1. Primers used for detecting the target genes and the melting temperatures (Tm) used for primers.

Gene name	Primers				Amplification efficiency	
	Forward	Reverse	Tm °C	Ref.	Assay 1	Assay 2
16S rRNA gene	5′-AGA GTT TGA TCC TGG CTC AG-3′			[60]		
		5′-CTG CTG CSY CCC GTA GGA-3′	60	[61] modif.		82–100
		5′-CTG CTG CCT CCC GTA GG-3′	60	[61]	87	
tetC	5′-TGC GTT GAT GCA ATT TCT ATG C-3′	5′-GGC GCC TAC AAT CCA TG-3′	64	[62]	80	93–101
tetM	5′-GCA ATT CTA CTG ATT TCT GC-3′	5′-CTG TTT GAT TAC AAT TTC CGC-3′	60	[61]	90–93	89–108
sul1	5′-CGG CGT GGG CTA CCT GAA CG-3′	5′-GCC GAT CGC GTG AAG TTC CG-3′	64	[63]	90–93	81–88
sul2	5′-GCG CTC AAG GCA GAT GGC ATT-3′	5′-GCG TTT GAT ACC GGC ACC CGT-3′	64	[62]	102	91–103
blactx-m-32	5′-CGT CAC GCT GTT GTT AGG AA-3′	5′-CGC TCA TCA GCA CGA TAA AG-3′	64	[59]	87	89–101
blashv-34	5′-GCG TTA TTT TCG CCT GTG TA-3′	5′-AGG TGC TCA TCA TGG GAA AG-3′	60	[63]	92–94	97–108
blaoxa-58	5′-GCA ATT GCC TTT TAA ACC TGA-3′	5′-CTG CCT TTT CAA CAA AAC CC-3′	60	[63]	90	97–111

qPCR amplification efficiency is given for 16S RNA gene and for ARGs, R^2 of the linear range of standards was always >0.99.

were used. Primers and annealing temperature, Tm (°C), are given in Table 1. Number of technical replicates in both qPCR assays was 3.

1.5. Standards used for quantification

A plasmid vector and fragments of ARGs were constructed and used as standards for quantifying the raw qPCR results. The standard plasmids were checked for the correct inserts by sequencing. In Assay 1 for the three Helsinki time points (Dec 2010; March 2011 and June 2011) the 16S rRNA gene quantification standard was genomic DNA from *E. coli* K12 (genome size 4.6 Mbp with seven copies of the rRNA operon). In Assay 2, used for all other samples, the standard for quantifying 16S rRNA gene was constructed from a 16S rRNA gene fragment from the natural aquatic *Chryseobacterium* strain isolated from Emajõgi River, which receives WWTP effluent water, and was cloned into a plasmid and validated by sequencing. Information about the plasmids used and Genbank accession numbers are given in Table 2.

1.6. Quantification and normalisation of ARGs

Standard curves (Ct per log copy number) for 16S rRNA gene and ARG quantification were obtained for each run using the plasmid constructs (or genomic *E. coli* DNA for 16S rRNA gene in Assay 1) (Table 2) in ten-fold serial dilutions. The gene copy number of a standard was determined from the plasmid/genomic DNA concentration (measured using a NanoDrop in Assay 1 or by fluorescent staining with PicoGreen (Invitrogen) and using VICTOR X3 Multilabel Plate Readers (Perkin-Elmer) in Assay 2). The ARG levels in the sample were calculated using the standard curve equation and measured Ct value, the quality control of raw Ct values for standard curve and unknown samples was done before further analysis. The limit of quantification (LOQ) was defined as the lowest point of the linear part of standard curve: Assay 1, ARGs 100 and 16S rRNA gene 1000 gene copy number per reaction; Assay 2, ARGs and 16S rRNA gene 100, except *sul1* with 10 copy numbers per reaction. The negative control was the reaction mix with nuclease-free water instead of the template DNA. The negative control had always a Ct value at least 3.3 cycles lower than the smallest standard used

Table 2. Plasmids and PCR fragments used as standards.

Gene name	Standard constructs	Accession number	Reference
16S rRNA gene	**Assay 1**		
	Genomic DNA from *E. coli* K12 genome size 4.6 Mbp with 7 copies of rRNA operon		
	Assay 2		
	PCR product cloned in plasmid PGEM-T Easy Vector System (Promega)	KF737394	present study
tetC	pDrive (Qiagen)		[62]
tetM	pDrive (Qiagen)		[62]
sul1	R388		[64]
sul2	RSF1010		[64]
$bla_{ctx\text{-}m\text{-}32}$	pUC19	KF737395	present study
$bla_{shv\text{-}34}$	PGEM-T Easy Vector System (Promega)	KF737397	present study
$bla_{oxa\text{-}58}$	pUC19	KF737396	present study

for calculation of LOQ. Technical replicates were incorporated in the statistical analysis to reflect differences in quantification (see below for details).

1.7. Water quality variables

WWTPs in Europe analyse water quality parameters according to the EU directive for urban wastewater treatment 91/271/ECC. These parameters are monitored regularly in accredited laboratories according to standard methods. The parameters used in this study were: biochemical oxygen demand (BOD_7) - standard method EN 1899-2; total suspended solids (SS) - EN 872; total phosphorus (Ptot) – EN ISO 6878; total nitrogen (Ntot) – EN ISO 11905. In addition, the automated measurements of flow rate and water temperature in the process were recorded. We received the data for water variables from the staff of each WWTP.

1.8. Statistics

Linear mixed models were fitted to the data using the functionality of the package lme4 [42]. the statistical software used was R version 3.0.1 [43]. Average levels of gene copy numbers (16S rRNA gene or ARGs) were modelled using the location of the WWTP (Helsinki, Tallinn, Tartu), sample source (IF/EF) and method protocol (Assay1/Assay2) as fixed effects and technical and biological replicates as random effects. Significance of fixed effects was assessed by an F-test using a (pre-specified) significance level of 1%. In addition, the combined effect of each of the environmental variables BOD_7, SS, Ptot, Ntot, flow rate (i.e water discharge - WD) and temperature in inflowing versus effluent waste on gene copy numbers was described by a linear mixed model version of an analysis of covariance model. Model checking was based on residual plots and normal probability plots using the raw residuals. Models were reduced using the likelihood ratio test. A 1% or 5% significance level was used. Pairwise comparisons were evaluated based on adjusted p-values obtained using the single-step method [44]. Values below LOQ were not included in the analyses.

Principal component analysis (PCA) and its extensions to between groups (BGA) and within groups (WGA) analyses (ade4 package in R) was used to analyse the grouping of inflowing/effluent samples by monitored water quality variables (nutrients, suspended solids, biological oxygen demand, water discharge and temperature).

Results and Discussion

2.1. Selection of ARGs for the study

Initially, we made a screening for 25 selected genes that could spread with higher probability from any WWTPs into environment. Initial selection based on four major criteria: (i) clinically relevant genes (risk to human health) previously detected in WWTPs (e.g. [29]); (ii) genes found in various mobile elements, demonstrating their potential for transfer between bacteria [45–48]. (iii) high consumption antibiotics–sulphonamides, tetracyclines, and beta-lactams; (iv) incorporation of long-used antibiotics (tetracycline, sulfonamides) and the newer extended spectrum beta-lactams (carbapenems and 3rd and 4th generation cephalosporins) and their resistance genes. Altogether 12 *bla*, 6 *tet*, 3 *sul*, 3 *qnr* (floroquinolones) and 1 vancomycin resistant gene were tested using traditional PCR. Finally, two resistance genes for sulphonamide (*sul1* and *sul2*), two for tetracycline (*tetM* and *tetC*), and three for extended spectrum beta-lactams ($bla_{oxa\text{-}58}$, $bla_{shv\text{-}34}$, and $bla_{ctx\text{-}m\text{-}32}$) remained in the study based on their abundance and frequency in screening study. It was necessary to choose a small number of relevant target genes because the variety of different ARGs is large. In the Comprehensive Antibiotic Resistance Database [49] (http://arpcard.mcmaster.ca/accessed Sept. 23, 2013), the number of ARGs is 2153 and testing all these genes quantitatively in one study would be prohibitively difficult.

2.2. Abundance of genes (16S rRNA gene and ARGs)

To evaluate the abundance of the total bacterial community, we quantified 16S rRNA gene in the samples using qPCR. The amplification efficiency is given in Table 1. Initially, the raw gene copy numbers were used to estimate the general changes of bacterial levels during wastewater purification. The copy number of 16S rRNA gene was several orders of magnitude lower in the effluent (EF) than the inflow (IF) (Figure 1A, Table S2 in File S1). The differences between the IF and EF samples were statistically significant in all cities (adjusted $p<0.01$) (Figure 1A).

The raw gene copy numbers of ARGs/ml decreased during processing in the WWTP water phase. The levels of ARGs detected in the EF were lower than IF in all three plants (Figure 1B, Table S2 in File S1). The decrease of abundance from IF to EF was statistically significant ($p<0.01$) for all ARGs in the large (Helsinki) WWTP. In the medium (Tallinn) WWTP, for all ARGs except $bla_{shv\text{-}34}$ ($p>0.05$), the decrease was statistically significant ($p<0.01$). In the small WWTP (Tartu), the raw abundance of ARGs decreased after purification but the decrease was statistically significant only for $bla_{ctx\text{-}m\text{-}32}$ and *tetM* ($p<0.01$), and weakly significant for $bla_{oxa\text{-}58}$ ($p = 0.03$). Earlier studies have demonstrated the large variation of treatment plant efficiency in removing microorganisms [50] and micropollutants [51]. which also depends on the capacity of the WWTP plant. Low capacity treatment facilities are more vulnerable to changes in inflowing wastewater composition and flow rates. In addition, the WWTP in Tartu did not have biological post-filtration at the time of sampling.

ARGs are surprisingly rarely quantified directly using community DNA in both IF and EF water samples from WWTPs. As in our study (Figure 1B), a few other studies have found from 100 to 1000 fold reductions of raw ARG copy numbers during the purification process; e.g. *sul1, tetW*; [52]; *tetC, tetA*; [53]. *tetG, tetQ* [54]. In one study, the relative abundance of *sul1* increased, while *sul2* decreased slightly in WWTP effluent [31]. In our study, resistance genes for "older" antibiotics (with exception of *tetC*) were more commonly detected. *sul1, sul2,* and *tetM* were present above LOQ in all sites and samples (Table 3 and Figure S2 in File S1). High abundances of various tetracycline resistance genes and sulphonamide resistance genes were also demonstrated in other studies [21,23]. Quantitative studies that target resistance to the newer beta-lactams in community DNA in WWTPs are almost completely absent–only bla_{TEM} by Lachmayr et al. [38]. in addition, *bla* genes were quantified in the river water under the influences of wastewater but not directly in the WWTP effluent [55].

2.3. Normalised/Relative abundances of ARGs

To avoid inconsistencies among qPCR assays, including suboptimal efficiency in some cases, we used 16S rRNA gene-normalised values, and the different protocol (Assay1/Assay2) was added as an additional fixed effect into the statistical models. This type of data analysis allows one to quantify the relative changes in ARG abundances, whether more or fewer ARGs appear per microbial genome. When relative abundances were compared, statistically significant ($p<0.01$) differences between IF and EF were detected in only four cases (Figure 2). For $bla_{oxa\text{-}58}$ and *tetC* in the Helsinki plant we observed a relative decrease after purification. In addition, *tet*M decreased in the Tallinn WWTP EF

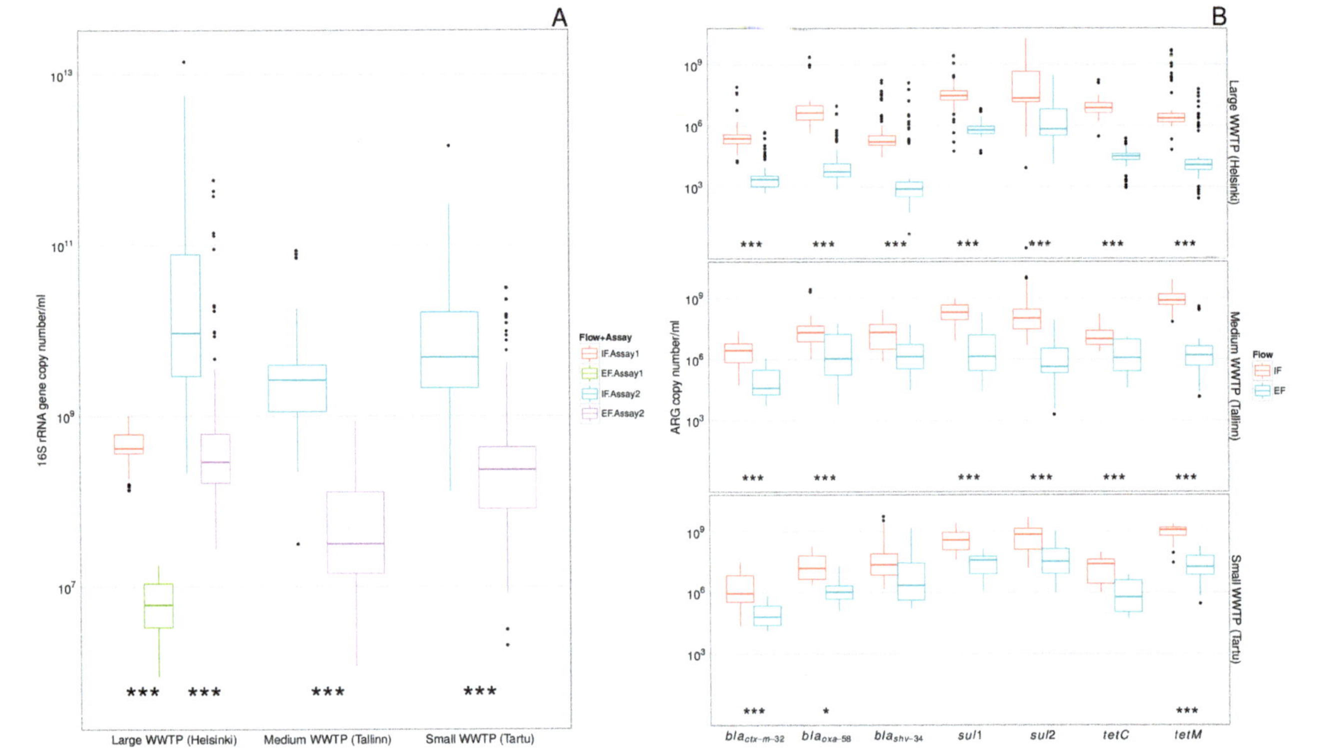

Figure 1. Raw gene copy numbers detected in a WWTP sample (copy number/ml). A - 16S rRNA gene in inflow (IF) and effluent (EF). Assay1 was used only for samples from large WWTP (Helsinki) from Winter 2010 to Autumn 2011; B–Antibiotic resistance genes (ARGs). Statistical significance between inflow wastewater and effluent samples: *** - $p<0.01$; *$0.03>p>0.01$. For the pairs not marked the statistical difference between inflow and outflow was statistically insignificant. The line in each box marks the median and boxes: 25th and 75th percentiles; whiskers: 5th and 95th percentiles and outliers ±1.5 * IQR. See Figure S2 in File S1 for abundances of same genes presented by each sampling event.

samples. The only increase in EF was observed for $bla_{shv\text{-}34}$ in the Tallinn WWTP. In all other cases the purification process had no significant effect on the relative abundances of resistance genes.

We conclude from our study that there is neither considerable enrichment (selection) nor purification of ARGs during processes in WWTPs (Figure 2) at the whole community level. Effective selection would be assumed when there are appropriate conditions, i.e. increased concentration of ABs occur. In two WWTPs studied, the measured levels of AB concentrations of several compounds were very low but measurable compared to highly labile beta-lactams [56]. This suggests that conditions could favour enrichment of at least *tet* and *sul* genes within studied WWTPs. Although, quantitative enrichment of ARGs (*sul*) responsible for resistance against refractory ABs with longer history of usage has

Table 3. Detection of ARGs in different WWTPs (total of all analyses per gene, n = 15).

ARG	City	IF %(number) detected	EF %(number) detected
tetC	Helsinki	93 (14)	80 (12)
	Tallinn	67 (10)	27 (4)
	Tartu	67 (10)	73 (11)
$bla_{oxa\text{-}58}$	Helsinki	100 (15)	87 (13)
	Tallinn	100 (15)	47 (7)
	Tartu	87 (13)	80 (12)
$bla_{shv\text{-}34}$	Helsinki	100 (15)	100 (15)
	Tallinn	87 (13)	87 (13)
	Tartu	100 (15)	100 (15)
$bla_{ctx\text{-}m\text{-}32}$	Helsinki	100 (15)	100 (15)
	Tallinn	87 (13)	40 (6)
	Tartu	80 (12)	47 (7)

Only genes that were sometimes not detected are given. sul1, sul2 and tetM were detected 100% in all IF and EF samples from the WWTPs.

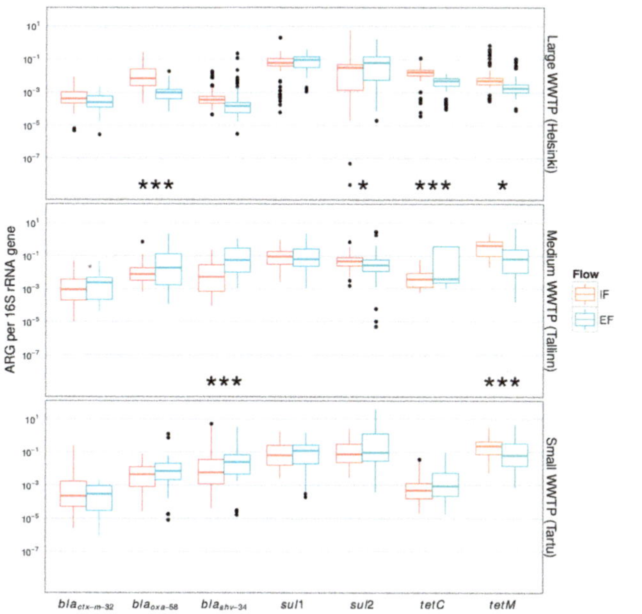

Figure 2. Normalised ARG abundances. Antibiotic resistance gene copy numbers normalised to 16S rRNA gene copy numbers. The results are given for all samples for one gene for a WWTP, no seasonal comparison. Statistically significant comparison results are marked with *** at $p<0.01$, $*0.03>p>0.01$. The line in each box marks the median and boxes: 25th and 75th percentiles; whiskers: 5th and 95th percentiles and outliers ±1.5 * IQR.

been demonstrated in some studies [31]. reduction or no change has been observed in most studies [21,23,40]. To date, only one study demonstrated similar case of positive selection for newer ARG possessing organisms in a WWTP, which suggests that bacteria harbouring bla_{TEM} are released more from effluent water compared to wastewater [38].

2.4. Abundance of ARGs and treatment efficiency of wastewater

In EU countries, treatment efficiency of WWTPs is estimated by monitoring a few water quality measures, according to European directive 271/1991/EC. The compulsory parameters monitored are total nitrogen (Ntot) and phosphorus (Ptot), Biological Oxygen Demand (BOD_7), and suspended solids (SS). In addition, a few generic background parameters are measured in all WWTPs e.g. Water Discharge (WD) and temperature. Such water quality parameters are good for evaluating wastewater purification in the traditional sense–removal of excess nutrients, labile organic compounds etc. Obviously, the wastewater was purified of excess nutrients and organic compounds in the WWTPs studied because all parameters were up to an order of magnitude lower in EF than IF. We observed a change of between 6 to 45 fold depending on the parameter. Averages in IF: BOD_7 - 231; P_{tot} 8; N_{tot} 51 and SS 306 mg/l and in EF: BOD_7 5; P_{tot} 0.6; N_{tot} 8 and SS 8 mg/l (Table S1B in File S1). Volumetric concentrations of nutrients, SS and labile organic compounds were higher in the IF of the small (Tartu) than in medium (Tallinn) and large (Helsinki) WWTP (Figure S4 in File S1; IF samples are strongly associated with these variables, permutation test, 1000 replicates, $p<0.01$). This could be caused by shorter solid retention time in smaller plants [57]. At the same time, the efficiency of purification in the traditional sense did not differ dramatically among plants (Figure S3 in FileS1; residual differences among WWTPs disappear after decomposing the IF/EF level differences, permutation test, 1000 replicates, $p>0.05$). A relationship between the change of ARG abundances and the efficiency of nutrient removal and temperature has been reported previously (e.g. [58,59]). These studies suggest that changes in ARG abundance could depend on processes and conditions in the WWTP. However, none of these parameters are designed to estimate threats associated with the spread of either ARB or ARGs. Indeed, in our study, temperature, water discharge and concentrations of nutrients did not help in estimating the efficiency of ARG removal. This was demonstrated by the absence of the combined effect of monitored environmental variables and abundances of ARGs (Figure S4 in File S1). Moreover, a new treatment step (installation of biological post-filtration for final effluent treatment mainly for nitrogen removal) was added during the study period in the Tallinn WWTP (July 2011). However, no statistically significant effect was observed on ARG removal after this event. The Helsinki WWTP had the biological post-filtration installed throughout the study period.

In conclusion, these results force us to reject our original hypothesis. All ARGs were detected in most wastewater and effluent samples, however, the conventional WWTPs under study seem not to be important sites for changes in the relative abundance of ARGs at the whole community level: no enrichment in relative abundance was observed. Furthermore, no additional reduction of ARGs occurred; raw abundance changed in proportion to the decrease of bacterial abundance. We conclude that many unknown factors may influence the biological purification processes in conventional WWTPs and the evaluation of their relationship to ARG removal or selection requires more complex case studies.

Acknowledgments

We acknowledge the staff from HSY at Viikinmäki, Tartu Veevärk AS and Tallinna Vesi AS for providing samples, and Dr. David Schryer for language revision.

Author Contributions

Conceived and designed the experiments: TT MV VK. Performed the experiments: ML AK VV. Analyzed the data: ML AK VV CR VK. Contributed reagents/materials/analysis tools: TT MV. Contributed to the writing of the manuscript: ML AK VV CR TT MV VK.

References

1. Levy SB, Marshall B (2004) Antibacterial resistance worldwide: causes, challenges and responses. Nat Med 10: S122–9.
2. Lindberg RH, Wennberg P, Johansson MI, Tysklind M, Andersson BA V (2005) Screening of Human Antibiotic Substances and Determination of Weekly Mass Flows in Five Sewage Treatment Plants in Sweden. Environ Sci Technol 39: 3421–3429.
3. Le-Minh N, Khan SJ, Drewes JE, Stuetz RM (2010) Fate of antibiotics during municipal water recycling treatment processes. Water Res 44: 4295–4323.
4. Novo A, André S, Viana P, Nunes OC, Manaia CM (2013) Antibiotic resistance, antimicrobial residues and bacterial community composition in urban wastewater. Water Res 47: 1875–1887. Available: http://www.sciencedirect.com/science/article/pii/S0043135413000274. Accessed 23 January 2013.

5. Khachatourians GG (1998) Agricultural use of antibiotics and the evolution and transfer of antibiotic-resistant bacteria. Can Med Assoc J 159: 1129–1136.
6. Durso LM, Miller DN, Wienhold BJ (2012) Distribution and Quantification of Antibiotic Resistant Genes and Bacteria across Agricultural and Non-Agricultural Metagenomes. PLoS One 7: e48325.
7. Schwartz T, Kohnen W, Jansen B, Obst U (2003) Detection of antibiotic-resistant bacteria and their resistance genes in wastewater, surface water, and drinking water biofilms. FEMS Microbiol Ecol 43: 325–335.
8. Kümmerer K (2004) Resistance in the environment. J Antimicrob Chemother 54: 311–320.
9. Baquero F, Martínez J-L, Cantón R (2008) Antibiotics and antibiotic resistance in water environments. Curr Opin Biotechnol 19: 260–265.
10. Allen HK, Donato J, Wang HH, Cloud-Hansen KA, Davies J, et al (2010) Call of the wild: antibiotic resistance genes in natural environments. Nat Rev Microbiol 8: 251–259. Available: http://dx.doi.org/10.1038/nrmicro2312. Accessed 21 March 2014.
11. Stoll C, Sidhu JPS, Tiehm A, Toze S (2012) Prevalence of clinically relevant antibiotic resistance genes in surface water samples collected from Germany and Australia. Environ Sci Technol 46: 9716–9726.
12. Brown KD, Kulis J, Thomson B, Chapman TH, Mawhinney DB (2006) Occurrence of antibiotics in hospital, residential, and dairy effluent, municipal wastewater, and the Rio Grande in New Mexico. Sci Total Environ 366: 772–783.
13. Duong HA, Pham NH, Nguyen HT, Hoang TT, Pham HV, et al (2008) Occurrence, fate and antibiotic resistance of fluoroquinolone antibacterials in hospital wastewaters in Hanoi, Vietnam. Chemosphere 72: 968–973.
14. Chang X, Meyer MT, Liu X, Zhao Q, Chen H, et al (2010) Determination of antibiotics in sewage from hospitals, nursery and slaughter house, wastewater treatment plant and source water in Chongqing region of Three Gorge Reservoir in China. Environ Pollut 158: 1444–1450.
15. Reinthaler F., Posch J, Feierl G, Wüst G, Haas D, et al (2003) Antibiotic resistance of E. coli in sewage and sludge. Water Res 37: 1685–1690.
16. Kim S, Aga DS (2007) Potential Ecological and Human Health Impacts of Antibiotics and Antibiotic-Resistant Bacteria from Wastewater Treatment Plants. J Toxicol Environ Heal Part B 10: 559–573.
17. Figueira V, Vaz-Moreira I, Silva M, Manaia CM (2011) Diversity and antibiotic resistance of Aeromonas spp. in drinking and waste water treatment plants. Water Res 45: 5599–5611.
18. Michael I, Rizzo L, McArdell CS, Manaia CM, Merlin C, et al (2013) Urban wastewater treatment plants as hotspots for the release of antibiotics in the environment: A review. Water Res 47: 957–995.
19. Łuczkiewicz A, Jankowska K, Fudala-Książek S, Olańczuk-Neyman K, Łuczkiewicz A, et al (2010) Antimicrobial resistance of fecal indicators in municipal wastewater treatment plant. Water Res 44: 5089–5097. Available: http://www.sciencedirect.com/science/article/pii/S0043135410005713. Accessed 23 January 2013.
20. Li D, Qi R, Yang M, Zhang Y, Yu T (2011) Bacterial community characteristics under long-term antibiotic selection pressures. Water Res 45: 6063–6073.
21. Munir M, Wong K, Xagoraraki I (2011) Release of antibiotic resistant bacteria and genes in the effluent and biosolids of five wastewater utilities in Michigan. Water Res 45: 681–693.
22. Zhang X-X, Zhang T (2011) Occurrence, Abundance, and Diversity of Tetracycline Resistance Genes in 15 Sewage Treatment Plants across China and Other Global Locations. Environ Sci Technol 45: 2598–2604.
23. Gao P, Munir M, Xagoraraki I (2012) Correlation of tetracycline and sulfonamide antibiotics with corresponding resistance genes and resistant bacteria in a conventional municipal wastewater treatment plant. Sci Total Environ 421–422: 173–183.
24. Guardabassi L, Lo Fo Wong DM., Dalsgaard A (2002) The effects of tertiary wastewater treatment on the prevalence of antimicrobial resistant bacteria. Water Res 36: 1955–1964.
25. Moura A, Henriques I, Smalla K, Correia A (2010) Wastewater bacterial communities bring together broad-host range plasmids, integrons and a wide diversity of uncharacterized gene cassettes. Res Microbiol 161: 58–66.
26. Jury KL, Khan SJ, Vancov T, Stuetz RM, Ashbolt NJ (2011) Are Sewage Treatment Plants Promoting Antibiotic Resistance? Crit Rev Environ Sci Technol 41: 243–270.
27. Courvalin P (1994) Transfer of antibiotic resistance genes between gram-positive and gram-negative bacteria. Antimicrob Agents Chemother 38: 1447–1451.
28. Schlüter A, Szczepanowski R, Pühler A, Top EM, Schlüter A, et al (2007) Genomics of IncP-1 antibiotic resistance plasmids isolated from wastewater treatment plants provides evidence for a widely accessible drug resistance gene pool. FEMS Microbiol Rev 31: 449–477.
29. Szczepanowski R, Linke B, Krahn I, Gartemann K, Guetzkow T, et al (2009) Detection of 140 clinically relevant antibiotic-resistance genes in the plasmid metagenome of wastewater treatment plant bacteria showing reduced susceptibility to selected antibiotics. Microbiology-Sgm 155: 2306–2319.
30. West BM, Liggit P, Clemans DL, Francoeur SN (2010) Antibiotic Resistance, Gene Transfer, and Water Quality Patterns Observed in Waterways near CAFO Farms and Wastewater Treatment Facilities. Water, Air, Soil Pollut 217: 473–489.
31. Czekalski N, Berthold T, Caucci S, Egli A, Bürgmann H (2012) Increased Levels of Multiresistant Bacteria and Resistance Genes after Wastewater Treatment and Their Dissemination into Lake Geneva, Switzerland. Front Microbiol 3.
32. Pruden A, Arabi M, Storteboom HN (2012) Correlation between upstream human activities and riverine antibiotic resistance genes. Environ Sci Technol 46: 11541–11549.
33. Rizzo L, Manaia C, Merlin C, Schwartz T, Dagot C, et al (2013) Urban wastewater treatment plants as hotspots for antibiotic resistant bacteria and genes spread into the environment: A review. Sci Total Environ 447: 345–360.
34. Da Silva MF, Tiago I, Verssimo A, Boaventura RAR, Nunes OC, et al (2006) Antibiotic resistance of enterococci and related bacteria in an urban wastewater treatment plant. FEMS Microbiol Ecol 55: 322–329.
35. Levy SB (2002) Factors impacting on the problem of antibiotic resistance. J Antimicrob Chemother 49: 25–30. Available: http://jac.oxfordjournals.org/content/49/1/25.
36. Jury KL, Khan SJ, Vancov T, Stuetz RM, Ashbolt NJ (2011) Are Sewage Treatment Plants Promoting Antibiotic Resistance? Crit Rev Environ Sci Technol 41: 243–270.
37. Auerbach EA, Seyfried EE, McMahon KD (2007) Tetracycline resistance genes in activated sludge wastewater treatment plants. Water Res 41: 1143–1151.
38. Lachmayr KL, Kerkhof LJ, DiRienzo AG, Cavanaugh CM, Ford TE (2009) Quantifying Nonspecific TEM β-Lactamase (blaTEM) Genes in a Wastewater Stream. Appl Environ Microbiol 75: 203–211.
39. Zhang T, Zhang M, Zhang X, Fang HH (2009) Tetracycline Resistance Genes and Tetracycline Resistant Lactose-Fermenting Enterobacteriaceae in Activated Sludge of Sewage Treatment Plants. Environ Sci Technol 43: 3455–3460. Available: http://dx.doi.org/10.1021/es803309m. Accessed 27 March 2013.
40. Börjesson S, Mattsson A, Lindgren P-E (2010) Genes encoding tetracycline resistance in a full-scale municipal wastewater treatment plant investigated during one year. J Water Health 08: 247.
41. Boom R, Sol CJ, Salimans MM, Jansen CL, Dillen PMW, et al (1990) Rapid and simple method for purification of nucleic acids. J Clin Microbiol 28: 495–503.
42. Bates D, Bolker MM and Ben (2013) lme4: Linear mixed-effects models using S4 classes.
43. R Development Core Team (2013). R: A language and environment for statistical computing. R Foundation for Statistical Computing, Vienna, Austria. URL http://www.R-project.org/.
44. Hothorn T, Bretz F, Westfall P (2008) Simultaneous inference in general parametric models. Biom J 50: 346–363.
45. Toleman MA, Bennett PM, Walsh TR (2006) Common regions e.g. orf513 and antibiotic resistance: IS91-like elements evolving complex class 1 integrons. J Antimicrob Chemother 58: 1–6.
46. Bennett PM (2008) Plasmid encoded antibiotic resistance: acquisition and transfer of antibiotic resistance genes in bacteria. Br J Pharmacol 153: S347–S357.
47. Roberts AP, Mullany P (2009) A modular master on the move: the Tn916 family of mobile genetic elements. Trends Microbiol 17: 251–258.
48. Liu M, Zhang Y, Yang M, Tian Z, Ren L, et al (2012) Abundance and Distribution of Tetracycline Resistance Genes and Mobile Elements in an Oxytetracycline Production Wastewater Treatment System. Environ Sci Technol 46: 7551–7557.
49. McArthur AG, Waglechner N, Nizam F, Yan A, Azad MA, et al (2013) The comprehensive antibiotic resistance database. Antimicrob Agents Chemother 57: 3348–3357.
50. Koivunen J, Siitonen A, Heinonen-Tanski H (2003) Elimination of enteric bacteria in biological/chemical wastewater treatment and tertiary filtration units. Water Res 37: 690–698. doi:10.1016/S0043-1354(02)00305-6.
51. Luo Y, Guo W, Ngo HH, Nghiem LD, Hai FI, et al (2014) A review on the occurrence of micropollutants in the aquatic environment and their fate and removal during wastewater treatment. Sci Total Environ 473<unicod: 619–641. doi:10.1016/j.scitotenv.2013.12.065.
52. Gao P, Munir M, Xagoraraki I (2012) Correlation of tetracycline and sulfonamide antibiotics with corresponding resistance genes and resistant bacteria in a conventional municipal wastewater treatment plant. Sci Total Environ 421–422: 173–183.
53. Zhang T, Zhang M, Zhang X, Fang HH (2009) Tetracycline Resistance Genes and Tetracycline Resistant Lactose-Fermenting Enterobacteriaceae in Activated Sludge of Sewage Treatment Plants. Environ Sci Technol 43: 3455–3460.
54. Auerbach EA, Seyfried EE, McMahon KD (2007) Tetracycline resistance genes in activated sludge wastewater treatment plants. Water Res 41: 1143–1151.
55. Graham DW, Olivares-Rieumont S, Knapp CW, Lima L, Werner D, et al (2011) Antibiotic Resistance Gene Abundances Associated with Waste Discharges to the Almendares River near Havana, Cuba. Environ Sci Technol 45: 418–424.
56. Lillenberg M (2011) Residues of some pharmaceuticals in sewage sludge in estonia, their stability in the environment and accumulation into food plants via fertilizing residues of some pharmaceuticals in sewage sludge in estonia, their stability in the environment and accumul Estonian University of Life Sciences. Available: http://dspace.emu.ee/xmlui/bitstream/handle/10492/146/Thesis_Lillenberg_2011.pdf.
57. Clara M, Kreuzinger N, Strenn B, Gans O, Kroiss H (2005) The solids retention time-a suitable design parameter to evaluate the capacity of wastewater treatment plants to remove micropollutants. Water Res 39: 97–106.
58. Novo A, Manaia CM (2010) Factors influencing antibiotic resistance burden in municipal wastewater treatment plants. Appl Microbiol Biotechnol 87: 1157–1166. Available: http://www.ncbi.nlm.nih.gov/pubmed/20396880.

59. Nõlvak H, Truu M, Tiirik K, Oopkaup K, Sildvee T, et al (2013) Dynamics of antibiotic resistance genes and their relationships with system treatment efficiency in a horizontal subsurface flow constructed wetland. Sci Total Environ 461–462: 636–644.
60. Edwards U, Rogall T, Blöcker H, Emde M, Böttger EC (1989) Isolation and direct complete nucleotide determination of entire genes. Characterization of a gene coding for 16S ribosomal RNA. Nucleic Acids Res 17: 7843–7853.
61. López-Gutiérrez JC, Henry S, Hallet S, Martin-Laurent F, Catroux G, et al (2004) Quantification of a novel group of nitrate-reducing bacteria in the environment by real-time PCR. J Microbiol Methods 57: 399–407.
62. Tamminen M, Karkman A, Lohmus A, Muziasari WI, Takasu H, et al (2010) Tetracycline Resistance Genes Persist at Aquaculture Farms in the Absence of Selection Pressure. Environ Sci Technol 45: 386–391.
63. Pei R, Kim S-C, Carlson KH, Pruden A (2006) Effect of River Landscape on the sediment concentrations of antibiotics and corresponding antibiotic resistance genes (ARG). Water Res 40: 2427–2435.
64. Heuer H, Smalla K (2007) Manure and sulfadiazine synergistically increased bacterial antibiotic resistance in soil over at least two months. Environ Microbiol 9: 657–666.

22

Coliform Bacteria as Indicators of Diarrheal Risk in Household Drinking Water

Joshua S. Gruber*, Ayse Ercumen, John M. Colford, Jr.

Division of Epidemiology, University of California, Berkeley, California, United States of America

Abstract

Background: Current guidelines recommend the use of *Escherichia coli* (EC) or thermotolerant ("fecal") coliforms (FC) as indicators of fecal contamination in drinking water. Despite their broad use as measures of water quality, there remains limited evidence for an association between EC or FC and diarrheal illness: a previous review found no evidence for a link between diarrhea and these indicators in household drinking water.

Objectives: We conducted a systematic review and meta-analysis to update the results of the previous review with newly available evidence, to explore differences between EC and FC indicators, and to assess the quality of available evidence.

Methods: We searched major databases using broad terms for household water quality and diarrhea. We extracted study characteristics and relative risks (RR) from relevant studies. We pooled RRs using random effects models with inverse variance weighting, and used standard methods to evaluate heterogeneity and publication bias.

Results: We identified 20 relevant studies; 14 studies provided extractable results for meta-analysis. When combining all studies, we found no association between EC or FC and diarrhea (RR 1.26 [95% CI: 0.98, 1.63]). When analyzing EC and FC separately, we found evidence for an association between diarrhea and EC (RR: 1.54 [95% CI: 1.37, 1.74]) but not FC (RR: 1.07 [95% CI: 0.79, 1.45]). Across all studies, we identified several elements of study design and reporting (e.g., timing of outcome and exposure measurement, accounting for correlated outcomes) that could be improved upon in future studies that evaluate the association between drinking water contamination and health.

Conclusions: Our findings, based on a review of the published literature, suggest that these two coliform groups have different associations with diarrhea in household drinking water. Our results support the use of EC as a fecal indicator in household drinking water.

Editor: Sudha Chaturvedi, Wadsworth Center, United States of America

Funding: The authors have no funding or support to report.

Competing Interests: The authors have declared that no competing interests exist.

* Email: jsgruber@gmail.com

Introduction

Globally, drinking water has been established as a primary transmission pathway for diarrhea pathogens. [1,2] In industrialized countries, centrally treated drinking water distribution systems have largely eliminated outbreaks of waterborne diseases, such as typhoid fever and cholera. [3] In developing countries, there is a large body of evidence that improving the microbial quality of drinking water by household treatment and safe storage reduces diarrhea. [4–6] Yet, evidence directly linking diarrheal illness to measured fecal contamination in drinking water remains inconclusive. [3,7,8]

In general, it is not feasible to test water for all known waterborne pathogens to assess whether it is safe for drinking. [2,9–11] Instead, since the early 1900s there has been heavy reliance on fecal indicator organisms as measures of drinking water quality. [1,12] Current World Health Organization (WHO) guidelines recommend *Escherichia coli* (EC) and/or thermotolerant ("fecal") coliforms (FC) as indicators of the effectiveness of disinfection processes, and as index organisms for the potential presence of fecal contamination and waterborne pathogens; [2,10,12,13] previous WHO guidelines use categories of EC and FC concentrations to define levels of disease risk from drinking water. [14] While EC are considered the most suitable indicator organism due to their specificity to fecal sources of contamination, FC are also recommended as an acceptable surrogate; this recommendation exists despite the recognition that the FC group includes coliform species of environmental origin, and is therefore not likely specific to fecal contamination. [2,12,13] Despite heavy reliance on EC and FC to assess the microbiological safety of drinking water, it has yet to be shown that either of these specific indicators is associated with waterborne illness. [8]

A previous systematic review and meta-analysis evaluated the evidence for the link between household drinking water quality, measured by fecal indicator organisms, and health; [8] the authors found no evidence of an association between diarrhea and indicators of drinking water contamination (EC, FC and *fecal streptococci*). This review, however, was limited by the availability, size and quality of articles published at the time the search was conducted (2001). [8] Specifically, the previous review included only three relatively small studies that reported using EC, and therefore focused mostly on FC as a measure of water quality. Given the limited number of studies available to the authors, they were also unable to evaluate the performances of EC and FC separately. Since the previous review, several studies have been published using EC as a measure of water quality (n = 6) [15–21], motivating our updated review with this newly available evidence. In addition, this larger body of evidence allows us to evaluate the performance of EC and FC separately; we hypothesized that the two coliform groups would have different associations with diarrhea given their different specificities for fecal contamination. Given the widespread use of EC and FC as measures of water quality by researchers and policy makers, and the limited evidence regarding the association of these proxy measures with actual disease outcomes, we believe that our systematic review and meta-analysis will make a significant contribution to the body of knowledge on the performance of indicator organisms in household drinking water, and have implications for their recommended use for future guidelines and research. In addition, based on the findings of our systematic review, we are able to make several recommendations for the design and reporting of future studies that assess the relationship between drinking water contamination and health.

Methods

Search Strategy

We searched MEDLINE, EMBASE, and Web of Knowledge databases for relevant articles. We used broad search terms for water, water quality and indicator organisms, diarrhea and household (point-of-use) sampling (Table 1). Titles and abstracts from the search were examined, and the full texts of relevant articles were reviewed. The bibliographies of relevant articles identified from full-text reviews were scanned to identify additional relevant studies. Searches were limited to articles published in peer-reviewed journals, in English, Spanish, German or Turkish (languages spoken by the authors). No restrictions were placed on date of publication. The final search was conducted on February 27, 2013. A protocol was not registered for this systematic review.

Inclusion and Exclusion Criteria

We reviewed articles that reported data on the association between exposure to *Escherichia coli* (EC) or thermotolerant ("fecal") coliforms (FC) in household drinking water, and the occurrence of diarrhea under non-outbreak conditions; we excluded other, less commonly used indicator organisms (e.g., total coliforms, *fecal streptococci*). Our inclusion criterion was that the study collected exposure and outcome data at the household level or at the point of use; studies were excluded if they only reported measures of source water quality (e.g., shared wells, distribution system "nodes"), or used ecologic outcome data (e.g., health ministry reports). No further restrictions were placed on: definitions of diarrhea, recall period, study design, age groups, study location (developed/developing country), study setting (urban/rural), or drinking water source (as long as exposure to indicator bacteria was measured at the point of use).

Study Reporting and Data Extraction

For each relevant study two authors (JG and AE) independently extracted basic study characteristics, including year of publication, sample size, age groups, study location, setting and design, indicator organism and enumeration methods, and diarrhea definition and enumeration methods into standardized forms, and resolved any discrepancies. Where possible, a relative risk (RR) such as odds ratio, cumulative incidence ratio, incidence density ratio or prevalence ratio was extracted from relevant studies. When no effect measure was reported by the authors, raw data were extracted to calculate an appropriate RR and confidence interval using standard methods. [22] If authors reported both raw data and unadjusted RRs we confirmed the reported estimates using the raw data; we used the RRs reported by the authors, unless otherwise noted. If authors reported results for both indicators (EC and FC) the results were treated as two separate studies in the meta-analysis. All RRs were extracted such that a value greater than unity indicated increased risk of illness among the group exposed to contaminated water. We also noted whether authors controlled for additional variables (by study design, stratified analysis or regression methods) and whether they accounted for correlation (clustering) of outcomes in their analysis.

Exposure Thresholds

There are multiple ways to define comparison groups in water quality studies. For example, authors often use a threshold to define exposed and non-exposed groups based on indicator organism counts measured in 100 ml samples of water (i.e., exposed: ≥10 EC/100 ml; non-exposed: <10 EC/100 ml; with RRs calculated to compare these two groups); however, threshold levels can vary between studies. Current WHO guidelines recommend that drinking water that is safe for human consumption should have no detectable EC or FC in any 100 ml sample (i.e., a 1 EC/FC threshold); however, there is also evidence from the literature that much higher indicator organism concentrations are required to cause disease. For example, a study conducted in the Philippines on source water quality [23] previously reported a "threshold effect", where no association with diarrhea was found at the 1 EC, 10 EC or 100 EC cutoffs, but an association was found at the 1000 EC cutoff; it is plausible that a similar threshold effect could be present for point-of-use drinking water. Given the limited number of studies included in the previously published systematic review and meta-analysis, the authors made no attempt to differentiate indicator organism-diarrhea associations at different threshold levels. [8] In our review, we attempted to extract RRs for different thresholds at the 1 EC or FC, 10 EC or FC, 100 EC or FC and 1000 EC or FC levels, when the authors reported these data.

Categorical Exposures

As an alternative approach to exposure categorization, previous WHO guidelines defined disease risk from drinking water based on categories of indicator organism counts measured in 100 ml samples of water: 0 EC or FC, safe; 1–10 EC or FC, low risk; 11–100 EC or FC, intermediate risk; 101–1000 EC or FC, high risk; > 1000 EC or FC, very high risk. [14] Investigators commonly use these guidelines to define exposure categories in studies of drinking water quality; each category is treated as an independent exposure group, and RRs are calculated for each elevated category relative to the 0 EC or FC exposure group. In our review, we extracted RRs for these categorical exposures to explore the presence of a dose-response (increased disease risk associated with increasing exposure levels) and to evaluate previous WHO guidelines that defined health risk based on these exposure categories. Within

Table 1. Systematic Review Search Terms. [a]

Water Exposure	Household Sampling	Disease Outcome
water quality	household	Diarrhea
water microbiology	point of use	Diarrheoa
water pollution	Pou	gastrointestinal illness
water contamination	point-of-use	gastrointestinal disease
water supply	drinking water	gastrointestinal infection
drinking water	consumption	Dysentery
"indicator bacteria"	tap water	"HCGI"
"indicator organism"	well water	"highly credible gastrointestinal illness"
"thermotolerant coliforms"	domestic	"AGI"
"thermo-tolerant coliforms"		"AGII"
"fecal coliforms"		"acute gastrointestinal illness"
escherichia coli		
"e. coli"		
"fecal bacteria"		
"fecal contamination"		
"microbiological indicators"		

[a]Search terms combined using Boolean logic: within column terms combined using "or" statements; "and" statements were used to combine terms between columns.

studies that evaluated dose-responses, we extracted *p*-values for tests of trend. In studies that did not report *p*-values for tests of trend but reported results for different categorical exposure levels, we extracted the appropriate data, where possible, and conducted our own tests of trend using previously published methods that estimate a log-linear dose-response while accounting for correlations between RRs from a single study (*glst* command: STATA 12, STATA Corp, College Station Texas). [24]

Meta-analysis

For consistency with the previously published review, we conducted meta-analyses to calculate a summary measure pooled across all studies which provided an extractable RR, combining all EC and FC studies. [8] The previous review was not explicit about which results were used for meta-analyses if data could be extracted at multiple thresholds. In our review, we estimated a summary measure across all studies using the lowest extractable threshold from each study (e.g., if a study reported raw data that could be combined at the 1 EC, 10 EC and 100 EC thresholds, we included the RR calculated for the 1 EC threshold in the main meta-analysis). In addition, we conducted a sub-group analysis to explore the type of indicator organism used in each study as a source of heterogeneity in the main analysis; for this sub-group analysis we stratified studies by indicator used (EC and FC) and re-pooled summary measures separately for each indicator. In a secondary meta-analysis, we explored evidence for a threshold effect in household drinking water (see "Exposure Thresholds" above). For this analysis, we restricted meta-analyses to studies that reported extractable RRs at the 1 EC, 10 EC and 100 EC thresholds, respectively – this analysis was only feasible for studies that reported extractable data for all three threshold levels.

Summary measures were calculated by meta-analysis using random effects models with inverse variance weighting (STATA 12). [25] Heterogeneity across studies was evaluated using the Mantel-Haenztel χ^2 test; we considered a *p*-value on the χ^2 statistic <0.2 as evidence of heterogeneity. We used funnel plots to evaluate publication bias; plot asymmetry was interpreted as evidence of "small study bias", for which publication bias is a likely contributor. [26,27]

Results

Systematic Review

Database searches were completed on February 27, 2013, from which we identified 5,801 titles and abstracts; 20 relevant articles were identified for review (Figure 1). Of the 20 relevant articles, 14 presented data in an extractable format (see Methods) and were included in meta-analyses. One article [28] presented results for both EC and FC; these results were treated as separate studies in the meta-analyses (resulting in 15 total studies).

Study Characteristics: All Studies

The 20 relevant studies were published between 1977 and 2012 (Table 2). Two studies were conducted in a developed country (Canada), the rest in low-resource settings; no studies were conducted among populations with access to centralized water-treatment and distribution systems. The study settings included a mix of urban/peri-urban (n = 7) and rural (n = 13), and the study populations included a range of age groups (Table 2). The included articles used cohort (n = 14) and case-control (n = 6) study designs (Tables 3 and 4). We identified one randomized control trial [16] that reported data on the direct association between water quality and diarrhea; we considered this study a cohort design for the purposes of the review since water quality was not randomized. We note that, while cohort designs traditionally measure an exposure and then follow a population over time to measure outcomes, the cohort designs we identified in this review measured exposure (EC or FC): (i) simultaneously with outcomes through cross-sectional sampling; (ii) at intervals during ongoing diarrhea surveillance, or; (iii) the timing and analysis of water samples relative to diarrhea measurement was mixed or not clearly reported (Tables 3 and 4). The included case-control studies measured exposure after disease occurrence by design; however one study [19] utilized a nested case control design with

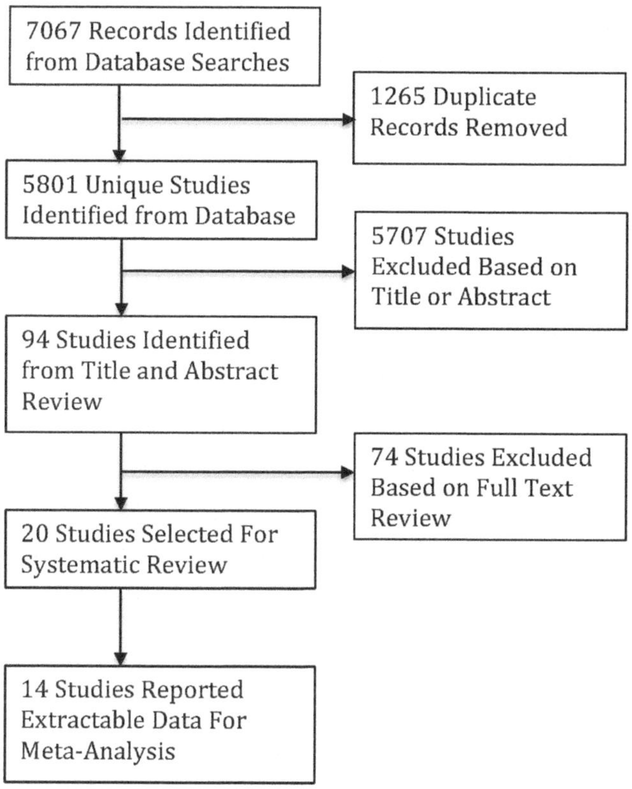

Figure 1. Systematic review flow chart.

risk set sampling in which disease and EC exposure were measured on the same day.

Included studies measured EC (n = 9), FC (n = 10), or both (n = 1) in drinking water samples collected at the point of use. Overall, studies varied in their reporting of enumeration and incubation methods, and we could not always confirm the indicator organism enumerated based on the reported information (Tables 3 and 4). In at least one study, authors reported using EC, but a description of their methods suggested they measured FC. [29] With this exception, we relied on the authors' reporting or deferred to the categorization used by the previous review. [8]

Most studies used a definition of diarrhea that is consistent with current recommendations: three or more loose or watery stools in a 24-hour period (Tables 3 and 4). [30,31] Four studies included the presence of blood or mucus in stool as part of their definition of diarrhea, [15,17,18,32] and two studies [15,17] reported dysentery as an outcome separate from "general" diarrhea; we did not include dysentery as a separate outcome in our review. Cohort studies in low-resource settings relied primarily on household surveillance and self-reported symptoms (e.g., maternal recall) for outcome assessment, using a range of recall periods (daily, up to one month; Tables 3 and 4); the two Canadian studies relied on monthly calendars maintained in the household and reported by telephone. Case-control studies mostly identified cases through hospitals and clinics, and selected controls from community samples; however, two studies identified cases and controls based on concurrent [19] or prior [33] household diarrhea surveillance conducted by the investigators.

Among the studies identified in our review, eight studies reported results that controlled for potential confounders; the other 12 studies found no evidence for confounding, or did not report attempting to control for confounding (Tables 5 and 6) – we did not differentiate between these studies in meta-analyses. In general, reporting on the independence of observations or methods to account for clustered outcomes was poor (Tables 5 and 6). Only five of the 20 studies reported using methods to account for correlated outcomes (e.g., generalized estimating equations; GEE). In general, the RRs and confidence intervals we calculated from available raw data were consistent with those reported by the authors, whether or not they reported accounting for correlated outcomes. However, in one of these studies, that reported using GEE, we calculated confidence intervals from extracted raw data that were substantially more conservative than those reported by the authors; [15] we included the more conservative confidence intervals for this study in our meta-analyses (Table 5).

Disease-Indicator Associations

Tables 5 and 6 summarize results from EC and FC studies, respectively. Among the ten EC studies, eight reported data that quantified the relationship between indicator exposure and diarrhea (Table 5). Of the eight studies with quantitative results, six reported threshold results at the 1 EC/100 ml cutoff [15,18,20,21,28,33] and all six reported RRs greater than unity; however, only one study was able to rule random error out as an explanation for their results (Table 5). [15] The remaining two studies, among the eight that quantified a relationship between EC and diarrhea, reported significant relationships between diarrhea and linear increases in exposure to mean log10 EC concentrations; the results were very consistent across these two studies (Table 5). [16,19] The two studies that did not have extractable, quantifiable results reported that no "association" or "correlation" was found between EC and diarrhea. [17,34] However, we could not confirm whether the authors' conclusions were based on the size of the effect estimate (i.e., similar to unity), precision (i.e., "non-significant" confidence intervals or *p*-values), or both.

Among the 11 FC studies (including FC results from Genthe et al. 1997 as separate from their EC results), nine studies reported data that quantified the relationship between indicator exposure and diarrhea (Table 6). Results from these nine FC studies were less consistent than results from EC studies. Using the lowest extractable threshold, five studies reported estimates that suggest increases in the risk of diarrhea with exposure to FC [32,35–38], and four reported effects equivalent to unity or reduced diarrhea risk with exposure to FC. [28,39–41] Henry and Rahim (1990) reported FC effects for two distinct study areas differentiated by "adequate" versus "poor" sanitation infrastructure; while a single summary measure for this study suggested an increased risk of diarrhea with exposure to FC, the results from the two separate study populations were very different (Table 6; FC counts were associated with an increase in illness in the area with improved sanitation and a reduction in illness in the area with unimproved sanitation). Of the two FC studies that did not report extractable data, results were also mixed: one reported no difference in FC isolation from water in case and control households, [29] the other reported that households with diarrhea had higher overall geometric mean FC levels compared to control households. [42]

Summary Measures

For consistency with the previous review, we conducted meta-analyses across all studies (combining both EC and FC studies), using the lowest extractable threshold for each study (Tables 5 and 6). Across all studies, our summary estimate suggests a non-significant increase in diarrhea risk from exposure to drinking water contaminated with EC or FC (RR 1.26 [95% CI: 0.98,

Table 2. Selected Characteristics of Included Studies.

First Author	Year	Location	Setting	Study Size[a]	Ages	Indicator
deAceituno	2012	Honduras	Rural	1020	All	EC
Levy	2012	Ecuador	Rural	115	NR	EC
Gundry	2009	S. Africa/Zimbabwe	Rural	254	1–2 y	EC
Brown	2008	Cambodia	Rural	1196	All	EC
Jensen	2004	Pakistan	Rural	209	<5 y	EC
Bhargava	2003	Bangladesh	Rural	99	1–10 y	FC
Strauss	2001	Canada	Rural	647	All	EC
Raina	1999	Canada	Rural	531	All	EC
Genthe	1997	S. Africa	Urban	316	"Pre-school"	Both
Jagals	1997	S. Africa	Urban	100	NR	FC
Vanderslice	1993	Philippines	Urban	254	<2 y	FC
Knight	1992	Malaysia	Rural	196	4–59 m	FC
Han	1991	Myanmar	Urban	208	6–29 m	FC
Henry	1990	Bangladesh	Rural	92	6–18 m	FC
Henry and Rahim	1990	Bangladesh	Urban	137	1–6 y	FC
Echeverria	1987	Thailand	Urban	NR	<5 y	FC[b]
Esrey	1986	Lesotho	Rural	545	1–60 m	FC
Lloyd-Evans	1984	Gambia	Urban	20	6–36 m	EC
Black	1982	Bangladesh	Rural	40	5–18 m	EC
Rajasekaran	1977	India	Rural	1091	<5 y	FC

EC: Escherichia Coli; FC: Fecal Coliform.
[a]May not reflect the analytic sample used by the authors;
[b]Authors report EC but methods suggest FC.

1.63]; Figure 2). However, we also found evidence of significant heterogeneity ($X^2 = 36.13$, degrees of freedom (d.f.) $= 12$, $p < 0.001$), and evidence of publication bias (Figure 3).

In order to explore a potential source of heterogeneity across all studies, we stratified results by indicator organism (EC vs. FC) and re-estimated summary measures in these sub-groups. Across all EC studies, the summary measure suggests a significant association between exposure to EC in household drinking water and diarrhea (RR: 1.54 [95% CI: 1.37, 1.74]); Figure 4). While we found no evidence of heterogeneity across EC studies ($X^2 = 1.65$, d.f. = 5, p = 0.90), and minimal evidence of publication bias (Figure 5), we note that the EC summary measure was disproportionately influenced by one study [15] ("% weight", Figure 4). We therefore conducted a sensitivity analysis by repeating the EC analysis without the study by Brown et al. Removing this study did not impact the magnitude of the summary measure, but did reduce the precision of the estimate (RR 1.48 [95% CI: 1.02, 2.15]); Figure S1); there was no evidence of heterogeneity after removing Brown et al. from the analysis ($X^2 = 1.59$, d.f. = 4, p = 0.81). Across all FC studies, our summary effect measure suggests no association between FC exposure in drinking water and diarrhea (RR: 1.07 [95% CI: 0.79, 1.45]); Figure 4); we also found evidence of significant heterogeneity ($X^2 = 2.30$, d.f. = 6, p = 0.06), and publication bias (Figure 6).

Threshold Effects

Two EC studies provided extractable data that allowed us to explore threshold effects at three different levels (1 EC, 10 EC, and 100 EC per 100 ml; Table 5) [15,18]. RRs were calculated by comparing outcomes in the exposed group (≥EC threshold) to outcomes in the control group (<EC threshold). In one study, [15] increasing the threshold from 1EC to 10EC suggested a slight increase in diarrhea risk; there was no further increase in risk when the threshold was increased to 100EC (Table 5). In the other study, increasing the threshold beyond 1EC did not increase the magnitude of the effect estimates (Table 5). [18] We pooled estimates across these two studies, and found no evidence that increasing thresholds beyond WHO drinking water guidelines (1 EC/100 mL) was associated with an increased risk of diarrhea (Figure 7). [2] A third EC study reported finding no threshold effect at 1000 EC/100 ml, but did not report a RR. [19]

Categorical Exposure and Dose-Response

Four EC studies reported investigating a dose-response relationship. [15,18,19,21] Only three of these studies evaluated the presence of a dose-response using similar exposure categories (<1 EC (referent), 1–10 EC, 11–100 EC, 100+, or 101–1000 EC and 1000+; Table 5). Of these three studies, one did not report extractable data, but reported finding no evidence of a dose-response effect. [19] Results from the other two studies [15,18] suggest a qualitative dose response (increased risk of diarrhea with increasing exposure categories; Table 5); formal tests of trend, however, had mixed results (Brown et al: $p<0.001$ provides evidence of a dose response; Jensen et al: p = 0.26 does not provide evidence of a dose response). A fourth study [21] used different EC exposure categorizations. The results from this study qualitatively suggested increasing risk across 0.1–1.5 EC and 1.6–700 EC exposure categories; however, the authors found no evidence of a linear trend (p = 0.45; Table 5).

Table 3. Summary of EC Study Exposure and Outcome Measurements.

First Author	Year	Study Design	Outcome [a]	Recall	Measured EC Relative to Outcome	Enumeration Method	Medium	Temp (°C)[f]
deAceituno	2012	RCT	Diarrhea	7-day	After (<1 week)	MPN	IDEXX Colilert	NR
Levy	2012	CC	Diarrhea	Daily	Same Day	MF	BD Difco ml agar[e] (EPA 1604)	30
Gundry	2009	Cohort	Diarrhea	Daily	Before (<2 weeks)	MPN	IDEXX Colilert	NR
Brown	2008	Cohort	Diarrhea	7-day	After (<1 week)	MF	"selective medium"[d]	NR
Jensen	2004	Cohort	Diarrhea	7-day	Before/After (<3 days)	MF	Hach m-ColiBlue 24 Broth PourRite	35
Strauss	2001	Cohort	AGII[b]	Monthly Diary	Twice over 28-d surveillance	MF	m-FC-BCIG agar	44.5
Raina	1999	Cohort	GI Illness[c]	Monthly Diary	5 times over 1-y surveillance	NR	"standard methods" (MOE 1992)	NR
Genthe	1997	CC	Diarrhea	Hospital	After	NR	"standard methods" (APHA 1989)	NR
Lloyd-Evans	1984	CC	Diarrhea	7-day	After (1–2 months)	MF	Millipore (DHSS 1969)	NR
Black	1982	Cohort	Diarrhea	NR	After (<1 month)	Plating	MacConkey's Agar	37

EC: *Escherichia Coli*; CC: Case Control; MF: Membrane Filtration; MPN: Most Probable Number; NR: Not Reported; C: Celsius; DHSS: Department of Health & Social Security: *The bacteriological examination of water* supplies; APHA: American Public Health Association: *Standard Methods for the Examination of Water and Wastewater*; MOE: *Ministry of Environment: Ontario's Drinking Water Objectives*.
[a]Unless noted, used a definition consistent with World Health Organization recommendation of 3 or more loose/watery stools in 24 hour period;
[b]Acute Gastrointestinal Illness: 1) vomiting or liquid diarrhea or 2) nausea or soft, loose diarrhea combined with abdominal cramps;
[c]Diarrhea with or with out vomiting;
[d]selective medium: "containing chromogenic and fluorogenic b-glucuronide and b-galactoside substrates";
[e]Multiple EC enumeration methods used – we report the method with the largest sample size and for which the authors report an extractable RR;
[f]Incubation temperature.

Table 4. Summary of FC Study Exposure and Outcome Measurements.

First Author	Year	Study Design	Outcome [a]	Recall	Measured FC Relative to Outcome	Enumeration Method	Medium	Temp (°C)[e]
Bhargava	2003	Cohort	GI Morbidity [b]	1-month	After (<1 month)	MF	*DIFCO* m-FC media	45
Genthe	1997	CC	Diarrhea	Hospital	After	NR	"standard methods" (APHA 1989)	NR
Jagals	1997	CC	Diarrhea [c]	Clinic	After	MF	"standard methods" (APHA 1992)	44.5
Vanderslice	1993	Cohort	Diarrhea [c]	7-day	2–5 times over 1 year	MF	m-FC agar (APHA 1985)	44.5
Knight	1992	CC	Diarrhea	Clinic	After (4–14 days)	MF	*Paqualab* (ELE International 1986)	NR
Han	1991	Cohort	Diarrhea	7-day	Averaged over surveillance	MF	m-FC agar (WHO 1983)	NR
Henry	1990	Cohort	Diarrhea	7-day	Averaged over surveillance	NR	"standard methods" (APHA 1981)	NR
Henry and Rahim	1990	Cohort	Diarrhea	14-day	Unclear (<6 months)	dip-slides	*ORION Hygicult* Agar Slides [d]	37
Echeverria	1987	CC	Diarrhea	Hospital	After	MF	m-FC agar (APHA 1975)	44
Esrey	1986	Cohort	Diarrhea [c]	24-hour	Unclear	MF	Faecal Coliform Broth (AJPH 1975)	NR
Rajasekaran	1977	Cohort	Diarrhea	NR	Unclear (<15 days)	MPN	IMViC (APHA 1965)	44

FC: Fecal Coliform; CC: Case Control; GI: Gastrointestinal; MF: Membrane Filtration; MPN: Most Probable Number; NR: Not Reported; C: Celsius; APHA: American Public Health Association: *Standard Methods for the Examination of Water and Wastewater*; WHO: World Health Organization: *Guidelines for Drinking Water Quality*; IMViC: Indole, methyl red, Voges–Proskauer and citrate [a] [a] Unless otherwise noted, study used a definition consistent with current World Health Organization recommendation of 3 or more loose/watery stools in a 24 hour period; [b] Morbidity included: diarrhea, dysentery, vomiting, stomachache, acidity, typhoid, cholera; [c] Did not report a definition; [d] Report enumerating *Enterobacteriaceae* – included as FC for consistency with previous review. [e] Incubation temperature

Table 5. EC Study Results.

First Author	Year	Adjusted Analysis	RR	Threshold Results [a] RR (95% CI)	Categorical/Dose Response: [b]RR (95% CI)	Alternate Results: RR (95% CI)
deAceituno	2012	Yes	OR [c]	-	-	Log10 increase: 1.26 (1.08,1.46)
Levy	2012	NR	OR [c]	1000 EC: "no effect"	no "dose-response" (1–10 EC, 11–100 EC, 101–1,000 EC, 1,000+ EC)	Log10 increase: 1.29 (1.02,1.65)
Gundry	2009	Yes	NR	-	-	"no association"
Brown [d]	2008	No	PR [c]	1 EC: 1.55 (1.36,1.76) 10 EC: 1.70 (1.52,1.90) 100 EC: 1.67 (1.51,1.84) 1000 EC: 1.37 (1.21,1.54)	1–10 EC: 0.98 (0.81–1.20) 11–100 EC: 1.35 (1.13–1.61) 101–1000 EC: 1.83 (1.58–2.11) 1001+EC: 1.81 (1.54–2.11) [e]	-
Jensen	2004	No	IDR	1 EC: 1.32 (0.77,2.27) 10 EC: 1.25 (0.83,1.88) 100 EC: 1.23 (0.85,1.80)	1–10 EC: 1.17 (0.55–2.47) 11–100 EC: 1.26 (0.70–2.28) 100+EC: 1.45 (0.81–2.59) [f]	-
Strauss	2001	Yes	OR [c]	1 EC: 1.52 (0.33,6.92)	0.1–1.5 EC: 0.85 (0.10,7.19) 1.6–700 EC: 2.69 (0.34,21.56) [g]	-
Raina	1999	Yes	OR [c]	1 EC: 2.11 (0.90,4.94)	-	-
Genthe	1997	Yes	OR	1 EC: 1.26 (0.59,2.72)	-	-
Lloyd-Evans	1984	No	OR	1 EC: 3.86 0.33,45.57)	-	-
Black	1982	NR	NR	-	-	"no correlation found"

EC: *Escherichia Coli*; RR: Relative Risk; OR: Odds Ratio; PR: Prevalence Ratio; IDR: Incidence Density Ratio; NR: Not Reported; CI: Confidence Interval;
[a]Compares risk in groups exposed to ≥EC value, to <EC value;
[b]Reference group is <1 EC for all categories;
[c]Authors report using methods to account for correlated (clustered) outcomes;
[d]RRs and 95% CIs calculated from raw data;
[e]*p*-value for linear trend <0.01;
[f]p-value for linear trend =0.26;
[g]p-value for linear trend =0.45.

Table 6. FC Study Results.

First Author	Year	Adjusted Analysis	RR	Threshold Results [a] RR (95% CI)	Alternate Results RR (95% CI)
Bhargava	2003	Yes	NA	-	ML Coefficient: 0.204 (SE: 0.08) *p-value* <0.05
Genthe	1997	Yes	OR	1 FC: 0.84 (0.46,1.52)	
Jagals	1997	NR	NA	-	Geometric mean FC higher in homes with diarrhea (compared to control homes)
Vanderslice	1993	Yes	NA	-	Probit coefficient: −0.002 *p-value* >0.10
Knight	1992	Yes	OR	1 FC: 1.45 (0.57,4.76)	-
Han	1991	No	PR	Medium: 0.72 (0.56,0.94) [b]	Medium-High: 0.73 (0.52,1.02) [c] High: 0.72 (0.52,1.00) [c]
Henry	1990	No	IDR	Low: 1.03 (0.75,1.42) [b]	-
Henry and Rahim	1990	No	CIR	10^4 FC: 1.26 (0.59,2.72) [d]	Improved Area: 2.58 (0.70,9.54) [d] Unimproved Area: 0.86 (0.34–2.18) [d]
Echeverria	1987	NR	NR	-	FC isolated from water as often in homes with diarrhea as in control homes
Esrey	1986	No	PR	10 FC: 1.53 (0.69,3.40) 100 FC: 1.46 (0.62,3.45)	10–100 FC: 1.41 (0.53,3.76) [e]>100 FC: 1.64 (0.64,4.18) [e]
Rajasekaran	1977	No	PR	10 FC: 2.93 (1.10,7.81)	-

FC: Fecal Coliform; RR: Relative Risk; OR: Odds Ratio; PR: Prevalence Ratio; IDR: Incidence Density Ratio; NR: Not Reported; CI: Confidence Interval; ML: maximum likelihood
[a]Compares risk in groups exposed to ≥FC value, to <FC value;
[b]Based on median of samples or arbitrary cutoffs;
[c]Compared to "Low" group – Authors do not report specific FC concentrations used to determined categories;
[d]Reference group is <1000 colony forming units FC per gram;
[e]Reference group <10 FC.

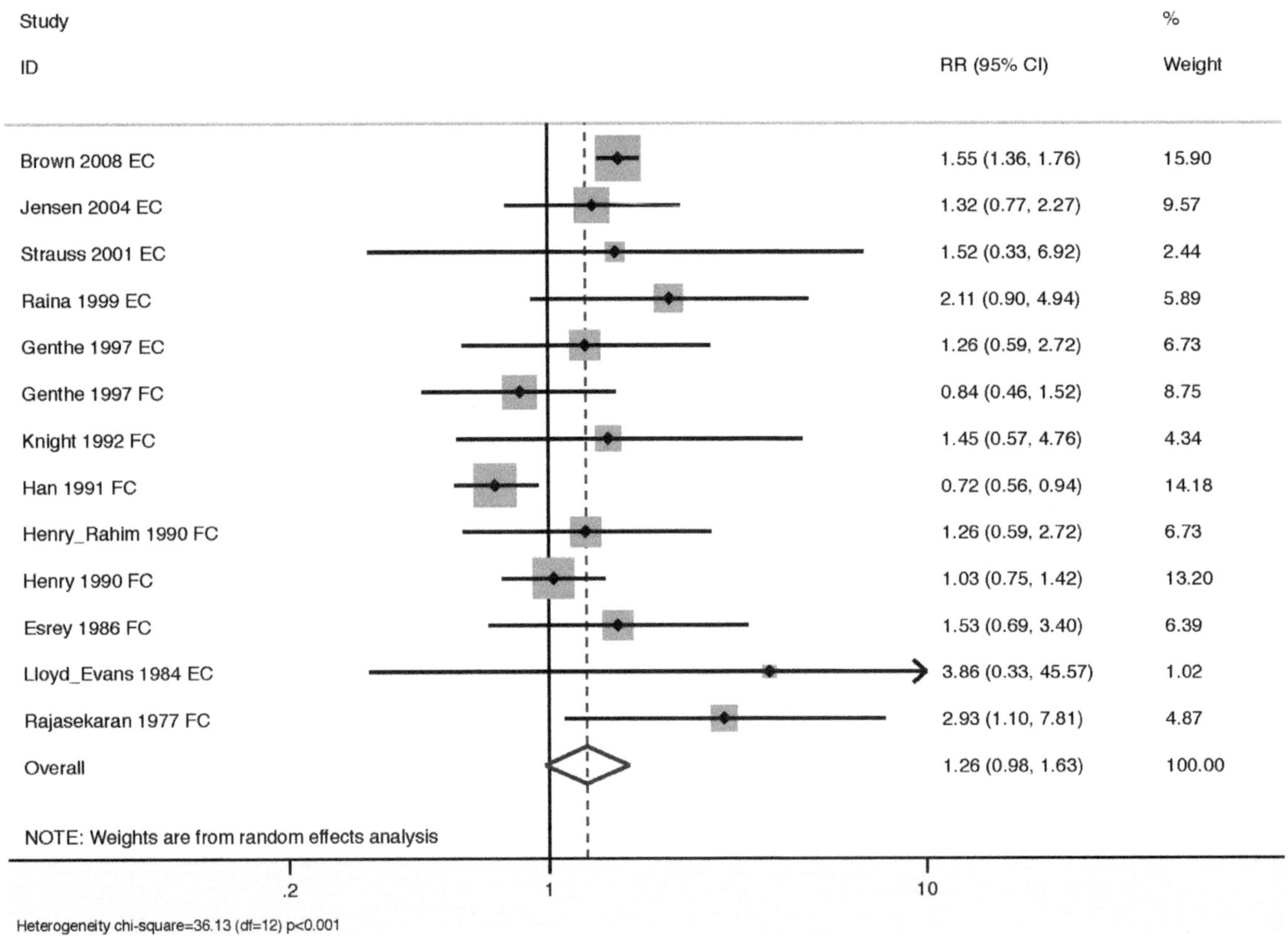

Figure 2. Forest plot for all included studies.

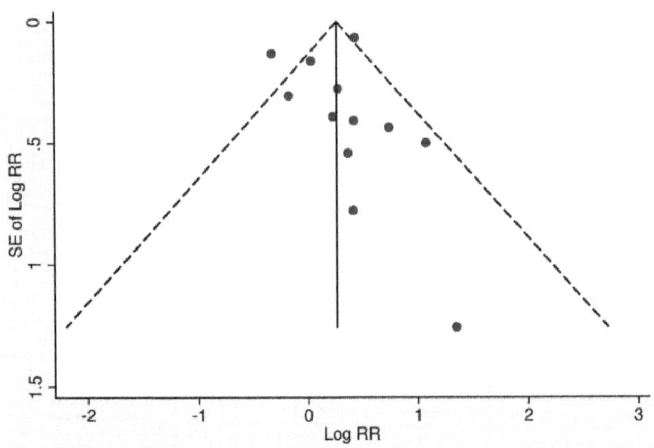

Figure 3. Funnel plot for all included studies (EC and FC studies). Studies are plotted relative to size (SE of Log RR), and reported log relative risk (Log RR); studies higher on the vertical axis are larger, and studies further to the right have larger RRs. The solid vertical line is the log of the summary RR for all studies, and the dotted lines are "pseudo 95% confidence intervals" for the summary measure. The absence of smaller studies with small or null RRs (lower left region of the plot) creates plot asymmetry, and provides evidence of possible publication bias.

Discussion

Summary of Findings

We conducted a systematic review and meta-analysis to evaluate the relationship between diarrheal illness and the presence of EC or FC indicators in household drinking water. Consistent with a prior review, [8] we did not find conclusive evidence for an association between contaminated drinking water and diarrhea when EC and FC results were combined (previous review: RR 1.12 [95% CI: 0.85, 1.48]; our review: RR 1.26, [95% CI: 0.98, 1.63]; Figure 2). We also found evidence of significant heterogeneity and publication bias across all studies (Figure 2, Figure 3). After stratifying studies by EC and FC to explore the choice of indicator organism as a source of heterogeneity, we found that EC studies reported consistent effect estimates that suggest an increased risk of diarrhea with exposure to contaminated drinking water with no evidence of heterogeneity (pooled RR 1.54 [95% CI: 1.37, 1.74]; Table 5, Figure 4). In contrast, FC studies reported mixed results with significant heterogeneity and provided minimal evidence of an association with diarrhea (pooled RR 1.07 [95% CI: 0.79, 1.45]; Table 6, Figure 4). We found no evidence that exposure thresholds greater than 1 EC/100 mL (e.g., 10 EC/100 mL, 100 EC/100 mL) were associated with increased diarrheal disease, but this evidence is based on a limited number of studies. Results regarding a dose-response relationship between EC in household drinking water and diarrhea were limited and

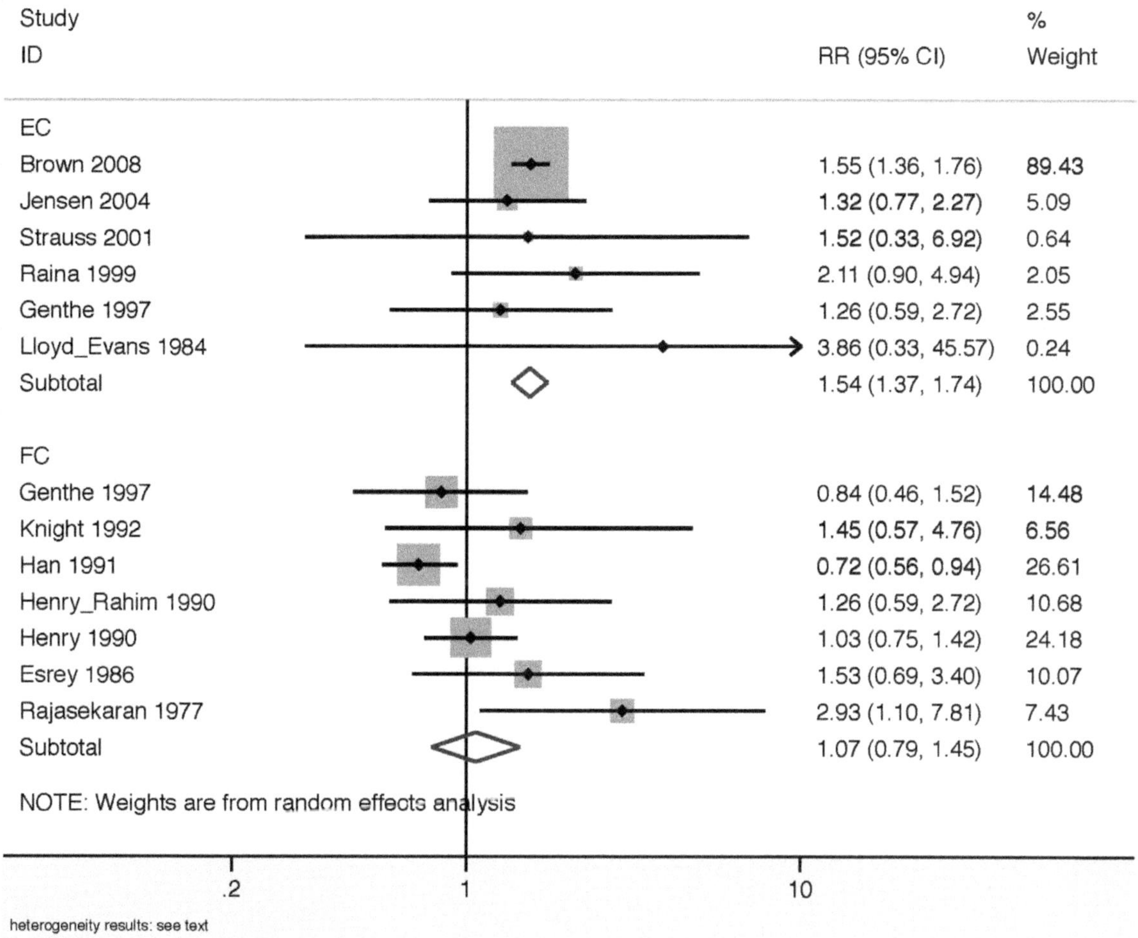

Figure 4. Forest plot for all studies stratified by indicator organism. Top panel, EC: *Escherichia coli*; Bottom panel, FC: thermotolerant ("fecal") coliforms.

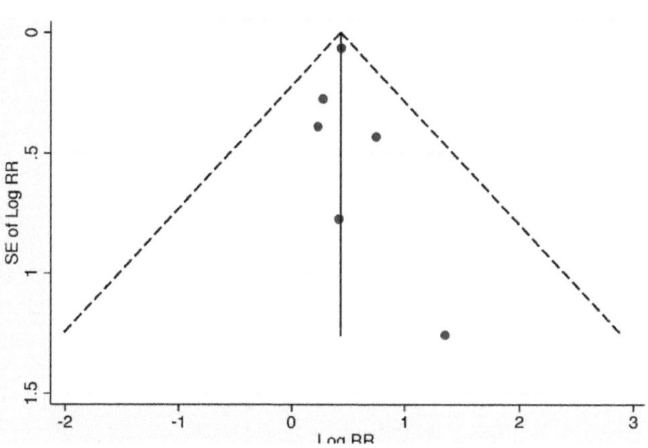

Figure 5. Funnel plot for EC studies. Studies are plotted relative to size (SE of Log RR), and reported log relative risk (Log RR); studies higher on the vertical axis are larger, and studies further to the right have larger RRs. The solid vertical line is the log of the summary RR for all studies, and the dotted lines are "pseudo 95% confidence intervals" for the summary measure. Studies on this plot are generally symmetrical, with the exception of one small study with a large RR, providing minimal evidence of publication bias.

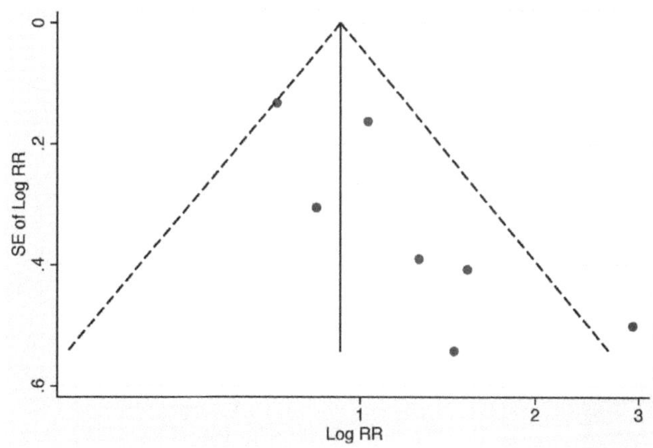

Figure 6. Funnel plot for all FC studies. Studies are plotted relative to size (SE of Log RR), and reported log relative risk (Log RR); studies higher on the vertical axis are larger, and studies further to the right have larger RRs. The solid vertical line is the log of the summary RR for all studies, and the dotted lines are "pseudo 95% confidence intervals" for the summary measure. The absence of smaller studies with small or null RRs (lower left region of the plot) creates plot asymmetry, and provides evidence of possible publication bias.

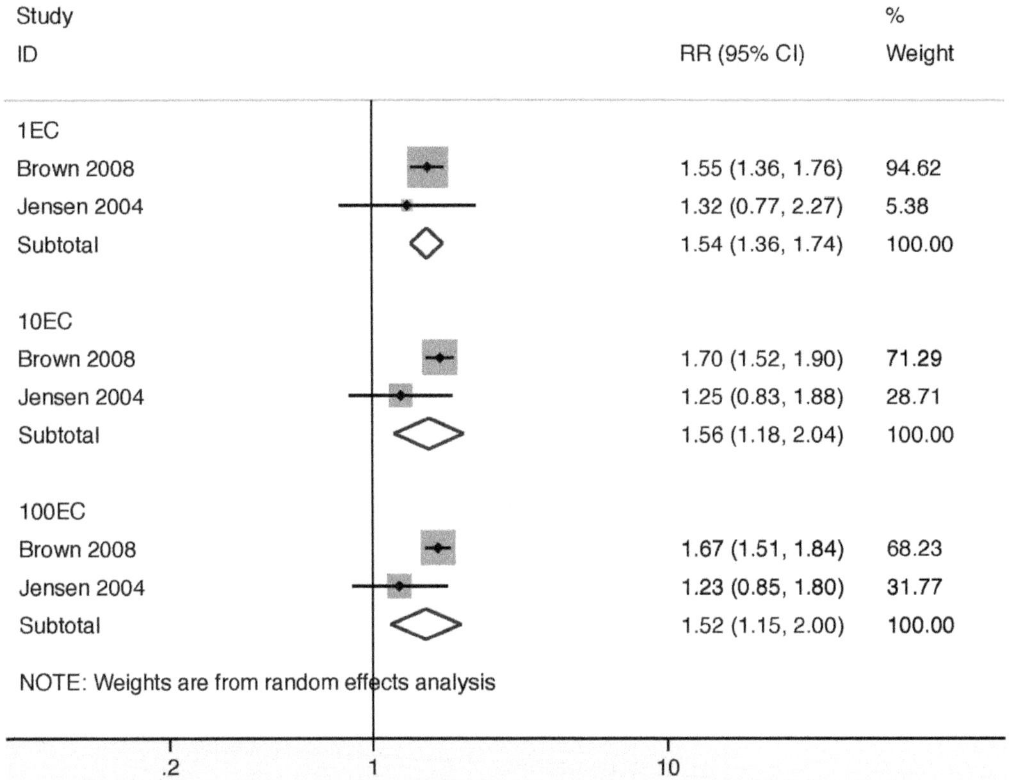

Figure 7. Forest plot for two studies with extractable data at multiple thresholds. Data were pooled across the two studies at the 1 EC/100 ml threshold (top panel), 10 EC/100 ml threshold (middle panel), and 100 EC/100 ml threshold (bottom panel). EC: *Escherichia coli.*

inconclusive. In addition to the results of the meta-analysis, we identified several elements of study design and reporting that could improve future studies evaluating the relationship between water quality and health.

Implications for Choice of Indicator Organism to Assess Microbiological Safety of Drinking Water

Current guidelines recommend two specific uses for indicator organisms: (i) fecal indicators (index organisms), used to indicate the presence of fecal contamination, potential pathogens, and possible health risk, and; (ii) process indicators used to evaluate the effectiveness of a water treatment process. [1,2,13]. While the performance of coliform bacteria as process indicators has been described previously, [2,43] and is outside the scope of this review, the results from our review have implications for the use of EC and FC as index organisms in household drinking water. Currently, EC is considered the "most suitable" fecal indicator; however, FC is generally recognized as a suitable surrogate. [1,2,12,13] Our findings support the continued use of EC as a fecal indicator (index organism) to assess water quality at the point of use, but do not support the use of FC as a surrogate as it does not demonstrate a clear association with diarrheal disease outcomes.

Our findings that EC and FC perform differently as indicators of diarrhea risk from contaminated drinking water are not surprising. While EC is generally of fecal origin, [10] and likely to be present simultaneously with diarrheagenic pathogens in recently contaminated drinking water, FC are known to include coliforms of environmental origin (e.g., *Klebsiella*), in addition to EC. [1,2,13] The consistency with which FC methods will be specific to human pathogens will therefore vary with the environments in which water samples are being collected. Indeed, our results show that the magnitude and direction of the effect estimates for FC were highly inconsistent (Figure 4, Table 6); even within a single study [37] the authors found highly divergent results using FC in two separate study areas (Table 6). In contrast, the magnitude and direction of effect estimates across the EC studies included in our review were consistent across diverse study regions, and when pooled provided evidence of a significant association with diarrhea; these results are consistent with the higher specificity of EC for fecal contamination, compared to FC (Table 5, Figure 4). The presence of a dose-response between EC and diarrhea could have provided further evidence to support our findings that drinking water contaminated with EC is associated with diarrhea. However, dose-response results from our review were inconclusive, and represent an avenue for future research.

While our summary measures suggest that the presence of EC in household drinking water is significantly associated with diarrhea, it is important to note that only one individual EC study in our review was able to rule out chance as a possible explanation of their findings [15]. It has been demonstrated that indicator levels can vary considerably even over short periods of time in both household and source water, [44,45] and that pathogens and indicator organisms have inconsistent correlations. [46] The studies in this review relied on intermittent grab samples to classify household water quality. Even daily grab samples likely provided only a crude assessment of a household's exposure to fecal contamination and diarrheagenic pathogens, which likely contributed to the lack of precision in the individual studies included in our review. While the imprecision of results from individual studies supports a previous assertion that fecal indicator organisms are "blunt" tools to characterize water quality, [19] our findings suggest that they have continued utility in health and water research. Indeed, while recent WHO drinking water guidelines have shifted focus from recommendations using specific indicator levels to define water safety to a more integrated approach centered on water safety plans, these recommendations are largely directed towards the management of water distribution systems. [2,13] In practice, regional organizations and researchers working in areas without access to drinking water distribution systems rely heavily on indicator organisms to assess the risk of disease from drinking water, and to measure the effectiveness of household water treatment and safe storage interventions; this is particularly true in resource poor settings with limited access to advanced laboratory facilities. Our results suggest that EC have continued value in both of these applications.

Threshold Effects for Diarrhea Risk

The findings from our review do not support the existence of a "threshold effect" in contaminated household drinking water. A previous study evaluating the effect of contaminated *source* water on diarrhea [23] found no evidence of an association between EC and diarrhea at the 1 EC threshold level; however, significant associations with diarrhea were observed at the 1000 EC threshold level. The authors concluded that their findings of a "threshold effect" supported earlier assertions that relaxing drinking water guidelines (beyond the 1 EC threshold) might be acceptable in developing countries. [47] Contrary to these findings, in our review of household drinking water we found consistently elevated risks of diarrhea at the 1 EC threshold, with a significant summary measure when pooled across all studies. These results support current WHO guidelines recommending that all drinking water intended for human consumption contain zero EC/100 ml, regardless of location. [2] Beyond the 1 EC threshold, our results did not suggest increased risk of diarrhea at elevated cutoffs, but this evidence was based on a limited number of studies.

Recommendations for Future Water Quality Studies

The studies included in our review covered several decades, and varied considerably with respect to methodological quality and completeness of reporting. In this section, we highlight several findings from our systematic review that could improve the design and reporting of future studies that seek to evaluate the relationship between water quality and health.

The studies in our systematic review used very consistent definitions of diarrhea, relying primarily on self- or caregiver reporting of diarrhea symptoms (Tables 3 and 4). However, we note that diarrhea recall periods and identification strategies varied greatly (e.g., daily recall, one-month recall, cases presenting to clinics; Tables 3 and 4). Reliance on self-reported diarrhea symptoms as a health measure could have introduced bias into individual studies through two primary mechanisms, which have implications for the use of self-reported diarrhea as an outcome in future studies. First, as with any self-reported, subjective outcome, there is a potential for differential reporting relative to exposure status [48]. If study-specific reporting biases were differential with respect to drinking water contamination, it is possible that the results of individual studies (and therefore our review) could be biased. However, there is no reason to suspect that individuals would be aware of the indicator status of their drinking water when reporting diarrhea symptoms, which would decrease the potential for this bias. Second, longer follow-up periods can bias diarrhea reporting due to poor outcome recall, which could have biased findings towards the null if non-differential with respect to exposure, and/or compounded biases associated with differential reporting in individual studies. [49] The inclusion of objective health outcome measures (e.g., pathogen-specific antibody responses) could improve future studies by reducing reliance on self-reported diarrhea symptoms and their associated biases. [50,51] In addition, while our review focused exclusively on diarrhea, expanding the range of water-related health outcomes studied and reviewed (e.g., indicators of growth and malnutrition) could also improve our understanding of the health impacts of drinking water contamination beyond gastrointestinal illnesses.

Very few studies in our review were able to establish temporality between exposures (indicator organism) and outcomes (diarrhea), and therefore mostly relied on the assumption that drinking water samples were representative of an exposure period that was relevant to measured disease outcomes. While this assumption might be reasonable, or the only option given the practicalities of field conditions, it does leave open the possibility of reverse causality (i.e., diarrhea caused by other environmental pathways led to increased fecal contamination of drinking water). The design of future water quality studies should attempt to measure exposure in advance of disease outcomes, and all studies should be explicit about the timing of exposure and disease measurements in their reporting. These recommendations would all but preclude the use of "traditional" case-control studies, where cases and controls (i.e., disease status) are identified first, followed by visits to households to collect water samples to ascertain exposure status. Instead, a cohort design that measures exposure to contaminated drinking water first, followed by surveillance visits after an appropriate incubation period to measure health outcomes, is a more suitable study design. Decisions regarding follow-up frequency, recall period, and measures of disease occurrence (i.e., incidence, prevalence) will vary depending on study context and logistics, but should be carefully considered and explicitly reported. [49,52]

In studies evaluating the relationship between drinking water quality and health, confounding bias is a concern: there are numerous factors that could cause both drinking water contamination and diarrhea in a given study setting. Overall, the studies in our review were not consistent in reporting adjusted results (Tables 5 and 6), and there was no consistent set of factors used across studies when authors did report attempting to control for confounding. Incomplete control of confounding in the included studies could have biased individual study results, and therefore the findings of our review. Future studies should be explicit about their attempts to measure and control for confounding factors. Additionally, several studies did not account for correlated outcomes in their analyses, which could have influenced the precision of summary estimates. However, for EC studies we note the consistency of the direction and magnitude of the associations reported across diverse study locations, and across studies that did and did not adjust for confounding and/or account for clustering. Regardless, reporting on these issues could be improved upon in future studies.

Our systematic review included published, peer-reviewed studies. As with any systematic review, publication bias, where smaller studies with small or null ("non-positive") results are systematically excluded from the published literature, could have impacted our findings. In our review, we found evidence of publication bias among FC studies (smaller studies with smaller effect sizes appear to be absent from the funnel plot; lower left region, Figure 6); however, inclusion of smaller FC studies with RRs at or less than unity would have strengthened our findings of no association between FC and diarrhea. We found minimal evidence to suggest publication bias among EC studies (Figure 5);

however, we cannot rule out publication bias as a possible explanation of our findings – the funnel plot suggests that most studies were balanced around the summary measure with the exception of one small study with a large positive finding. Taken together, there is evidence that smaller studies with small ("non-positive") effects sizes are systematically missing from the literature on the relationship between water quality and diarrheal illness.

In our review, we pooled RRs across relevant studies using meta-analysis to investigate the association of fecal indicator measures in drinking water with diarrhea. The limitations of meta-analysis have been discussed [53]; we caution against interpreting the summary measures from our review as "true" underlying causal effects, but rather consider them additional pieces of evidence regarding the relationship between fecal indicators in drinking water and diarrhea. The evidence from our pooled summary measures, taken together with biologic plausibility and the consistency of results across studies, provides evidence to support a relationship between diarrhea and the presence of EC in point-of-use drinking water; we do not, however, find evidence to support an equivalent relationship between diarrhea and FC.

Conclusions

We reviewed the available literature to identify studies that evaluated the relationship between microbial indicators of drinking water quality and diarrhea. We found that studies using *Escherichia coli* (EC) as an indicator of household drinking water quality reported consistent effect estimates, that when pooled suggested a significant association with increased diarrheal illness. Results from studies using thermotolerant ("fecal") coliforms (FC), on the other hand, were inconsistent, and suggested no association with diarrhea when pooled. In this review, we also note several areas where the design and reporting of the included studies could have been improved, and make recommendations for future studies. The results from our review suggest that EC has value as a fecal indicator organism, but that use of FC should be considered carefully in contexts where an association with diarrheal disease outcomes is important.

Acknowledgments

The authors would like to acknowledge Professor Kara Nelson for generously reviewing this manuscript, and providing key feedback.

Author Contributions

Conceived and designed the experiments: JG JC. Performed the experiments: JG AE. Analyzed the data: JG. Contributed reagents/materials/analysis tools: JG AE. Wrote the paper: JG AE JC.

References

1. Dufour AP (2003) Assessing microbial safety of drinking water: Improving approaches and methods: International Water Assn.
2. WHO (2011) Guidelines for drinking-water quality, fourth edition. Geneva: World Health Organization.
3. Cutler D, Miller G (2005) The role of public health improvements in health advances: The twentieth-century United States. Demography 42: 1–22.
4. Arnold BF, Colford JM Jr (2007) Treating water with chlorine at point-of-use to improve water quality and reduce child diarrhea in developing countries: a systematic review and meta-analysis. The American journal of tropical medicine and hygiene 76: 354.
5. Clasen T, Schmidt WP, Rabie T, Roberts I, Cairncross S (2007) Interventions to improve water quality for preventing diarrhoea: systematic review and meta-analysis. bmj 334: 782.
6. Fewtrell L, Kaufmann RB, Kay D, Enanoria W, Haller L, et al. (2005) Water, sanitation, and hygiene interventions to reduce diarrhoea in less developed countries: a systematic review and meta-analysis. The Lancet infectious diseases 5: 42–52.
7. National Research Council (2004) Indicators for Waterborne Pathogens. Washington, DC: The National Academies Press.
8. Gundry S, Wright J, Conroy R (2004) A systematic review of the health outcomes related to household water quality in developing countries. J Water Health 2: 1–13.
9. Gerba C (2000) Indicator Microogranisms. Environmental Microbiology. San Diego, CA: Academic Press.
10. Leclerc H, Mossel D, Edberg S, Struijk C (2001) Advances in the bacteriology of the coliform group: their suitability as markers of microbial water safety. Annual Reviews in Microbiology 55: 201–234.
11. Reynolds KA, K.D Mena, Gerba CP (2008) Risk of waterborne illness via drinking water in the United States. Reviews of Environmental Contamination and Toxicology 192: 117–158.
12. Tallon P, Magajna B, Lofranco C, Leung KT (2005) Microbial indicators of faecal contamination in water: a current perspective. Water, Air, and Soil Pollution 166: 139–166.
13. Ashbolt NJ, Grabow WOK, Snozzi M (2001) Indicators of microbial water quality. In: Fewtrell L, Bartram J, editors.Water Quality: Guidelines, Standards and Health. London, UK: IWA Publishing,.
14. WHO (1997) Guidelines for drinking-water quality, 2nd edition. Geneva: World Health Organization.
15. Brown JM, Proum S, Sobsey MD (2008) Escherichia coli in household drinking water and diarrheal disease risk: evidence from Cambodia. Water Science and Technology 58: 757–763.
16. de Aceituno AMF, Stauber CE, Walters AR, Sanchez REM, Sobsey MD (2012) A Randomized Controlled Trial of the Plastic-Housing BioSand Filter and Its Impact on Diarrheal Disease in Copan, Honduras. American Journal of Tropical Medicine and Hygiene 86: 913–921.
17. Gundry SW, Wright JA, Conroy RM, Du Preez M, Genthe B, et al. (2009) Child dysentery in the Limpopo Valley: A cohort study of water, sanitation and hygiene risk factors. Journal of Water and Health 7: 259–266.
18. Jensen PK, Jayasinghe G, van der Hoek W, Cairncross S, Dalsgaard A (2004) Is there an association between bacteriological drinking water quality and childhood diarrhoea in developing countries? Trop Med Int Health 9: 1210–1215.
19. Levy K, Nelson KL, Hubbard A, Eisenberg JN (2012) Rethinking indicators of microbial drinking water quality for health studies in tropical developing countries: case study in northern coastal Ecuador. Am J Trop Med Hyg 86: 499–507.
20. Raina PS, Pollari FL, Teare GF, Goss MJ, Barry DA, et al. (1999) The relationship between E. coli indicator bacteria in well-water and gastrointestinal illnesses in rural families. Can J Public Health 90: 172–175.
21. Strauss B, King W, Ley A, Hoey JR (2001) A prospective study of rural drinking water quality and acute gastrointestinal illness. BMC Public Health 1: 8.
22. Rothman K, Greenland S, (1998) Modern Epidemiology. Philadelphia: Lippincott-Raven Publishers.
23. Moe C, Sobsey M, Samsa G, Mesolo V (1991) Bacterial indicators of risk of diarrhoeal disease from drinking-water in the Philippines. Bulletin of the World Health Organization 69: 305.
24. Greenland S, Longnecker MP (1992) Methods for trend estimation from summarized dose-response data, with applications to meta-analysis. American Journal of Epidemiology 135: 1301–1309.
25. Egger M, Smith G Davey, Altman D (2001) Systematic Reviews in Healthcare. Meta-analysis in Context. London: BMJ Books.
26. Egger M, Smith GD, Schneider M, Minder C (1997) Bias in meta-analysis detected by a simple, graphical test. bmj 315: 629–634.
27. Sterne JAC, Egger M (2001) Funnel plots for detecting bias in meta-analysis: Guidelines on choice of axis. Journal of Clinical Epidemiology 54: 1046–1055.
28. Genthe B, Strauss N, Seager J, Vundule C, Maforah F, et al. (1997) The effect of type of water supply on water quality in a developing community in South Africa. Water Science and Technology 35: 35–40.
29. Echeverria P, Taylor DN, Seriwatana J (1987) Potential sources of enterotoxigenic Escherichia coli in homes of children with diarrhoea in Thailand. Bulletin of the World Health Organization 65: 207–215.
30. Baqui AH, Black RE, Yunus M, Hoque ARA, Chowdhury H, et al. (1991) Methodological issues in diarrhoeal diseases epidemiology: definition of diarrhoeal episodes. International journal of epidemiology 20: 1057.
31. WHO (2013) Diarrhoeal disease; fact sheet.

32. Rajasekaran P, Dutt PR, Pisharoti KA (1977) Impact of Water Supply on the Incidence of Diarrhoea and Shigellosis Among Children in Rural Communities in Madurai. Indian J Med Res 66: 189–199.
33. Lloyd-Evans N, Pickering HA, Goh SGJ, Rowland MGM (1984) Food and Water Hygiene and Diarrhoea in Young Gambian Children: a Limited Case Control Study. Transactions of the Royal Society of Tropical Medicine and Hygiene 78: 209–211.
34. Black RE, Brown KH, Becker S, Alim AR, Merson MH (1982) Contamination of weaning foods and transmission of enterotoxigenic Escherichia coli diarrhoea in children in rural Bangladesh. Trans R Soc Trop Med Hyg 76: 259–264.
35. Bhargava A, Bouis HE, Hallman K, Hoque BA (2003) Coliforms in the water and hemoglobin concentration are predictors of gastrointestinal morbidity of Bangladeshi children ages 1–10 years. American journal of human biology 15: 209–219.
36. Esrey SA, Habicht J, Casella G, Miliotis MD, Kidd AH, et al. (1986) Infection, Diarrha, and Growth Rates of Young Children Following the Installation of Village Water Supplies in Lesotho. Interanation Symptosium on Water-Related Health Issues November: 11–16.
37. Henry FJ, Rahim Z (1990) Transmission of diarrhoea in two crowded areas with different sanitary facilities in Dhaka, Bangladesh. J Trop Med Hyg 93: 121–126.
38. Knight SM, Toodayan W, Caique WC, Kyi W, Barnes A, et al. (1992) Risk factors for the transmission of diarrhoea in children: a case-control study in rural Malaysia. Int J Epidemiol 21: 812–818.
39. Han AM, Oo KN, Aye T, Hlaing T (1991) Bacteriologic studies of food and water consumed by children in Myanmar: 2 Lack of association between diarrhoea and contamination of food and water. J Diarrhoeal Dis Res 9: 91–93.
40. Henry F, Huttly S, Patwary Y, Aziz K (1990) Environmental sanitation, food and water contamination and diarrhoea in rural Bangladesh. Epidemiology and infection London, New York NY 104: 253–259.
41. Vanderslice J, Briscoe J (1993) All Coliforms Are Not Created Equal - a Comparison of the Effects of Water Source and in-House Water Contamination on Infantile Diarrheal Disease. Water Resources Research 29: 1983–1995.
42. Jagals P, Grabow WOK, Williams E (1997) The effects of supplied water quality on human health in an urban development with limited basic subsistence facilities. Water Sa 23: 373–378.
43. Sobsey MD (1989) Inactivation of health-related microorganisms in water by disinfection processes. Water Science & Technology 21: 179–195.
44. Levy K, Hubbard AE, Nelson KL, Eisenberg JN (2009) Drivers of water quality variability in northern coastal Ecuador. Environmental science & technology 43: 1788–1797.
45. Levy K, Nelson KL, Hubbard A, Eisenberg JN (2008) Following the water: a controlled study of drinking water storage in northern coastal Ecuador. Environmental health perspectives 116: 1533.
46. Wu J, Long S, Das D, Dorner S (2011) Are microbial indicators and pathogens correlated? A statistical analysis of 40 years of research. J Water Health 9: 265–278.
47. Feachem RG, McGarry MG, Mara DD (1977) Water, wastes and health in hot climates: John Wiley and Sons Ltd.
48. Wood L, Egger M, Gluud LL, Schulz KF, Jüni P, et al. (2008) Empirical evidence of bias in treatment effect estimates in controlled trials with different interventions and outcomes: meta-epidemiological study. bmj 336: 601–605.
49. Arnold BF, Galiani S, Ram PK, Hubbard AE, Briceño B, et al. (2013) Optimal recall period for caregiver-reported illness in risk factor and intervention studies: A multicountry study. American journal of epidemiology 177: 361–370.
50. Crump JA, Mendoza CE, Priest JW, Glass RI, Monroe SS, et al. (2007) Comparing serologic response against enteric pathogens with reported diarrhea to assess the impact of improved household drinking water quality. American Journal of Tropical Medicine and Hygiene 77: 136.
51. Schmidt W-P, Cairncross S (2009) Household Water Treatment in Poor Populations: Is There Enough Evidence for Scaling up Now? Environmental science & technology 43: 986–992.
52. Schmidt W-P, Arnold BF, Boisson S, Genser B, Luby SP, et al. (2011) Epidemiological methods in diarrhoea studies—an update. Int J Epidemiol 40: 1678–1692.
53. Berk RA, Freedman DA (2003) Statistical assumptions as empirical commitments: Law, Punishment, and Social Control: Essays in Honor of Sheldon Messinger, 2nd edn. Aldine de Gruyter.

Permissions

All chapters in this book were first published in PLOS ONE, by The Public Library of Science; hereby published with permission under the Creative Commons Attribution License or equivalent. Every chapter published in this book has been scrutinized by our experts. Their significance has been extensively debated. The topics covered herein carry significant findings which will fuel the growth of the discipline. They may even be implemented as practical applications or may be referred to as a beginning point for another development.

The contributors of this book come from diverse backgrounds, making this book a truly international effort. This book will bring forth new frontiers with its revolutionizing research information and detailed analysis of the nascent developments around the world.

We would like to thank all the contributing authors for lending their expertise to make the book truly unique. They have played a crucial role in the development of this book. Without their invaluable contributions this book wouldn't have been possible. They have made vital efforts to compile up to date information on the varied aspects of this subject to make this book a valuable addition to the collection of many professionals and students.

This book was conceptualized with the vision of imparting up-to-date information and advanced data in this field. To ensure the same, a matchless editorial board was set up. Every individual on the board went through rigorous rounds of assessment to prove their worth. After which they invested a large part of their time researching and compiling the most relevant data for our readers.

The editorial board has been involved in producing this book since its inception. They have spent rigorous hours researching and exploring the diverse topics which have resulted in the successful publishing of this book. They have passed on their knowledge of decades through this book. To expedite this challenging task, the publisher supported the team at every step. A small team of assistant editors was also appointed to further simplify the editing procedure and attain best results for the readers.

Apart from the editorial board, the designing team has also invested a significant amount of their time in understanding the subject and creating the most relevant covers. They scrutinized every image to scout for the most suitable representation of the subject and create an appropriate cover for the book.

The publishing team has been an ardent support to the editorial, designing and production team. Their endless efforts to recruit the best for this project, has resulted in the accomplishment of this book. They are a veteran in the field of academics and their pool of knowledge is as vast as their experience in printing. Their expertise and guidance has proved useful at every step. Their uncompromising quality standards have made this book an exceptional effort. Their encouragement from time to time has been an inspiration for everyone.

The publisher and the editorial board hope that this book will prove to be a valuable piece of knowledge for researchers, students, practitioners and scholars across the globe.

List of Contributors

Qi Liu, Shiliang Liu, Haidi Zhao, Li Deng, Cong Wang, Qinghe Zhao and Shikui Dong
School of Environment, State Key Laboratory of Water Environment Simulation, Beijing Normal University, Beijing, China

Bora Demiri
Great Lakes Institute for Environmental Research, University of Windsor, Windsor, Ontario, Canada

Teresa J. Crease
Department of Integrative Biology, University of Guelph, Guelph, Ontario, Canada

Ianina Altshuler
Department of Biology, McGill University, Montreal, Quebec, Canada

Melania E. Cristescu
Great Lakes Institute for Environmental Research, University of Windsor, Windsor, Ontario, Canada
Department of Biology, McGill University, Montreal, Quebec, Canada

Tamara García-Armisen, Özgül I˙nceoğlu, Nouho Koffi Ouattara, Adriana Anzil and
Pierre Servais
Ecology of Aquatic Systems, Université Libre de Bruxelles, Campus de la Plaine, Brussels, Belgium

Michel A. Verbanck
Department of Water Pollution Control, Université Libre de Bruxelles, Campus Plaine, Brussels, Belgium

Natacha Brion
Analytical and Environmental Chemistry, Vrije Universiteit Brussels, Brussels, Belgium

Lu Cai
Collaborative Innovation Center for Geo-Hazards and Eco-Environment in Three Gorges Area, Hubei Province, Yichang, PR China

Lei Chen and Yong Gao
Institute of Chinese Sturgeon Research, China Three Gorges Project Corporation, Yichang, PR China

Prashant Mandal
Engineering Research Center of Eco-environment in Three Gorges Reservoir Region, Ministry of Education, China Three Gorges University, Yichang, PR China

David Johnson
Engineering Research Center of Eco-environment in Three Gorges Reservoir Region, Ministry of Education, China Three Gorges University, Yichang, PR China
School of Natural Sciences and Mathematics, Ferrum College, Ferrum, Virginia, United States of America

Min Fang, Zhiying Tu and Yingping Huang
Collaborative Innovation Center for Geo-Hazards and Eco-Environment in Three Gorges Area, Hubei Province, Yichang, PR China
Engineering Research Center of Eco-environment in Three Gorges Reservoir Region, Ministry of Education, China Three Gorges University, Yichang, PR China

Peter Houk
University of Guam Marine Laboratory, UOG Station, Mangilao, Guam
Pacific Marine Resources Institute, Saipan, Northern Mariana Islands

David Benavente, John Iguel, Steven Johnson and Ryan Okano
CNMI Bureau of Environmental and Coastal Quality, Saipan, Northern Mariana Islands

Samuel Kersey Sturdivant
Division of Marine Science and Conservation, Nicholas School of the Environment, Duke University, Beaufort, North Carolina, United States of America

Mark J. Brush and Robert J. Diaz
Virginia Institute of Marine Science, College of William & Mary, Gloucester Point, Virginia, United States of America

Tara Chestnut
Oregon State University, Environmental Science Graduate Program, Corvallis, Oregon, United States of America
US Geological Survey, Oregon Water Science Center,Portland, Oregon, United States of America

Chauncey Anderson
US Geological Survey, Oregon Water Science Center,Portland, Oregon, United States of America

Radu Popa
Biological Sciences, University of Southern California, Los Angeles, California, United States of America

Andrew R. Blaustein
Oregon State University, Department of Integrative Biology, Corvallis, Oregon, United States of America

Mary Voytek
Astrobiology Program, National Aeronautics and Space Administration Headquarters, Washington DC, United States of America

Deanna H. Olson
US Forest Service, Pacific Northwest Research Station, Corvallis, Oregon, United States of America

Julie Kirshtein
US Geological Survey, National Research Program, Reston, Virginia, United States of America

Euan D. Reavie
Center for Water and the Environment, Natural Resources Research Institute, University of Minnesota Duluth, Duluth, Minnesota, United States of America

Adam J. Heathcote
Groupe de Recherche Interuniversitaire en Limnologie, Département des Sciences Biologiques, Université du Québec à Montréal, Montréal, Québec, Canada

Victoria L. Shaw Chraïbi
Department of Earth and Atmospheric Sciences, University of Nebraska-Lincoln, Lincoln, Nebraska, United States of America

Heather K. Patterson and Ruth H. Carmichael
Department of Marine Sciences, University of South Alabama, Mobile, Alabama, United States of America
Dauphin Island Sea Lab, Dauphin Island, Alabama, United States of America

Anne Boettcher
Department of Biology, University of South Alabama, Mobile, Alabama, United States of America

Noah R. Lottig
Center for Limnology Trout Lake Station, University of Wisconsin Madison, Boulder Junction, Wisconsin, United States of America

Tyler Wagner
U.S. Geological Survey, Pennsylvania Cooperative Fish and Wildlife Research Unit, Pennsylvania State University, University Park, Pennsylvania, United States of America

Emily Norton Henry
Department of Fisheries and Wildlife, Michigan State University, East Lansing, Michigan, United States of America
Department of Outreach and Engagement, Oregon State University, Tillamook, Oregon, United States of America

Kendra Spence Cheruvelil
Department of Fisheries and Wildlife, Michigan State University, East Lansing, Michigan, United States of America
Lyman Briggs College, Michigan State University, East Lansing, Michigan, United States of America

Katherine E. Webster
School of Natural Sciences, Trinity College, Dublin, Ireland

John A. Downing
Department of Ecology, Evolution, and Organismal Biology, Iowa State University, Ames, Iowa, United States of America

Craig A. Stow
Great Lakes Environmental Research Laboratory, National Oceanic and Atmospheric Administration, Ann Arbor, Michigan, United States of America
Bart Van der Bruggen
ProcESS, Department of Chemical Engineering, KU Leuven, Leuven, Belgium

Gomotsegang F. Molelekwa
ProcESS, Department of Chemical Engineering, KU Leuven, Leuven, Belgium
Department of Environmental Health, Tshwane University of Technology, Pretoria, South Africa

Murembiwa S. Mukhola
Department of Environmental Health, Tshwane University of Technology, Pretoria, South Africa

Patricia Luis
Materials and Process Engineering (iMMC-IMAP), Université Catholique de Louvain, Louvain-la-Neuve, Belgium

Yu Xia, Xianghua Wen and Yunfeng Yang
Environmental Simulation and Pollution Control State Key Joint Laboratory, School of Environment, Tsinghua University, Beijing, China

Xiaohui Wang
Environmental Simulation and Pollution Control State Key Joint Laboratory, School of Environment, Tsinghua University, Beijing, China
Department of Environmental Science and Engineering, Beijing University of Chemical Technology, Beijing, China

Jizhong Zhou
Environmental Simulation and Pollution Control State Key Joint Laboratory, School of Environment, Tsinghua University, Beijing, China
Institute for Environmental Genomics and Department of Microbiology and Plant Biology, University of Oklahoma, Norman, Oklahoma, United States of America
Earth Sciences Division, Lawrence Berkeley National Laboratory, Berkeley, California, United States of America

John M. Guinotte
Marine Conservation Institute, Seattle, Washington, United States of America

Andrew J. Davies
School of Ocean Sciences, Bangor University, Menai Bridge, Anglesey, United Kingdom

Pieter Boets and Peter L. M. Goethals
Aquatic Ecology Research Unit, Department Applied Ecology and Environmental Biology, Ghent University, Ghent, Belgium

Gert Everaert
Aquatic Ecology Research Unit, Department Applied Ecology and Environmental Biology, Ghent University, Ghent, Belgium
Environmental Toxicology Research Group, Department Applied Ecology and Environmental Biology, Ghent University, Ghent, Belgium

Jan De Neve
Department of Mathematical Modelling, Statistics and Bioinformatics, Ghent University, Ghent, Belgium

Luis Dominguez-Granda
Department of Chemical and Environmental Sciences, Escuela Superior Politécnica del Litoral (ESPOL), Guayaquil, Ecuador

Seid Tiku Mereta and Argaw Ambelu
Department of Environmental Health Science and Technology, Jimma University, Jimma, Ethiopia

Thu Huong Hoang
School of Environmental Science and Technology, Hanoi University of Science and Technology, Hanoi, Vietnam

Olivier Thas
Department of Mathematical Modelling, Statistics and Bioinformatics, Ghent University, Ghent, Belgium
National Institute for Applied Statistics Research Australia (NIASRA), School of Mathematics and Applied Statistics, University of Wollongong, Wollongong, Australia

Marisa Mazari-Hiriart
Laboratorio Nacional de Ciencias de la Sostenibilidad, Departamento de Ecología de la Biodiversidad, Instituto de Ecoléogía, Universidad Nacional Autónoma de México, México, D.F., México

Gustavo Pérez-Ortiz
Laboratorio Nacional de Ciencias de la Sostenibilidad, Departamento de Ecología de la Biodiversidad, Instituto de Ecoléogía, Universidad Nacional Autónoma de México, México, D.F., México
Departamento de Biología, Facultad de Ciencias, Universidad Nacional Autónoma de México, México, D.F., México

Isaura Yañez-Noguez and María Teresa Orta-Ledesma
Coordinación de Ingeniería Ambiental, Instituto de Ingeniería, Universidad Nacional Autónoma de México, México, D.F., México

Felipe Armas-Vargas
Posgrado en Ciencias de la Tierra, Instituto de Geología, Universidad Nacional Autónoma de México, México, D.F., México

Marco A. Tapia
Departamento de Biología, Facultad de Ciencias, Universidad Nacional Autónoma de México, México, D.F., México
Posgrado en Ciencias Bioló gicas, Facultad de Ciencias-Instituto de Ecología, Universidad Nacional Autónoma de México, México, D.F., México

Rosa Solano-Ortiz
Posgrado en Ciencias Bioló gicas, Facultad de Ciencias-Instituto de Ecología, Universidad Nacional Autónoma de México, México, D.F., México

Miguel A. Silva
Posgrado de Ciencias Bioquímicas, Departamento de Ingeniería Celular y Biocatá lisis, Instituto de Biotecnología, Universidad Nacional Autónoma de México, Cuernavaca, Morelos, México

Yolanda López-Vidal
Programa de Inmunología Molecular Microbiana, Departamento de Microbiología y Parasitología, Facultad de Medicina, Universidad Nacional Autónoma de México, México, D.F., México

Carlos Díaz-Ávalos
Departamento de Probabilidad y Estadística, Instituto de Investigaciones en Matemáticas Aplicadas y en Sistemas, Universidad Nacional Autónoma de México, México, D.F., México

Arijit Ganguly, Ranita Chakravorty, Dipak K. Mandal and Parimalendu Haldar
Department of Zoology, Visva-Bharati University, Santiniketan, West Bengal, India

Angshuman Sarkar
Department of Statistics, Visva-Bharati University, Santiniketan, West Bengal, India

Julieta Ramos-Elorduy and Jose Manuel Pino Moreno
Departamento de Zoología, Universidad Nacional Autonoma de Mexico, Mexico City, Distrito Federal, México

Carlos Iñiguez-Armijos
Departamento de Ciencias Naturales, Universidad Técnica Particular de Loja (UTPL), Sección de Ecología, Loja, Ecuador
Institute for Landscape Ecology and Resources Management (ILR), Research Centre for Bio Systems, Land Use and Nutrition (IFZ), Justus-Liebig-University Giessen, Giessen, Germany

Adrián Leiva
Departamento de Ciencias Naturales, Universidad Técnica Particular de Loja (UTPL), Sección de Ecología, Loja, Ecuador

Hans-Georg Frede and Lutz Breuer
Institute for Landscape Ecology and Resources Management (ILR), Research Centre for Bio Systems, Land Use and Nutrition (IFZ), Justus-Liebig-University Giessen, Giessen, Germany

Henrietta Hampel
Universidad de Cuenca, Quinta Balzaín, Cuenca, Ecuador

Jiabo Chen
Department of Natural Resources, College of Environment and Natural Resources, Zhejiang University, Hangzhou, Zhejiang Province, China

Jun Lu
Department of Natural Resources, College of Environment and Natural Resources, Zhejiang University, Hangzhou, Zhejiang Province, China
China Ministry of Education Key Lab of Environment Remediation and Ecological Health, Zhejiang University, Hangzhou, Zhejiang Province, China

Kade Small
School of Environmental Sciences, Charles Sturt University, Albury, New South Wales, Australia

Robyn J. Watts
School of Environmental Sciences, Charles Sturt University, Albury, New South Wales, Australia
Institute for Land, Water and Society, Charles Sturt University, Albury, New South Wales, Australia

R. Keller Kopf
Institute for Land, Water and Society, Charles Sturt University, Albury, New South Wales, Australia

Julia Howitt
Institute for Land, Water and Society, Charles Sturt University, Albury, New South Wales, Australia
School of Agricultural and Wine Sciences, Charles Sturt University, Wagga Wagga, New South Wales, Australia

Jason W. Marion, Cheonghoon Lee, Chang Soo Lee and Timothy J. Buckley
Division of Environmental Health Sciences, College of Public Health, The Ohio State University, Columbus, Ohio, United States of America

Qiuhong Wang and Linda J. Saif
Food Animal Health Research Program, Ohio Agricultural Research and Development Center, Department of Veterinary Preventive Medicine, The Ohio State University, Wooster, Ohio, United States of America

Stanley Lemeshow
Division of Biostatistics, College of Public Health, The Ohio State University, Columbus, Ohio, United States of America

Jiyoung Lee
Division of Environmental Health Sciences, College of Public Health, The Ohio State University, Columbus, Ohio, United States of America
Department of Food Science and Technology, The Ohio State University, Columbus, Ohio, United States of America

Mailis Laht, Veiko Voolaid, Tanel Tenson and Veljo Kisand
Institute of Technology, University of Tartu, Tartu, Estonia

Marko Virta and Antti Karkman
Department of Food and Environmental Sciences, University of Helsinki, Helsinki, Finland

Christian Ritz
Department of Nutrition, Exercise and Sports, Faculty of Science, University of Copenhagen, Copenhagen, Denmark

Joshua S. Gruber, Ayse Ercumen and John M. Colford, Jr.
Division of Epidemiology, University of California, Berkeley, California, United States of America

Index